CORN Publication Series 13

ISBN 978-2-503-54058-0 (print)
ISBN 978-2-503-54695-7 (online)
D/2013/0095/112

Landscapes or seascapes?

The history of the coastal environment in the North Sea area reconsidered

Edited by Erik Thoen, Guus J. Borger
Adriaan M.J. de Kraker, Tim Soens, Dries Tys,
Lies Vervaet & Henk J.T. Weerts

BREPOLS

CONTENTS

PART II LANDSCAPES VERSUS SEASCAPES?
SOCIAL ORGANISATION AND CHANGING COASTAL LANDSCAPES

LIST OF CONTRIBUTORS

CECILE BAETEMAN	Geological Survey of Belgium (RBINS), Brussels Free University (VUB), B
FRANS BEEKMAN	Den Haag, NL
GUUS J. BORGER	University of Amsterdam and VU University Amsterdam, NL
MICHELE CAMPOPIANO	University of York, GB
ADRIAAN M.J. DE KRAKER	VU University Amsterdam, NL
JAMES A. GALLOWAY	Institute of Historical Research, University of London, GB
SJOERD J. KLUIVING	VU University Amsterdam, NL
OTTO S. KNOTTNERUS	Zuidbroek, NL
MICHEL A. LASCARIS	Cultural Heritage Agency, NL
ANTHONY LONG	University of Durham, GB
DIRK MEIER	Working group Coastal Archaeology and History Schleswig-Holstein, DE
JOHANNES A. MOL	Leiden University and Fryske Akademy, NL
RICHARD ORAM	University of Stirling, GB
STEPHEN RIPPON	University of Exeter, GB
HANS RENES	VU University Amsterdam and Utrecht University, NL
TIM SOENS	University of Antwerp, B
STEVEN A. SOETENS	VU University Amsterdam, NL
ERIK THOEN	Ghent University, B
DRIES TYS	Brussels Free University (VUB), B
BERT VAN DER VALK	Deltares, Delft, NL
ANTON M.M. VAN HAPEREN	Staatsbosbeheer Regio Zuid, NL
LIES VERVAET	Ghent University, B
MARTYN WALLER	Kingston University London, GB
HENK J.T. WEERTS	Cultural Heritage Agency, NL

LIST OF FIGURES

LIST OF TABLES

1 The North Sea coastal plains over the last two millennia: landscapes or seascapes? A new collection of essays

Erik THOEN, Ghent University
Guus J. BORGER, University of Amsterdam and VU University Amsterdam
Adriaan M.J. DE KRAKER, VU University Amsterdam
Tim SOENS, University of Antwerp
Dries TYS, Brussels Free University (VUB)
Henk J.T. WEERTS, Cultural Heritage Agency, the Netherlands

I. The goal of the book

This volume is, to a large extent, the result of a gathering organised by its editors in Ghent, which took place April 22–24, 2010, and dealt with the geographical evolution of the coastal areas adjacent to the North Sea, with a focus upon the last two thousand years. This conference was actually the third in a series of symposiums on the same broad theme. The first took place more than fifty years ago, in 1958, and the second in 1978. However, the recent gathering differed from the previous two in one principal way: it was not limited to the Low Countries alone but included other countries in the North Sea area such as Scotland, England and Germany, and also northern Italy. Recognised specialists were invited to present their research in a variety of fields relating to the subject at hand, and this book represents their work.

Our purpose in amassing and presenting the 'state of the art' in this volume is to stimulate future investigations. In addition, the various disciplines in which the coastal plains are studied too often remain within their own borders, and so we have set out to thoroughly interweave them in the hope that this will spur greater interdisciplinary cooperation. All of the texts contained herein underwent a process of peer review and editorial oversight in order to ensure quality and readability because this collection is intended to appeal not just to experts in historical geography, but to historians and scientists working in any field who wish to gain insights into the present 'state of play'. With that in mind, specialist terminology has been simplified on the editors' discretion in order to facilitate cross-disciplinary understanding and research.

II. The genesis of this volume

The first, rather modest, symposium in 1958 was the initiative of two soil scientists working at the University of Ghent, R. Tavernier and J. Ameryckx. Combining their

Landscapes or seascapes? The history of the coastal environment in the North Sea area reconsidered, ed. by Erik THOEN, Guus J. BORGER, Adriaan M.J. DE KRAKER, Tim SOENS, Dries TYS, Lies VERVAET & Henk J.T. WEERTS, Turnhout, 2013 (*CORN Publication Series*, 13), pp. 1–7.

BREPOLS PUBLISHERS DOI: 10.1484/M.CORN.1.101545

latest findings regarding different chronological deposits, with the new research by their Dutch colleagues, C.H. Edelman and J. Bakker, they proposed that there were successive changes in the level of the North Sea during the Holocene period, although the precise dates of these fluctuations were, as yet, unknown. In order to try and track this evolution through the historical era, Tavernier enlisted the help of the historian Jan Dhondt, also from the University of Ghent, who co-organised the conference. In turn, Dhondt inspired another co-worker and historian, Adriaan Verhulst, to compare the soil scientists' results with historical source material and the literature by Dutch colleagues such as S.J. Fockema Andreae, J.F. Niermeyer, and E. Gottschalk. Verhulst added further to the conversation by involving archaeologists and place-name specialists, such as Maurits Gysseling.

The dating of the deposits in what were believed to be alternating phases of 'transgression' and 'regressions' in sea level, called 'Calais' or 'Dunkirk' for the pre-historic and historic periods respectively, were at the heart of these thought-provoking discussions. However, the results of the workshop were only published at a regional level[1]. And while many unexplored written sources in both Belgium and the Netherlands were thought to contain information that could lead to a more accurate chronology of these sea level changes in coastal evolution, many questions and uncertainties remained. It was already clear, for example, that the theories promoted by the soil scientists had been applied too indiscriminately.

However, it took another twenty years before a second conference was held. This took place in 1978, and was the initiative of Verhulst of Ghent University and Elisabeth Gottschalk of the University of Amsterdam[2]. Another meeting was essential because of the huge advances in research made during the previous two decades. First, it had become clear that soil scientists, historians, and archaeologists could not simply use each other's data to support their own chronologies and interpretations of coastal geomorphology. Historical investigations had concluded that the direct causes of large scale flooding in the south-west Netherlands were primarily individual storm surges, rather than Dunkirk-transgressions, as had been assumed. Even though the underlying reasons remained unknown, by 1977 a reliable chronology of storm surge events had been established[3].

For their part, archaeologists had demonstrated that the transgression-regression cycle needed to be delineated from mean sea level rise; earth scientists, thanks to many new techniques allowing them to greatly increase their knowledge of long and short term coastal sea level evolutions, had shown that this had drastically slowed from *c.* 1000 BC onwards. Historical climatology also entered the fray, as attempts were made to

---

1 Ameryckx, J. and Verhulst, A. (1958) 'Enkele historisch-geografische problemen in verband met de oudste geschiedenis van de Vlaamse kustvlakte. Een colloquium van bodemkundigen en historici', *Handelingen der Maatschappij voor Geschiedenis en Oudheidkunde te Gent*, 12, pp. 1–24.

2 Verhulst, A. and Gottschalk, M.K.E (1980) *Transgressies en occupatiegeschiedenis in de kustgebieden van Belië en Nederland*, Gent (Belgisch Centrum voor Landelijke Geschiedenis; 66).

3 Gottschalk, M.K.E. (1971–1977) *Stormvloeden en rivieroverstromingen in Nederland,* 3 vols, Assen.

associate storm surges with warmer climatic periods[4], this way re-introducing in fact short term changes in sea level to the debate. In addition to which, the 1978 conference strongly appealed for more investigation of the so-called 'Dark Ages', the period prior to 1000 AD, primarily because archaeological research was not yet able to confirm human occupation of the coastal plain between the fourth and tenth centuries. The lack of evidence was regarded as an *argumentum ex nihilo* in support of the hypothesis that the early medieval coastal plain was almost entirely abandoned during that period because of large scale flooding.

To summarise: in 1978, apart for essays to adapt their chronology, there was a growing interest in the underlying reasons for different stages of occupation and reclamation. However, many of the conference's participants left rather unsatisfied; particularly with regards to the duration and nature of phases in coastal plain evolution because one of the primary outcomes of the symposium was a growing scepticism towards simplistic patterns and phases of transgressions and regressions. From that point onwards, these concepts lost all meaning for many researchers. Nevertheless, a small minority of specialists, and many non-specialists, persisted in this approach and are still using its terminology to this day[5].

The three decades following the second conference have yielded a wealth of knowledge from an ever-expanding body of research. Many of the issues then raised in 1978 have since been addressed, but only through the salutary dispersal of investigations across even more disciplines. Above all, the latest research on deposits was incorporated into attempts to explain coastal phenomena in a more profound and nuanced way than ever before. Detailed soil sampling in many areas provided new data, with peat layers proving to be especially excellent materials for pollen analysis and radiocarbon dating. By using such methods, researchers gradually gained a better understanding of the close relationship between the processes of deposition, sea-level change, and land formation taking place across multiple regions. In particular, earth scientists increasingly came to realise that the formation and loss of land in coastal environments must be understood as the result of the complex interplay of many natural mechanisms. Furthermore, we now know that during the last millennia of the Holocene, semi-long-term small-scale changes in the absolute sea level (formerly incorrectly called 'transgressions' or 'regressions') either did not occur or were so subtle as to be immeasurable[6], while regional land loss and gain occurred widely but on different temporal and spatial scales, and resulted in local, *relative* changes in sea level.

Historical and archaeological research also evolved during this time. Most significantly, ideas regarding the chronology of human occupation have changed, and since 1978 we have become much better informed on this aspect of the 'Dark Ages'. This has been achieved through the combined efforts of archaeologists, historical geographers, and

---

[4] Lamb, H.H. (1980) 'Climatic fluctuations in historical times and their connexion with transgressions of the sea, storm floods and other coastal changes', in: A. Verhulst en M.K.E. Gottschalk (eds), *Transgressies en occupatiegeschiedenis*, pp. 251–290.

[5] For more details regarding the recent research by earth scientists in particular, see Baeteman, this volume.

[6] For a summary of the most important natural processes influencing *relative* sea level changes, see Weerts, this volume.

place name specialists. Generally speaking, current research into the Roman occupation demonstrates that there was a much denser population than previously thought, while the Early Middle Ages witnessed a decline in occupation. Thereafter, from the Carolingian period onwards, there was a clear revival in the population density of many coastal areas. Indeed, the eleventh and twelfth centuries show an increase in population and settlement, the very same era previously believed to have had higher sea levels[7].

Until about the twelfth century AD, it appears that, there was overall relatively little divergence between the levels of occupation in coastal areas and the majority of 'inland' areas in the Low Countries and Western Europe, and this despite regional and local variations due to drainage problems near the coast. However, this changed starting in the thirteenth century. New research on what followed indicates that the coasts were suffering the consequences of long-term over-exploitation. Some of the articles in this book explore this theme and demonstrate that after a period of intensive cultivation, an ever-increasing effort was required in order to work coastal regions, thereby pushing many farmers into commercial production. This line of investigation demonstrates how historians have recently increasingly taken social structures into account when explaining coastal geographical changes and local evolutions.

III. Facilitating further research

Despite the progress made in the last few decades, much remained to be investigated. The findings of earth scientists, historians, and archaeologists were not widely reported on within their respective circles. One of the purposes of the 2010 conference was to foster *an improvement in the mutual knowledge of different disciplines*, and it succeeded. As a result of this gathering, and the research presented here, we now know that while many studies have underlined the importance of 'accommodation space' in mitigating the risks of flooding, this margin has been heavily influenced by man made changes to the environment. Many, for example, underlined the importance and scale of peat compaction as a result of peoples' activities, the consequences of which were flooding, long-term land loss, and an increase in 'accommodation space'. Others focussed on the role of the human impacted hinterland in sediment transport and drainage into the open sea, together with the opening up of inlets. Still others drew attention to how the process of dike building carried with it an increased risk of flooding.

This scope of the research collected in this volume is also important because it has increasingly become evident that land loss and gain were the results of *regional factors*. Moreover, it is now clear that humans devised survival strategies, and thus organised their activities in relation to the environment, on a regional basis, which means that the causes of local changes must have been both natural and socio-historical[8]. The two were interwoven. Yet recent investigations, and many of the articles presented here, stress that

[7] At one time, the eleventh century was thought to have seen a third 'Dunkirk transgression', and the presumed general flooding was supposed to have been linked to semi-long-term small-scale changes in the absolute sea level.

[8] See Thoen, this volume.

the human factor is still largely underestimated in the history of coastal landscapes; this is probably to a large extent responsible for the fact that the last two to three thousand years of the Holocene are referred to as a 'period of instability' by Martyn Waller. While we may still misjudge this element in relation to the pre-Roman era, the activities of mankind were hugely influential during the Roman period and increased further after the seventh and eighth centuries.

It has now become clearer than ever that there is no single chronological scheme capable of explaining the coastal evolution across the entirety of the North Sea area. Earth scientists have shown that there was a very small, gradual, relative rise in level of the southern portion of the North Sea during the last millennia, although this change differed from one area to another. Semi-short-term fluctuations over decades or centuries, the so-called 'transgressions' and 'regressions', have now been completely disproven, and cannot have been the cause of periods of changed marine influence in an area. As was underlined many years ago by C. Baeteman, P. Vos, H. Weerts, A. Waller and others, there is no evidence for it in the Late-Holocene stratigraphy; nor, as D. Tys, B. Augustyn, P. Henderikx, G. Borger and others have shown, is there any indication of it in the archaeological and historical data. The same is true for storm surges; although they did bring about some alterations, their influence was varied too much by locale to have been decisive or explicative.

Yet, the dismissal of an overall pattern of natural change, does not rule out the changed importance of human influence on a regional scale, as highlighted by S. Rippon.[9] Indeed, instead of natural, successive phases of transgressions and regressions, sea-level fluctuations in other words, coastal land loss and land gain were likely caused by *social cycles*, as O. Knottnerus aptly dubbed them at the conference. Schematically one can discern four stages in such a cycle:

Stage 1: initial, 'low impact' exploitation and modification the natural environment.

Stage 2: transformation of the natural environment into a cultural environment via increased colonisation and more intensive land use.

Stage 3: accumulative environmental adaptation through extensive alterations. Overexploitation subsequently often led to environmental problems including land subsidence and higher water levels in the estuaries due to the embankment of adjacent saltmarshes.

Stage 4: resulting, at times, in the abandonment of land and occupation sites and the emergence of new natural environmental conditions. This could lead back to stage one.

Interestingly, this pattern of human behaviour in marine areas shows similarities with Malthusian cycles. However, just as the Malthusian model cannot be applied everywhere in the same fashion, the four-stage schema, as it is described above, also has its limitations. These phases did not occur everywhere in the same fashion, or with the same negative consequences. Indeed, to explain regional differences one must also take into account the different social structures of regionally organised rural societies, or *social agro-systems*[10], which to a large extent determined the sustainability of (coastal) areas.

---

[9] See Rippon, this volume.
[10] See Thoen, this volume.

Moreover, it seems that there were both *cumulative* and *lateral spatial* effects. This means that local phenomena, such as overexploitation or problems arising from drainage issues, in one area could have had a huge influence on an adjacent area in the same period, or further down the road. H. Weerts demonstrates that coastal land gain or land loss differed both regionally and temporally; while D. Meier shows how locale specific alterations to creek systems changed the surrounding environment. Poor maintenance of dike and drainage infrastructure could also cause problems in adjacent areas. At the same time, however, a disastrous loss of land on one side of an estuary could mean safety for the other side for centuries to come.

The book underscores the limits of current research in an effort to stimulate future investigations to fill some of the lacunas. While we currently have a vague picture of the most important natural and human influences on changes in marine areas and how they function, it is clear that we still have not undertaken enough regional studies either to fully understand the interplay of the factors at work, or to explain the underlying mechanisms of change. Some periods clearly deserve more research, such as that spanning the third and the seventh centuries AD. There is also still much to study regarding infrastructure, water management in general, the reclamation of peat areas and wetlands, and land use, of the dunes for example. Not to mention the techniques used to make alterations, the social organisations and administrations, and the settlement structures. We also need to improve our ability to measure the physical effects of such changes, like land subsidence. Despite the deeper insights afforded us by research undertaken over the last few decades, our models are still to some extent based on fragmented documentation in several areas.

Finally, we must ask ourselves how past societies responded to environmental changes or challenges and in what way these impacted social relations among different coastal regions. Finding answers to these questions could aid us in dealing with current ecological transformations. They touch upon recent global discussions regarding flooding hazards in terms of how societies and governments respond to this issue, effectively or otherwise. At the dawn of the twenty-first century, the belief in our ability to find technological, 'sustainable' solutions for countering hazards such as coastal flooding has waned. 'Environmental risk' is itself increasingly seen as a cultural concept, highly dependent upon time and place specific relations of power and property. Fully interdisciplinary studies of the marine landscapes around the North Sea, undertaken by extensive teams of researchers, are necessary for understanding the evolution of these coastal areas, both in the past and in the future.

IV. Outline

The book is split into two parts. The first focuses on the influence of natural phenomena on coastal evolution and closes with an article by H. Weerts, which provides a general literature review of our present state of knowledge regarding these occurrences, including natural processes unintentionally triggered by mankind. The second discusses direct human influence upon the evolution of the coastal areas and finishes with a general overview from E. Thoen of the most up-to-date literature on the subject. We have refrained from providing an overall synthesis for the entire North Sea area, primarily because the research conducted thus far has been too fragmentary to allow it. Maybe – hopefully – this will be possible after the next conference, although we certainly do not want to wait another twenty or thirty years for that happen!

Figure 1.1 Regions discussed in the different contributions in this volume. The numbers refer to the chapters in this volume

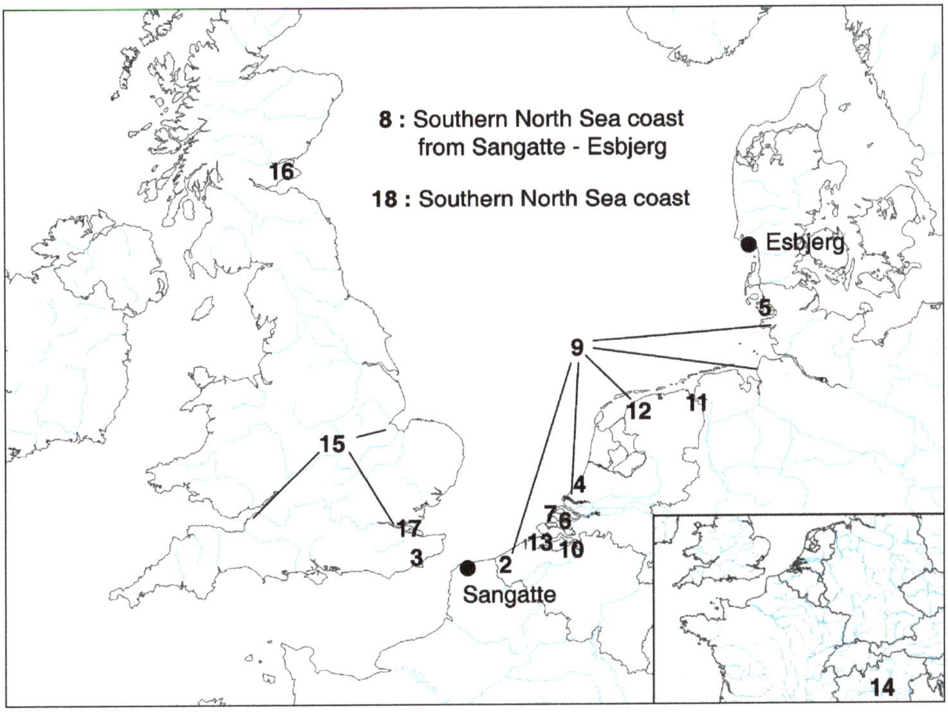

V. Acknowledgements

Since this volume is not a 'classical' publication of conference proceedings but a collection of restyled, peer reviewed, and extended articles with quite a few illustrations, it required a great deal of nuts and bolts attention to detail. This was carried out primarily through the initiative of dra. Lies Vervaet, assistant in the Department of History of Ghent University, and co-editor of this book. She organised the peer reviewing process, which was performed with the help of both the Flanders-Netherlands publishing team and specialist colleagues. We are also grateful to Bart De Wit, of the Ghent University, who painstakingly reformatted many of the illustrations, as well as to dr. Alasdair Ross, Senior Lecturer in Environmental and Medieval History at University of Stirling, who proved the ideal native English speaker for correcting the language in the texts.

Finally, we are indebted to the institutions that financed this publication; namely, the Research Foundation Flanders (FWO Flanders), the Netherlands Organisation for Scientific Research (NWO), the Faculty of Arts and Philosophy of Ghent University, and the Cultural Heritage Agency of the Netherlands, which made some maps, including Figure 1.1 above, which provides an overview of the regions dealt with in this book.

Part I

Seascapes versus landscapes? The natural context for human survival

2 History of research and state of the art of the Holocene depositional history of the Belgian coastal plain

Cecile BAETEMAN, Geological Survey of Belgium (RBINS), Brussels Free University (VUB)

Abstract

This paper evaluates the major issues that surround two different concepts of Holocene coastal evolution in the Belgian coastal plain with reference to the possibility of human occupation. The first concept, which was generally accepted until the 1980s, only considered late Holocene deposits. The coastal evolution was then conceived of as a series of successive transgressions (in the sense of a sea-level rise) that breached the dune belt and flooded the entire plain. The coastal plain was geographically divided into different landscapes according to the different transgressions that were chronologically defined on the basis of archaeological and historical sources. From the 1980s a complete different view on coastal development has been developed on the basis of intensive new geological fieldwork, whereby the entire Holocene sediment succession is investigated. Here, the development of the coastal plain is considered in the context of the dynamic character of tidal sedimentary environments. Sedimentology, radiocarbon datation, and diatom analyses provide the framework to explain the driving mechanisms and processes of coastal change. These results allow archaeologists and historians to better interprete their findings in a palaeoenvironmental and chronological context. In effect, this paper presents a summary of our current understanding of the Holocene coastal development.

I. Introduction

Knowledge of Holocene sedimentary successions and coastal changes is important for archaeologists and historians who work in coastal plains and who need to interpret their findings within a context of palaeoenvironmental change. As non-specialists of coastal change, archaeologists and historians have to rely on existing publications and maps. They do not hold detailed information about the site-specific stratigraphy, sea-level history or processes of coastal change. Consequently, they often are not sufficiently aware of possible misconceptions in older publications and maps. This is particularly true for investigations in the Belgian coastal plain. The late Holocene deposits were for most of the twentieth century interpreted within the context of successive transgressions with the emphasis on their chronology and lateral extension. Therefore, the history of research in the Belgian coastal plain will be used as an illustration of these misconceptions that have dominated literature for a long time.

In Belgian literature it was widely believed that the Holocene deposits were formed during two transgressions, i.e. the Calais and Dunkerque transgressions, separated by a

Landscapes or seascapes? The history of the coastal environment in the North Sea area reconsidered, ed. by Erik THOEN, Guus J. BORGER, Adriaan M.J. DE KRAKER, Tim SOENS, Dries TYS, Lies VERVAET & Henk J.T. WEERTS, Turnhout, 2013 (*CORN Publication Series*, 13), pp. 11–29.

BREPOLS PUBLISHERS DOI: 10.1484/M.CORN.1.101546

1 to 2 m thick peat bed (called surface peat). The latter was interpreted as being formed behind a continuous dune belt during a fall in sea level, or according to others, a stable sea level. This thick peat bed was accepted as the stratigraphic boundary between the Calais and Dunkerque transgressions/deposits. During this early period of research in the Belgian coastal plain, the Calais deposits received little attention. They were described as sandy deposits with some clay lenses at the top. The Dunkerque deposits, on the other hand, have been subject of investigation by soil scientists and historians. The concept of transgressions that flooded the entire plain lasted until the 1980s when, for the Belgian coastal plain, a complete different view on coastal development was introduced. This latter view resulted from new geological field work and new concepts developed by coastal geologists in northern Germany. It took about twenty years before the traditional Calais-Dunkerque stratigraphy was also abolished in the Netherlands (Ebbing *et al.*, 2003).

To demonstrate the misconception this paper first presents a short overview of the pre-1980s literature. The overview focuses on how scientists conceived the formation of the coastal plain and the possibility of human occupation rather than on the frequently changing stratigraphy and chronology. For a critical analysis of the latter see Baeteman 1981 and 1983. In the second part of the paper an overview of the results of the geological research of the last thirty years is presented. The dynamic character of the tidal sedimentary environments building up a coastal plain, and the driving mechanisms of coastal change are explained to provide a context for the interpretation of coastal archaeology and human behaviour.

II. History of the research – The major issues

The Belgian coastal plain (Figure 2.1) is the cradle of the stratigraphy and in particular of the chronology of late Holocene deposits. Research in the coastal plain started as early as 1927 with Cornet who introduced the term 'Dunkerque transgression' in the Belgian stratigraphy following Dubois (1924) who defined the 'Assise de Calais' and 'Assise de Dunkerque' in northern France. Both authors used the term 'Dunkerque' to name the sediments. It was Briquet (1930) who associated the terms with a transgression in the sense of a sea-level rise and a period of sedimentation. However, Halet (1931) could not agree with the stratigraphical subdivision. He considered that it was impossible to distinguish specific units in the deposits of the coastal plain because of a lack of distinct boundaries that could be detected over the entire area. Halet was well aware of the complexity of the development and evolution of the coastal plain. However, his accurate observations and sound interpretations were not adopted by researchers during the following decades. Tavernier (1938) studied in particular the upper deposits of the plain and he followed the ideas of Briquet. Tavernier put the emphasis on the concept of a transgression in the sense of a sea-level rise, causing breaching of the dunes and flooding of the plain.

A profound change in ideas concerning the evolution of the coastal plain was introduced by Tavernier in his publication of 1948. The author stated that the upper deposits were caused by inundations since Roman times. These inundations breached the dunes several times and in several places. Therefore, he subdivided the sea breaches into different phases and introduced three Dunkerque transgressions (DI, DII and DIII), which he placed chronologically on the basis of comparison with archaeological findings from

Figure 2.1 Location map of the coastal plains of Northern France, Belgium and Zeeland

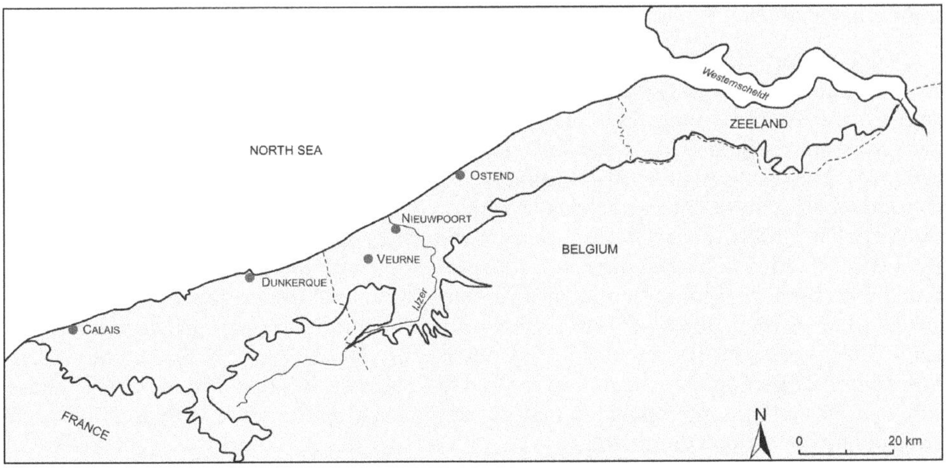

Zeeland (The Netherlands). Because of the absence of archaeological findings and historical sources from particular periods, it was considered that the area was not suitable for human occupation because of the inundation. This idea was in support of the introduction of the three Dunkerque transgressions. Geological evidence, apart from vague descriptions such as sand on clay and/or clay on sand, was not provided to document the tripartite subdivision. This publication became a landmark that was frequently quoted during at least the next thirty years.

Since the early 1950s research in the coastal plain was carried out in the context of soil mapping with Moormann and Ameryckx carrying out fieldwork. Both soil mappers used the tripartite subdivision. However, its chronology and in particular the duration of the transgressions changed frequently during their investigations (see Baeteman, 1983 for an overview). These changes most probably were the result of a confrontation with the complexity of the deposits of the plain during their detailed mapping. From their descriptions it is obvious that both soil mappers had problems with the tripartite subdivision. Moormann (1955), for example, wrote that he did not understand the puzzle of sand above clay and clay above sand in the same area. He also had to introduce a new transgression (D0) in order to explain the sediment succession in the same area. Ameryckx (1959) had problems with the DIII-transgression so he consequently distinguished two sub-phases. In the same publication, Ameryckx introduced the term 'regression' because of the presence of a vegetation horizon. This later became the frequently used 'Roman regression' following the DI transgression. Despite the discrepancy between the field data and the tripartite Dunkerque subdivision, the latter was maintained. The legend of the soil map was established according to the tripartite subdivision, and the time and financial pressure to produce maps was so high that there was no possibility of properly interpreting the field data.

Sedimentary processes of infill were hardly considered, except for the phenomenon of relief inversion explained by compaction of the peat. It was stated that the creeks associated with the Dunkerque inundations eroded the peat. Sand was deposited in the creeks and clay above the peat outside the creeks. After the embankment the peat compacted, resulting in a lowering of the surface of these areas while the sand-filled creeks remained at their original level.

Collaboration with historians during the soil mapping phase was mainly focused on the chronology of the transgressions linked to the possibility of human settlement based on historical sources alone. This resulted in a change of the relative chronology of the transgressions as proposed by the soil mappers into an absolute chronology (Verhulst, 1980: 11). The results of this collaboration were presented in 1958 at a colloquium in Gent, focused on the geographical-historical evolution of the Belgian coastal plain (Ameryckx and Verhulst, 1958). These results, as with the major issues surrounding the tripartite Dunkerque subdivision, the concept of transgressions with breaching of the dunes associated with the formation of a branching creek system, and the explanation of the relief inversion, were maintained in literature between 1960 and 1976 (see e.g. Maréchal *et al.*, 1964; Tavernier & Ameryckx, 1970; De Breuck *et al.*, 1970; Ozer, 1976). Actually, there was no proper geological research in the coastal plain during that whole period. These publications were no more than a repetition of the existing literature without significant new findings or critical commentary.

The first questions came from the archaeologist Hugo Thoen (1978), who was investigating the Roman period. He had doubts about the geographical extent of the DI and the chronology of the DII transgression. More importantly, he proved that the archaeological findings in Zeeland used to establish the chronology of the Dunkerque transgressions, were not reliable because they were not found in situ.

The concept of the Dunkerque transgressions was also adopted in the Netherlands for the legend of the new geological map (de Jong and Hageman, 1960). Afterward, it assumed a life of its own and, depending on the author, the context of the term Dunkerque changed. It was used as a lithostratigraphic, a chronostratigraphic, or a geomorphologic unit, or used to name transgressions or phases of inundations. The sedimentological characteristics of the deposits and their indicative meaning, however, were not considered. Moreover, the possibility of absolute dating by radiocarbon led to a further chronological enhancement of the rigid framework of the transgressions. For instance Roeleveld (1974) defined five Calais and six Dunkerque transgressions in Groningen purely on that basis. The author also confirmed that the Dunkerque II transgression took place in northern France between 300 and 550AD (Roeleveld, 1977).

However, during the same period, Streif (Barckhausen *et al.*, 1977; Streif, 1971, 1972, 1978) made a clean sweep of the Calais-Dunkerque stratigraphy and proposed a totally new and ingenious classification system for mapping the coastal plain deposits. The Holocene deposits were subdivided according to a hierarchic system on the basis of the vertical succession and the lateral interfingering of clastic sediments on one hand, and peat beds on the other (Figure 2.2). This classification system accurately reflects the depositional history of the Holocene coastal plain deposits in relation to the rate of relative sea-level (RSL) rise, sediment budget and accommodation space (see below).

In 1978 another congress was organised to present a synthesis of twenty years of 'intensive research' (sic, Verhulst, 1980: 17; Verhulst and Gottschalk, 1980) in the coastal

Figure 2.2 Schematic representation of the linkage of the Streif classification system with the effect of the changes in the rate of RSL rise on the Holocene depositional history (source: Baeteman, 2005b)

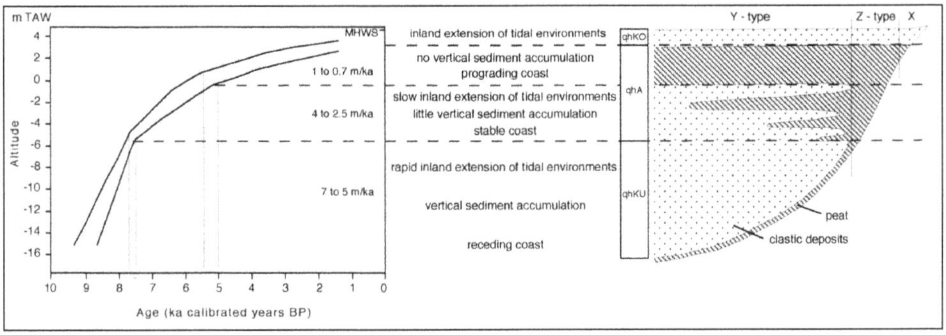

plain by soil scientists, archaeologists, and historians. Basically, however, the working hypothesis presented in 1958 was only repeated and confirmed. The major difference was that transgressions were now considered as storm surges related to climate. The huge inventory undertaken by Gottschalk formed the basis for this new idea. Essentially storm surges are occasional events that happened irregularly in space and time and so they cannot be fixed in a rigid chronological framework (Gottschalk, 1980: 27). What remained, however, was the concept of the breaching of the dunes and catastrophic flooding of the plain. The D-II transgression was considered as being the most catastrophic and long-lasting event.

III. Results of the geological research in the Belgian coastal plain of the past thirty years

The research was initially focused on the deeper sediments, i.e. the early and middle Holocene deposits. This was essential because the understanding of the development and evolution of a coastal plain requires knowledge of the complete vertical sediment succession to determine the driving mechanisms and ongoing processes of coastal change. The characteristics of the present landscape are an inheritance of what happened before, in particular in terms of coastal processes. A coastal plain superficially records only terminal geomorphic surfaces, but not the steps necessary to develop those surfaces. These steps are found in the subsurface.

In the period between the late 1970s and the 1990s about 1150 undisturbed hand-augered gouge cores, with a large number of them reaching the Pleistocene/Holocene boundary, and about hundred undisturbed mechanically-drilled cores which reached the Tertiary boundary, were collected in the western part of the plain (Figure 2.3). Over two hundred fifty radiocarbon datings were obtained from organic material and shells. The data sets of radiocarbon dates are published in Baeteman and Van Strydonck (1989), Baeteman and Declercq (2002), Baeteman *et al.* (2002) and Baeteman (2005a). The characteristics and ages of the late-Holocene shallow deposits were primary taken from temporary outcrops. The investigation focused mainly on the sedimentological aspects

Figure 2.3 Map of the western coastal plain of Belgium with location of the boreholes (source: Baeteman, 2005a)

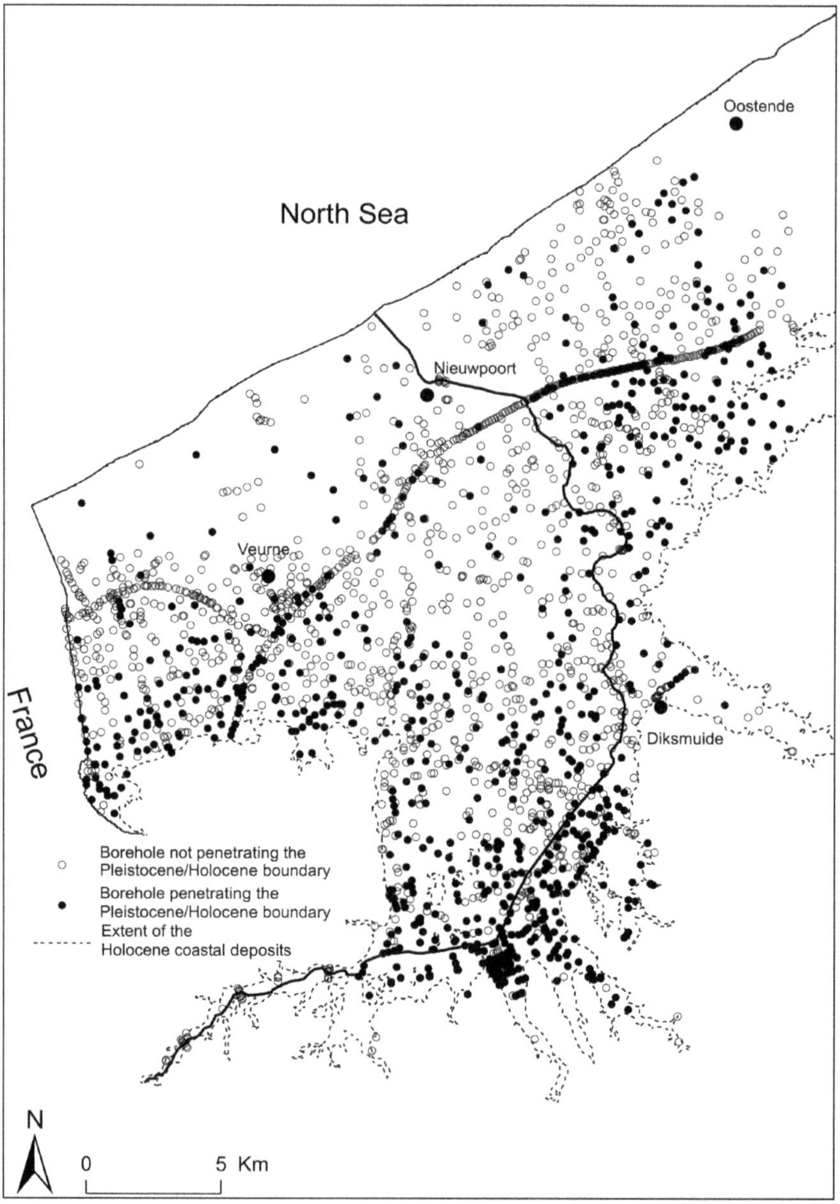

of the deposits to determine the sedimentary processes. The borehole database of the systematic mapping (Figure 2.3) enabled an integrated reconstruction of the coastal changes throughout the Holocene.

Figure 2.4 Schematical representation of the different sedimentary environments in a tidal flat in relation to water levels.
MHW: mean high water spring; MTL: mean tide level; MLW: mean low water spring

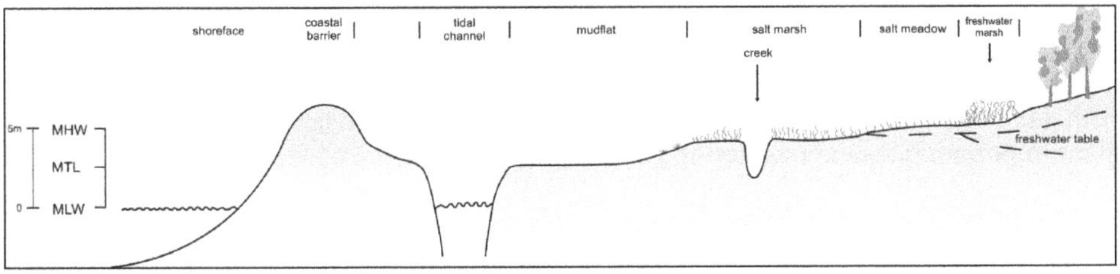

The deposits were interpreted in terms of depositional sedimentary environments (coastal barrier, tidal channel, mudflat, salt marsh, Figure 2.4). This interpretation was enhanced by detailed diatom analyses by Denys (1985, 1989, 1993a, 1993b, 1993c, 1994, 1995, 1999). Such environmental interpretation is essential to document coastal changes because each tidal sedimentary environment has its specific relation to sea level, and their position to one another shows a typical zoning pattern in relation to the position of the tidal channels. Moreover, they form a dynamic system capable of shifting either landward or seaward, and one over another, in relation to the following controlling factors: rate of RSL rise, sediment supply, and accommodation space (the height difference between a sediment surface and high tide). The evolution of one environment into another cannot be generalised for the entire plain. It is related to the position of the tidal channels.

The tidal flat is dissected by a very extensive network of tidal channels and creeks in the supratidal zone (salt marsh), which act as local sediment suppliers. When the sediment supply surpasses the accommodation space created by the relative sea-level rise, the area will rapidly silt up to the high-water level. Consequently, the frequency of tidal inundation decreases and a part of the drainage network serving that particular area becomes redundant and silts up. Eventually the area falls out of the reach of tidal inundation and the salt marsh encroaches upon the mudflat, followed by peat accumulation if sufficient time is available for the formation of a freshwater lens in the shallow subsoil. However, at the same time, intertidal deposition continues in an adjoining part until, in turn, this area will also build up to above high-water level. Then the same evolution is repeated there. If no sufficient time is available for the freshwater lens to form (which was the case in the period characterised by a rapid rate of RSL rise), only vegetation horizons can develop. Consequently, vegetation horizons or peat beds are not synchronous and have no stratigraphical implication, as their occurrence only depends on whether sufficient sediment was supplied to allow silting up of a particular part of the tidal flat.

The boundary between the open sea and the tidal flat is formed by the coastal sand barrier, a dynamic feature controlled by sand supply in relation to RSL rise. The coastal barrier, a huge sand body extending both below and above sea level, includes the shoreface, tidal deltas and inlets, washovers, and the beach and dunes. Coastal barriers are essentially transitory features that maintain themselves in dynamic equilibrium with rising sea level by the landward transfer of sand, eroded from the shoreface, to the tidal flat via tidal inlets and overwash processes. Landward shifting of the barrier is fairly continuous while sea level is rising relatively rapidly but may become intermittent when it slows down. In addition, the rate of landward shifting is reduced as the accommodation space for sediment in the tidal flat decreases, i.e. when the tidal flat has silted up to high-water level. The role of sediment supply then becomes dominant. With a positive sediment budget, landward shifting of the barrier is reversed and the barrier starts shifting seaward; this is also called progradation (Roy *et al.*, 1995). The dynamic character of the sedimentary environments implies that at any time throughout the Holocene all the sedimentary environments existed next to one another, even over short distances. The breaching of the dunes associated with a complete flooding of the plain during a particular time never occurred. This is of great importance when considering the possibility of human occupation.

The different approach of the geological investigation, considering the sediment characteristics and the evolving sedimentary environments together with age determination, resulted in significant differences in the understanding of the development and evolution of the coastal plain. The major steps of the development will be recalled here, summarised from Baeteman (1985, 1991, 1999, 2004a, 2005a), Baeteman and Van Strydonck (1989), Baeteman *et al.* (1999), Baeteman and Declercq (2002). The development of a coastal plain is a function of the following controlling factors: rate of RSL rise, sediment budget and accommodation space, the latter effected by sediment and peat compaction. During the infill of the area, initially caused by the RSL rise, the relative importance of the individual factors changed in the course of time.

Knowledge of the morphology of the pre-Holocene surface that reflects the original topography at the beginning of the post-glacial sea-level rise, is essential to understanding the distribution of the various sediment successions in the coastal plain. The analysis of the borehole database showed that in the western part of the Belgian coastal plain the landscape prior to the marine flooding consisted of a drainage pattern of four relatively small and shallow rivers that joined in the central part of the western plain and formed a southeast-northwest depression (Figure 2.5). This depression is the palaeovalley of an ancient river IJzer formed during the glacial sea-level lowstand (Baeteman, 1999). This contradicts the idea that the formation of the valley was formed as a result of erosion in the Boreal period (Maréchal, 1953). The palaeovalley, however, is located several kilometers to the west of the present-day position of the river. It has a gentle slope with depths of -12 m and -18 m TAW in the central and seaward part of the plain, respectively. The relief of the pre-Holocene surface in the western part contrasts greatly with the eastern part of the coastal plain, which had consequences for its later development. The eastern part has a flat morphology with a gentle slope and shows a much higher elevation between +5 and -5 m TAW. Therefore, it was influenced much later by the Holocene transgression (Baeteman, 2008a). The RSL rise is recorded in the western plain as from *c.* 9500 calBP (calendar years before present, i.e. 1950). This contradicts the general view in literature where it is said that sea level starts to rise as from *c.* 5500 calBP (e.g. Maréchal, 1992; Jacobs and De Ceukelaire, 2002). The rising sea invaded the area by way of the palaeovalley which developed into an estuary with tidal channels and flats. As groundwater level rose with sea level, the flooded area was fringed by freshwater marshes in which peat accumulated. This peat is the basal peat, a transgressive unit shifting landward with the rising sea level (see Figure 2.2). It forms the basis of the Holocene sediment succession where it is not eroded by late Holocene tidal channels. Age and altitude of the basal peat were used to construct the relative sea-level curve of the area (Denys and Baeteman, 1995).

Figure 2.5 Contour map of the pre-Holocene surface at a 2 m interval relative to TAW
(TAW is the Belgian ordinance datum referring to mean low water spring which is about 2 m below mean sea level).
Because of the vertical exaggeration the map gives a wrong impression of the relief of the valleys. In reality they are shallow with gentle slopes (redrawn from Baeteman and Declercq, 2002)

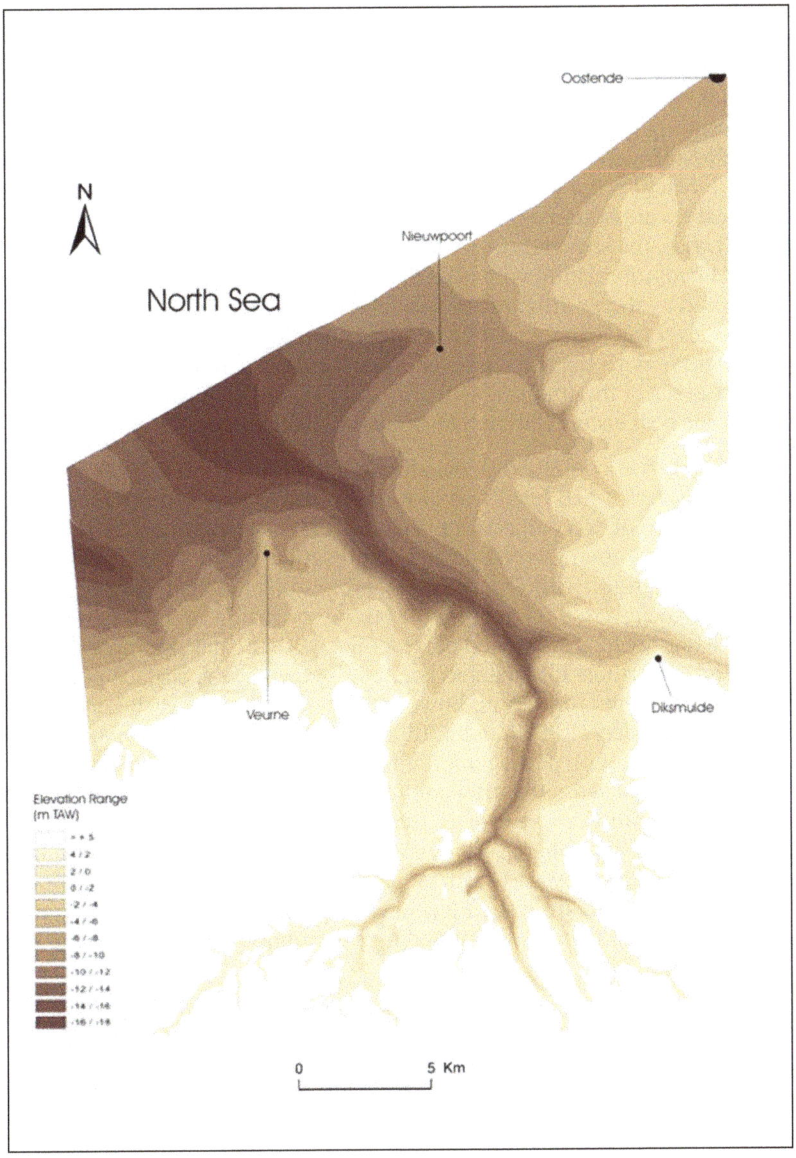

The general trend of the relative sea-level curve shows that initially the RSL rise was very rapid, rising at an average rate of 7 m/ka in the period before *c.* 7500 calBP. This resulted in a rapid shift of the sedimentary environments across the continental shelf towards a position close to the present-day boundary of the coastal plain, associated with a significant vertical sediment accretion with the development of vegetation horizons (see Figure 2.2). Erosion occurred at the shoreface by waves resulting in a landward shift of the coastal barrier. At *c.* 7500–7000 calBP, the relative sea-level curve shows a distinct retardation to an average of 2.5 m/ka. Consequently, the rapid landward shift of the environments stopped and the position of the coastal barrier became more or less stable. Sediment supply was now in balance with the accommodation space created by the RSL rise and parts of the tidal flat were infilled. Periods of emergence lasted much longer and freshwater conditions prevailed for short periods (in the order of about 200–300 years). Salt-marsh vegetation evolved into reed growth resulting in peat accumulation. Initially these freshwater marshes were short-lived and developed locally in the relatively higher silted-up areas which, for a certain time, were out of the reach of the tidal flooding and so deprived of sediment. However, by *c.* 6800 calBP, peat accumulation developed on a more regional scale. As a result of the lower rate of the RSL rise, the sequence deposited between *c.* 7800 and 5500 calBP consists of successive peat beds alternating with tidal flat deposits (see Figure 2.2). The formation of this was governed by the position of the tidal channels and thus by local sediment supply and not by sea-level fluctuations as claimed by Gullentops and Broothaers (1996: 26) in their sea-level curve whereby the sea level drops by about 1m for every intercalated peat bed. The depressions of the small palaeovalleys in the landward part of the plain are filled with a continuous peat accumulation that can attain thicknesses of 4 m.

The rate of RSL rise continued to decelerate and after *c.* 5500–5000 calBP it fell to an average of 0.70 m/ka. Sea level was close to its maximum and sediment supply had exceeded the creation of accommodation space. Landward migration of the sedimentary environments stopped completely, and the stabilisation of the coastal barrier changed into barrier progradation, with the coastline prograding beyond the present-day one. Periods of peat growth lasted longer and the lateral extension of the freshwater marshes became more widespread. This period corresponds with the development of the (now) 1 to 3 m thick and uppermost intercalated peat bed known as the surface peat (see Figure 2.2). The beginning of its formation, however, dates back to as early as *c.* 6200 calBP in certain areas. Between *c.* 5500 and 4500 calBP almost the entire plain had changed into a freshwater marsh with peat accumulation. In the very western and seaward part of the plain, however, tidal sedimentation continued. This peat accumulation, which lasted 2000–3000 years, could keep pace with the RSL rise. Traces of tidal influence in the thick peat bed in the landward area (Denys, 1993a) suggest that the major tidal channels remained open, although they were almost silted up and served mainly as drainage for the peat bog and the runoff from the Pleistocene higher grounds outside the plain. This also implies that the plain was not bounded by a continuous and closed dune belt.

The late Holocene deposits of the coastal plain were formed by a renewed expansion of the tidal environment. This resulted in a profound change in the coastal landscape whereby brackish clastic sediments replaced the peat accumulation. From the available data it is known that the end of the peat accumulation is not synchronous. However, precise dates for the end of the process are not known because the determination of accurate dates for the end of the peat formation is extremely difficult. This is because some erosion and reworking of the surface of the peat is always likely during the initial stages of inundation. Moreover, the surface of the peat might have been affected by a limited degree of intrusion, oxidation, or surface reworking making radiocarbon dates less reliable (Waller *et al.*, 2006). In the western plain only a few locations show a gradual transition with the overlying sediments giving reliable dates. The youngest age of all for the top of the peat taken from the very landward part of the plain where no erosion or oxidation is assumed, is about 1525 calBP. The other dates from the surface of the peat range from 3370 to 1525 calBP, with a cluster occurring between 2250 and 2000 calBP.

The renewed tidal expansion entered the area via the mid-Holocene tidal channels that were meandering and landward branching across the peat bog to almost reach the landward edge of the plain. The channels were almost completely silted up and the major ones served as drainage for the peat bog and the higher-lying hinterland. It is suggested that an excessive runoff from the hinterland was caused by an abrupt climate change *c.* 2800 calBP. This involved an increase in precipitation, most probably combined with tree cutting in the hinterland during the Iron Age period, which eroded the upper part of the mid-Holocene channels. Tree cutting in this period has not yet been documented in the study area, but is most likely via analogy with the observations of Smyth and Jennings (1990) in England. Once the channels were superficially 'cleaned', tidal waters could enter them again. This re-entrance, however, happened in successive steps. At first the tidal waters inundated the peat in the areas adjacent to the channels. This happened with modest erosion of the peat along the channel banks and this first step was restricted to small crevasse channels. At two locations, Veurne and Raversijde nearby Oostende, the formation of the crevasse channels with shallow erosion could be dated to *c.* 2400 and 2700–2400 calBP, respectively. At Raversijde, the local flooding by a crevasse channel was enhanced by Iron-age peat extraction that lowered the surface because of de-watering and compaction of the peat (Baeteman and Pieters, 2013).

In addition, in the areas adjacent to the channels seawater stopped the peat growth, which consequently lost its potential of retaining water, resulting in compaction of the bog (in the sense of a reduction in volume) and subsidence of the land adjacent to the channel. Peat extraction during Roman times had the same effect (Baeteman, 2007b). The land subsidence provided accommodation space for the deposition of intertidal sediments. Interlaminated mud and sand with peat detritus, indicative of tidal deposition, was formed. Although the erosion was modest during the first step, it had far-reaching consequences. Little by little, it caused drainage and de-watering of the peat bog with compaction and collapse of it in the vicinity of the channels. This resulted in a gradual increase of the tidal prism (the area the channels drain) to which the channels adapted by enlarging their cross-section (depth-width). Because the peat has a great erosion resistance, the channels scoured deeply into the easily erodable sand of their predecessors (the sand-filled early and mid-Holocene channels). The detailed mapping (Baeteman, 2004b, 2005a) showed that the major late Holocene channels reoccupied the same location as their mid-Holocene predecessors and sometimes eroded even into the Pleistocene deposits to depths of 25 m. However, tidal currents in the channels detached large masses of peat that slumped and eventually became buried in the channel fill. The radiocarbon dates of peat boulders in the channel fills indicate that they originate from the upper peat bed, even those in the lags at the base of the channels. During this phase, the channel network enlarged erosively further responding to the continuously increasing tidal prism. Progressively, larger parts of the

Figure 2.6 Map of the western part of the Belgian coastal plain showing the location of the major late Holocene sand-filled channels (redrawn from Baeteman, 2007b)

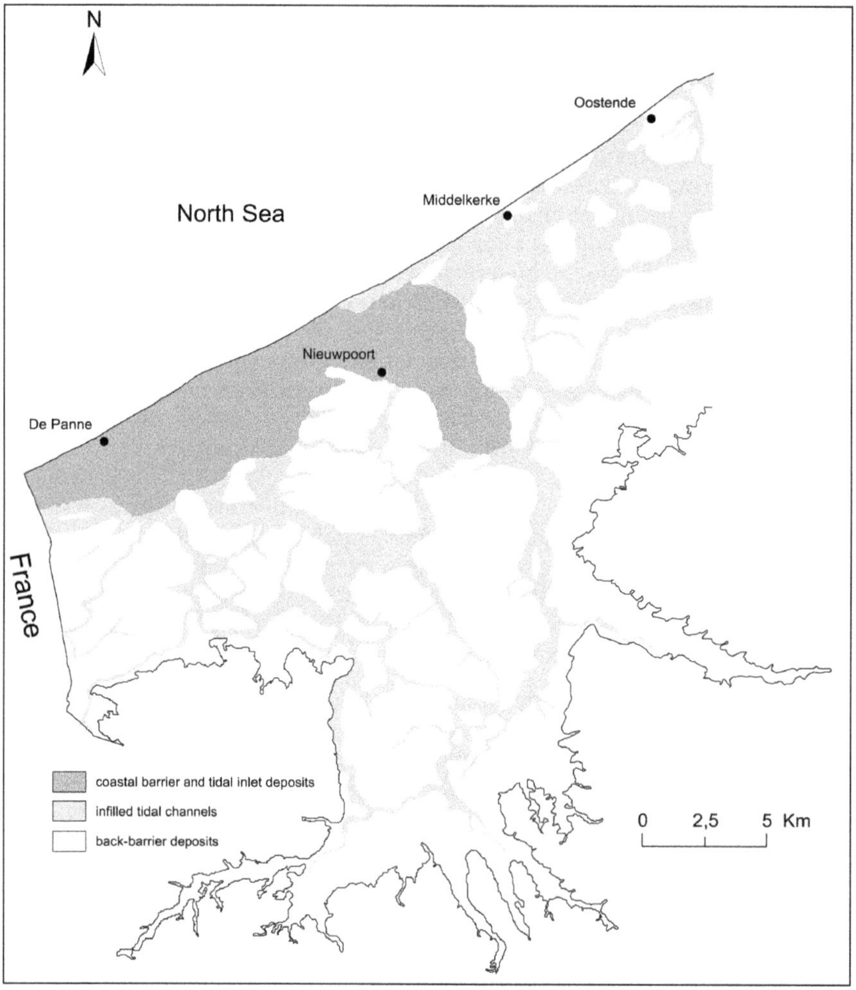

peat bog were attained and collapsed because of de-watering. This eventually contributed to the expansion of an increasingly dense channel network in the plain (Figure 2.6).

In the major part of the plain where the upper peat bed has not been entirely eroded, not all deposits overlying the peat bed show the same sediment characteristics. In the most distal reaches of the channels, little or no sediment differentiation is discernible vertically in the 1 to 2 m thick clay. Surfaces indicative of breaks in the sedimentation are lacking. This at a first glance implies that the clay was deposited by one single process and at the onset of the tidal inundation. In the proximity of the sand-filled channels, however, sediment changes in these deposits are frequent. The sedimentological study, together with radiocarbon dating of shallow outcrops in the proximity of sand-filled tidal channels, enabled to document the mechanisms and processes that caused the complex sediment variations, together with the evolution of the changing coastal landscape during the last 2000 years (Baeteman 2007a, 2008b; Baeteman *et al.*, 2002).

A time and sediment gap between the top of the peat and the overlying sediments suggests that some areas experienced a long period of slow or no sediment deposition. The sites where the peat collapsed came most probably in a subtidal position with a minimum of sediment deposition. It is assumed that this is the result of a negative sediment balance (cf. Beets and van der Spek, 200). All available sediments were used to fill the deeply scoured tidal channels. Moreover, during the 2–3 ka years of uninterrupted peat growth, sea level had risen by about 2 m. This, together with the collapse of the peat bog, implies that a huge accommodation space first had to be filled before a state of dynamic equilibrium was reached between the channel cross-section, tidal prism, sediment supply, and the sea level. A portion of the necessary sediment came from the early and mid-Holocene channel fills. However, since such a large volume was needed, the seaward area also had to supply sediment. This explains the landward migration of the shoreline due to erosion of the tidal deltas and the shoreface.

Once the dynamic equilibrium between the controlling factors was attained, the sediment surface reached an intertidal position. The youngest ages of all for the top of the peat indicate that the peat accumulation came to an end and was replaced by a tidal flat in the entire plain between *c.* 1600 and 1500cal BP (*c.* 350–400 AD). While the major part of the plain changed into an intertidal flat and silted up into a supratidal flat, most of the channels eventually filled until they had reached an intertidal position. This happened in the period between *c.* 1400–1200 calBP. This is about the same age the soil mappers and historians dated the catastrophic D-II flooding to. As described above, they considered that during the D-II transgression (300–800 AD) the plain was completely flooded. However, the sediment and age determinations indicate that in this period the plain had already changed into an intertidal and supratidal flat. Hence, the D-II transgression as such never existed.

Because of the silting up of the area and the very weak RSL rise, accommodation space was no longer created. Therefore, the channels started to migrate laterally due to meandering. This caused shallow erosion and reworking of the upper part of the channel fill and the adjacent supratidal flats and explains the sediment variations. The available dates indicate that this reworking of the upper channel fill did not happen simultaneously across the entire area. Not every channel reacted in a similar way because of the interplay of local factors that controlled their dynamics. Therefore, the sediment variations and the time periods are not identical. This implies that sea-level fluctuations were not responsible for the sediment variations recorded in the late Holocene deposits.

Finally, the channels and tidal inlets silted up to a large extent and the channel networks contracted. Archaeological and historical evidence documents that from the tenth century AD onwards, the major part of the plain consisted of salt marshes evolving into salt meadows, which were progressively embanked as well as some of the channels that had become sufficiently shallow (Ervynck *et al.*, 1999; Loveluck and Tys, 2006). Because of the embankment, man-made drainage with ditches and sluices was necessary. This caused compaction of the tidal-flat deposits overlying the peat bed, resulting in lowering of the surface (subsidence) in the embanked areas. Consequently, the reduced cross-sectional area of some of the major channels, mainly because of continued sedimentation and/or the dikes, could no longer accommodate a large volume of tidal water. When storms occurred, the storm-induced currents during the return flow after the storm set-up, eroded and reworked the upper channel fill again. The effect of storms in terms of sediment variation, however, was restricted to the channels and nearby areas.

IV. Conclusions

The geological research in the Belgian coastal plain has clearly shown that our understanding of the depositional history of a coastal plain requires investigation of 1) the entire sediment sequence and 2) the dynamic character of the tidal sedimentary environments that eventually formed the plain. The formation of the coastal plain, which in fact is the infilling of palaeovalleys, was controlled by palaeotopography, a decelerating RSL rise, sediment supply, and accommodation space in turn affected by sediment and peat compaction. With the exception of the early Holocene period, sedimentary processes, whereby tidal channels played a major role, caused the sediment variations reflecting changes in the coastal landscape. Sediment variations should not necessarily be associated with transgressions and regressions or sea-level fluctuations. The record of changes does not involve rising and falling sea levels. Landscape changes were not synchronous over the entire plain. This is particularly true for the last 2000 years. For archaeologists and historians it is of importance that they are aware that at any time the coastal landscape consisted of all the different sedimentary environments next to another, even over short distances. The configuration of the tidal flat, especially the proximate and distal position of the tidal channel in relation to a particular site must also be considered. By itself, such an approach provides an understanding of the complex sediment variations and their context is required for the interpretation of the archaeological findings and human activities.

Considering the overview of the research history together with a summary of our current understanding of Holocene coastal development, it is strongly suggested that the terms Calais and Dunkerque should no longer be used because they are too much loaded with the erroneous concept of the development of the coastal plain and its chronology. Despite the overwhelming field evidence supported by absolute age determinations, the Calais-Dunkerkque stratigraphy and the concept introduced by the soil mapping are still being used in Belgium with moreover systematically disregarding the research papers published since the 1980s (e.g. De Moor and Pissart, 1992; Maréchal, 1992; Augustyn, 1992, 1995; Houthuys *et al.*, 1993; Gullentops and Broothaers, 1996; Louwye and Declercq, 1998; De Moor *et al.*, 2010). The invalid stratigraphy is also adopted as a generally accepted view and as such used for the official lithostratigraphic legend of the Quaternary map of Belgium (Gullentops and De Moor, 2001). Also in the international literature, the Calais and Dunkerque transgressions, separated by regressions, together with the concept of a strongly fluctuation sea level, have been advocated recently (Behre, 2007). However, in order to better inform both archaeologists and historians this paper has critically demonstrated that the supposed transgressions and regressions are non-existent and that the accompanying sea-level fluctuations are highly unlikely to be real features of the sea-level history of the southern North Sea (Baeteman *et al.*, 2011).

Bibliography

Ameryckx, J. (1959) 'De ontstaansgeschiedenis van de zeepolders', *Biekorf*, 60, pp. 1–26.

Ameryckx, J. and Verhulst, A. (1958) 'Enkele historisch-geografische problemen in verband met de oudste geschiedenis van de Vlaamse kustvlakte', *Handelingen van de Maatschappij voor Geschiedenis en Oudheidkunde te Gent,* 12, pp. 1–24.

Augustyn, B. (1992) *Zeespiegelrijzing, transgressiefasen en stormvloeden in maritiem Vlaanderen tot het einde van de 16e eeuw. Een landschappelijke, ecologische en klimatologische studie in historisch perspectief*, Brussels.

Augustyn, B. (1995) 'Evolution of the dune ecosystem in Flanders during the Middle Ages: anthropogenic factors versus sea level change theory', *Historisch-geografisch Tijdschrift,* 13, pp. 9–19.

Baeteman, C. (1981) 'An alternative classification and profile type map applied to the Holocene deposits of the Belgian coastal plain', *Bulletin Belgische Vereniging voor Geologie,* 90, pp. 257–280.

Baeteman, C. (1983) 'De Holocene sedimenten van de westelijke kustvlakte: een analyse van de Belgische litteratuur', *Professional Paper, Belgian Geological Survey,* 9/204, pp. 1–45.

Baeteman, C. (1985) 'Development and evolution of sedimentary environments during the Holocene in the western coastal plain of Belgium', *Eiszeitalter und Gegenwart,* 35, pp. 23–32.

Baeteman, C. (1991) 'Chronology of the coastal plain development during the Holocene in West Belgium', *Quaternaire,* 2, pp. 116–125.

Baeteman, C. (1999) 'The Holocene depositional history of the IJzer palaeovalley (western Belgian coastal plain) with reference to factors controlling peat beds', *Geologica Belgica,* 2, pp. 39–72.

Baeteman, C. (2004a) 'The Holocene development of a tide-dominated coastal lowland. Western coastal plain of Belgium', in: *Field Guide. The QRA Third International Postgraduate Symposium Fieldtrip, September 17, 2004*, Brussels.

Baeteman, C. (2004b) *Geologische kaart van België 1/25.000. Profieltypenkaart van de Holocene Kustafzettingen*, Brussels.

Baeteman, C. (2005a) 'How subsoil morphology and erodibility influence the origin and pattern of late Holocene tidal channels: case studies from the Belgian coastal lowlands', *Quaternary Science Reviews,* 24, pp. 2146–2162.

Baeteman, C. (2005b) 'The Streif classification system: a tribute to an alternative system for organising and mapping Holocene coastal deposits', *Quaternary International,* 133–134, pp. 141–149.

Baeteman, C. (2007a) 'De laat holocene evolutie van de Belgische kustvlakte: sedimentatieprocessen versus zeespiegelschommelingen en Duinkerke transgressies', *Geoarchaeological and Bioarchaeological Studies,* 8, pp. 1–17.

Baeteman, C. (2007b) 'Roman peat-extraction pits as possible evidence for the timing of coastal changes. An example from the Belgian coastal plain', in: J.J. J.M. Beenakker, F.H. Horsten, A.M.J. de Kraker and H. Renes (eds), *Landschap in ruimte en tijd*, Amsterdam, pp. 16–25.

Baeteman, C. (2008a) 'De Holocene geologie van de Belgische kustvlakte', *Professional Paper – Geological Survey of Belgium* 2/304, pp. 1–36.

Baeteman, C. (2008b) 'Radiocarbon-dated sequences from the Belgian coastal plain: testing the hypothesis of fluctuating or smooth late-Holocene relative sea-level rise', *The Holocene*, 18, pp. 1219–1228.

Baeteman, C., Waller, M., Kiden, P. (2011) 'Reconstructing middle to late Holocene sea-level change: A methodological review with particular reference to "A new Holocene sea-level curve for the southern North Sea" as presented by K.-E. Behre', *Boreas,* 40, pp. 557–572.

Baeteman, C. and Strydonck, M. Van (1989) 'Radiocarbon dates on peat from the Holocene coastal deposits in West-Belgium', *Professional Paper, Geological Survey of Belgium,* 6/241, pp. 59–91.

Baeteman, C. Beets, D.J. and Strydonck, M. Van (1999) 'Tidal crevasse splays as the cause of rapid changes in the rate of aggradation in the Holocene tidal deposits of the Belgian Coastal Plain', *Quaternary International,* 56, pp. 3–13.

Baeteman, C., and Declercq, P-Y. (2002) 'A synthesis of early and middle Holocene coastal changes in the western Belgian lowlands', *Belgeo,* 2, pp. 77–107.

Baeteman, C., Scott, D.B. and Strydonck, M. Van (2002) 'Changes in coastal zone processes at a high sea-level stand: a late Holocene example from Belgium', *Journal of Quaternary Science,* 17, pp. 547–559.

Baeteman, C. and Pieters, M. (2013), 'De laat Holocene afzettingen van de site Raversijde', in: M. Pieters (ed.), *Raversijde: vuurstenen artefacten, een Romeinse dijk, een 14de-eeuws muntdepot, een 15de-eeuwse sector van een vissersnederzetting en bewijs van 17–18de-eeuwse belegeringen van Oostende. Opgravingsverslag van te Raversijde uitgevoerd archeologisch bodemonderzoek (Oostende, West-Vlaanderen, België) in 1992–1999, in 2003 en in 2005,* Relicta Monografieën 8: Raversijde, pp. 29–51.

Barckhausen, J., Preuss, H. and Streif, H. (1977) 'Ein lithologisches ordnungsprinzip für das Küstenholozän und seine Darstellung in Form von Profiltypen', *Geologisches Jarhbuch,* A44, pp. 45–74.

Beets, D.J., and Spek, A.J.F. van der (2000) 'The Holocene evolution of the barrier and the back-barrier basins of Belgium and the Netherlands as a function of late Weichselian morphology, relative sea-level rise and sediment supply', *Geologie en Mijnbouw/ Netherlands Journal of Geosciences,* 79, pp. 3–16.

Behre K.-E. (2007) 'A new Holocene sea-level curve for the southern North Sea', *Boreas*, 36, pp. 82–102.

Briquet, A., (1930) *Le littoral du nord de la France et son évolution morphologique*, Paris.

Cornet, J., (1927) *Leçon de Géologie*, Brussels.

De Breuck, W., De Moor, G. and Maréchal, R. (1970) 'Lithostratigrafie van de kwartaire sedimenten in het Oostelijk kustgebied (België)', *Natuurwetenschappelijk Tijdschrift,* 51, pp. 125–137.

de Jong, J.D. and Hageman, B.P. (1960) 'De legenda voor de Holocene afzettingen op de nieuwe geologische kaart van Nederland, schaal 1:50.000', *Geologie en Mijnbouw*, 39, pp. 644–653.

De Moor, G. and Pissart, A. (1992) 'Het reliëf', in: *Geografie van België*, Brussels, pp. 129–215.

De Moor, G., Ozer, A. and Heyse, I. (2010) 'Belgium', in: E. Bird (ed.), *Encyclopedia of the world's coastal landforms*, Dordrecht, pp. 669–671.

Denys, L. (1985) 'Diatom analysis of an Atlantic-Subboreal core from Slijpe (western Belgian coastal plain)', *Review of Palaeobotany and Palynology,* 46, pp. 33–53.

Denys, L. (1989) 'Observations on the transition from Calais deposits to surface peat in the western Belgian coastal plain – results of a palaeoenvironmental diatom study', *Professional Paper, Geological Survey of Belgium,* 6/241, pp. 20–43.

Denys, L. (1993a) *Palaeoecologisch diatomeeënonderzoek van de Holocene afzettingen in de westelijke Belgische kustvlakte,* Unpublished Dissertation, Universitaire Instelling Antwerpen.

Denys, L. (1993b) 'The diatom record of Holocene deposits near Oostkerke and the development of the Avekapelle gully', in: C. Baeteman and W. de Gans (eds), *Quaternary Shorelines in Belgium and The Netherlands*, pp. 25–42 (Excursion Guide of the 1993 Fieldmeeting of the INQUA Subcommission of Shorelines of Northwestern Europe).

Denys, L. (1993c) 'Holocene valley fills at the southern limit of the western Belgian coastal plain – general palaeoecological outline', in: C. Baeteman and W. de Gans (eds), *Quaternary Shorelines in Belgium and The Netherlands*, pp. 56–73 (Excursion Guide of the 1993 Fieldmeeting of the INQUA Subcommission of Shorelines of Northwestern Europe).

Denys, L. (1994) 'Diatom assemblages along a former intertidal gradient: a palaeoecological study of a Subboreal clay layer (western coastal plain, Belgium)', in: C. Bakker, H. Van Dam, C. Davids and H. De Wolf (eds), *Memorial Volume A. Van der Werff, Netherlands Journal of Aquatic Ecology,* 28, pp. 85–96.

Denys, L. (1995) 'The diatom record of a core from the seaward part of the coastal plain of Belgium', in: D. Marino and M. Montresor (eds), *Proceedings of the thirteenth International Diatom Symposium*, Bristol, pp. 471–487.

Denys, L. (1999) 'A diatom and radiocarbon perspective of the palaeoenvironmental history and stratigraphy of Holocene deposits between Oostende and Nieuwpoort (western coastal plain, Belgium)', *Geologica Belgica,* 2/3, pp. 111–140.

Denys, L. and Baeteman, C. (1995) 'Holocene evolution of relative sea-level and local mean high water spring tides in Belgium – a first assessment', *Marine Geology,* 124, pp. 1–19.

Dubois, G. (1924) 'Recherche sur les terrains quaternaires du Nord de la France', *Mémoires de la Société Géologique du Nord,* 8.

Ebbing, J.H.J., Weerts, H.J.T., Westerhoff, W.E. (2003) 'Towards an integrated land-sea stratigraphy of the Netherlands', *Quaternary Science Reviews*, 22, pp. 1579–1587.

Ervynck, A., Baeteman, C., Demiddele, H., Hollevoet, Y., Pieters, M., Schelvis, J., Tys, D., Strydonck, M. Van and Verhaeghe, F. (1999) 'Human occupation because of a regression, or the cause of a transgression? A critical review of the interaction between geological events and human occupation in the Belgian coastal plain during the first millennium AD', *Probleme der Küstenforschung im südlichen Nordseegebied,* 26, pp. 97–121.

Gottschalk, M.K.E. (1980) 'Subatlantische transgressiefasen en stormvloeden', in: A. Verhulst and M.K.E. Gottschalk (eds), *Transgressies en occupatiegeschiedenis in de kustgebieden van Nederland en België,* Ghent, pp. 21–27.

Gullentops F, and Broothaers L. (1996) 'Overzicht van de geologie van Vlaanderen', in: F. Gullentops and L. Wouters (eds), *Delfstoffen in Vlaanderen,* Lier, pp. 1–27.

Gullentops, F. and De Moor, G. (2001) 'Quaternary lithostratigraphic units (Belgium). 2.2 Remaining marine-estuarine deposits', *Geologica Belgica*, 4, pp. 153–164.

Halet, F., (1931) 'Contribution à l'étude du Quaternaire de la plaine maritime belge', *Bulletin de la Société Géologique*, 32, pp. 152–162.

Houthuys, R., De Moor, G. and Somme, J. (1993) 'The shaping of the French-Belgian North Sea Coast throughout recent geology and history', in: R. Hillen and H.J. Verhaegen (eds), *Coastlines of the Southern North Sea,* New York, pp. 27–40.

Jacobs, P. and De Ceukelaire, M. (2002) *Toelichting bij de Geologische Kaart van België – Vlaams Gewest – Kaartblad Veurne-Roeselare (19–20).*

Loveluck, C. and Tys, D. (2006) 'Coastal societies, exchange and identity along the Channel and southern North Sea shores of Europe, AD 600–1000 AD', *Journal of Maritime Archaeology,* 1, pp. 1–30.

Louwye, S. and Declercq, E. (1998) 'Relative water level change in the intracoastal zone of Belgium and Northern France over the last 2500 years', *Boreas*, 27, pp. 162–177.

Maréchal, R. (1953) 'Ontstaan en morfologie van de Golf van Lo', *Natuur Wetenschappelijk Tijdschrift,* 35, pp. 159–166.

Maréchal R. (1992) 'Kwartairgeologie, Lithologie van de oppervlakkige lagen', *Tweede atlas van België*, Blad II 3, pp. 1–25.

Maréchal, R., De Breuck, W., De Moor, G. and Verheye, W. (1964) *Geologie. Survey West-Vlaanderen*, Geologisch Instituut Rijksuniversiteit Gent.

Moormann, F.R. (1955) 'Over het ontstaan en de opbouw van De Grote Moeren', *Biekorf,* pp. 79–84, and pp. 111–115.

Ozer, A. (1976) 'La morphologie des polders – les dépôts côtiers holocènes', in: A. Pissart (ed.),

Géomorphologie de la Belgique, Liège, pp. 17–27.

Roeleveld, W. (1974) 'The Groningen Coastal Area: A study in Holocene geology and lowland physical geography', *Berichten Rijksdienst Oudheidkundig Bodemonderzoek 1970–1971*, 20–21, pp. 7–25; 1974, 24, pp. 5–132.

Roeleveld, W. (1976) 'De Dunkerque II transgressie langs de Noordfranse en Noordnederlandse kust', *Groningse Volksalmanak*, pp. 151–160.

Roy, P.S., Cowell, P.J., Ferland, M.A. and Thom, B.G. (1995) 'Wave-dominated coasts', in: R.W.G. Carter and C.D. Woodroffe (eds), *Coastal Evolution: Late Quaternary shoreline Morphodynamics*, Cambridge, pp. 121–186.

Smyth, C. and Jennings, S. (1990) 'Late Bronze Age-Iron Age valley sedimentation in East Sussex, Southern England', in: J. Boardman (ed.), *Soil erosion on Agricultural Land*, Chichester, pp. 273–284.

Streif, H. (1971) 'Stratigraphie und Faziesentwicklung im Küstengebied von Woltzeten in Ostfriesland', *Beihefte zum Geologischen Jahrbuch,* 119.

Streif, H. (1972) 'The results of stratigraphical and facial investigations in the coastal Holocene of Woltzeten/Ostfriesland, Germany', *Geologiska Föreningen i Stockholm Förhandlingar,* 94, pp. 281–299.

Streif, H. (1978) 'A new method for the representation of sedimentary sequences in coastal Regions', in *Proceedings 16th Coastal Engineering Conference,* Hamburg, pp. 1245–1256.

Tavernier, R. (1938) 'De geologische ontwikkeling van de Vlaamse kust', *Wetenschap in Vlaanderen*, 4, pp. 41–48.

Tavernier, R. (1948) 'De jongste geologische geschiedenis der Vlaamse kustvlakte', *Handelingen der Maatschappij voor Geschiedenis en Oudheidkunde te Gent*, 3, pp. 107–115.

Tavernier, R. and Ameryckx, J. (1970) *Kust, Duinen, Polders*. Atlas van België, Blad 17, Brussels.

Thoen, H. (1978) *De Belgische kustvlakte in de Romeinse Tijd. Bijdrage tot de studie van de landelijke bewoningsgeschiedenis*, Brussels (Verhandelingen van de Koninklijke Academie voor Wetenschappen, Letteren en Schone Kunsten van België. Klasse der Letteren, 40(88)).

Verhulst, A. (1980) 'De bewoningsgeschiedenis van de kustgebieden van Nederland en België in het licht van de nieuwe opvattingen over transgressie, stormvloeden en klimaatswijziging', in: A. Verhulst and M.K.E. Gottschalk (eds), *Transgressies en occupatiegeschiedenis in de kustgebieden van Nederland en België*, Ghent, pp. 9–19.

Verhulst, A. and Gottschalk, M.K.E. (eds) (1980) *Transgressies en occupatiegeschiedenis in de kustgebieden van Nederland en België*, Ghent.

Vos, P., and Zeiler, F. (2008) 'Overstomingsgeschiedenis van zuid-west-Nederland, interactie tussen natuurlijke en antropogene processen', *Grondboor en hamer*, 3–4, pp. 86–95.

Waller, M. P., Long, A. J. and Schofield, J. E. (2006) 'Interpretation of radiocarbon dates from the upper surface of late-Holocene peat layers in coastal lowlands', *The Holocene,* 16, pp. 51–61.

3 The Holocene coastal deposits of Sussex: a re-evaluation*

Martyn WALLER, Kingston University London
Antony LONG, University of Durham

Abstract[1]

The Holocene deposits of the Sussex coast have been described both as being relatively uniform and as exhibiting considerable differences, with regional (relative sea-level) and local (particularly barrier formation and breaching) processes respectively seen as controlling influences. We reopen this debate detailing the evidence collected over the last twenty five years. Knowledge of the key processes is reviewed along with the stratigraphic data available from the Romney Marsh depositional complex, Pevensey Levels and the Eastbourne area and the valleys of East (Combe Haven, Cuckmere, Ouse) and West (Adur, Arun) Sussex. The valleys and levels of East Sussex contain tripartite sedimentary sequences, with mid-Holocene peats separating clastic sediments of marine/brackish origin. Differences in the thickness of the peat beds can be attributed to a number of factors, including stratigraphic architecture and post-depositional compaction as well as the influence of local processes. Variation in the timing of the onset of peat formation at sites along the East Sussex coastline can be attributed to differences in sediment supply. Evidence is presented that much of the sand and gravel within the Romney Marsh depositional complex was transported landward with rising sea level from offshore sources close to Romney Marsh. The redistribution of gravel along the coast by longshore processes only seems to have become significant in the late Holocene. The valleys of West Sussex do not appear to contain thick mid-Holocene deposits of peat, though further work is required to clarify their Holocene stratigraphies. It is argued that size of the Romney Marsh depositional complex has played an important role in the formation and preservation of a record which reflects a regional driving process, the rate of relative sea-level rise. Smaller systems, in particular the valleys of West Sussex, due to a combination of confined topography and high site exposure, seem more prone to local influences.

---

* This contribution was first published in the volume M.P. Waller, E. Edwards, L. Barber (eds), *Romney Marsh, Persistence and Change in a Coastal Lowland,* Sevenoaks, 2010. The editors sincerely thank the authors and the Romney March Trust for the authorization to republish the text.
[1] Our interest in the coastal deposits of Sussex have been stimulated and enhanced by many individuals, but we would particularly like to thank Michael Tooley, Jill Eddison, Richard Bryant, Paul Burrin, Andy Plater and the late Simon Jennings. We also thank Claire Ivison (Kingston University) for the production of figures 3.1 and 3.3 and Jason Kirby for his insightful comments on a draft of the paper. We would like to thank the officers of the Romney Marsh Trust (in particular the late Terry Burke, David Williams and Sarah Pearson) for their support in the production of the paper.

Landscapes or seascapes? The history of the coastal environment in the North Sea area reconsidered, ed. by Erik THOEN, Guus J. BORGER, Adriaan M.J. DE KRAKER, Tim SOENS, Dries TYS, Lies VERVAET & Henk J.T. WEERTS,Turnhout, 2013 (*CORN Publication Series*, 13), pp. 31–59.

BREPOLS PUBLISHERS DOI: 10.1484/M.CORN.1.101547

I. Introduction

The early 1980s marks the beginning of attempts to understand the Holocene evolution of the Sussex coastline by applying modern techniques of litho-, bio- and chronostratigraphy to the coastal deposits of the region. Differences soon emerged between workers, both as to the degree of stratigraphic uniformity between sites along the coast and their interpretation, with Jennings and Smyth (1982a; 1982b) emphasising differences and invoking local factors (particularly breaching of gravel barriers) and Burrin (1982; 1983) advocating a more uniform stratigraphic model for the region as first proposed by Jones (1971).

Following these early publications, further details of many of the sequences discussed by Jennings and Smyth and Burrin have been published (e.g. Jennings and Smyth, 1987; Smyth and Jennings, 1988; Burrin and Jones, 1991) and new work has been undertaken, not only from the Romney Marsh area (e.g. Waller *et al.*, 1988; Long and Innes, 1995; Plater *et al.*, 1999; Long *et al.*, 2007), but also from Pevensey Levels (Lake *et al.*, 1987; Jones, 2002), Willingdon Levels (Jennings *et al.*, 2003) and the Ouse valley (Waller and Hamilton, 2000). In addition, knowledge of a number of key processes, in particular sea-level change (Long and Shennan, 1993; Waller and Long, 2003), the development of the gravel beaches (Jennings and Smyth, 1990; Nicholls, 1991) and sediment input from terrestrial sources (e.g. Smyth and Jennings, 1990; Waller, 1994; Waller and Schofield, 2007) has been advanced. Now is therefore an opportune time to re-evaluate the coastal deposits of this region and the evidence they contain for the evolution of the Sussex coastline during the Holocene. In terms of the Romney Marsh area, this wider context is important in understanding the longshore supply of sediment as well as helping to establish the degree to which the stratigraphic sequences recorded here reflect local or regional processes.

The early debate on the Sussex coastal deposits focussed on the Ouse valley, the Eastbourne area and Combe Haven (Figure 3.1). However, it is logical to include in any review not only the whole of the Romney Marsh deposition complex but also the coastal deposits as far west as Selsey Bill; together these all fall within a single regional sediment cell (Nicholls, 1991). The Holocene coastal deposits of this coastline include those within a series of confined valleys (Arun, Ader, Ouse, Cuckmere and Combe Haven), which are relatively narrow at their mouths, and the infill of two extensive former embayments; Pevensey Levels and Romney Marsh with their associated valleys. In as much as the data allows, comparisons are made with reference to their three-dimensional stratigraphic architecture, as a number of factors mitigate against attempting to characterise the deposits in any one of these areas in terms of a single or 'type' sequence. At all of the above locations, inland fluvial sediments interdigitate with marine/brackish deposits, while facies variations in the latter are ubiquitous. In addition, in the Romney Marsh depositional complex, west to east progradation and retreat during the mid- and late-Holocene has produced diachronous surfaces, as is evident in the formation of the main marsh peat (see *The Romney Marsh Depositional Complex*).

Radiocarbon dates are presented in calendar years BP. For consistency, all dates have been converted to calendar ages using the calibration program OxCal v.4.0 (Bronk Ramsey, 1995; 2001) and the INTCAL04 dataset (Reimer *et al.*, 2004). The material used for

Figure 3.1 The Sussex coast and Romney Marsh depositional complex showing the sites mentioned in the text

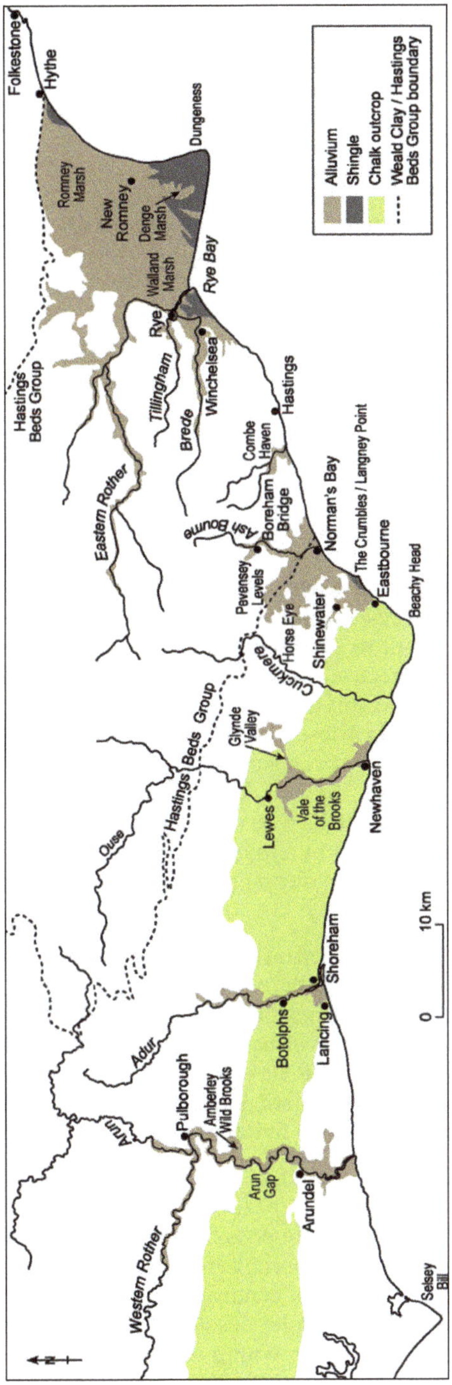

the radiocarbon dates has varied, along with thickness of the samples and methods used (conventional/AMS). The dates in the text have therefore been rounded up to nearest fifty years from the mid-point of the 2 σ calibrated age range.

II. Controls on sedimentation

We start our review by introducing the main processes which have influenced coastal development.

II.1. Relative sea-level change

Accumulations of Holocene sediment >20 m thick exist along the Sussex coast (e.g. at Langney Point, Tilling Green Rye, Dungeness, Newhaven). Relative sea-level (RSL) rise has provided much of the accommodation space (the height difference between a sediment surface and high tide) required, with the contribution from crustal movements (subsidence) estimated at 0.9 mm/yr^{-1} (Gehrels, 2010). The Holocene RSL history of the Sussex coast can be divided into three periods (Waller and Long, 2003). Rapid rise occurred during the early Holocene with an average rate (assuming no significant compaction or lag between RSL rise and sedimentation) of 12 mm/yr^{-1} between *c.* 9200 and 7800 calBP (Waller and Kirby, 2002). During the mid Holocene (*c.* 7800–3700 calBP) the rate of RSL rise declined to between 4 and 2 mm/yr^{-1}. Trends from the late Holocene are difficult to establish (Figure 3.2 from Waller and Long, 2003, with post 2003 data added) because most of the sea-level index points are from the top of peat beds and have been lowered by varying amounts from their original elevation by sediment compaction. Data from regressive contacts (from peat that overlies clastic material and is less affected by compaction) from the Romney Marsh area indicate RSL continued to rise during this interval, but at reduced rate of <2 mm/yr^{-1} (Long *et al.*, 2006a; 2006b). In addition to RSL rise, the accommodation space required for sediment accumulation can be produced by erosional phases and also sediment compaction, as in the Rye area after inundation during the thirteenth century AD (Long *et al.*, 2006b; Long *et al.*, 2007).

While emphasis is usually placed on the nature and rate of sediment accretion, the rate of RSL rise can also influence gravel-barrier dynamics by controlling the supply of sediment to, and stability of, these landforms (Orford *et al.*, 1991). Jennings *et al.* (1998) indicate conditions are favourable for gravel-barrier consolidation with a RSL rise of between 2 and 8 mm/yr^{-1}; below 2 mm/yr^{-1} RSL is too slow for sediment replenishment resulting in the longshore reworking of barriers, while above 8 mm/yr^{-1} the high-tide shoreline is moving so fast that barriers will be overtopped and fail to consolidate.

As a result of the vertical scatter in the sea-level index points (Figure 3.2), caused largely by the differential effects of peat compaction but also potentially by changes in local or regional tidal range, it is not possible to identify smaller amplitude, metre-scale fluctuations in RSL. This is an important limitation to current knowledge since fluctuations of this scale over timescales of centuries to millennia can cause significant changes in coastline position by, for example, destabilising coastal barriers of sand and gravel (Orford *et al.*, 1991) and also by drowning coastal wetlands.

Figure 3.2 The sea-level index points from the Romney Marsh depositional complex and the Sussex coast.
These index points are from transgressive or regressive contacts, which approximate to the former position of mean high water spring tides (MHWST). To allow for variations in present tidal range the data have reduced to mean sea level (MSL).

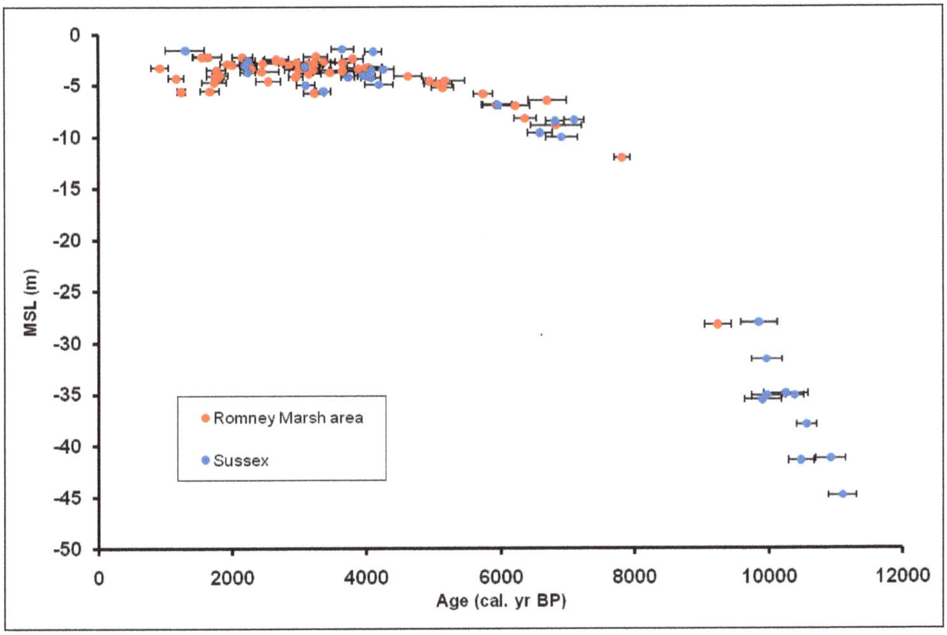

II.2. Coastal processes

Water and sediment movement along the Sussex coast is largely controlled by tidal-wave conditions at the eastern end of the English Channel. Prior to the opening of the Strait of Dover *c.* 8200 calBP this region is likely to have been a low energy, low tidal-range embayment. The subsequent shift to a meso-macro tidal environment, with high tidal-stream velocities through the Strait of Dover, most probably initiated the easterly nearshore drift of sediment (Austin, 1991; Long *et al.*, 1996). Nicholls (1991) argued that under modern conditions sediment moves predominately eastward from Selsey Bill to a regional depocentre at Dungeness (the Sussex 'master cell').

Located at the downwind end of the English Channel, the Sussex coast experiences frequent storms originating in the adjacent North Atlantic (Lozano *et al.*, 2004). Storms are potentially important in enhancing the rate of littoral drift (Nicholls, 1991) and can therefore promote episodes of barrier initiation, consolidation and breakdown depending on whether particular stretches of coastline are sources or sinks of sediment. Both variations in storm frequency and the magnitude of storm events will influence sediment movement. Attempts to use sand-blowing as a proxy indicator of storminess in the North Atlantic suggest variations in frequency may be related to major climatic shifts (Clarke and Rendell, 2009). Episodes of sand drift appear to correspond with periods of Northern Hemisphere cooling and might therefore be expected to coincide with Bond events (approximately cyclical cooling episodes recorded at *c.* 1500 yr intervals; Bond *et al.*, 1997). Historical records also provide valuable information regarding periods of high storm magnitude/frequency, such as during the thirteenth century AD when storms ravaged much of the south coast of England and the North Sea Low Countries (Lamb, 1991; 1995).

It is not certain whether the modern longshore drift system is typical of the past, even the late Holocene. Jennings and Smyth (1990; 1991) noted the coastline would have been more crenulated in the early Holocene and that the presence of large estuaries (such as Pevensey Levels) would have resulted in numerous closed sediment cells even in the late Holocene. However, while acknowledging short-term interruptions, Nicholls (1991) regards longshore transport as being fundamental in determining gravel supply during the Holocene, with the episodic movement of gravel around headlands occurring during stormy periods.

II.3. The supply of gravel

The modern Sussex shoreline consists mostly of sand and gravel, with flint the dominant component of the latter. There are three potential sources of flint, the first of which is from erosion of the chalk during the Holocene. Dornbusch *et al.* (2006), from investigations of proportion of flint in the chalk and measurements of cliff erosion, calculate that 7700 m^3/yr^{-1} of flint is produced by cliff erosion and platform downwearing along the East Sussex chalk coastline. This value is lower than the approximations made by earlier authors (e.g. the 25,000 m^3/yr^{-1} calculated in Jennings and Smyth 1990) and if supplied at a constant rate over the last 5000 years would only account for approximately one fifth of the volume of shingle on Dungeness Foreland (Dornbusch *et al.,* 2006). The volume of flint released is therefore insufficient to account for the amount presently stored in the main gravel depositories which also include The Crumbles, Rye Harbour and Hythe. A second source is from sediment transported by rivers to the floor of the English Channel during Pleistocene cold phases when RSL was low. Jennings and Smyth (1990) proposed that such material was reworked into the nearshore zone by rising sea level and then subsequently moved onshore by storm waves. A third potential source (Nicholls, 1991) is the reworking of gravel stored in the Pleistocene raised beach deposits of West Sussex, although this material would mainly have been accessible during the late Holocene when RSL was close to present.

A decline in supply of gravel as the result of the exhaustion of one or more of these sources can be expected to have led to barrier reworking and potentially changes in the position and permanence of tidal inlets. Changes to the position of mouths of the Sussex rivers are well-documented for the historic period (e.g. Lake *et al.*, 1987; Woodcock, 2003). It would also have altered the morphological and sedimentary characteristics of the barriers, which in itself may have made them more vulnerable to overtopping in major storms (Jennings *et al.*, 1998). A critical question is whether there is a relationship in the timing between phases of initiation, consolidation and breakdown of different barriers along the Sussex coast. Cannibalisation of older barriers may, for example, correlate to episodes of down-drift barrier initiation and consolidation, assuming sediment pathways, including those around headlands, were open.

II.4. Catchment influences and human activity

The catchment characteristics of the levels/valleys within the study area differ significantly, in terms of their area, their topography, as well as their geology and geomorphology (Figure 3.1). These differences will have combined to influence the supply of clastic sediment and freshwater to the coastal zone and, thereby potentially patterns of coastal evolution. Analysis of alluvial sediment sequences across Britain suggests a strong link between episodes of flood-related sedimentation and the Northern Hemisphere cooling events noted above (Macklin and Lewin, 2003; Macklin *et al.*, 2005). The number of flood units increases markedly after *c.* 4500 calBP, suggesting clearance and larger-scale agriculture increased sediment supply from river catchments; indeed Macklin and Lewin (2003) argue that land-use change made catchments more responsive to climate change. The quantity and quality (in terms of dissolved mineral nutrients) of freshwater supplied via the river systems has been shown to influence the development of wetland vegetation in the coastal zone (e.g. Waller *et al.*, 1999; Long *et al.*, 2007). The supply of water from the river catchments is likely to have increased following deforestation due to a reduction in evapotranspiration and the destabilisation of soils.

Inter-tidal environments, particularly in protected back-barrier settings, have long been prime targets for reclamation. Reclamation has two main geomorphological consequences (Long *et al.*, 2007). Firstly, the accompanying reduction in the tidal prism promotes sediment infilling and the ultimate closure of tidal inlets. Secondly, the space available to accommodate storm water is reduced and this raises flood-water levels. When combined with the lowering of drained land surfaces by compaction, this can greatly increase the likelihood of sudden and extensive inundation.

III. Models of coastal evolution

A number of conceptual models have been applied to the coastal deposits of southern England and more specifically to the Sussex coast. The model of Jennings and Smyth (1990) emphasises barrier formation and variations in the quantity and nature of the sediment supply as the key local control on the nature of back-barrier and valley-fill coastal sedimentation. They envisage, for the latter part of the Late Glacial, rising RSL reworking sand and gravel landwards across the floor of the English Channel and into the river channels. During the early Holocene through to *c.* 6000 calBP, this sediment moved progressively shoreward, with numerous river channel/tidal inlets preventing the development of an open littoral drift system until the offshore river valleys were infilled with sediment and overtopped. From *c.* 6000–300 calBP, relatively stable RSL and shallow nearshore environments encouraged the landward transfer of sediment, particularly gravel, resulting in extensive barrier formation and shoreline progradation. Jennings and Smyth (1990) argue that the last 300 years has seen the longshore redistribution of sediment.

A broadly similar three-phase model of barrier development was applied by Long and Innes (1995) to Romney Marsh/Dungeness, although the chronology differs significantly to that of Jennings and Smyth (1990). Based on an interpretation of barrier and back-barrier records, they proposed that barrier initiation occurred *c.* 6500–5000 calBP as a result of an initial influx of sediment, with stabilisation and consolidation *c.* 5500–2000 calBP indicating continued or increasing sediment supply. Barrier breakdown, the segmentation and reorientation of the barrier, in the last 2000 calBP resulted from a decline in the supply of sediment, an increase in storm magnitude/frequency or accelerated RSL rise (Long and Innes, 1995).

Lastly, Long *et al.* (2000) proposed a three-phase model of estuary development from work in Southampton Water, which by emphasising regional changes (RSL), may be applicable to southern England, including the Sussex and Kent coasts (Long, 2001). During the early Holocene, estuary expansion, driven by rapidly rising RSL, resulted in the formation of basal peats and thick deposits of estuarine sediment. Estuary contraction occurred in the mid Holocene as peat formation, dependent upon continued but slowly rising RSL, became widespread. Wetland inundation, coinciding with relatively stable RSL in the late Holocene, was the result of conditions being unfavourable for the continued accumulation of organic material due to either the negligible RSL rise or peat compaction (Long *et al.*, 2000).

IV. The coastal deposits

We now outline the stratigraphic information available (summarised in Figure 3.3) on the Holocene deposits of the valleys and former embayments of Sussex, starting at the eastern end of the regional sediment cell and working westwards.

Figure 3.3 Schematic stratigraphies of the Holocene deposits of the Sussex coast and the Romney Marsh depositional complex.
Age ranges (calBP) are given for the peat contacts where lithostratigraphic continuity has been established between sites

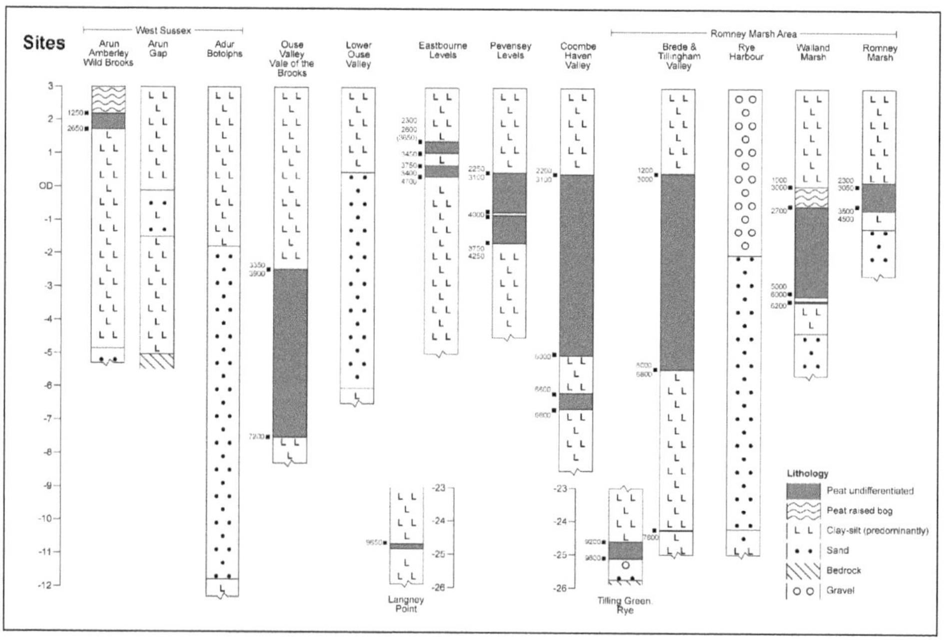

IV.1. The Romney Marsh depositional complex

The Romney Marsh depositional complex (Figure 3.1) comprises the now reclaimed back-barrier wetlands of Walland Marsh, Romney Marsh proper and Denge Marsh, a series of valleys which feed into the western side of the marshland (Eastern Rother, Tillingham and Brede), and the gravels of a degraded barrier system (at Rye Harbour, Dungeness and Hythe).

Away from the valleys, few boreholes have fully penetrated the back-barrier deposits of the area and knowledge of the pre-inundation land surface is therefore limited. Investigations from the Tillingham near Rye indicate the presence of a palaeo-valley at a depth of *c.* 25 m from the modern surface (Waller and Kirby, 2002). The upper contact of a thin basal peat from Tilling Green Rye provides the earliest sea-level index point (*c.* 9200 calBP at -24.68 m OD) from the complex (Figure 3.2). This confirms that tidal waters had extended into this part of the eastern English Channel by this date. The overlying sediment, indeed the bulk of the fill in the valleys (e.g. Waller *et al.* 1988), consists of blue-grey silty-clays, which microfossil evidence suggests was deposited in a low-energy intertidal marsh or tidal-flat environment. A thin peat, dated to *c.* 7800 calBP at -10.24 m OD, is only recorded from the lower Brede and Tillingham and seems to record a local episode of wetland expansion and then retreat (Waller and Kirby, 2002). In the lower valleys a coarsening-upwards sequence is apparent with marine sands of tidal-flat or tidal channel origin deposited after *c.* 8000 calBP. In most of the boreholes sunk across Walland Marsh and Romney Marsh proper (e.g. Long and Innes, 1995; Long *et al.*, 1998) sand (impenetrable by hand-coring) is recorded at a depth of *c.* -3 m OD. Termed the 'Midley Sand' by Green (1968), these deposits are now known to be stratigraphically distinct from the surface sand deposits regarded by Green as part of the same unit (Long and Innes 1993). Blue-grey silts and clays above these lower sands indicate a decline in environmental energy before peat formation, most probably due to the development of a protective coastal barrier across Rye Bay.

Across the western side of Walland Marsh (from Horsemarsh Sewer to Pett Level) a thin (*c.* 10 cm) layer of peat occurs, intercalated between clay layers, prior to the formation of the main marsh peat (Tooley and Switsur, 1988; Waller *et al.*, 1988; Long and Innes, 1995). The latter is a laterally persistent peat bed which varies in thickness from *c.* 6 m in the valleys to <1 m on Romney Marsh proper, though locally in one of the small tributary valleys (the Pannel), it is thicker and extends to the surface (Waller, 1993). Radiocarbon dates from the base of this layer (where it overlies clays and silts of marine/brackish origin) range from *c.* 6800 calBP in the Brede valley (at Brede Bridge) to *c.* 3500 calBP on Romney Marsh proper near Newchurch (Long *et al.*, 2007). After a short-lived transitional reedswamp phase, eutrophic fen carr communities became established in the valleys, with alder the dominant taxa (Waller, 2002). In the late Holocene, the peat-forming communities on Walland Marsh indicate progressively more acidic conditions, with an area likely to have been isolated from nutrient-rich waters eventually (*c.* 2750 calBP) developing into a raised bog (Waller *et al.*, 1999; Long *et al.*, 2007). Dates from the upper surface of the peat range from *c.* 3200 to 1300 calBP in the Brede valley alone, and are not, in general terms, thought to be a reliable guide to the timing of the end of peat formation or the beginning of the deposition of overlying marine/brackish sediments (Waller *et al.*, 2006).

The post-peat sediments of the Romney Marsh depositional complex, the 'Young Alluvium' of Green (1968), accumulated in tidal inlets with entrances at Hythe, New Romney and Rye. They comprise a complex sequence of grey or brown clays, silts and sands and are frequently laminated. The Hythe inlet existed during the Roman period from c. 1800 calBP with reclamation certainly well advanced by c. 1200 calBP (Long *et al.*, 2007), though the Rumensea Wall (enclosing Romney Marsh proper in the southwest) may date from as early as c. 1400 calBP (Allen, 1999). The tidal area associated with the New Romney inlet, which dates from c. 1400 calBP, appears to have been relatively small, confined by the Rumensea Wall and the Walland Marsh raised bog. A large tidal channel, the Cheyne Channel, extends in a north-westerly direction (Long *et al.*, 2007), and the tidal channels in which the near-surface 'Midley Sands' were deposited (Kirby *et al.*, 2010) seem also to be related to this inlet. The first breach of the coastal barrier at Rye probably occurred as early as *c.* 1200 calBP, though it was certainly considerably widened by the storms of the thirteenth century AD. Here, barrier failure is associated with widespread peat erosion (see below) and the movement of gravel and sand inland. Sediment, including gravel, appears to have been driven into the area which became the Rother estuary and the gravel was subsequently redistributed eastward along the northern arm of the Wainway Channel to form small beaches as observed at Moneypenny Farm (Long *et al.*, 2007). Both reclamation and the regrowth of the coastal barrier in the Rye Harbour area from *c.* 600 calBP (Lovegrove, 1953; Long *et al.*, 2007) are implicated in the reduction of the tidal area of the Rye inlet and the consequence demise of New Winchelsea and Rye as major ports.

The post-peat sediments might be expected to have been derived from terrestrial catchment sources. However, primary woodland clearance in the river catchments occurred largely in the period from 4000 to 3300 calBP (Waller and Schofield 2007). With no immediate response to land-cover changes evident, it appears that any sediment released either passed into temporary sediment stores or was transported directly to the coastal zone. However, a phase of arable activity during the Roman period around Rye resulted in the deposition of thin (<0.5 m) slopewash deposits between the peat and overlying marine/brackish sediments at a number marshland edge locations (Long *et al.*, 2007).

Recent work on the back-barrier deposits of the Romney Marsh area (e.g. Long *et al.*, 2007; Plater *et al.*, 2009) cautions against assuming continuous incremental sediment accumulation and gradual coastal change. For example, the peat-accumulation rate in the Brede valley declines after *c.* 4000 calBP and accumulation appears to have ceased *c.* 1000 cal. yr before inundation and the deposition of the bulk of the post-peat sediments (Waller *et al.*, 2006). Phases of exceptionally rapid sedimentation are also evident, with the analysis of tidal rhythmites (finely laminated beds of sand and mud) suggesting that in certain circumstances (e.g. tidal channels) sediment accumulation rates of >0.5 m/yr^{-1} may be attained (Stupples, 2002; Long *et al.*, 2007). In addition, post-depositional factors, notably erosion and sediment compaction, need to be considered. For example in the Rye Harbour area, the thirteenth century AD breach resulted in the erosion of peat from a 4 km wide corridor and in peripheral locations subsequent compaction lowered the peat surface by at least 3 m (Long *et al.*, 2006b). Dating the upper surface peat layers is considered particularly problematic, as the sediment dated may have accumulated slowly, experienced compaction or the peat layer may have been erosionally truncated (Waller *et al.*, 2006).

In addition to work in the back-barrier area, optically stimulated luminescence (OSL) dating has provided a direct chronology for the evolution of Dungeness Foreland (Long *et al.*, 2007; Roberts and Plater, 2007). Dates from the underlying shoreface sands constrain the timing of the emplacement of the overlying the gravel. These shoreface sands accumulated as an inter-tidal or subtidal spit, under conditions of high sediment supply, with the overlying gravel deposited soon after their emergence under the influence of storms and waves. The most westerly exposed beaches at Broomhill Level are the oldest and date to *c.* 4100 yr BP, probably forming as part of a bay bar (Long *et al.*, 2007). The beaches between Broomhill and Holmstone accumulated between 4000 and 2000 yr BP. From *c.* 2000 yr BP, there was a change in the rate and direction of progradation with rapid easterly growth, which may be a result of the updrift cannibalisation of sediment. A second phase of north-easterly extension occurred from *c.* 1300 to 400 yr BP with barrier reorientation and further cannibalisation probably due to the updrift breach in the barrier system at Rye (Roberts and Plater, 2007; Plater *et al.*, 2009).

IV.2. The Combe Haven valley

The deposits of the lower Combe Haven valley (Figure 3.1) have been subject to extensive investigation (Jennings and Smyth, 1987; Smyth and Jennings, 1988). Three main Holocene units are recognised. The deepest, silty clay (from above *c.* -19 m OD) accumulated in a marine/brackish environment. An overlying peat is, at mid valleys sites, divided into two by an intervening wedge of the silty clay. The lower thin (<0.8 m) peat layer formed between *c.* 6800 and 6600 calBP and the upper layer after *c.* 6000 calBP. The latter thickens (maximum *c.* 5 m) upstream and is overlain by silty clays (<1 m thick) or, adjacent to the coast, gravel. Dates from the transition between the peat and the upper silty clay range from *c.* 3100 to *c.* 2200 calBP. Pollen evidence suggests that the upper silty clay can be divided into an upstream freshwater facies and downstream marine facies, with a shift to freshwater conditions at downstream locations recorded *c.* 0.25 m below the surface (Jennings and Smyth 1987).

Smyth and Jennings (1988; 1990) report further investigations of the peat/upper silt transition, noting the coincidence between changes in pollen stratigraphy (declines in tree pollen, particularly alder, and rises in herbaceous pollen) and the switch to minerogenic sedimentation. At the upstream locations this shift was attributed to forest clearance *c.* 3100 calBP, with the later (*c.* 2200 calBP) inundation at downstream sites seen as a consequence of this activity; increased river discharge following clearance would have produced a widening in the estuary mouth and promoted an extension of the tidal limit up-valley (Jennings and Smyth, 1987; Smyth and Jennings, 1988; 1990).

IV.3. Pevensey Levels

Lake *et al.* (1987) outline the alluvial deposits of Pevensey Levels (Figure 3.1) which are described as being 15 m thick (and deeper near the coast) overlying a complex basal surface. Bedrock 'islands' occur adjacent to the coast (Hooe Level) and at a number of inland locations (e.g. Horse Eye). The alluvial deposits typically comprise four units. The basal deposit consists of firm clayey silts with various inclusions including peat fragments. A basal peat was recorded near Normans' Bay at a depth of -6.4 m OD. The overlying clay with sand (maximum thickness >9 m) is of marine/brackish origin at New Bridge (Horse

Eye Level) where foraminifera indicate sand-flat associations initially, tending towards a high salt-marsh environment in the upper levels (Lake *et al.*, 1987). A peat bed follows with a maximum thickness of 1.8 m, though it is absent from the former river courses. North and west of Horse Eye peat occurs extensively at 0.9 to 1.5 m below ground surface and locally outcrops at the surface in several small tributary valleys. *Cerastoderma* shells occur in the uppermost deposit, clay. In addition, sand bodies within the sequences are recorded to the south and east of Horse Eye and extending northwards towards the Ash Bourne valley. Lake *et al.* (1987) suggest they represent channel fills though they could not exclude the possibility that some originated as barrier-bars.

Moffat (1984; 1986) and Jones (2002) report palaeoenvironmental investigations from the main peat bed and adjacent sediments. In pollen diagrams from eastern Pevensey Levels alder appears to be dominant, with sedge values increasing in the overlying sediment, though a diagram from a more central location indicates more open and wetter conditions. At the latter site foraminifera indicate the under and overlying silty clays are of marine/brackish origin along with a thin (8 cm) intercalated clay layer which was also recorded in adjacent boreholes (within *c.* 1 km).

Given the architecture of the deposits of the valleys feeding into the western side of the Romney Marsh depositional complex and Combe Haven, the Pevensey Levels peat bed could be expected to thicken considerably in the lower Ash Bourne valley. This appears not to be the case. A borehole (sunk to test this hypothesis) at Boreham Bridge, revealed a sequence of blue-grey clay (between 8 m and 3.79 m below the surface), overlain by 2.17 m of peat with clayey intercalations and 1.62 m of silty clay.

Six radiocarbon dates from lower (regressive) contacts of the peat bed range from *c.* 4250 to 3750 calBP (Moffat, 1984; Jones, 2002), and although from geographically disparate locations including a tributary valley (Court Lodge), towards the coast (near Normans' Bay) and inland (Horse Eye), show no obvious spatial trends. The thin intercalated clay reported by Jones (2002) appears to have been deposited over a short period (<100 cal. yr) at *c.* 4000 calBP. Three dates from the upper (transgressive) contact of the peat bed range from *c.* 3100 to 2250 calBP (Jones 2002). Doubts as to whether the latter accurately reflect the age at which the bulk of the upper clays were deposited are raised by the founding grant (AD 1180) of Otham Abbey (on the western edge of the marshland) which gave the Abbey the right to take '60 cart loads of peat yearly from the manor of Pevensey' (Salzmann, 1910). The caveat 'so long as the moor may last' suggested to Salzmann (1910) realisation of a gradual extension of tidal conditions at this time, though might alternatively reflect recognition that the resource was dwindling under the pressure of peat cutting. Two dates from organic detritus within the upper clay (within 0.4 m of the surface) of *c.* 500 calBP were obtained by Moffat (1984), although the dated material may be reworked or have been contaminated by the vertical penetration of roots.

The reclamation history of Pevensey Levels is detailed by Salzmann (1910), Dulley (1966) and Rippon (2000). Although reclamation appears to have started relatively late (Rippon, 2000), by the mid-thirteenth century AD most of the wetland had been drained and three major tidal inlets embanked; Pevensey Haven, Old Haven/Waller's Haven (into which the river Ash Bourne flowed) and Esthaven. The storms of AD 1287 caused flooding at a number of locations and were followed by several attempts to improve the sea defences and drainage system, which included damming Pevensey Haven. Subsequently, and presumably as consequence, the tidal flow appears to have been insufficient to keep the mouth of Ash Bourne (Old Haven) open and the lower Ash Bourne (e.g. Hooe Levels) experienced extensive freshwater flooding from the mid-fourteenth century AD. Old Haven was then abandoned and several cuts made through to Esthaven, with the current outflow at Normans' Bay dating to AD 1455 (Dulley, 1966; Lake *et al.*, 1987).

IV.4. The Eastbourne area

The Holocene deposits of the Eastbourne area occupy a short valley/basin (Willingdon Levels/Lottbridge Drove) between the Chalk in the west and, in the east, a ridge of Weald Clay which rises to a maximum 35 m OD and separates this area from Pevensey Levels (Figure 3.1). At the coast is an accumulation of gravel known as The Crumbles. Deep boreholes indicate bedrock at a depth of 33.4 m beneath The Crumbles at Langney Point and *c.* 12 m at Lottbridge Drove. Stratigraphic investigations from Langney Point, Eastbourne Level and Southbourne Level (Lottbridge Drove) and Willingdon Levels are reported in Jennings (1985), Jennings and Smyth (1985; 1987) and Jennings *et al.* (2003).

The Langney Point sequence starts with a basal peat (with dates of *c.* 10,850 and 9800 calBP; Harkness and Wilson, 1979) or minerogenic sediment overlain by a thin peat (Jennings 1985). The transgressive contact of the latter is dated (at -24.7 m OD) to *c.* 9850 calBP (Jennings, 1985). The overlying clay is replaced by sand at -14.2 m OD which in turn is followed by *c.* 8 m of gravel above -3.7 m OD. The biostratigraphy of the clay suggests an estuarine origin with more fully marine conditions during the deposition of the sand (Jennings and Smyth, 1987). At Lottbridge Drove, valley gravels are overlain above -6.08 m OD by silty clay, which again contains a biostratigraphy indicative of estuarine conditions. An overlying peat occurs over most of the Lottbridge Drove/Willingdon Levels area and is a maximum 1.17 m thick. At some locations it is divided in two by estuarine clay. An upper silty clay (1.75 m thick at Lottbridge Drove) was divided by Jennings and Smyth (1987) into a lower estuarine and an upper freshwater facies.

As a result of the discovery of a late Bronze Age occupation platform within the upper part of the peat at Shinewater in 1995 (Greatorex, 2003), further investigations were undertaken from the peat from a number of sites across Willingdon Levels (Jennings *et al.*, 2003). The four dates obtained from lower (regressive) contacts range from *c.* 4700 calBP on the western edge of Willingdon Levels at Hydneye to *c.* 3400 calBP at Mornings Mill Farm on the northern edge. The thin intercalated clay was dated at Shinewater to between *c.* 3750 and 3450 calBP. The four dates from upper (transgressive) contacts on Willingdon Levels (Jennings *et al.*, 2003) are relatively consistent (*c.* 2600–2300 calBP), while a much earlier date (*c.* 3650 calBP) was obtained from Lottbridge Drove (Jennings and Smyth, 1987). The pollen stratigraphy from the Willingdon Levels peat is dominated by taxa indicative of open and wet conditions. Only at Mornings Mill Farm does alder attain frequencies indicative of the development of fen woodland. Jennings *et al.* (2003) attribute the open conditions to anthropogenic activity.

Jennings and Smyth (1987) suggest development of the gravel foreland at The Crumbles occurred from the twelfth century AD and initiated the freshwater facies in the upper silty clay. Cartographic evidence suggests The Crumbles has experienced sediment depletion and shoreline retreat since *c.* AD 1600 (Jennings and Smyth, 1990).

IV.5. The Ouse and Cuckmere valleys

The deposits of the Ouse valley (Figure 3.1) were first reported by Jones (1971) and Thorley (1971), with further details of these investigations provided in Thorley (1981) and Burrin and Jones (1991) and additional information available from Lake *et al.* (1987) and Waller and Hamilton (2000). The bedrock surface in the lower valley, shown in detail by Burrin and Jones (1991), extends below -25 m OD at Newhaven, shallowing to *c.* -10 m OD in the Vale of the Brooks. Basal sands and gravels (a maximum 3 m thick) of uncertain origin are widespread. Burrin and Jones (1991) divide the overlying fill into an inland sequence and, from the Chalk outcrop downstream, a perimarine sequence. The former comprise fluvial and colluvial derived deposits and locally peat (Robinson and Williams, 1983). The latter include estuarine sediments clays, with interbedded peats, which are thickest (*c.* 10 m) in the Vale of the Brooks. Overlying sand and clay units thicken downstream of the Vale of the Brooks. Above the basal unit, the only gravel encountered was that forming the contemporary beach (estimated a *c.* 5 m thick) across the valley mouth.

Bio- and chronostratigraphic information is available from the Vale of the Brooks and from a tributary, the Glynde valley. The deepest sediment (at *c.* -9 m OD) investigated by Thorley (1981) is a peaty clay at Lewes I which appears from the pollen stratigraphy (Waller and Hamilton, 2000) to have been deposited under salt marsh or tidal-flat conditions. This is followed by a phase of freshwater peat formation which began at Lewes I *c.* 7200 calBP (Thorley, 1981) and prior to *c.* 7100 calBP in the Glynde valley (Waller and Hamilton, 2000). Peat formation was evidently extensive during the mid Holocene, from *c.* 6500 to 3350 calBP at Lewes II and spanning *c.* 5800 to 4900 calBP from a sequence reported from the Vale of the Brooks by Lake *et al.* (1987). The peat is consistently described as consisting of a basal woody unit and an upper detrital peat, though alder pollen values are high throughout the three published sequences. Two dates have been obtained from top of the peat; at Lewes II *c.* 3350 calBP and *c.* 3900 calBP from the Glynde valley. At both these sites the pollen stratigraphies suggest that contacts with the overlying marine/brackish deposits are undisturbed and continuous, though the variation in the altitude of the peat surface across the Vale of the Brooks (Thorley, 1981; Waller and Hamilton, 2000) suggests that at least some locations that the overlying clastic unit (largely clayey silts in the Vale of the Brooks) sits unconformably on the peat.

Knowledge of the deposits of the lower Cuckmere valley is limited. Burrin (1983) reports boreholes from Cuckmere Haven and Litlington that suggest a similar stratigraphy to the lower Ouse. At Cuckmere Haven basal gravels are replace *c.* 28 m below the surface by silty clays (to *c.* 20 m), then sands (to *c.* 3 m) followed by an upper silty clay.

IV.6. The Adur and Arun valleys (West Sussex)

The Holocene deposits of the lower Adur are detailed in Young and Lake (1988) from borehole records held by the British Geological Survey. The deepest part of the bedrock channel (overlain by basal gravels of unknown origin) at the mouth of the Adur (below -23 m OD) lies mid-way between Lancing and Shoreham (Figure 3.1). Where the A27 crosses the valley and near Botolphs (*c.* 2 and 6 km inland respectively) a tripartite Holocene sequence can be distinguished. Lower silts and clayey sands (maximum *c.* 10 m thick) are replaced by fine- to medium-grained sands (maximum *c.* 10 m thick) and by further silts and clayey sands (at the A27) and clays (at Botolphs). In a borehole log from Lower Beeding (just north of Botolphs) traces of peat are reported from the sand unit and the overlying clay is described as organic (Young and Lake 1988). A section immediately inland of the coastal outcrop of gravel, shows a *c.* 10 m gravel bed resting on bedrock and sand, overlain by a few metres of sand or clay. Gravel (<1 m thick) was also recorded within and overlying the sand unit in two boreholes near the A27. Details of the development of the gravel spit at the Adur mouth from the mid-seventeenth century AD are provided in Ward (1922) and Young and Lake (1988).

Litho-, chronostratigraphic and seismic data are available for the offshore palaeo-Arun valley (Gupta *et al.*, 2004; 2007; Bayliss *et al.*, 2007). The Holocene infill includes thin (up to 20 cm thick) organic layers, some of which are intercalated within estuarine deposits. Radiocarbon dates from these peats range from 11,300–9750 calBP with peat age related to altitude (Figure 3.2). Brief details of the stratigraphy of the lower Arun valley are provided in Aldiss (2002). Boreholes indicate the buried valley extends to a depth of 36 m below surface at the mouth of Arun and 31 m at Arundel. The overlying sediments are said to comprise silty clays, silts and fine sands, which is consistent with the schematic stratigraphies presented in Castleden (1982).

Boreholes sunk for the purpose of palaeoenvironmental reconstruction from near the edge of the floodplain in the Arun gap at North Stoke reveal an 8 m sequence above the chalk comprising clayey silts with a sandier unit *c.* 3 to 4 m below the surface (Figure 3.3). Organic material (wood etc.) is scarce and clearly not *in situ*. Although investigations focussed on the near-surface peat deposits, Waton (1983) provides some details of the deeper stratigraphy of Arun upstream of the chalk at Amberley Wild Brooks. On the western side, peat is underlain directly by sand. To the east peat is underlain by clay (the upper layers of which are rich in *Phragmites*) which extends to depth of *c.* 8 m beneath the surface where sand was encountered. The pollen assemblage from the upper metre of the clay beneath the peat suggests deposition in a marine/brackish environment, with the onset of peat formation dated to *c.* 2650 calBP (Godwin and Willis 1964). The presence of *Sphagnum* and *Molinia* layers indicate that after *c.* 1250 calBP (Waton, 1983) the peat formed in an oligotrophic environment, a raised bog (Godwin1943). Tributary valleys, the Western Rother and the Chilt (near Pulborough), contain organic fills, with peat forming in the Chilt from *c.* 4300 calBP to the present (Groves 2008) and over 5 m of peat reported in the Western Rother (Aldiss 2002).

V. Comparison of the coastal sequences

The stratigraphies detailed above are represented schematically in Figure 3.3. The altitudes are indicative; where detailed lithostratigraphic information is available the altitude of peat beds is shown to be locally influenced by the thickness of overburden and the depth of the Holocene sequence (e.g. Long *et al.* 2006b).

Both basal and, below *c.* -24 m OD, intercalated peats are common both offshore and inland in the buried valley systems found along the Sussex coast. The latter, reported from Langney Point and the Arun, are thin and appear to be short-lived episodes (invariably only one radiocarbon date is available) dating prior to *c.* 9000 calBP. The sediments deposited from *c.* -24 to -10 m OD appear to be largely fine-grained clastics, clays and silts, though the data available are largely derived from commercial boreholes at locations inland. These sediments are compatible with low energy, probably low tidal-range environments, before the opening of the Strait of Dover. The coarsening upwards sequence that follows, giving rise to the deposition of sands in the Romney Marsh area after *c.* 8000 calBP, may also be evident in Adur valley and possibly the lower Ouse (though see below).

The peats which are consistently recorded in the coastal deposits of East Sussex began to accumulate *c.* 7200 calBP, though interruptions during the early stages of peat growth appear common with marine conditions returning to Combe Haven and western side of the Romney Marsh depositional complex. Neither the onset of peat formation nor the age of the intercalated clays appears consistent along the coast. Distinctions between the valleys and levels (e.g. Jennings and Smyth, 1987) are difficult to justify since where data are available the continuity of the sequences is clear. Intercalations within the main peat bed have only been recorded from Pevensey Levels and the Eastbourne levels where clay layers of marine/brackish origin dating to *c.* 4000–3500 calBP have been consistently recorded from some areas.

During the mid Holocene, alder-dominated vegetation occurs in valleys and is replaced spatially by communities indicative of wetter conditions on central Pevensey Levels (Jones, 2002) and by more acidic conditions on Walland Marsh (Waller *et al.*, 1999). At the seaward end of some sequences, peat is absent. In the case of the Rye Harbour area, widespread erosion occurred during a phase of post-peat inundation, destroying the peat and creating the accommodation space for the deposition of sands. The relationship between the peat deposits of Vale of the Brooks and the sands recorded by Burrin and Jones (1991) in the lower Ouse is unclear and the latter unit could also have been deposited after extensive peat erosion.

In contrast to East Sussex, thick mid-Holocene peat layers have not been recorded from the lower Adur and Arun valleys. There are three possible explanations. Firstly, peat may in fact exist but has yet to be discovered, either due to the relative scarcity or nature (mostly commercial borehole) of the lithostratigraphic data. Secondly, as seems possible in the lower Ouse, peat may have been removed by erosion. Thirdly, peat may never have formed, with the lower valleys remaining as open estuaries characterised by intertidal and subtidal conditions into the late Holocene (see *Coastal processes and evolution*).

In the late Holocene, marine-brackish conditions returned to the valleys and levels of East Sussex. Dates from upper (transgressive) contacts range from *c.* 3650 to 1000 calBP and are locally inconsistent, as has been most clearly demonstrated in the Brede valley (Waller *et al.*, 2006). There is no reason to regard the problems associated with the dating of transgressive contacts to be confined the Brede and such dates can only be regarded as providing a maximum age for the deposition of the overlying sediments. Archaeological and historical evidence in some cases provides additional dating evidence (Long *et al.*, 2006a). Peat formation continued at some locations during the late Holocene, including the edges of both the Romney Marsh complex (e.g. the Pannel valley; Waller 1993) and Pevensey Levels, and inland valley locations in the Ouse and Arun. In addition, peat initiation occurred on Amberley Wild Brooks (the Arun valley) with a raised bog developing here (*c.* 1250 calBP) and over southern Walland Marsh (*c.* 2700 calBP), though peat formation ceased in the latter area following marine inundation probably associated with the thirteenth-century AD storms.

VI. Coastal processes and evolution

In this section we reappraise the roles of regional and local processes in controlling the Holocene evolution of the Sussex coast using the stratigraphic information previously detailed.

VI.1. Relative sea-level change

The data now available (updated from Waller and Long, 2003) to reconstruct RSL changes in the study area are shown in Figure 3.2. The 65 index points divide roughly evenly between the Romney Marsh depositional complex (35 dates) and the rest of the Sussex coast (30 dates). The deepest dates are mostly from the palaeo-Arun (Gupta *et al.*, 2004; Bayliss *et al.*, 2007) and indicate RSL rose from -45 m to -25 m mean sea level (MSL) between *c.* 11,500–10,000 calBP. The RSL plot assumes constancy in tidal range between the present and the time of sample formation. As noted above, the tidal range in the eastern English Channel is likely to have increased after the opening of the Strait of Dover and, for this reason, these older dates probably plot slightly above their true elevation (though by not more than 2 m). However, these deeper samples also lack detailed biostratigraphic supporting data from which to determine their precise elevation with respect to contemporaneous sea level, being derived from a range of materials including peat, leaf and plant-stem fragments as well as herbaceous plant remains (Gupta *et al.,* 2004). They probably formed at or above contemporaneous mean high-water spring tide (MHWST), but the precise reference water level is not known. In our analysis we assume that they formed at MHWST but if they formed below or above this level, their elevations would require adjustment. In practice, the rate of RSL rise (*c.* 20 m in less than 2000 cal. yrs) was so fast during this interval that these corrections are probably of secondary importance.

There is a significant gap in data between the deepest population of samples, below *c.* -25 m MSL, and the second group above *c.* -12 m MSL. This gap records the division between the offshore and deeply buried samples and the organic sediments that are within easy coring reach from terrestrial settings across the region. The majority of the data after this interval are derived from the Romney Marsh depositional complex and, as noted previously, they record an upwards rise in sea level at a decelerating rate until *c.* 3700 calBP, after which the rate slowed significantly. Much of the height and age scatter in the dates after this interval reflect the effects of sediment compaction (all data points are from intercalated deposits) and for this reason the trend in the RSL data does not pass through 0 m at the present day.

The coastal deposits of East Sussex, with thick mid-Holocene peats present, are broadly consistent with the three-phase model of estuary development of Long *et al.* (2000), with the availability of accommodation space, governed by changes in RSL, being the regional underlying controlling process. During the early Holocene the creation of accommodation space generally outstripped sediment supply and terrestrialisation and peat formation were therefore rare. During the mid Holocene, although the rate of RSL rise declined, the continued creation of accommodation space resulted in the accumulation of peats, with the eutrophic alder dominated communities indicative of high groundwater levels. During the late Holocene, the rate of RSL rise and sedimentation rates decline and thick accumulations of sediment are rare except for locations where accommodation space was created by erosion and compaction or where peat formation occurred independently of the groundwater table.

However, variations in the occurrence and timing of peat formation in East Sussex and in particular the apparent absence of mid-Holocene peats from the West Sussex valleys suggest that other processes, which operated variably in time and space, were able to override the influence of RSL. Given the pattern of RSL rise, it might be expected that such processes would become increasingly influential as the Holocene progressed. The processes which may locally have determined the nature and timing of sedimentation include the supply clastic sediment, the degree of protection (the evolution of gravel barriers), the morphological and size characteristics of catchments, climate change/ storm activity and human activity. These are considered separately below though many of these processes are interlinked and separating their respective contributions through time is not yet possible.

VI.2. Sediment supply

Basal peats and thick sequences of clastic sediment, as recorded in many of the offshore samples and deeper boreholes from the region (e.g. the palaeo-Arun, Tilling Green Rye, Langney Point), are consistent with rapidly rising RSL. However, the presence of intercalated peats during a period of rapid RSL rise, at Langney Point and offshore in the Arun, is noteworthy. It is possible that these formed under temporary slow-downs in RSL, not evident from the rather coarse RSL data available at this time. Evidence from beyond the study area indicates that the rate of 'eustatic' sea-level rise was not constant during the early Holocene, with for example a possible abrupt rise in sea level associated with the drainage of glacial lakes that surrounded the former Laurentide Ice Sheet at *c.* 8200 calBP recorded in the coastal deposits of the Netherlands (Törnqvist *et al.*, 1998; Hijma and Cohen, 2010). However, the early Holocene peats observed in the Sussex coastal lowlands are generally older than this event and inconsistent in age suggesting that they reflect local and not regional or global processes. These layers seem likely to have formed behind ephemeral barriers reflecting both an abundance of clastic sediment (the sediment transported by rivers to the floor of the English Channel during the Devensian) and the presence of channels associated with the braided rivers of the Late Glacial.

The timing of the onset of widespread peat formation in the East Sussex valleys varies and is therefore unlikely to be the product of fluctuations in RSL. In these protected locations the point at which the accumulation of clastic material (which became the platform necessary for peat growth) was able to exceed the creation of accommodation space will have been strongly influenced by sediment supply. Fine-grained sediments will have ultimately derived from the Wealden catchments, however, sediment may have been supplied directly from the catchment and reworked in the tidal estuaries or have originated from the movement of material onshore. Investigations of the fluvial deposits of the Ouse valley led Burrin and Scaife (1984) to suggest woodland clearance resulted in increased sedimentation from the Mesolithic. However, these deposits were not directly dated and later work on the vegetation history of Weald (e.g. Waller, 1993; 1994; Waller and Schofield, 2007; Groves 2008) suggests catchments remained well-wooded and the yield of sediment was probably therefore low until at least the early Bronze Age (*c.* 4000 calBP). This is consistent with an increase in flood-related sedimentation noted across Britain after *c.* 4500 calBP by Macklin and Lewin (2003) and Macklin *et al.* (2005) and suggests offshore sources of sediment were important well into the mid Holocene. Spatial differences in the supply of sediment from offshore, across any coastal barrier into the back-barrier areas/valleys, therefore provide the most likely explanation for variations in the timing of the onset of peat formation.

The proportion of clastic sediment derived from the catchments is likely to have increased in the late Holocene and Jennings and Smyth (1987) refer to the Combe Haven floodplain acting as a major sediment store following late Bronze Age/Iron Age clearance. However, not only are these clastic sequences difficult to date, but gross sedimentation rates both in the Combe Haven valley and parts of the Brede valley were considerably lower than the early and mid Holocene. The upper silts and clays of these valleys often form a thin veneer (<1 m) over thick sequences (>10 m) of early Holocene clastics and mid-Holocene peats. The contribution of catchment derived material to estuary infill (e.g. Pevensey Levels, Romney Marsh proper) may have increased in the late Holocene, though thick clastic sequences which formed in the Rye Harbour area following barrier failure are likely to contain material derived from reworking. The relative contributions of offshore and catchment sources to the sediment budget are therefore difficult to determine even during the late Holocene. However, the likely increase in sediment supply in the late Holocene from the catchments did not lead to gross increases in sedimentation rates on the floodplains. Moreover, in some valley locations peat formation continued uninterrupted, suggesting that the supply of catchment sediment was not the underlying reason behind the widespread switch from the deposition of peat to clastic material in the late Holocene.

Amberley Wild Brooks may provide an exception to this pattern and an example of an increase in sediment supply from the catchment in the late Holocene leading to peat formation. The Wild Brooks occupy a protected setting beyond the chalk and if the absence of peat in the mid Holocene was the result of a failure to fill accommodation space, then terrestrialisation and peat formation did not occur (*c.* 2650 calBP) until the catchment had been extensively deforested (Groves 2008).

VI.3. The degree of protection/gravel-barrier evolution

Gravel has been found beneath but not within the early Holocene marine deposits of the region. RSL rise is likely to have been too rapid for gravel-barrier consolidation and the wave and tidal conditions, together with the crenulated coastline created, unsuitable for the long-shore transfer of gravel during this period (Jennings and Smyth 1990).

The only accumulations of beach gravel dated from the region to mid Holocene are those at Broomhill (from *c.* 4700 calBP; Roberts and Plater, (2007). Indeed gravels have only been recorded within the Holocene sequences inland of the present coastline from the Romney Marsh area (where the early beaches have been protected by the easterly development and extension of the cuspate foreland) and from two boreholes in the Adur. Elsewhere mid-Holocene shorelines (and any gravel beaches) are likely to have been seaward of the present coastline, which is confirmed east of Beachy Head by exposure of peats on the foreshore at various locations. Longshore sediment movement during the late Holocene appears to have been sufficient to remove any other direct evidence of these mid-Holocene barriers.

Headlands can act as anchor points for barriers (Jennings *et al.*, 1998) and Pevensey Levels and, in particular the Eastbourne area, lying immediately east of Beachy Head (itself a potential source of gravel) would also be expected to record early peat formation. However, the stratigraphic review above shows that this is not the case with peats here dating from after *c.* 4700 calBP. The limited extent of peat formation at these sites probably reflects site exposure, though that the input of freshwater is likely to have been relatively low (see below) may also be important. The lack of open-water lagoonal deposits, such as those deposited behind other south coast barriers (e.g. Slapton Ley, Devon) certainly suggest that barriers, if they existed at all, failed to exert a strong control on patterns of sedimentation. The similarity of the sedimentary sequences of Eastbourne and Pevensey Levels and the presence of estuarine conditions for long periods (Jennings and Smyth, 1987) may also record a connection between these areas prior to the late Holocene landward advance of the coastline.

The apparent absence of mid-Holocene peats from the West Sussex valleys may be attributed to more exposed conditions caused by open-coast conditions in this area. This would imply that at this time the Pleistocene raised beaches of West Sussex made a minor contribution to the supply of gravel along the coast. Although the chalk cliffs begin east of Brighton, the recent studies on rates of erosion also indicate that the gravel sourced from cliff erosion may not be a major source. Much of the gravel currently stored along the East Sussex coast appears then to have originated from sources further east, either reflecting the distribution of sediments stored during the sea-level lowstand of the Late Devensian, presumably largely in the palaeovalleys of the English Channel which are now empty of sediment (Hamblin *et al.,* 1992), or possibly from a precursor of Dungeness formed by the transfer of gravel eastward during the last interglacial.

In broad terms the history of the back-barrier environments and historical coastal changes along the East Sussex coast conform to the pattern that might be expected given the major controls on gravel-barrier evolution. The mid Holocene was a period of peat formation and infilling with the rate of rising RSL providing conditions suitable for barrier consolidation. Late Holocene instability can be attributed to the slowing down in the rate of RSL rise and the exhaustion of the supply of (stored) gravel. The evidence of updrift cannibalisation and downdrift initiation of barriers during this period comes not only from Dungeness Foreland (Long *et al.*, 2007; Roberts and Plater, 2007; Plater *et al.*, 2009) but also the Rye Harbour area. The latter gravel body accumulated rapidly at the western end of the Romney Marsh depositional complex from the fourteenth century AD until the middle of the nineteenth century AD (Lovegrove, 1953; Long *et al.*, 2007). Although the cartographic evidence for depletion of the gravel foreland of The Crumbles at Eastbourne dates only dates from the early seventeenth century AD (Jennings and Smyth, 1990), the problems experienced in keeping the mouth of the Ash Bourne open from the mid-fourteenth century AD (Lake *et al.*, 1987) suggests depletion commenced earlier, with the storms of the thirteenth century AD probably initiating the movement of the gravel which was eventually deposited at Rye Harbour.

VI.4. Catchment characteristics

Along with the supply of sediment and the degree of site exposure, the supply of freshwater from the catchments could potentially have been a local influence in determining the timing the shift between clastic sedimentation and peat formation during the mid and late Holocene in the East Sussex valleys. Unfortunately insufficient data on the Holocene discharges of the rivers (present-day data are unreliable due to groundwater extraction) and sub-alluvial geometry of the valleys, makes it difficult to determine whether differences in freshwater discharge could have had significant influence. The lack of freshwater input (aside from the Ash Bourne) may be a contributory factor in the late development of peat on Pevensey Levels and in the Eastbourne area. However, that extensive peat formation occurred in relatively small catchments such Combe Haven, while apparently not in Arun and Adur valleys, argues against the supply of freshwater being a major factor. Valley morphology is potentially also of interest; the valleys in the west of the study area tend to have rather large catchments with narrow lower valleys, resulting in greater potential for channel migration and sediment reworking compared with the wider coastal valleys in the east, notably Pevensey Levels and the Romney Marsh depositional complex.

VI.5. Climate change/storm activity

The lack of consistency in the timing of major changes in sedimentation in the coastal deposits and in the vegetation changes which occurred in the peat forming communities (Waller *et al.*, 1999) argues against a strong climatic influence. However, it may be possible to detect some climatic influences. Increased run-off, due to climate change and clearance activity, may be responsible for the eventual formation of peat in the Eastbourne area and at Amberley Wild Brooks. Ombrotrophic bog development over southern Walland Marsh *c.* 2800 calBP certainly coincides with wetter conditions as recorded at many sites around north-western Europe (van Geel *et al.* 1996), though both vertical and spatial isolation from base-rich water appear to be prerequisites for this occur (Waller *et al.*, 1999). The importance of local factors in determining the shift to ombrotrophic conditions is, however, demonstrated by bog development occurring much later at Amberley Wild Brooks (*c.* 1250 calBP). Western Pevensey Levels might also be expected to have become isolated from base-rich water in the late Holocene. Unfortunately stratigraphic information is particularly lacking from this area, though the records of peat cutting may be significant in this respect (ombrotrophic peat is likely to have been preferred).

Clear evidence for barrier breaching as a result of storm activity during the Holocene is only available from the Rye and New Romney areas. Here recognition of barrier failure is facilitated by the historical record. Nevertheless the stratigraphic evidence from Rye, which includes the erosion of peat from a 4 km wide corridor and deposition of a substantial gravel deposit at Moneypenny Farm, demonstrates the scale of this event. The thick sand unit found in the lower Ouse appears to sit unconformably on the underlying sediment (Burrin and Jones, 1991) and as noted previously this deposit may be the product of a similar event. However, the sand deposits of Pevensey Levels appear to be associated with the tidal channels and comparable sediments are also absent from Combe Haven and the Eastbourne Levels. The thin estuarine clay (dated *c.* 4000 calBP) found intercalated in the peat in both the Eastbourne and Pevensey Levels appears not to occur elsewhere and could represent an overtopping event or possibly even the landward extension of clastic sedimentation after breaching during a storm event. However, if linked to the presence of a barrier it might be more simply explained by a change in the mouth of the Ash Bourne as a gravel barrier migrated eastward. That the only unequivocal example of barrier breaching as a result of a storm is from the late Holocene may be a preservation issue. However, it is likely that during this period, even when storms can be demonstrated to be the proximate cause, the diminishing supplies of gravel are likely to have been the ultimate cause of barrier failure.

VI.6. Human activity

In addition to influencing sediment and water supply through changes in the vegetation of the catchment, human activity may have had a direct influence on floodplain sedimentation. As well as peat cutting, activities such as vegetation cutting and grazing, for which there is evidence from the Bronze Age onwards (Waller and Schofield, 2007), may have retarded the accumulation of peat and promoted a shift from organic to clastic sedimentation in the East Sussex valleys. Lowering the water table by drainage, would have caused peat compaction, a process which may also have occurred naturally as a result of tidal creek extension following inundation (Baeteman *et al.*, 2002). These processes could all have provided accommodation space in the late Holocene, though the absence of thick post-peat deposits in valley's such as the Brede and Combe Haven, in contrast to the some of the levels, suggests that coastal processes (inundation followed by erosion and sediment compaction), were of greater significance in this respect. As noted in the introduction the reclamation of inter-tidal areas is likely to have enhanced the impact of inundation. It is unlikely to be purely coincidental that widespread inundation in the thirteenth century AD in both the Romney Marsh area and Pevensey Levels was preceded by episodes of reclamation.

VI.7. Implications for the evolution of the Romney Marsh area

During the early Holocene the Sussex coast evolved in a comparatively uniform way, the rapid rate of RSL rise providing a unifying force. Differences began to emerge from *c.* 7000 calBP with the onset of peat formation in some areas, notably in the east. From the perspective of Romney Marsh area, topography might appear to have played a significant role in this, with the protected east facing palaeovalleys providing locations where peat forming vegetation could become established. However, that peat formation occurred in some of the apparently more exposed valleys (the Ouse, Combe Haven) earlier suggests that the supply of fine-grained sediment, in terms of the infilling of accommodation space and providing a platform for salt marsh and then freshwater peat formation, was crucial.

During the mid Holocene, the wetlands of Romney Marsh extended in an easterly direction, enclosed by and protected from the high energy of the English Channel by a coastal barrier. The size of the wetlands created far exceeds those created in the topographically restricted bedrock valleys and levels to the west. Again the supply of sediment, though now sand and gravel, appears to have been crucial. With the earliest evidence for the barrier formation coming from Rye Bay and progradation beyond the current shoreline seemingly limited to the eastern end of coastline (beyond Eastbourne), the source of much this gravel appears to been offshore and closer to the Romney Marsh area than previously suspected. The growing importance of longshore supply of sediment during the late Holocene can be attributed to many of the valleys to the west of the Romney Marsh area being infilled with sediment, with the coastline between them straightened and less crenulated. This change in regional coastal morphology would have favoured increased longshore drift. Moreover, the reduction of offshore sediment sources of sand

and gravel during the late Holocene made the reworking of onshore sediments and the easterly transport more likely.

In the Romney Marsh area, the main marsh peat bed is testament both to the spatial extent of, and the length of time over which back-barrier areas, were protected. These wetlands would have acted as traps for precipitation, effectively doubling the catchment area of the Eastern Rother. Moreover, although traversed by large rivers and tidal channels, the development over large areas of acidophilous vegetation suggests that these water courses were peripheral to the much of the marshland, indeed this is confirmed by stratigraphic studies (Long and Innes, 1995). In West Sussex, the confined topography, combined with relatively large catchment areas, meant that the propensity for sediment reworking or non-deposition due to channel processes was considerably higher. The message that emerges is that size matters; large depositional complexes, especially those unconstrained by topography, are instrumental in creating and maintaining conditions conducive to prolonged periods of wetland development.

In the late Holocene, the slowing down in the rate of RSL rise is likely to have created the necessary underlying conditions for the return of tidal environments to the valleys and levels of East Sussex by triggering a decline in peat accumulation rendering them increasingly susceptible to inundation. This process was later aggravated by human activities. The difficulties in accurately dating the upper contact of the peat layers make it difficult to determine precisely when marine conditions returned. In the Romney Marsh area deposition was related to the sequential opening of tidal inlets, with the overall resilience of the marshland and its capacity for internal readjustment a product of the size of the complex (Long *et al.*, 2006a). More widely, in the absence of a strong regional driving force (RSL), it is likely that other processes became increasingly influential. With the decline in sediment supply and reworking of gravel breaches, the Romney Marsh area became more vulnerable to influences, such as storms and variations in the longshore movement of sediment, which have left their imprint along the coastline.

Finally, size is probably also important in explaining why the deposits of the Romney Marsh area bear a greater resemblance, in terms of gross stratigraphic architecture, to the other large depositional complexes in southern England (the Thames Estuary and the Solent) and the southern North Sea (the Belgium coastal plain). Smaller systems, it would seem, are more prone to local influences in controlling their spatial and temporal patterns of sedimentation in response to variations in terrestrial and marine processes. From this perspective, perhaps it would be more appropriate to describe the coastal stratigraphies of West Sussex as atypical, reflecting a particular combination of topography/spatial extent and site exposure.

VII. Conclusions

A considerable body of data on the coastal deposits of the Romney Marsh area and Sussex has been amassed over the last twenty five years. Perhaps not surprisingly, comparison between sites does not support a stratigraphic model where there is a high degree of uniformity (Burrin, 1982) or one which can be explained largely by reference to local processes (Jennings and Smyth, 1982). For East Sussex/the Romney Marsh depositional complex a tripartite model (with mid-Holocene peats separating clastic sediments of marine/brackish origin) is applicable, over which a regional driving process, changes in the rate of RSL rise, provided the ultimate control. Here the main differences are in the thickness of the peat, which in part reflects local influences on the onset and end of peat formation. However, these differences in thickness are also the product of post-depositional changes and the stratigraphic architecture of the deposits (with peat formation occurring earlier in valleys and at upland margins).

From the Romney Marsh perspective, comparison of the Holocene stratigraphies of this region with areas to the east throws new light on the origin of the gravel. Certainly the source of the older (mid Holocene) gravel beaches here appears to have been offshore and closer to Romney Marsh than had previously been assumed. Many of the differences between Romney Marsh and the Sussex valleys/levels can be attributed to size. By virtue of its size, the Romney Marsh complex appears both to have been conducive to prolonged periods of wetland development and resilient, with large parts of the sedimentary protected from subsequent destruction.

Despite the considerable progress made in our understanding of the coastal deposits of Sussex over the last twenty five years, it remains possible that some of the apparent differences (e.g. in the presence, thicknesses and age range of the peat) and discrepancies may turn out to be the product of the uneven distribution of stratigraphic information (now heavily biased in favour of the Romney Marsh area). In particular, further work is required to clarify whether the Holocene stratigraphies of the Adur and Arun valleys are atypical.

Bibliography

Aldiss, D.T. (2002) *Geology of the Chichester and Bognor District: a Brief Explanation of the Geological Map*. Sheet explanation of the British Geological Survey. 1: 50 000 Sheet 317/332 Chichester and Bognor District (England and Wales), Keyworth.

Allen, J.R.L. (1999) 'The Rumensea Wall and the early settled landscape of Romney Marsh (Kent)', *Landscape History,* 21, pp. 5–18.

Austin, R.M. (1991) 'Modelling Holocene tides on the NW European Continental shelf', *Terra Nova,* 3, pp. 276–288.

Baeteman, C., Scott, D.B. and Strydonck, M., Van (2002) 'Changes in coastal zone processes at a high sea-level stand: a late Holocene example from Belgium', *Journal of Quaternary Science*, 17, pp. 547–559.

Bayliss, A., Bronk Ramsey, C., Cook, G. and Plicht, J., van der (2007) *Radiocarbon dates from samples funded by English heritage under the aggregates levy sustainability fund 2002–4*, Swindon.

Bond, G., Showers, W., Cheseby, M., Lotti, R., Almasi, P., deMenocal, P., Priore, P., Cullen, H., Hajdas, I. and Bonani, G. (1997) 'A pervasive millennial scale cycle in North Atlantic Holocene and glacial climates', *Science*, 278, pp. 1257–1266.

Bronk Ramsey, C. (1995) 'Radiocarbon calibration and analysis of stratigraphy: the OxCal program', *Radiocarbon,* 37, pp. 425–430.

Bronk Ramsey, C. (2001) 'Development of the radiocarbon calibration program', *Radiocarbon,* 43, pp. 355–363.

Burrin, P.J. (1982) 'The coastal deposits of the southern Weald', *Quaternary Newsletter,* 38, pp. 16–24.

Burrin, P.J. (1983) 'On the coastal deposits of East Sussex: a further comment', *Quaternary Newsletter,* 39, pp. 29–31.

Burrin, P.J. and Jones, D.K.C. (1991) 'Environmental processes and fluvial responses in a small temperate zone catchment: a case study of the Sussex Ouse valley, southeast England', in: L. Starkel, K.J. Gregory and J.B. Thornes (eds), *Temperate Palaeohydrology*, Chichester, pp. 217–252.

Burrin, P.J. and Scaife, R.G. (1984) 'Aspects of Holocene valley sedimentation and floodplain development in southern England', *Proceedings of the Geologists' Association*, 95, pp. 81–96.

Castleden, R. (1982) *Classic Landforms of the Sussex Coast,* Sheffield (Landform Guides 2).

Clarke, M.L. and Rendell, H.M. (2009) 'The impact of North Atlantic storminess on western European coasts: a review', *Quaternary International*, 195, pp. 31–41.

Dornbusch, U., Robinson, D.A., Moses, C.M. and Williams, R.B.G. (2006) 'Chalk coast erosion and its contribution to the shingle budget in East Sussex', *Zeitschrift für Geomorphology N.F. Supplement*, 144, pp. 215–230.

Dulley, A.J.F. (1966) 'The level and port of Pevensey in the Middle Ages', *Sussex Archaeological Collections*, 104, pp. 26–45.

Geel, B., van, Buurman, J. and Waterbolk, H.T. (1996) 'Archaeological and palaeoecological indications of an abrupt climatic change in The Netherlands, and evidence for climatological teleconnections around 2650 BP', *Journal of Quaternary Science*, 11, pp. 451–461.

Gehrels, W.R. (2010) 'Late Holocene land- and sea-level changes in the British Isles: implications for future sea-level predictions', *Quaternary Science Reviews*, 29, pp. 1648–1660.

Godwin, H. (1943) 'Coastal peat beds of the British Isles and North Sea', *Journal of Ecology,* 31, pp. 199–247.

Godwin, H. and Willis, E.H. (1964) 'Cambridge University natural radiocarbon measurements VI', *Radiocarbon,* 6, pp. 116–137.

Greatorex, C. (2003) 'Living on the margins? The late Bronze Age landscape of the Willingdon Levels', in: D. Rudling (ed.), *The Archaeology of Sussex to AD2000*, King's Lynn, pp. 89–100.

Green, R.D. (1968) *Soils of Romney Marsh*, Harpenden (Soil Survey of Great Britain 4).

Groves, J.A. (2008) *Late Quaternary Vegetation History of the Acidic Lithologies of South East England*, Unpublished Dissertation, Kingston University.

Gupta, S., Collier, J., Palmer-Felgate, A., Dickinson, J., Bushe, K. and Humber, S. (2004) *Submerged Palaeo-Arun River: Reconstruction of Prehistoric Landscapes and Evaluation of Archaeological Resource Potential* (Integrated Projects 1 and 2). Final Project Report for English Heritage.

Gupta, S., Collier, J.S., Palmer-Felgate, A. and Potter G. (2007) 'Catastrophic flooding origin of the shelf valley systems in the English Channel', *Nature*, 448, pp. 342–345.

Hamblin, R.J.O., Crosby, A., Balson, P.S., Chadwick, R.A., Penn, I.E. and Arthur, M.J. (1992) *The Geology of the English Channel, United Kingdom*, London (Offshore Regional Report 10).

Harkness, D.D. and Wilson, H.W. (1979) 'Scottish Universities Research and Rector Centre radiocarbon measurements III', *Radiocarbon*, 21, pp. 203–256.

Hijma, M.J. and Cohen, K.M. (2010) 'Timing and magnitude of the sea-level jump preluding the 8200 year event', *Geology*, 38, pp. 275–278.

Jennings, S. (1985) *Late-Quaternary Environmental Change at Eastbourne, East Sussex*, Unpublished Dissertation, The Polytechnic of North London.

Jennings, S.C. and Smyth, C. (1982a) 'A preliminary interpretation of coastal deposits from East Sussex', *Quaternary Newsletter*, 37, pp. 12–19.

Jennings, S.C. and Smyth, C. (1982b) 'A reply to the coastal deposits of the southern Weald', *Quaternary Newsletter*, 38, pp. 24–9.

Jennings S.C. and Smyth, C. (1985) 'The origin and development of Langney Point: a study of Flandrian coastal and sea-level change', *Quaternary Newsletter*, 45, pp. 12–22.

Jennings, S.C. and Smyth, C. (1987) 'Coastal sedimentation in East Sussex during the Holocene', *Progress in Oceanography*, 18, pp. 205–241.

Jennings, S.C. and Smyth, C. (1990) 'Holocene evolution of the gravel coastline of East Sussex', *Proceedings of the Geologists' Association*, 101, pp. 213–224.

Jennings, S.C. and Smyth, C. (1991) 'Holocene evolution of the gravel coastline of East Sussex: reply', *Proceedings of the Geologists' Association*, 102, pp. 306–308.

Jennings, S.C., Orford, J.D., Canti, M., Devoy, R.J.N and Straker, V. (1998) 'The role of relative sea-level rise and changing sediment supply on Holocene gravel barrier development: the example of Porlock, Somerset, UK', *The Holocene,* 8, pp. 165–81.

Jennings, S.C., Greatorex, C., Smyth, C. and Spurr, G. (2003) 'The environmental archaeology of the Late Bronze Age occupation platform at Shinewater, near Eastbourne', in: A.J. Howard, M.G. Macklin and D.G. Passmore (eds), *Alluvial Archaeology in Europe*, Rotterdam, pp. 93–110.

Jones, D.K.C. (1971) 'The Vale of the Brooks', in: R.B.G. Williams (ed), *Guide to Sussex Excursions*, Brighton, pp. 43–46 (Institute of British Geographers. Special Publication).

Jones, S.L. (2002) *Sea-level Changes in Southeast England and Northern France*, Unpublished Dissertation, Queen Mary and Westfield College London.

Lake, R.D., Young, B., Wood, C.J. and Mortimore, R.N. (1987) *Geology of the Country around Lewes*, England and Wales (Memoir of the British Geological Survey, Sheet 319).

Lamb, H.H. (1991) *Historic Storms of the North Sea, British Isles and Northwest Europe,* Cambridge.

Lamb, H.H. (1995) *Climate, History and the Modern World*, London.

Long, A.J. (2001) 'The mid and late Holocene evolution of Romney Marsh and the Thames Estuary', *Archaeology in the Severn Estuary,* 11, pp. 55–68.

Long, A.J. and Innes, J.B. (1993) 'Holocene sea-level changes and coastal sedimentation in Romney Marsh, south-east England', *Proceedings of the Geologists' Association,* 104, pp. 223–237.

Long, A.J. and Innes, J.B. (1995) 'The back-barrier and barrier depositional history of Romney Marsh, Walland Marsh and Dungeness, Kent, England', *Journal of Quaternary Science*, 10, pp. 267–283.

Long, A.J. and Shennan, I. (1993) 'Holocene sea-level and crustal movements in southeast and northeast England, UK', *Quaternary Proceedings,* 3, pp. 15–19.

Long, A.J., Plater, A.J., Waller, M.P. and Innes, J.B. (1996) 'Holocene coastal sedimentation in the eastern English Channel: data from the Romney Marsh region, United Kingdom', *Marine Geology*, 136, pp. 97–120.

Long, A.J., Waller, M.P., Hughes, P. and Spencer, C. (1998) 'The Holocene depositional history of Romney Marsh proper', in: J. Eddison, M. Gardiner and A.J. Long (eds), *Romney Marsh: Environmental Change and Human Occupation in a Coastal Lowland,* Oxford, pp. 45–63 (Oxford University Committee for Archaeology Monograph 46).

Long A.J., Scaife, R.G. and Edwards, R.J. (2000) 'Stratigraphic architecture, relative sea level, and models of estuary development in southern England: new data from Southampton Water', in: K. Pye and J.R.L. Allen (eds), *Coastal and Estuarine Environments: Sedimentology, Geomorphology and Geoarchaeology*, Bath, pp. 253–79 (Geological Society Special Publication; 175).

Long, A.J., Waller, M.P. and Plater, A.J. (2006a) 'Coastal resilience and late Holocene tidal inlet history: the evolution of Dungeness Foreland and the Romney Marsh (UK) depositional complex', *Geomorphology*, 82, pp. 309–330.

Long, A.J., Waller, M.P. and Stupples, P. (2006b) 'Driving mechanisms of coastal change: sediment autocompaction and the destruction of late Holocene coastal wetlands', *Marine Geology*, 225, pp. 63–84.

Long A.J., Waller, M.P. and Plater, A.J. (2007) *Dungeness and the Romney Marsh: Barrier Dynamics and Marshland Evolution*, Oxford.

Lovegrove, H. (1953) 'Old shorelines near Camber Castle', *Geographical Journal,* 109, pp. 200–207.

Lozano, I., Devoy, R.J.N., May, W. and Andersen, U. (2004) 'Storminess and vulnerability along the Atlantic coastlines of Europe: analysis of storm records and of a greenhouse gases induced climate scenario', *Marine Geology,* 210, pp. 205–225.

Macklin, M.G. and Lewin, J. (2003) 'River sediments, great floods and centennial-scale Holocene climate change', *Journal of Quaternary Science,* 18, pp. 101–105.

Macklin, M.G., Johnstone, E. and Lewin, J. (2005) 'Pervasive and long-term forcing of Holocene river instability and flooding in Great Britain by centennial-scale climate change', *The Holocene,* 15, pp. 937–943.

Moffat, B. (1984), *An Evaluatory Study of the Methods used in the Reconstruction of Historical Vegetation and Land-use*, Unpublished Dissertation, The Polytechnic of North London.

Moffat, B. (1986) 'The environment of Battle Abbey estates (East Sussex) in medieval times; a re-evaluation using analysis of pollen and sediments', *Landscape History,* 8, pp. 77–93.

Nicholls, R.J. (1991) 'Holocene evolution of the gravel coastline of East Sussex: discussion', *Proceedings of the Geologists' Association,* 102, pp. 301–306.

Orford, J.D., Carter, R.W.G. and Jennings, S. (1991) 'Coarse clastic barrier environments: evolution and implications for Quaternary sea level interpretation', *Quaternary International*, 9, pp. 87–104.

Plater, A.J., Long, A.J., Spencer, C.D. and Delacour, R.A.P. (1999) 'The stratigraphic record of sea-level change and storms during the last 2000 years: Romney Marsh, southeast England', *Quaternary International,* 55, pp. 17–27.

Plater, A.J., Stupples, P. and Roberts, H.M. (2009) 'Evidence for episodic coastal change during the late Holocene: the Dungeness barrier complex, SE England', *Geomorphology,* 104, pp. 47–58.

Reimer, P.J., Baillie, M.G.L., Bard, E., Bayliss, A., Beck, J.W., Bertrand, C.J.H., Blackwell, P.G., Buck, C.E., Burr, G.S., Cutler, K.B., Damon, P.E., Edwards, R.L., Fairbanks, R.G., Friedrich, M., Guilderson, T.P., Hogg, A.G., Hughen, K.A., Kromer, B., McCormac, F.G., Manning, S.W., Ramsey, C.B., Reimer, R.W., Remmele, S., Southon, J.R., Stuiver, M., Talamo, S., Taylor, F.W., Plicht, J., van der, and Weyhenmeyer, C.E. (2004) 'IntCal04 terrestrial radiocarbon age calibration, 26–0 ka BP', *Radiocarbon,* 46, pp. 1029–1058.

Rippon, S. (2000) *The Transformation of Coastal Wetlands*, Oxford.

Roberts, H.M. and Plater, A.J. (2007) 'Reconstruction of Holocene foreland progradation using optically stimulated luminescence (OSL) dating: an example from Dungeness, UK', *The Holocene*, 17, pp. 495–505.

Robinson, D.A. and Williams, R.B.G. (1983) 'The soils and vegetation history of Sussex', in: The Geography Editorial Committee, University of Sussex (eds) *Sussex: Environment, Landscape and Society*, Gloucester, pp. 109–112.

Salzmann, LF. (1910) 'The inning of Pevensey Levels', *Sussex Archaeological Collections*, 53, pp. 32–60.

Smyth, C. and Jennings, S. (1988) 'Mid- to late-Holocene forest composition and the effects of clearances in the Combe Haven valley, East Sussex', *Sussex Archaeological Collections*, 126, pp. 1–20.

Smyth, C. and Jennings, S. (1990) 'Late Bronze Age–Iron Age valley sedimentation East Sussex, southern England', in: J. Boardman, I.D.L. Foster and J.A. Dearing (eds), *Soil Erosion on Agricultural Land*, Chichester, pp. 273–284.

Stupples, P. (2002) 'Tidal cycles preserved in late Holocene tidal rhythmites, the Wainway Channel, Romney Marsh, southeast England', *Marine Geology*, 182, pp. 231–246.

Thorley, A. (1971) 'Vegetation history in the Vale of the Brooks', in R.B.G. Williams (ed.) *Guide to Sussex Excursions*, Brighton, pp. 47–50 (Institute of British Geographers, Special Publication).

Thorley, A. (1981), 'Pollen analytical evidence relating to the vegetation history of the Chalk', *Journal of Biogeography*, 8, pp. 93–106.

Tooley, M.J. and Switsur, R.V. (1988) 'Water level changes and sedimentation during the Flandrian Age in the Romney Marsh area', in: J. Eddison and C. Green (eds), *Romney Marsh: Evolution, Occupation, Reclamation*, Oxford, pp. 53–71 (Oxford University Committee for Archaeology Monograph; 24).

Törnqvist, T.E., Ree, M.H.M., van, Veer, R., van and Geel, B., van (1998) 'Improving methodology for high-resolution reconstruction of sea-level rise and neotectonics by palaeoecological analysis and AMS 14C dating of basal peats', *Quaternary Research,* 49, pp. 72–85.

Waller, M.P. (1993) 'Flandrian vegetational history of south-eastern England: pollen data from Pannel Bridge, East Sussex', *New Phytologist,* 124, pp. 345–369.

Waller, M.P. (1994) 'Flandrian vegetational history of south-eastern England: stratigraphy of the Brede Valley and pollen data from Brede Bridge', *New Phytologist,* 126, pp. 369–392.

Waller, M.P. (2002) 'The Holocene vegetation history of the Romney Marsh region', in: A.J. Long, S. Hipkin and H. Clarke (eds), *Romney Marsh: Coastal and Landscape Change through the Ages*, Oxford, pp. 1–21. (Oxford University School of Archaeology Monograph; 56).

Waller, M.P. and Hamilton, S. (2000) 'The vegetation history of the English chalkland: a mid-Holocene pollen sequence from the Caburn, East Sussex', *Journal of Quaternary Science,* 15, pp. 253–272.

Waller, M.P. and Kirby, J. (2002) 'Late Pleistocene/Early Holocene environmental change in the Romney Marsh region: new evidence from Tilling Green, Rye', in: A.J. Long, S. Hipkin and H. Clarke (eds), *Romney Marsh: Coastal and Landscape Change through the Ages*, Oxford, pp. 22–39 (Oxford University School of Archaeology Monograph; 56).

Waller M.P. and Long, A.J. (2003) 'Holocene coastal evolution and sea-level change on the southern coast of England: a review', *Journal of Quaternary Science,* 18, pp. 351–359.

Waller, M.P. and Schofield, J.E. (2007) 'Mid to late Holocene vegetation and land-use history in the Weald of south-eastern England: multiple pollen profiles from the Rye area', *Vegetation History and Archaeobotany*, 16, pp. 367–384.

Waller, M.P., Burrin, P.J. and Marlow, A. (1988) 'Flandrian sedimentation and palaeoenvironments in Pett Level, the Brede and Lower Rother valleys and Walland Marsh', in: J. Eddison and C. Green (eds), *Romney Marsh: Evolution, Occupation, Reclamation,* Oxford, pp. 3–30 (Oxford University Committee for Archaeology Monograph; 24).

Waller, M.P., Long, A.J., Long, D. and Innes, J.B. (1999) 'Patterns and processes in the development of coastal mire vegetation: multi-site investigations from Walland Marsh, southeast England', *Quaternary Science Reviews*, 18, pp. 1419–44.

Waller, M.P., Long, A.J. and Schofield, J.E. (2006) 'The interpretation of radiocarbon dates from the upper surface of late Holocene peat layers in coastal lowlands', *The Holocene,* 16, pp. 51–61.

Ward, E.M. (1922) *English Coastal Evolution*, London.

Waton, P.V. (1983) *A Palynological Study of the Impact of Man on the Landscape of Central Southern England, with Special Reference to the Chalklands*, Unpublished Dissertation, University of Southampton.

Woodcock, A. (2003) 'The archaeological implications of coastal change in Sussex', in: D. Rudling (ed.), *The Archaeology of Sussex to AD 2000*, King's Lynn, pp. 1–16.

Young, B. and Lake, R.D. (1988) *Geology of the Country and Brighton and Worthing*, England and Wales (Memoir of the British Geological Survey, Sheets 318 and 333).

4 Potential and use of archaeological and historical data for a reconstruction of the sea level curve of the last 3000 years in the coastal zone of the southern North Sea. Results of a case study*

Sjoerd J. Kluiving, VU University Amsterdam
Michel A. Lascaris, Cultural Heritage Agency, The Netherlands
Adriaan M.J. de Kraker, VU University Amsterdam
Hans Renes, VU University Amsterdam and Utrecht University
Guus J. Borger, University of Amsterdam and VU University Amsterdam
Steven A. Soetens, VU University Amsterdam

Abstract

This paper demonstrates that methodologies from various disciplines can be utilised to explain the coastal development of the southern North Sea during the last 3000 years. The potential and uses of archaeological data are tested against the applicability of delivering data for reconstructing sea level or local water level curves. Geology, archaeology, and historical geography are disciplines that assess changes in absolute and regional sea levels, deposition, subsidence, or rises in surface level. In this paper a selection of the national archaeological database (ARCHIS2) from the western Netherlands is presented as archaeological proxy data that might help reconstruct the relationship between occupational history and landscape development.

The case study around Katwijk in the western Netherlands presents the known history and geology of the area, as well as the archaeological data present. Results show that data within ARCHIS2 is of too poor quality and that even with the help of excavation reports the required resolution for reconstructing local water level curves is currently not present in the recorded data. Archaeological data that has been recorded generally lacks precision in data location (x, y, and z) and generally does not have an absolute time control. The ideal database or model is the creation of a GIS (Geographic Information System) that can integrate data from the different disciplines. All data should be transparent and

* We thank Patrick Kiden (Deltares, the Netherlands), Henk Weerts (Cultural Heritage Agency, the Netherlands), as well as an anonymous reviewer for critical remarks which have substantially improved the quality of the paper. Marieke van Dinter (Utrecht University) kindly supplied us with unpublished data of her forthcoming dissertation. Dorit Sivan (Tel Aviv University, Israel) has made some helpful comments on an earlier draft of this paper.

Landscapes or seascapes? The history of the coastal environment in the North Sea area reconsidered, ed. by Erik Thoen, Guus J. Borger, Adriaan M.J. de Kraker, Tim Soens, Dries Tys, Lies Vervaet & Henk J.T. Weerts, Turnhout, 2013 (*CORN Publication Series*, 13), pp. 61–84.

BREPOLS PUBLISHERS DOI: 10.1484/M.CORN.1.101548

have a spatial and temporal resolution, so that data from different disciplines can corroborate each other. Its accuracy can then be compared so the better dataset to provide the optimal insight into landscape change associated with the sizes of coastal inlets and the local variations of tidal amplitudes.

I. Introduction

In Lowland areas human occupation in coastal zones is closely tied to changes in regional water tables and the (resulting) localities of water flow. Accordingly, the landscape development and occupation history of a regional coastal area may provide indications of fluctuations in Relative Sea Level (RSL). Relative Sea Level variations within the last 3000 years should be deduced from the most complete and precise dataset available. It appears that RSL rises in the last 3000 years in the southern North Sea is not only a result of natural factors but also of human impact. By establishing the natural signal of RSL rise, possible anthropogenic impact factors on rising sea levels can be detected. Since 3000 BP there is growing evidence from archaeological data that humans have occupied the coastal zones and have left an imprint on that landscape (Figure 4.1). For the past 1200 years historical geography can add data from archival sources. Lists of goods belonging to eighth-century monasteries, for example, give indications about the habitable status of a certain coastal area (Dekker, 1971). Landscape archaeology, by which we here mean the integration of geology, archaeology and historical geography, is proposed as a tool for the

Figure 4.1 Plough tracks, post holes (Iron age-Roman times: 800 BC – 450 AD) in 'Oerij' tidal basin in Velserbroek, Western Netherlands (source: Amsterdam Archaeological Centre (AAC), University of Amsterdam, http://dpc.uba.uva.nl/aac)

Figure 4.2 Instrumental data of high water level, mean sea level, low water level, and tidal range calculated over the time interval 1900–1980 at Vlissingen (Zeeland, the Netherlands) (source: van Malde, 1996a, b)

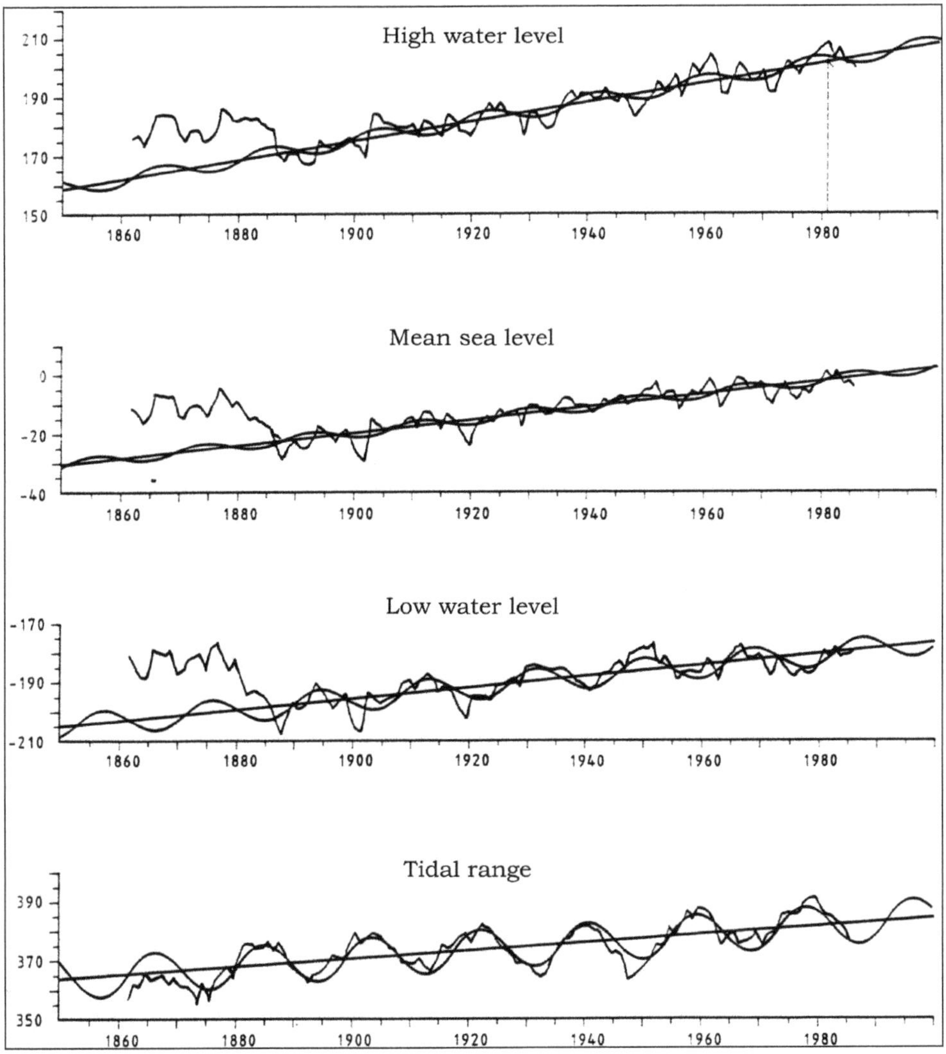

current research on the last 3000 years. After the end of the sixteenth century numerical data relating to local water levels can be found at several sites along the coast (Figure 4.2).

The aim of this paper is to address the standards to which the documentation of archaeological core and excavation data, such as found in databases like ARCHIS2, the Dutch national database in which all archaeological data is entered, should apply in order to contribute to the reconstruction of changes in regional water tables and resulting relative sea levels. However, is the proposed resolution of retrieving water level data from archaeological and historical geographical research sufficient to detect these small scale sea level variations? Does archaeological data from excavation reports generate proxy data with adequate resolution and reliability for local water level reconstruction in tidal basins?

Without adequate data to test the proposed research questions the Dutch national archaeological database ARCHIS2 is examined in this paper as a case study in retrieving RSL and local water level data from such databases. In this paper the content and use of the ARCHIS2 database will be evaluated with respect to the quantity and quality of data related to settlements in the coastal zone. The landscape development as well as the human occupation history around a former Rhine estuary is taken as a case study, to give answers in order to demonstrate the need of interdisciplinary data collection. Although this interdependency seems logical and straightforward it appears to be very complex when measured, observed, and recorded in the field. It appears that data quality has such low standards that hardly any conclusion can be drawn on local or regional water levels, let alone RSL. Baeteman (2007) has criticised the sea level curve of Behre (2003, 2005, and 2007) on the German North Sea coast for this reason.

Since the method of archaeological data recording can be regarded as highly systematic, due to national protocols and research agendas, like in the Netherlands the SIKB[1] and NOaA[2], the apparent omission in data precision quality can be viewed as a contradiction to the highly ordered nature of archaeological recordings in general. Therefore, this paper explores to what degree archaeological data have been reported on, and in what format archaeological data of soil cores and archaeological excavations have been recorded, in order to test if these datasets are suitable for the discussion on changes in RSL curves.

The novelty of the proposed methodology in this paper is to combine data from geology, archaeology, and historical geography, and to focus on the specific nature of that relation (Kluiving and Guttmann, 2012; Kluiving *et al.*, 2012). Combinations of archaeological research in studies of sea level reconstructions have proved to be fruitful in the Mediterranean (Sivan *et al.*, 2001, 2004; Toker *et al.*, 2011) and in the North Sea area (Parnell, 2005). Within the eastern Mediterranean, it has been demonstrated that the use of coastal water well data and a variety of archaeological remains, like bases, pavements or foundations of various fortified coastal structures combined with biological proxies, provide evidence for supporting a sea-level drop of about 50 ± 20 cm during the period 1000–1300 AD (Toker *et al.*, 2011). Along the English North Sea coast in the Humber estuary archaeological data points have been used to calculate reference water levels (Parnell, 2005). The advantage of archaeological data sets is that they correspond to periods where the geological evidence is sparse, and therefore are more effective in reducing RSL uncertainty.

---

[1] SIKB is a network, encompassing both the private and the public sector, set up to continuously and structurally enhance the standards of activities relating to soil management in The Netherlands. The Dutch government has required the market to develop national quality standards in archaeological research. SIKB builds and maintains the resulting quality assurance structure in a set of protocols.

[2] NOaA is the National Archaeological Research Agenda (NOaA). Within the agenda the NOaA clarifies its role in the system of quality management in archaeological heritage management in the Netherlands, devised by the archaeological sector itself (self-regulation).

II. Background of sea level studies

For the study of Late Holocene coastal changes of the southern part of the North Sea, more reliable RSL index points for the last 3000 years are needed and as a consequence additional sites in Belgium, the Netherlands and Germany are needed (Vink *et al.*, 2007) for sedimentary facies data. These are proxy data for changes in the evolution of coasts controlled by sedimentary processes and variations in sediment supply and accommodation space (Beets and Van der Spek, 2000). Moreover, the impact of regional variations in subsidence also need to be taken into account (0.1–1.0 m/ka) (Vink *et al.*, 2007). In addition, the human impact on sea level rise on coastal regions also needs to be included as an important influence for coastal formation. All of these variations may result in different landscapes and divergent coastal evolutions in neighboring regions (Gonzalez and Törnquist, 2009; Lambeck, 1997).

The tidal amplitude and range of extent of low and high tides reaching to the end of estuaries are most relevant for direct landscape transformations, traces of past cultures, and landscape archaeology. Changes in tidal range can be caused by estuary, floodbasin and/or river effects (Baeteman, 2007). In physiographic accommodation spaces such as estuaries and tidal basins the tidal amplitude is the most important feature on a decadal scale. Behind coastal barriers it is assumed that while the effect of RSL rise may be non-existent in a given accommodation space, former tidal amplitude can be reconstructed by reconstructing past landscapes in conjunction with archaeological sites. The focus needs to be on regional to local-scale landscape models with local tidal amplitude signals to define which impact RSL had at a given time and location (Figure 4.2).

In the coastal zone a highly dynamic estuary system is present that may show different states at short distances. At one place a gully system will incise, causing flood heights to increase and ebb lows to decrease, while nearby a gully system will silt up accompanied by decreasing tidal amplitude. Low ebb tides will improve natural drainage and better living conditions in the former situation, while in the latter silting up leads to swampy conditions unfavorable for settlement. A second characteristic of the natural spatial dynamics in the coastal area is that processes can act in a non-linear fashion across time: silted systems can rejuvenate again and a rejuvenated tidal gully may silt up again.

During the last 3000 years archaeological material was deposited and integrated in geological sediments, today providing witness to transgressions and regressions, as well as to storm sequences. It has been shown that the main elements which allow sedimentation in the coastal zone are the availability of sediment and the change in size of the tidal prism of the channels. More specifically, the compaction of peat resulted in an increase of accommodation space for sediments (Baeteman, 2007; Ervynck *et al.*, 1999). The suggested research and case studies will also address the full scale of correlations between archaeological periods and profound landscape changes. As such, the former 'Dunkirk transgression deposits', which suggested a 'fluctuating' sea level change, and which are (still) widely embedded in national (archaeological) research (Behre, 2003 and 2007) can be replaced by the nowadays more widely adopted 'smoothly rising' sea level (Vink *et al.*, 2007; Denys and Baeteman, 1995; Kiden, 1995; Van de Plassche, 1982; Van de Plassche *et al.*, 2005 and 2010; Weerts *et al.*, 2011) (Figure 4.3). Within the Netherlands, the former 'Dunkirk deposits' have been replaced by the 'Walcheren Member' in the 'Naaldwijk Formation' (De Mulder *et al.*, 2003; Weerts *et al.*, 2005).

Figure 4.3 The regional Mean Sea Level (MSL) curves represent the mid-lines of the respective MSL error bands (source: Vink et al., 2007).
The regional MSL curves are compared to the eustatic curves (source: Fleming et al., 1998; Milne et al., 2005)

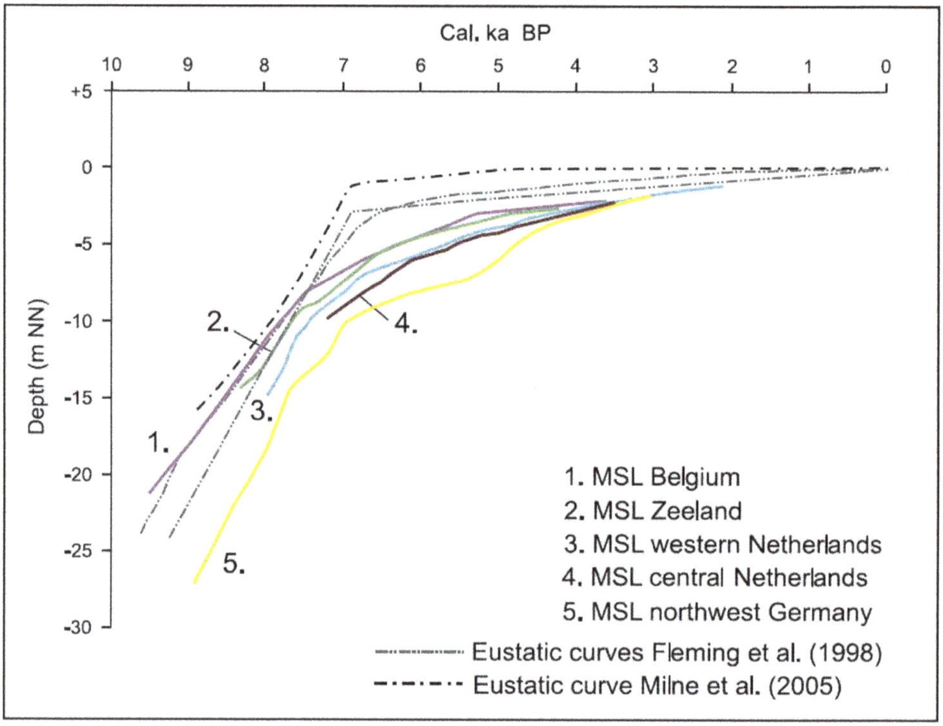

Further studies of RSL curves, more focused regional scale models, as well as landscape archaeological reconstructions in several tidal basins and estuaries along the Belgian, Dutch, and German coasts will establish new (regional) chronostratigraphical orders of sediments and cultural layers, positions of archaeological remains, historical landscapes, and (paleo-) coastlines.

This paper also addresses the issue of sea level rise due to present-day accelerated global warming. One basic question must be asked: to what degree is sea level variation the result of natural phenomena? A growing number of high resolution sea level studies from around the world are already available, and they show century-scale variations of at best a few decimeters (Gehrels, 2010; Gehrels *et al.*, 2002, 2005, 2006a, 2006b; Szkornik *et al.*, 2008; Van de Plassche, 1991 and 2000; Van de Plassche *et al.*, 1998 and 2010). The call for an interdisciplinary approach in this paper is further supported by the fact that historical geography in the Netherlands has already contributed to the issue of coastal landscape development and occupational history (van Veen, 1936 and 1950; Beenakker, 1987; Vos and van Heeringen, 1997; Schoorl, 1999; Beekman, 2007).

III. Methodology

An alternative source of proxy data for regional variations in RSL rise is the vast and still increasing amount of reports and other unpublished data of archaeological field research. The availability of archaeological soil traces and finds is a clear indication for favorable living conditions and indirectly for the ancient local water levels and the location of gullies in tidal basins. Information on the physical geography and the landscape context of the respective find spots is also published in these reports.

In the Netherlands all archaeological research, finds, and observations must be entered into the national database ARCHIS2 of the Cultural Heritage Agency (RCE). These also include data from excavations and chance finds of the pre-digital age. ARCHIS2 therefore contains multiple datasets, including the archaeological monument map (AMK). To generate proxy data for coastal development, the system parameter archaeological observation (*waarneming*) is of primary importance. If accurately recorded, each archaeological find spot also holds information on elevation, exact geographic location and age. Also references to full research reports and publications can be mentioned.

Disadvantages of the ARCHIS2 system can be summed up as follows. First, not all archaeological observations have been entered in the database because not all local archaeological authorities treat entering data into ARCHIS2 as a high priority. Second, when entered not all data is complete, usually only the archaeological find(s) and the nature of the research is mentioned. Finally, ARCHIS2 is a relatively closed system and access is restricted to archaeological specialists.

The archaeological observations in the ARCHIS2 database form the most extensive and heterogeneous dataset. Within a 40 km wide buffer from the Dutch coast to the inland over 25,000 archaeological observations are documented, for example. These observations have been inherited from both regular archaeological research and chance finds from non-archaeological digging activities. Potentially, ARCHIS2 forms a source of information to collect the necessary data to reconstruct the variety in development of coastal basins in the past 3000 years. To test the usability of data of ARCHIS2 for research on coastal development in this paper the archaeological observations of a specific coastal region, Katwijk (aan Zee), are evaluated (Figure 4.4).

The current Dutch coastline was shaped during the past 3000 years, when parts of the older sand bars were eroded and the younger coastal dunes were formed from 1000 BP onwards. Large peat cushions were already present in the lagoon behind the coastline. Within the same 3000 year period, this landscape also endured profound changes as a result of anthropogenic activities (such as peat extraction and land reclamations) and by inlets from the sea (such as the origin of the Zuiderzee and the Zeeland archipelago). Tidal basins in the hinterland of marine inlets were influenced by tidal amplitude and the sediment supply from rivers. The form and shape of the tidal and marine inlets (defining RSL) were not fixed, but changes occurred through the course of time (Figure 4.4a and 4.4b). These changes differed on a regional, intra-basinal scale, caused by complex interchanges between natural parameters ranging from sea level rise and coastal erosion to human interference like dike building and the lowering of the level of the land by draining peat lands. For this reason it should be clear that the coast of the southern North Sea today is a combination of a natural and a culturally constructed landscape, which can

Figure 4.4 Palaeogeographic maps of the Netherlands.
A. During the Iron Age (500 BC)
B. During Roman times (50 AD)
They show tidal inlets and basins at regularly spaced intervals along the coast while changing in form and shape during the course of time.
(source: RCE and TNO, 2008. Case study area of Katwijk is located within the rectangle. Developed for the 'Nationale Onderzoeksagenda Archeologie')

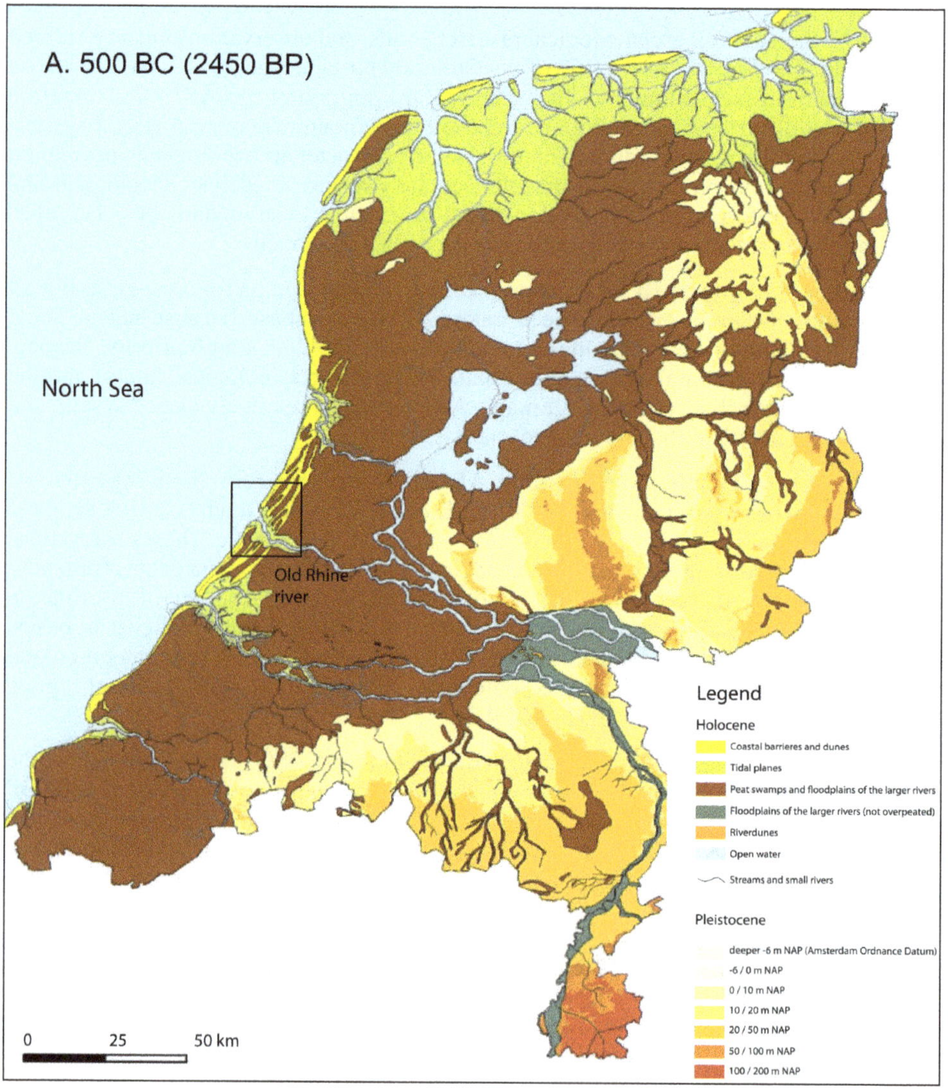

only be unraveled by interdisciplinary research. Archaeology is one of the disciplines that has not yet fully been employed in such an investigation.

IV. Case study Katwijk

Katwijk (surface area: 31.06 km²) is located in the position where the main River Rhine flowed into the North Sea during in the Bronze Age (Figure 4.4a). About 3000 years ago

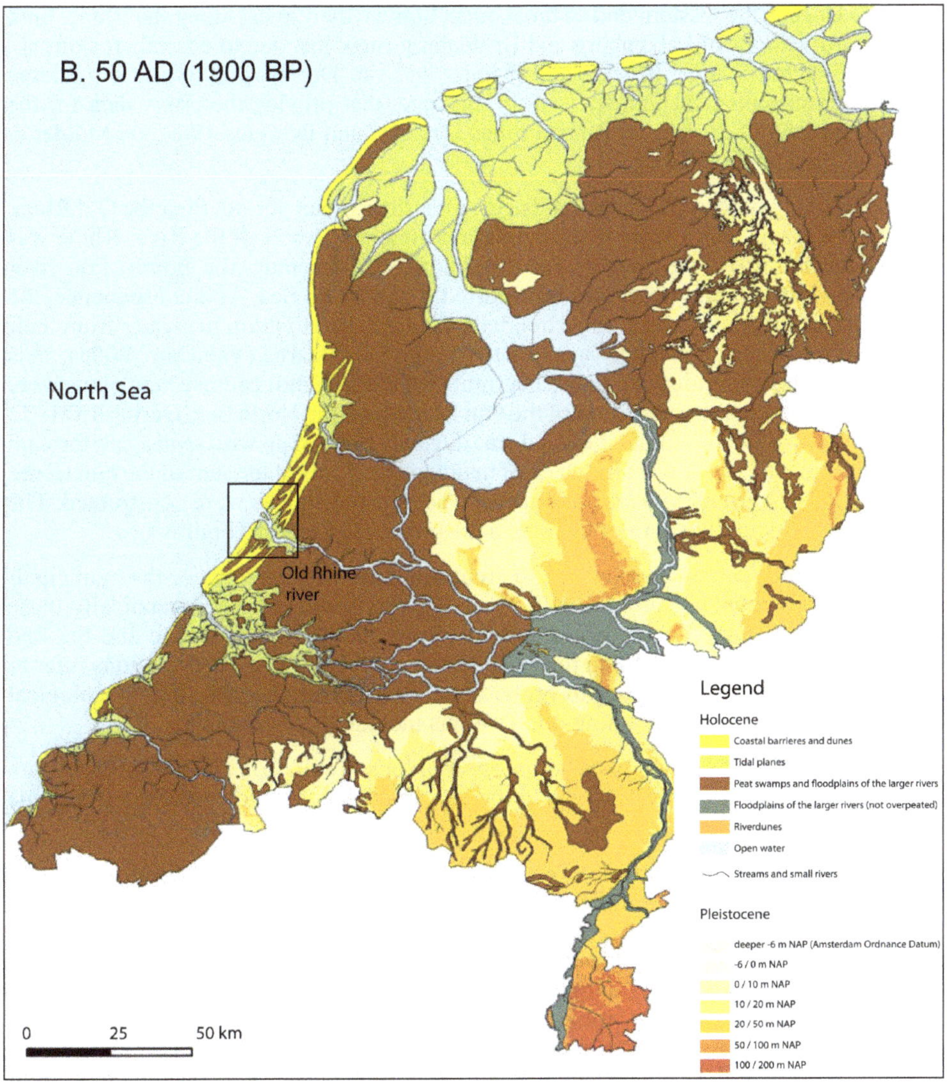

the coast was located a few kilometers landward of the current location (Pruisers and De Gans 1988; De Mulder *et al.*, 2003; van Heteren and van der Spek, 2008) and was aggrading in a seaward direction. A narrow zone of sand bars was located along the coast. In the Roman era parallel long-stretched dune rows with swampy strips in between them gave shape to the coastline. Behind these sand bars extended peat areas were present with limited accessibility. The marshy conditions of the peat area caused the relatively high and dry located sand bars to be much more attractive for settlement to the population of this area, particularly during the Iron Age and the Roman era. In the Katwijk area, which by no coincidence is the western end of the Roman border (the Limes) along the Rhine, both Roman fortifications of Valkenburg and Brittenburg (now lost due to coastal erosion) are considered archaeological benchmarks (Müller and van Doesburg 2008). It is the large number of archaeological find spots from this area, that provide the proxy data for the palaeogeography of the area in Roman times (Pruisers and De Gans 1988; De Mulder *et al.*, 2003; Van Dinter, 2013).

Drastic changes occurred in the landscape after the Roman period. First the Old Rhine of Roman Katwijk was a large river, but in later times the water of the River Rhine was diverted upstream to more southern courses. In the twelfth century the mouth of the river silted up and little if any water was transported to the North Sea. As a consequence, the inhabitants of the hinterland in the extended peat areas had to drain the water from their land to far more northern waterways like the IJ and the Spaarne (Van Dam, 2003). This situation was far from ideal, so at the beginning of the fifteenth century new plans were made to re-establish the connection of the Old Rhine with the North Sea. During 1571–72 at Katwijk a small modest drainage canal was dug, which filled up with sand again through the twin circumstances of war and a lack of maintenance. Only at the start of the nineteenth century was the current *Uitwateringskanaal* was dug and sluices were constructed. The canal is located slightly south of the original Roman river mouth (Figure 4.5).

The ARCHIS2 database contains 130 archaeological observations for the municipality of Katwijk (status in Spring 2010). It can be clearly seen that geographically these observations concentrated on former natural levees along the Old Rhine and the find spots are densely concentrated on the easternmost sand bar. Thirty observations refer to archaeological excavations, four to probe trenches and twenty-eigth to archaeological coring (Figure 4.5). The remaining observations include (chance) finds.

A query excluded all non-excavation observations (Table 4.1). The excavations are chosen as a first dataset to investigate critically, since they should be considered as the most complete dataset.

A few problems are immediately apparent: cryptic toponyms are used to refer to illustrious archaeological sites like the already mentioned *Praetorium Agrippinae* (Roman castellum) of Valkenburg. No elevation data is included. It also appears that these thirty excavations took place at fourteen geographical locations. The accuracy of the archaeological dates is deficient in that it remains unclear which is the most probable date and usually wide range chronologies are applied. Therefore, the ARCHIS2 database is considered not sufficient for purpose. Consulting excavation reports, other publications, or even the original field documentation remains a necessity. The archaeological site of Katwijk *the Zanderij* was thoroughly examined in the years 1996–2006 (Van der Velde *et al.*, 2008; Müller and Van Doesburg, 2008). The habitation and landscape history of

Figure 4.5 Archaeological observations in the municipality of Katwijk plotted on the palaeogeographical map in Roman times.
It shows the Rhine flowing directly into the North Sea (left) (source: van Dinter, map is available at DANS easy through http://www.persistent-identifier.nl/?identifier=urn:nbn:nl:ui:13–08qf-sf. This map goes with Dinter, M. van (2013)) and the same archaeological observations plotted on the current topographic map (right) (source: ARCHIS2), see also Table 4.1.

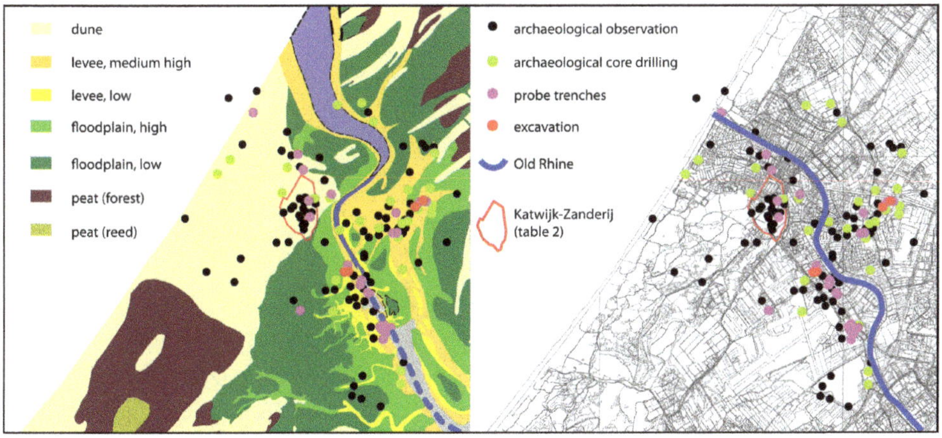

the region is determined by the coastal and fluviatile processes of the Old Rhine, which located the site *Zanderij* in the first millennium BC at the center of an extensive Rhine delta system (Figure 4.5). In the Late Bronze and Early Iron Ages the territory must have regularly flooded.

In periods with less river discharge and minor coastal activity the landscape consisted of a tidal flat with creeks and banks. During the first millennium BC the river mouth of the Old Rhine gradually changed into a much smaller delta shape with small creeks linked with the Old Rhine River. Even before the start of the first millenium the water retreated from the territory and small-scale aeolian activity resulted in a layer of sand covering the delta deposits, partly shown in the dune area in the palaeogeographic map (Figure 4.5, legend: dunes). In Roman times the landscape was inhabited but treeless with tidal flat vegetation. At the end of Roman times the habitation pattern diminished while large scale aeolian activity as well as the activity of the Old Rhine river increased. Between the fifth and eighth centuries habitation was reinitiated. In the High and Late Medieval Ages the land must also have been inhabited even though this is only indicated by a few finds lacking context. A single water-well of the sixteenth century is the only feature of the early modern times found in the *Zanderij* region (Van der Velde and Dijkstra, 2008). In the modern day landscape there is little remaining of the former coastal landscape, mainly because of the very intensive horticultural and agricultural industries, large scale sand mining, and modern development activities in the region.

This landscape history illustrates the dynamic and interactive character of habitation and landscape in the tidal basin of the *Zanderij* which is typical for large parts of the Dutch costal

Table 4.1 Archaeological excavations in the municipality of Katwijk (source: ARCHIS2 database)

NR	X_COORD	Y_COORD	Surface hight (NAP_)	MAX_DIEP-Tvondstdiepte	toponym	date	find
14807	89740	466130	?	0	VALKENBURG, MARKTVELD	12 BC-450 AD	castellum
14808	90240	465120	?	0	DE WOERD, ZONNEVELDSLA AN	12 BC – 150 AD; 525–725 AD	Pottery (TS); pottery & glass (burial gifts)
23990	88430	468840	0	0	KATWIJK, KLEIN DUIN	12 BC-450 AD; 525–725 AD	pottery, soil traces; (grave yard)
23994	89770	466170	0	0	VALKENBURGDORP	40–260 AD	castellum
24051	88550	468500	0	0	KATWIJK, CLEIJN DUIN	12 BC-450 AD	Traces of habitation, brick
24081	90200	465300	0	0	KATWIJK, DE WOERD	12 BC-270 AD	soil traces
24086	90150	465260	0	0	VALKENBURG, De Woerd (Limes)	70–270 AD	pottery
24116	90470	467230	0	0	RIJNSBURG, FORMER ABBEY	725–900 AD; 1050–1650 AD	Soil traces, pottery; monastery
24180	91040	467860	0	0	OEGSTGEEST, MOLENWEI / GAS-FABRIEK	525–725 AD	cemetery
24208	90200	465000	0	0	VALKENBURG, DE WOERD	12 vC-450 AD	soil traces
29706	90170	465200	0	0	VALKENBURG, DE WOERD	12 vC-450 AD	soil traces
30330	89120	468000	0	0	KATWIJK, Former AMK-terrain M7225 / CMA30E-012: Kerklaan; N.H. Kerk	1250–1500 AD	church
30811	90300	465230	0	0	DE WOERD	12 BC-450 AD; 525–900 AD; 1050–1250 AD	soil traces; zoological remains
32668	89890	465950	0	0	VALKENBURG, Veldzicht/Achter weg	12 BC-450 AD	soil traces
32879	89900	465930	0	0	VALKENBURG, Veldzicht area 2 (Limes)	12 BC-270 AD; 12 BC-450 AD	cremation, (also animal); soil traces

33238	89770	466170	0	0	VALKENBURG, DORPSHEUVEL	40–260 nC	castellum
33474	90470	467230	0	0	RIJNSBURG, FORMER ABBEY	12 BC-450 AD; 525–900 AD; 1050–1650 AD	potttery; foundations
33475	90450	467600	0	0	RIJNSBURG, DE WOERD-NARCISSENSTRA AT	525–900 AD	pottery and soil traces
33477	90450	467600	0	0	RIJNSBURG, DE WOERD-NARCISSENSTRA AT	525–900 AD	soil traces
33478	90470	467230	0	0	RIJNSBURG, FORMER ABBEY	1050–1650 AD	foundations
33483	90470	467230	0	0	RIJNSBURG, FORMER ABBEY	12th-16th cent. AD	Roman and Gothic building fragments.
33492	90470	467230	0	0	RIJNSBURG, FORMER ABBEY	12 BC - 450 AD; 450–1050 AD	pottery, soil traces
33525	90470	467230	0	0	RIJNSBURG, FORMER ABBEY	900–1050 AD	possible mound (Du.terp)
33596	90470	467230	0	0	RIJNSBURG, FORMER ABBEY	1050–1250 AD 1050–1500 AD; 1500–1950 AD	potterry (pingsdorf); monastery foundation; brick
36767	88680	467870	0	0	KATWIJK, ZANDERIJ WESTER-BAAN	12 BC-70 AD; 70270 AD; 4501050 AD	finds, soil traces
39577	90960	467840	0	0	RIJNSBURG, DE HORN	560–700 AD	graveyard
44657	88480	465600	0	0	VLIEGVELD VALKENBURG	1650–1850 AD	pottery, glass, brick
300465	89740	466200	0	0	Valkenburg	40–260 AD	castellum, various finds
401484	89740	466220	0	0	VALKENBURG, Dorpsheuvel (Limes)	40–260 AD	castellum, various finds
412353	88684	467540	0	0	KATWIJK, De Zanderij	70–270 AD; 450 1050 AD	soil traces

Table 4.2 Potential RSL proxy data from the archaeological excavations at Katwijk de 'Zanderij' (1996–2006) (source: van der Velde, 2008)

nr.	X	Y	topnym	feature	depth(MNAP)	date	dating-method	reference	reliability	accuracy
1	88700	467500	Katwijk, Zanderij	layer in excavation pit 68	-0,85	Iron age	ceramic	Van der Velde 2008, 109–111	1	5
2	88695	467540	Katwijk, Zanderij	building 4	0,1 (-0,2)	725–800 AD	ceramic	Muller & van Doesburg 2008, 77	4	1
3	88685	467518	Katwijk, Zanderij	Water well 52	0,1 (-0,2)	750 +/- 6 AD	dendro	Doesburg & Muller 2008, 186–187	1	1
4	88580	467910	Katwijk, Zanderij	Water well 29	0,95 (0,65)	550–600 AD	ceramic	Van der Velde 2008, 152	3	1
5	88700	467480	Katwijk, Zanderij	House 60	0,21 (-0,11)	550–650 AD	relative position	Van der Velde & Waldus 2008, 166, 179	4	4
6	88590	467380	Katwijk, Zanderij	House 58/59	-0,12 (-0,42)	550–650 AD	relative position	Van der Velde & Waldus 2008, 166, 178–179	4	4
7	88400	467390	Katwijk, Zanderij	Water well 30	0,12 (0,4)	672 +/- 8 AD	dendro	Dijkstra 2008, 202, 210	1	1

Figure 4.6 Katwijk the 'Zanderij'. Archaeological probe trenches and excavations 1996–2008 (adapted from van der Velde, 2008; legend, see table 4.2)

region. The essential elements for investigating the relation between coast and habitation are elevation and chronology of features, and the precision/range of these datasets. Elevation datasets are rare in both publications of the archaeological excavation (Van der Velde *et al.*, 2008; Müller and Van Doesburg, 2008). Nevertheless, by combining feature descriptions with maps and tables it was possible to generate several adequate datasets (Table 4.2; Figure 4.6).

With the exception of a layer with Iron Age finds (nr. 1) all depths are related to the excavation surface of the trenches and pits. The top surface height of the original soil of the accompanying archaeological feature is therefore reconstructed by adding 0.25–0.45 m to the measured depths. This 0.25–0.45 m is a rough estimate for the expected soil formation under conditions with relatively high ground water levels in sand. The accuracy and reliability of the archaeological data are also represented. Based on the above a range is given from 1, for the most accurate and reliable date, to 5 for the least accurate. The most accurate and probably most reliable dates were given to two water wells dated according to dendrochronology (Table 4.2, nrs. 3 and 7). Three features are dated on the basis of ceramic finds. The accuracy of dating and the reliability based on pottery is highly dependent on the recognisability (i.e. fragmentation and decoration) and the length of the period in which a ceramic type was in use. The amount of sherds does not influence the accuracy of chronology. If there is one sherd which can be dated and which is spatially linked to certain strata, the strata can be dated by *termini post* and *ante quem* dates: for example, a sherd of 20 BC dates the layer in which it is found to that period at the earliest. If a dated sherd occurs in the layer above it, the chronology can be locked to that layer (i.e. from 20 BC to 5 AD).

In Katwijk *de Zanderij* the dating of the pottery took place by period- and material-specialists (Van der Velde *et al.*, 2008; Dijkstra, 2008). Even so, the dating range of the features is rather diverse. The Iron Age dating of the layer in excavation pit 68 (Table 4.2, nr. 1), for example, has a range of at least 600 years. The wide range of time makes the dating of the layer not very accurate but relatively reliable. The chronologies of two buildings (nrs. 2 and 4), based on ceramic dating, appear to be more significant, with dating ranges of respectively seventy-five and fifty years. But these dates are classified as less reliable because they are based on only a few handcrafted and locally manufactured Iron Age sherds. These sherds are hard to assign a date to, and no correlation in an assembly is possible. Similar low reliabilities apply to building 60 as well as buildings 58/59, where the datings are only based on the lateral position in relation to other archaeological features (= relative position in Table 4.2).

To reconstruct water levels at the time of the archaeological finds in their context, it can be assumed that houses as well as ceramic finds in the *Zanderij* represent at least the surface elevation above the Mean High Water (MHW). The stratigraphy of water wells and related chronologies on the other hand can provide depth ranges within which we have certainty of the local water table level at the time of use. Care must be taken in using data from water wells in this wind-blown dune environment since they may reflect very local water tables that are even higher than MHW.

V. Discussion

It can be argued that archaeological data is not sufficiently precise, and if it were, the result is a local water level reconstruction rather than a RSL indicator. The analysis of the Katwijk case study area shows that it is hardly possible to a retrieve a selection of archaeological data that can be used at once for such a local water level reconstruction. It may be deduced that the outcome of the presented case study is not ideal for the purpose, while on the other hand it is clear that precise and absolute spatial referencing of archaeological features, as well as the precision of chronology and depth of these features, would greatly enlarge the value of archaeological data in the investigation of sea level changes. In addition, there is also a need for data on compaction and palaeotidal range evolution to make the step up to sea level reconstruction. If no information is available of compaction of strata, or if the palaeogeography of these sites remains only partly known, the fluctuations of sea levels of the last 3000 years may just as well be much larger than the estimates of the dm-scale today and regional divergence may become much more apparent. To solve this problem geological data and subsoil information have to be included in the reconstruction alongside archaeology. It is clear that archaeological data alone should not be used to attempt to resolve complex issues such as sea level variations in a tidal basin on a millennial time scale, or local water levels on centennial or decadal time scales. On the other hand, archaeological and historical geographical data provide timing of events in the last 3000 years that relate features with much more accurate chronologies to sea levels, which by geological data cannot be achieved. Geological layers, such as the peat deposits in this area, do allow for relevant RSL index points but their formation cannot be dated before 2800 BP (Van de Plassche, 1982). A strict selection of archaeological features directly related to the local ground water table may form the first step in an integrated approach to reconstruct the coastal development in the last 3000 years.

By adding archaeological data to geologically and statistically obtained RSL estimations, the precision in error in sea level fluctuations has been shown to increase by approximately 0.5 m (Parnell, 2005). Reference water levels are assumed to be closer to Highest Astronomical Tide (HAT) (Parnell, 2005; Waddelove and Waddelove, 1990). Estimations of HAT for the archaeological locations are used to calculate Relative Water Levels for the respective locations. It is worth considering if the investigation of existing (published) archaeological data has a benefit next to gathering new archaeological and geological data. The answer is probably yes, since the growing population of the southern North Sea coastal area in the last 3000 years did leave an imprint in houses, water wells, fish tanks, of which the latter approach good water level reference point data. The results of the case study area in the Katwijk/Old Rhine tidal basin show that archaeological data cannot be directly used as proxy data establishing relative sea level fluctuations in this specific geographical location. Next to database research, e.g. updating existing records, independent data retrieving and quality management strategies, as well as precise sampling methodologies, also need to be formulated and executed.

1. New archaeological observations should be entered in the National Database ARCHIS2 and standard formats for all data should be mandatory.
2. Each recorded archaeological find spot should hold information about elevation, exact geographic location and age, with precision of error margins on all data.
3. An open-system approach of database access within, for example, a controlled data entering procedure, is needed in order to enhance quality control.
4. Periodic peer review should be initiated and periodically undertaken to maintain standard quality protocols within the database contents.

Next to the improvement and updating of existing databases, like ARCHIS2 in the Netherlands, as well as recent progress in combining archaeological and geological data (e.g. Sivan *et al.*, 2001, 2004; Toker *et al.*, 2011), an interdisciplinary database is here proposed to model geological, archaeological, and historical geographical datasets of RSL and tidal range in the southern North Sea in absolute chronological ranges, so that they become mutually comparable (Figure 4.7). This model will allow researchers to either support or disprove whether, when, and where RSL (or associated local water level) did rise in the past 3000 years (the Late-Holocene). It will also show where and from which discipline we need additional observational data. Relative Sea Level changes from proxy methods at

Figure 4.7 Scheme of proposed geodatabase with 3000 years of coastal development of the southern North Sea

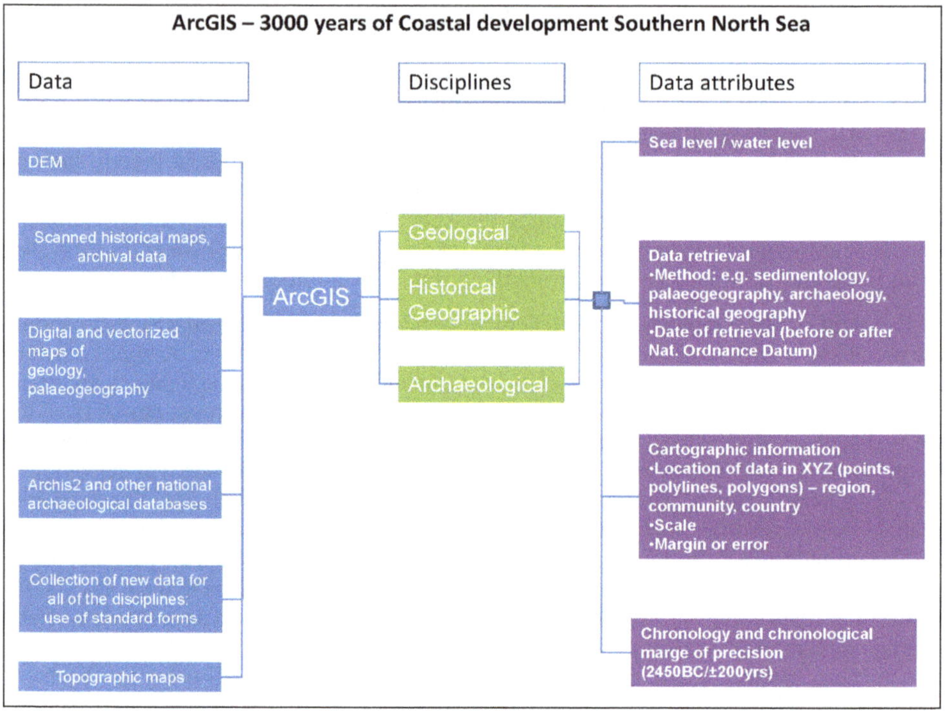

millennial to decadal timescales are different in archaeological, historical, geographical, and instrumental data. Data of RSL change collected before instrumental observations may have been homogenised into national Ordnance Datum in various countries.

Historical geographical data can be used to calibrate sea level movements for the last 300 years (Figure 4.2), but it is also feasible to use existing palaeogeographic reconstructions, which provide ever increasing details (e.g. Vos and Van Heeringen, 1997; Vos, 2006; Bazelmans *et al.*, 2011; Van Dinter, 2013). High resolution palaeogeography and adequate dating will result in detailed modeling of human occupation and its interaction with processes of erosion and sedimentation (Kluiving *et al.*, 2008). The Dutch government is currently developing a central database (BRO) with all soil and subsurface data that will be available in 2014 (http://www.dinoloket.nl) that will be very useful as input for the model.

The accuracy of elevation data from archaeological sites also needs to be verified. Errors should be quantified and provide elevation ranges. The errors in existing data (i.e. in ARCHIS2, excavations reports) can be quantified when they are compared to new field data, when collected with the particular aim to investigate RSL or local water level changes.

To accomplish this task, the model aims for a three step procedure:

1. Collection of data (old and new): elevation, chronology, accuracy and reliability
2. Analysing these data in a Geographical Information System (GIS)
3. Building a spatiotemporal model to evaluate RSL on regional and larger scales (Figure 4.7)

The main goal is to acquire a data set of high quality in number and spatial extent, coming from a variety of disciplines, to determine the highest achievable detail for both elevation and chronology. In view of comparability, standard formats for all data should be mandatory. The following groups of data should be collected:

1. Geology:
   a. sedimentological, palaeogeographical, and lithostratigraphical data, including data from (paleo)soils
   b. interpretation of marine fossils, shells, plant rests, foraminifera, diatoms, and other biological proxies
   c. C-14 dating on peat samples (sea-level index points, water level data) and shells in the coastal zone, OSL dating of silty and sandy sediments; geochronology of palaeosols

2. Archaeology:
   a. archaeological excavation data of benchmark sites regarding age, style, and duration of inhabitation, such as reconstruction of old surfaces, position of fireplaces, level of the arable land, water wells, or embankment levels
   b. C-14 on bone samples, archaeological materials (wood), OSL of soil pits, pottery, typological dating of pottery, numismatics
   c. x, y and z – coordinates from radiocarbon forms at relevant archaeological research locations

3. Historical geography:
   a. storm dates and sequences, historic storm levels and palaeoclimate reconstructions from archival data
   b. retrieval and integration of instrumental data on water level from State Surveys along the southern North Sea coast
   c. changes in inlets over longer historic time periods in terms of land-water and the geomorphology of inlets from historical data

The different phases of RSL or local water level changes can be visualised in GIS on a multi-layered/multi-scaled map, where evaluation of data quality will be of the utmost relevance. In steps, the existing (published) data should be compared to new (field) data to be collected under controlled conditions. Data from all disciplines should be referenced to one geographically referenced information system. Queries by region, discipline and error margins should provide insight in data quality and relevance, and be further validated by statistical analysis of sea- and water-level data for all disciplines. Finally with the highest achievable quality of RSL and water level data from the best resolution of information in time and scale, models can be built to visualise the sea level and local water level changes per region and through time.

VI. Conclusion

Generating proxy data for RSL based on data from archaeological excavations is challenging. Archaeological benchmark supporting dates constraining RSL rise have been increasingly recognised in the Mediterranean Sea and on the North Sea coast in the United Kingdom (e.g. Morhange, 1996; Parnell, 2005; Sivan *et al.*, 2001, 2004; Gehrels *et al.*, 2011; Token *et al.*, 2011). This study shows that through the use of the Dutch National Archaeological Database (ARCHIS2) there is access to a very large archaeological dataset, but it must be recognised that the use of data in ARCHIS2 in its present form is limited. A lot of data that potentially provides information on local water level reconstruction needs to be re-validated. An update of the ARCHIS2 database might therefore be necessary, particularly in establishing new supporting standard protocols that facilitate the accuracy of data recording. Archaeological studies are relevant but data should be more transparently presented and when applied to sea level research, geographically referenced in a fashion suitable to create index points and spatial range limits. Improvements can be made by accurate measuring of absolute elevations of the stratigraphy of an excavation or coring, or if absolute elevation cannot be retrieved it would also help if a (quantitative) margin of error and range of elevation can be determined (Parnell, 2005). Consulting excavation reports, other publications or even the original field documentation remains essential to investigate both RSL and local water levels. Only a combination of archaeological, geological, and historical geographical data will result in the optimal dataset to model RSL for the last 3000 years of sea level rise. It is hoped the outcome of this paper is an initiative to start a discussion in the archaeological community to improve the quality of data in national databases like ARCHIS2. Also the change from a closed to an open access system might enhance a procedure of directed peer review and critique by fellow scientists which in itself will enhance the quality of data stored in databases and reports.

The coast of the southern North Sea today is a combined natural and cultural constructed feature, which can only be unraveled by interdisciplinary research. The scarcity of geologically derived sea level index points directs the responsibility to retrieve these datasets elsewhere from archaeological and historical geographical sources. To achieve this aim, a database should integrate data with the best chronological and geographical definitions, both from existing as well as from newly acquired data from all three disciplines. The size and symmetry of the errors in the database should be established to determine the data quality. The best resolution data will then be used for the creation of models of former habitation patterns and how people interacted with Nature within diverse tidal basins areas along the southern North Sea coast. While the process of creating such a database may take a considerable amount of time, and is actually a goal for future research studies, we call for all (geo-)archaeologists, geologists, and historical geographers working in the coastal zone of the southern North Sea to present more key-sites and case studies with adequate archaeological bench-marks.

Bibliography

Baeteman, C. (2007) 'De laat holocene evolutie van de Belgische kustvlakte: sedimentatieprocessen versus zeespiegelschommelingen in Duinkerke transgressies', in: A.M.J. de Kraker and G.J. Borger (eds), *Veen-Vis-Zout. Landschappelijke dynamiek in de zuidwestelijke delta van de Lage Landen*, Amsterdam, pp. 1–17 (Geoarchaeological and Bioarchaeological Studies; 8).

Bazelmans, J., Meulen, M. van der and Weerts, H.J.T. (2011) *Atlas van Nederland in het Holoceen. Landschap en bewoning vanaf de laatste ijstijd tot nu*, Amsterdam.

Beekman, F. (2007) *De Kop van Schouwen onder het zand. Duizend jaar duinvorming en duingebruik op een Zeeuws Eiland*, Utrecht.

Beenakker, J.J.J.M. (1987) 'Een onderzoek naar de beheersing van het binnenwater: gevolgen voor landschap en bewoning', in: H. van der Wusten (ed.), *Postmoderne aardrijkskunde. De sociografische traditie voortgezet*, Muiderberg, pp. 212–224.

Beets, D.J. and Spek, A.J.F. van der (2000) 'The Holocene evolution of the barrier and the back-barrier basins of Belgium and the Netherlands as a function of late Weichselian morphology, relative sea-level rise and sediment supply', *Geologie en Mijnbouw/Netherlands Journal of Geosciences*, 79, pp. 3–16.

Behre, K.-E. (2003) 'Eine neue Meeresspiegelkurve für die südliche Nordsee: Transgressionen und Regressionen in den letzten 10.000 Jahren', *Probleme der Küstenforschung im südlichen Nordseegebiet*, 28, pp. 9–63.

Behre, K-E. (2005) 'Paläo-Ökologie an der Nordsee. Die Botanik als Schlüssel für die Umweltgeschichte', *Biologie unserer Zeit*, 35, pp. 320–330.

Behre, K.-E. (2007) 'A new Holocene sea-level curve for the southern North Sea', *Boreas*, 36, pp. 82–102.

Dam, P. van (2003) 'Rijnland onbegrensd: watergrenzen en kennisgrenzen', *Tijdschrift voor Waterstaatsgeschiedenis*, 12, pp. 50–58.

Dekker, C. (1971) *Zuid Beveland. De historische geografie en de instellingen van een Zeeuws eiland in de middeleeuwen*, Assen.

Denys, L., and Baeteman, C. (1995) 'Holocene evolution of relative sea level and local mean high water spring tides in Belgium – a first assessment', *Marine Geology*, 124, pp. 1–19.

Dijkstra, M.F.P. (2008) 'Aardewerk uit de Vroege-Middeleeuwen', in: H.M. van der Velde (ed.), *Cananefaten en Friezen aan de monding van de Rijn: tien jaar archeologisch onderzoek op de 'Zanderij'-Westerbaan te Katwijk (1996–2006)*, Amersfoort, pp. 251–263 (ADC Monografie; 5).

Dinter, M. van (2013), 'The Roman Limes in the Netherlands: how a delta landscape determined the location of the military structures', *Geologie en Mijnbouw/Netherlands Journal of Geosciences*, 92, pp. 11–32.

Ervynck, A., Baeteman C., Demiddele, H., Hollevoet, Y., Pieters, M., Schelvis, J., Tys, D., Strydonck, M., van and Verhaeghe, F. (1999) 'Human occupation because of a regression, or the cause of a transgression? A critical review of the interaction between geological events and human occupation in the Belgian coastal plain during the first millennium AD', *Probleme der Küstenforschung im südlichen Nordseegebiet*, 26, pp. 97–121.

Fleming, K., Johnston, P., Zwartz, D., Yokoyama, Y., Lambeck, K., Chappell J. (1998) 'Refining the eustatic sea-level curve since the Last Glacial Maximum using far- and intermediate-field sites', *Earth and Planetary Science Letters*, 163, pp. 327–342.

Gehrels, W.R. (2010) 'Sea-level changes since the Last Glacial Maximum: an appraisal of the IPCC Fourth Assessment Report', *Journal of Quaternary Science*, 25, pp. 26–38.

Gehrels, W.R., Belknap, D. F., Black, S. and Newnham, R. M. (2002) 'Rapid sea level rise in the Gulf of Maine, USA, since AD 1800', *The Holocene*, 12, pp. 383–389.

Gehrels, W.R., Kirby, J.R., Prokoph, A., Newnham, R.M., Achterberg, E.P., Evans, E.H., Black, S., Scott, D.B. (2005) 'Onset of recent rapid sea-level rise in the western Atlantic Ocean', *Quaternary Science Reviews*, 24, pp. 2083–2100.

Gehrels, W.R., Marshall, W.A., Gehrels, M.J., Larsen, G., Kirby, J.R., Eiriksson, J., Heinemeier, J., and Shimmield, T. (2006a) 'Rapid sea-level rise in the North Atlantic Ocean since the first half of the nineteenth century', *The Holocene*, 16, pp. 949–965.

Gehrels, W.R., Szkornik, K., Bartholdy, J., Kirby, J.R., Bradley, S.L., Marshall, W.A., Heinemeier, J., and Pedersen, J.B.T. (2006b) 'Late Holocene sea-level changes and isostasy in western Denmark', *Quaternary Research*, 66, pp. 288–302.

González, J.L., and Törnqvist, T.E. (2009) 'A new Late Holocene sea-level record from the Mississippi Delta: evidence for a climate/sea level connection?, *Quaternary Science Reviews*, 28, pp. 1737–1749.

Heteren, S. van and Spek, A. van der (2008) 'Waar is de delta van de Oude Rijn?', *Grondboor and Hamer*, 3/4, pp. 72–76.

Kiden, P. (1995) 'Holocene relative sea-level change and crustal movement in the southwestern Netherlands', *Marine Geology*, 124, pp. 21–41.

Kluiving, S.J., Kraker, A.M.J. de, Lascaris, M.A., Renes, J., Borger, G.J. and Soetens, S.A. (2008) 'Assimilating methodologies and measurements as integrated part of an interdisciplinary approach of analysing coastal development along the southern North Sea in the last 3000 years – a pre-proposal', *Geological Society, William Smith meeting, September 2008*, London.

Kluiving, S.J., and Guttmann E.B. (eds) (2012) 'Landscape archaeology between art and science. From a multi- to an interdisciplinary approach', Amsterdam (*Landscape & Heritage Series – Proceedings*).

Kluiving. S.J., Lehmkuhl, F. and Schuett, B. (2012) 'Landscape archaeology at the LAC2010 conference', *Quaternary International*, 251, pp. 1–6.

Malde, J. Van (1996a) 'Historical extraordinary water movements in the North Sea area', in D.J. Beets *et al.* (eds), *Coastal studies on the Holocene of the Netherlands*, pp. 27–39 (Mededelingen Rijksgeologische Dienst; 57).

Malde, J. Van (1996b) 'Relative rise of mean sea-levels in the Netherlands in recent times', in: M.J. Tooley and S. Jelgersma (eds), *Impacts of sea-level rise on European coastal lowlands*, Oxford, pp. 36–56.

Milne, G.A., Long, A.J. and Bassett, S.E. (2005) 'Modelling Holocene relative sea-level observations from the Caribbean and South America', *Quaternary Science Reviews*, 24, pp. 1183–1202.

Morhange, C., Laborel, J., Hesnard, A. and Prone, A. (1996) 'Variation of relative mean sea level during the last 4000 years on the northern shores of Lacydon, the ancient harbour of Marseilles (Chantier J. Verne)', *Journal of Coastal Research*, 12, pp. 841–849.

Mulder, E.F. De, Geluk, M.C., Ritsema, I.L., Westerhoff, W.E. and Wong, T.E. (2003) *De ondergrond van Nederland*, Groningen.

Müller, A. and Doesburg, J. van (2008) *Een archeologisch onderzoek naar een toevalsvondst in Katwijk-De 'Zanderij'. De opgraving van de 'Romeinse weg'*, Amersfoort (Rapportage Archeologische Monumentenzorg; 162).

Parnell, A.C. (2005) *The statistical analysis of former sea level*, Unpublished Dissertation, University of Sheffield.

Plassche, O. Van de (1982) *Sea-level change and water-level movements in the Netherlands during the Holocene*, 's-Gravenhage (Mededelingen Rijks Geologische Dienst; 36 (1)).

Plassche, O. Van de (1991) 'Late Holocene sea-level fluctuations on the shore of Connecticut inferred from transgressive and regressive overlap boundaries in salt-marsh deposits', *Journal of Coastal Research S.I.*, 11, pp. 159–180.

Plassche, O. Van de (2000) 'North Atlantic climate-ocean variations and sea level in Long Island Sound, Connecticut, since 500 cal yr AD', *Quaternary Research*, 53, pp. 89–97.

Plassche, O. Van de, Borg, K., van der and Jong, A.F.M. de (1998) 'Sea level-climate correlation during the past 1400 yr.', *Geology*, 26, pp. 319–322.

Plassche, O. Van de, Bohncke, S.J.P., Makaske, B. and Plicht, J., van der (2005) 'Water-level changes in the Flevo area, central Netherlands (5300–1500 BC): implications for relative mean sea-level rise in the Western Netherlands', *Quaternary International*, 133–134, pp. 77–93.

Plassche, O. Van de, Makaske, B., Hoek, W.Z., Konert, M., and Plicht, J., van der (2010) 'Mid-Holocene water-level changes in the lower Rhine-Meuse delta (western Netherlands): implications for the reconstruction of relative mean sea-level rise, palaeoriver-gradients and coastal evolution', *Geologie en Mijnbouw/Netherlands Journal of Geosciences*, 89, pp. 3–20.

Pruissens, A.P., and Gans, W. de (1988) *De bodem van Leidschendam*, Haarlem.

Schoorl, H. (1999) *De convexe kustboog. Texel-Vlieland-Terschelling. Bijdragen tot de kennis van het westelijk Waddengebied en de eilanden Texel, Vlieland en Terschelling. Deel 1: Het westelijk Waddengebied en het eiland Texel tot circa 1550*, Schoorl, pp. 14–34.

Sivan, D., Wdowinski, S., Lambeck, K., Galilid, E. and Rabana, A. (2001) 'Holocene sea-level changes along the Mediterranean coast of Israel, based on archaeological observations and numerical model', *Palaeogeography, Palaeoclimatology, Palaeoecology*, 167, pp. 101–117

Sivan, D., Lambeck, K., Toueg, R., Raban, A., Porath, Y. and Shirman, B. (2004) 'Ancient coastal wells of Caesarea Maritima, Israel, an indicator for relative sea level changes during the last 2000 years', *Earth and Planetary Science Letters*, 222, pp. 315–330.

Szkornik, K., Gehrels, R.W. and Murray, A.S. (2008) 'Aeolian sand movement and relative sea-level rise in Ho Bugt, western Denmark, during the 'Little Ice Age', *The Holocene*, 18, pp. 951–965.

Toker, E., Sivan, D., Stern, E., Shirman, B., Tsimplis, M. and Spada, G. (2011) 'Evidence for centennial scale sea level variability during the Medieval Climate Optimum (Crusader Period) in the eastern Mediterranean', *Earth and Planetary Science Letters*, 315, pp. 51–61.

Veen, J. van (1936) *Onderzoekingen in de Hoofden in verband met de gesteldheid der Nederlandsche kust, 's-Gravenhage* (Nieuwe verhandelingen van het Bataafsch Genootschap der Proefondervindelijke Wijsbegeerte te Rotterdam, Tweede Reeks, vol. 11).

Veen, J. van (1950) 'Eb- en vloedschaarsystemen in de Nederlandse getijdewateren', *Tijdschrift KNAG*, 67, pp. 303–325.

Velde, H.M. Van der (ed.) (2008) *Cananefaten en Friezen aan de monding van de Rijn: tien jaar archeologisch onderzoek op de 'Zanderij'-Westerbaan te Katwijk (1996–2006)*, Amersfoort (ADC Monografie; 5/ADC rapport; 1456).

Velde, H.M., Van der and Dijkstra, M.F.P. (2008) 'De monding van de Rijn gedurende de Romeinse tijd en de Vroege-Middeleeuwen', in: H.M. van der Velde (ed.) *Cananefaten en Friezen aan de monding van de Rijn: tien jaar archeologisch onderzoek op de 'Zanderij'-Westerbaan te Katwijk (1996–2006)*, Amersfoort, pp. 377–410 (ADC Monografie; 5/ ADC rapport; 1456).

Vink, A., Steffen, H., Reinhardt, L. and Kaufmann, G. (2007) 'Holocene relative sea-level change, isostatic subsidence and the radial viscosity structure of the mantle of northwest Europe (Belgium, the Netherlands, Germany)', *Quaternary Science Reviews*, 26, pp. 3249–3275.

Vos, P.C. and Heeringen, R.M. van (1997) 'Holocene geology and occupation history of the Province of Zeeland', in M.M. Fischer (ed.), *Holocene evolution of Zeeland (SW Netherlands)*, Haarlem, pp. 5–109.

Vos, P.C. (2006) 'Toelichting bij de nieuwe paleogeografische kaarten van Nederland', *Nationale Onderzoeksagenda Archeologie*, hoofdstuk 25 (versie 1.0) (www.noaa.nl), pp. 1–13.

Weerts, H.J.T., Westerhoff, W.E., Cleveringa, P., Bierkens, M.F.P., Veldkamp, J.G. and Rijsdijk K.F. (2005) 'Quaternary geological mapping of the lowlands of The Netherlands', *Quaternary International*, 133–134, pp. 159–178.

Weerts, H.J.T. (this volume) 'Holocene sea-level change, sedimentation, coastal change and palaeogeography in the southern North Sea lowlands – a 2012 geological literature overview'.

Waddelove, A. and Waddelove, E. (1990) 'Archaeology and research into sea-level during the Roman Era: towards a methodology based on highest astronomical tide', *Britannia*, 21, pp. 253–266.

5 From nature to culture: landscape and settlement history of the North-Sea coast of Schleswig-Holstein, Germany

Dirk MEIER, working group Coastal Archaeology and History Schleswig-Holstein

Abstract

This article gives an overview of the coastal geology, landscape, and settlement development from prehistoric times until the seventeenth century on the North-Sea coast of Schleswig-Holstein, Germany. The coastal population here was for thousands of years dependent upon the natural resources provided by the sea. Geo-archaeological investigations indicate that the earliest settlements were founded in the salt marshes of Schleswig-Holstein in the Roman period and date to the first century AD during a phase of a lowering storm flood level. Thereafter, c. 50 AD, higher storm surges forced the settlers to erect artificial Wurten (dwelling mounds).

In the seventh century a new occupation of the salt marshes took place in Dithmarschen, and a little bit later in North-Frisia, where large bogs limited the number of inhabitable areas to Pleistocene islands and smaller salt marshes in the western part of the area currently known as the North-Frisian Wadden Sea, as well as on the Eider estuary. Because of an increasing population during the high medieval period, new investment in farm land was necessary. Therefore, dikes were built and areas of peat were drained and cultivated. As a result, the natural environment changed more and more to a cultural landscape even though heavy storm surges sometimes disturbed this process. In the Uthlande *(outer lands) of North-Frisia large parts of the cultivated land were destroyed by heavy storm surges in 1362 and again in 1634. Remnants of these drowned settlements and fields can be observed as cultural traces in the tidal flats. These land losses in North-Frisia were partly the result of the human measures – like dike building and drainage – but geomorphology was perhaps more to blame. Accordingly, in order to fully understand why such disasters occurred geo-archaeological investigations are necessary.*

I. The geological development and early use of the landscape

The landscape and settlement history of the North-Sea coast of Schleswig-Holstein has been well established in the last fifty years. Initially, Albert Bantelmann excavated *Wurten* as artificial mounds of the Roman and early medieval periods. Subsequently, research into the landscape and settlement development of larger areas on the North Frisian Islands and the *Halligen* (small islands) were investigated by interdisciplinary research. This was based on the recording of the cultural heritage during the Norderhever-Project, run by the University of Kiel between 1975 and 1979 (Müller-Wille *et al.*, 1988). Since 1988

Landscapes or seascapes? The history of the coastal environment in the North Sea area reconsidered, ed. by Erik THOEN, Guus J. BORGER, Adriaan M.J. DE KRAKER, Tim SOENS, Dries TYS, Lies VERVAET & Henk J.T. WEERTS, Turnhout, 2013 (*CORN Publication Series*, 13), pp. 85–110.

BREPOLS PUBLISHERS DOI: 10.1484/M.CORN.1.101549

the Coastal-Archaeology unit of the Research and Technology Centre of the University of Kiel has continued recording such landscape and cultural elements in Eiderstedt and Dithmarschen. On the basis of this mapping, the landscape- and settlement history was described by the analysis of excavations of Roman- and Medieval period *terps* as artificial mounds (Meier, 2001a/b; 2004: 55–69; 2008: 18–30).

The landscape development of the coastal area in the Holocene was controlled both by sea-level changes and by the topography of the land surface which was inundated by the sea. From the south to north three zones of different morphology and landscape development can be distinguished along the North-Sea coast of Schleswig-Holstein: Dithmarshen, from the Elbe in the south to the Eider in the north; Eiderstedt, between the Eider estuary and the North-Frisian tidal flats, and North-Frisia. The eastern part of Dithmarschen and North-Frisia as well as the islands of Amrum, Föhr, and Sylt were formed by Pleistocene deposits of the Saalian glaciation. In general, the pre-Holocene relief slopes downwards from east to west and from north to south. In southern Dithmarschen, the basis of the Holocene coastal sediments dips from a present altitude of NN -10 m (NN: German Ordnance Datum) near by the Saale glacial cliff, to NN -30 m in the present day tidal flat area. During the Weichselian glacial period melt water from the River Elbe formed a valley which is deeply incised into the subsoil.

The Holocene transgression started in the southern part of Dithmarschen *c.* 6000 BC. At a level of NN -30 m a hilly landscape with height differences of 10 m was invaded by the sea. At that time, however, the global eustatic sea level had already risen higher than this low altitude suggests. What we see on the North Sea coast of Schleswig Holstein in general is the effect of the collapsing forebulge; due to isostatic subsidence the land-surface has been sinking since the beginning of the Holocene transgression and the basal fine grained sediments contain relicts of reeds. The elevation of the upper boundary of this layer ranges between a present level of NN -15 m in the western parts, and NN -10 m near the outcropping Pleistocene deposits in the east (Hoffmann, 2004: 31–32). Since *c.* 3000 BC, sand was primarily deposited and tidal flats came into being (Figure 5.1). This change of environment may be caused by strong tidal long shore currents which eroded the Pleistocene deposits and formed cliffs. As a consequence, a graded coastline consisting of cliffs and sand bars has been formed since 3000 BC (Hummel and Cordes, 1969; Menke, 1988). Sandy marine sediments were deposited on the seaward side of the beach ridges, and tidal flats developed. The valleys behind the beach ridges were covered with peat and westward salt marshes started to form *c.* 500 BC in Dithmarschen and the southern part of Eiderstedt.

In southern Dithmarschen, behind the natural levees of the River Elbe, a vegetation horizon, in depressions grading into peat, was formed at an altitude between NN -1 and -0.5 m. The lateral transition of the vegetation horizon into the peat layer was observed during the construction of a pipeline (Hoffmann, 2004: 32; Hoffmann *et al.*, 1997). The top of the peat was dated to *c.* 500 BC. In western Dithmarschen, near Marne, a peat layer on the top of this vegetation horizon was dated to the end of the last millennium BC. It is very likely that Roman period settlements founded in the beginning of the first millennium occupied this soil. This former land surface was subsequently covered with a vegetation horizon developed *c.* 500 AD. It is preserved only in a few places because it has mostly been reworked by ploughing. In contrast, peat did not develop during the last millennium BC in northern Dithmarschen. However, in the course of this period the sandy facies

Figure 5.1 North-Sea coast of Schleswig-Holstein. Reconstruction of the landscape *c.* 5000–3000 BC (source: Dirk Meier)

changed rapidly to a silty and clayey one. A marsh surface of the Roman period could only be demonstrated in one place and, moreover, the humic layer is weakly developed. Nevertheless, the great number of Roman-period settlements (Meier, 2001a) indicates that the salt marsh area was extensively occupied even though the sedimentation of clay had begun again by the second century AD. Gullies which drained the salt marsh were gradually filled up with mud before the formation of peat began. Thereafter, peat extended to the west, as far as the area of the early medieval *Wurten*. It is also noticeable that the clastic layers deposited on the top of the peat are rather thin. This may be a consequence of diking since the high medieval period. Remains of this peat, which was drained during both the high and late Medieval periods, are now preserved only below the medieval Wurten (Meier, 2001a: 128) and on top of the mud in the gullies (Hoffmann *et al.*, 1997; 2004: 32).

Northward of Dithmarschen (Figure 5.1), another system of beach ridges was formed *c.* 2000 BC in a west-east direction on the peninsula of Eiderstedt between the Eider estuary in the south and the North-Frisian tidal flats in the north (Menke, 1988). These sand walls do not extend as far as the outcropping Pleistocene deposits of the hinterland in the east. Dittmer (1952) supposed that the sand and gravel which had built up this spit system originated from the erosion of an old moraine shore (so called: Hever-Geestkern) located in the northwest of Eiderstedt in a place which belongs to the present offshore area. The interpretation was criticised by Austen (1992), who explained the bar as a system of sand banks with tidal channels between them. However, he did not explain the origin of the gravel. Moreover, the presence of a humic surface layer as well as a peat layer, dated at 800 BC (Menke, 1988), seem very unlikely to have developed on a sand bank. The beach ridges were visited by hunter gatherers of the Neolithic and Bronze Ages; later on – during the Roman period – settlements developed. South of the sand bars, salt marshes silted up *c.* 500 BC, and to the north peat formation started as a result of the protection offered by these sand bars against the sea (Meier, 2001b: 17–22).

The Holocene transgression entered the North-Frisan area later than in Dithmarschen. The present-day Pleistocene surface in this area is situated between NN -18 m, south of the island of Pellworm, and NN -5 m below the *Hallig* Langeness (Hoffmann, 1988). This is mirrored by the present-day topography of the landscape. The islands are located in places where the pre-Holocene surface occurs at higher elevations (e.g. Pellworm, Langeness, Hooge). The beginning of the Holocene transgression is not well dated, mainly because the organic layer at the base of the Holocene sequence is eroded. Initially, the depressions of the pre-Holocene surface were filled up by tidal clay. When, with the rising sea level, the higher areas were flooded and covered by sediments, the facies changed and became highly variable because sand, silt and clay were deposited over short distances. Later on, the facies of sediments became brackish in the southern part of North-Frisia. Humic clay mixed with reeds accumulated and in the first millenium peat developed. Palynological investigations on peat from Eiderstedt (Brande, 1988) and from the area between Pellworm and Nordstrand indicate a complete absence of marine influence *c.* 500 BC. It seems clear that the sea could not enter this region so it is supposed that a system of sand bars existed to the west of the North-Frisian area (Figure 5.1 and 5.2). It might have been built up by reworked sand material that came from the erosion of outcropping Pleistocene deposits which existed to the west of the present-day sand banks (in German the sands are called *Norderoogsand, Süderoogsand* and *Jaosand*). This barrier coast of sand bars and Pleistocene cores was partly breached by tidal channels. Unfortunately there are no remains of this barrier coast left today, neither from the outcropping Pleistocene nor from the sand bars themselves because they were later eroded by the invading sea. Cut off from the sea by these sand barriers, large areas of the present-day eastward part of the North-Frisian tidal flats were covered with peat bogs. The environment with the woodland on the Pleistocene areas, the coastal regions, and the large bogs provided good resources for hunter-gatherer groups. Several sites indicate that the coastal areas were used for fishing, hunting, and for sourcing flint by hunter-gatherers during the Mesolithic period. During the Neolithic and Bronze Age the forest on the Pleistocene areas were partly cleared by the first farmers but the coastal areas continued to be used for fishing, hunting, and obtaining flint.

The sequence of Holocene sediments in the tidal flats between the islands of Föhr and Sylt show that in large areas the pre-Holocene surface was covered with young Holocene tidal flat sand (Figure 5.1). Only parts of the basal peat are preserved. Three phases of erosion can be recorded in this area. The first phase began after the start of the transgression. At some locations the erosion removed only the upper part of the basal peat, at others the clay of the second sedimentary unit directly overlies the pre-Holocene deposits. The second phase of erosion occurred either before or during the sedimentation of the tidal flat sand, and the third phase of erosion is demonstrated by incision of gullies in the sandy deposits of the young tidal flats. In northern North-Frisia a complete sequence of the coastal Holocene deposits is only preserved in the part of the present-day marsh area of the island Sylt which is exposed to the north (Hoffmann, 1980; 1984: 34–35). *C.* 4000 BC, at a present altitude of NN -3.8 m, the basal peat was covered with clay which was then overlain by tidal flat sand. The top of this first unit is formed by clay changing into humic clay and peat. The deposition of this sequence began at the end of the sub-Boreal period at an altitude of NN -2.5 m. The third unit consists of silt and clay which have been were deposited by storm surges since 1300 AD. The Holocene sedimentary sequence shows periods of changing marine influence, which is also confirmed by palaeobotanical investigations of the basal peat (Averdiek, 1980). In the Holocene sequence in the southern marsh area of Sylt, only the second and the third sedimentary units were recorded. A settlement of the Roman period on top of these sediments above a vegetation horizon has been discovered by Kersten and La Baume (1958). The high medieval dike was subsequently built on top of this vegetation horizon (Hoffmann, 1980). Therefore, it can be concluded that no sedimentation occurred in the course of the first millennium AD.

II. The first step into the salt marshes: settlement history of roman period

The un-embanked salt marshes represent a particular settlement area with specific ecological conditions and their land use reveals a close correlation between topography, settlement, and agriculture. Near to the coast, meandering tidal channels formed a landscape with irregular parcelling and were used as transport routes between the sea and the hinterland. Whether arable farming was practised under these extreme ecological conditions has to be proved for each settlement individually. Farming was only carried out on a small scale and was restricted to the natural levees of the tidal channels. The settlements were totally dependent on the influence of the sea and, since the early first century AD, were mostly founded on elevated higher tidal channel banks in the river marshes along the Elbe and Stör, as well as in the salt marshes in Dithmarschen, the southern part of Eiderstedt, and the Wiedingharde in North-Frisia as well as around the island of Sylt and in some small parts of the today western part of the North-Frisian Wadden-Sea.

The settlements of the Roman period in the river marshes strictly adapted themselves to the natural conditions: high riverbanks, tidal-channel ridges, and the peat zone behind them. Archaeological finds, borings, and excavations in the *Stör*-marsh have demonstrated that the high river banks were densely populated in the first and second centuries AD but the low lying peat zone allowed no cattle grazing (Figure 5.2). A typical example of a river marsh settlement is *Hodorf*, where a stable-house of a ground settlement of the second century AD was unearthed in 1935 and 1936. By the fourth century AD the site had been heightened to become a *Wurt*. Possibly at the end of that phase of habitation the marsh was flooded (Haarnagel, 1937). The settlement of *Ostermoor* (Figure 5.2) in the south part Dithmarschen near the Elbe estuary is another similar example of a settlement in a river marsh of the first and second centuries AD (Bantelmann, 1957/58).

Though the settlements of the edge of the Pleistocene areas and sand ridges initially used the salt marsh only for grazing cattle, the old salt marshes of Dithmarschen were first permanently occupied by humans in the early first century AD (Figure 5.2). These marsh settlements are located along north-south oriented lines, mostly on marsh ridges. The landward part of the marshland was partly covered with peat (Meier, 2003: 285). In Süderbusenwurth, larger excavations between 1998–2002 (Meier, 2005a: 343–363) have pointed out the existence of single *Wurt* with stable-houses on higher ground (present altitude of NN +1.8 m) near a tidal channel. This *wurt* was likely created because the salt marshes around the settlement only reach a height of NN +0.8 m and were inundated at extreme high tides. After 150 AD the *Wurt* was further heightened with clay, but was abandoned *c.* 200 AD, probably because the low-lying marshland was becoming increasingly inundated at high tides. The main economy of this settlement was the husbandry of cows (fifty percent) and sheep (forty percent), but the ecological coastal resources were also used for hunting and fishing.

In the northern salt marshes of Dithmarschen, small settlements were arranged in two lines between 2 and 4km to the west of the edge of the Pleistocene hinterland on north-south orientated marsh ridges (Figure 5.2). Archaeological excavations have demonstrated the existence of a *Flachsiedlung* (ground settlement) from the first century AD in *Tiebensee* at a level of NN +1 to +1.3 m (Meier 2001a: 16–42). The house sites were raised a little in the second century AD before the settlers later abandoned the site, possibly because of higher ground water levels. Radiocarbon analysis of peat shows that bogs expanded over the hinterland during the late Roman period (Meier, 2001a: 128) and so new settlements were erected in a zone 2km further to the west in an area closer to the coastline. Here, in Haferwisch, single *Wurten* were built in 140 AD with sods of clay on a low lying salt marsh (Meier, 2001a: 43–67). After some subsequent height increases, the larger *Wurt,* which was formed from different single Wurten, was abandoned at the end of the fourth century AD.

Habitation of the high salt marshes along the meandering River Eider did not begin until the first or second century AD. The estuary of the Eider was smaller during the Roman period than today and it partly silted up, which led to lower Main High Water (MHW) than before. When the estuary got wider again, local MHW also rose. Changing topography at the mouth of an estuary can seemingly account for a fall or rise of the sea-level

Figure 5.2 North-Sea coast of Schleswig-Holstein. Reconstruction of the landscape *c.* AD 200 (source: Dirk Meier)

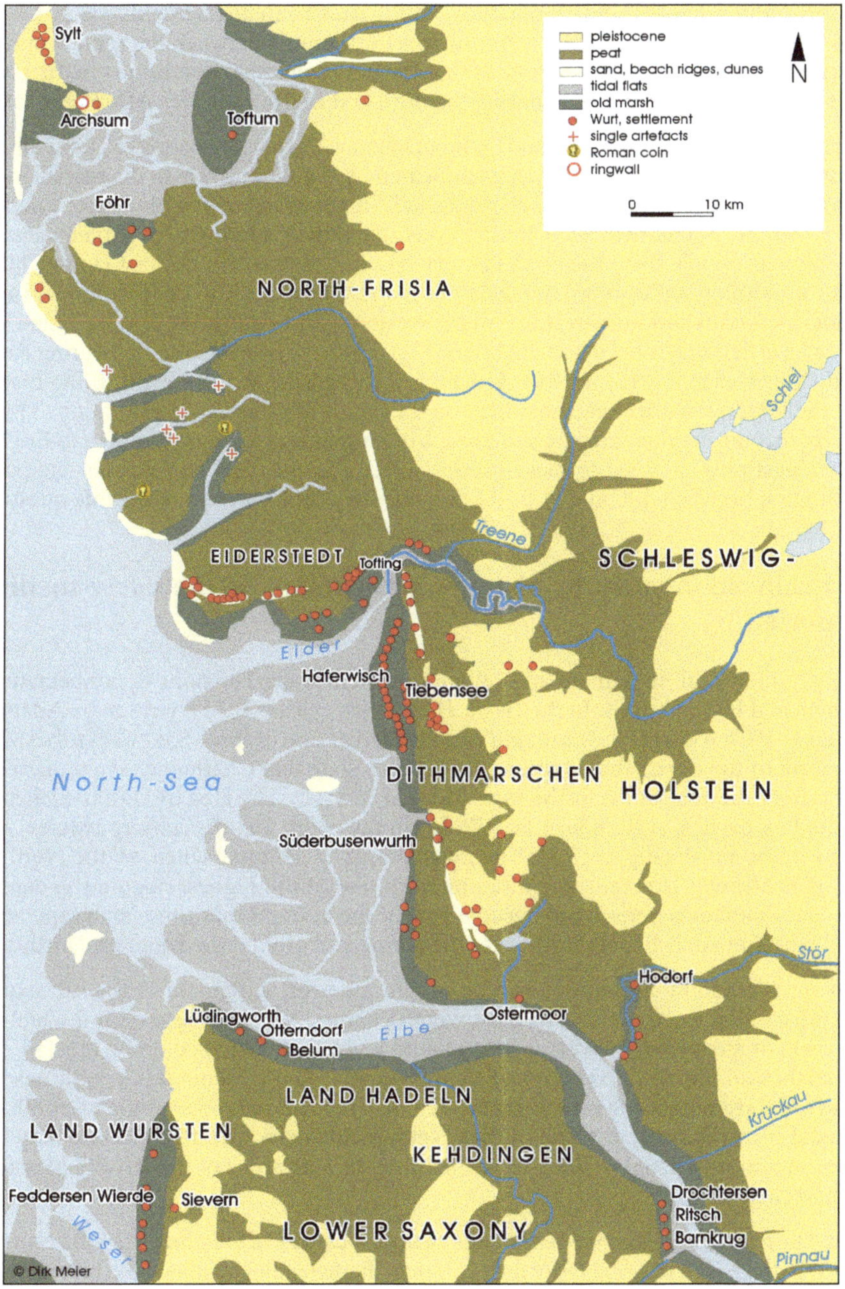

further upstream. What we see, however, are only fluctuations of local MHW that are not necessarily linked to sea level fluctuations at the coast. Because of renewed higher storm surges, which reached the higher silted-up river banks, the stable-houses of the excavated *Dorfwurt* (village hill/dwelling mound) of Tofting near Tönning (Bantelmann, 1955) were raised after a first short period of a ground level settlement between the second and fifth centuries AD. A further point to note is that the environmental conditions in the coastal marsh became increasingly salty during the third and fourth centuries AD (Behre, 1976).

The salt marshes of Eiderstedt and Dithmarschen were densely populated until the third and fourth centuries AD. Thereafter, as indicated by the settlements in the marshlands as well as by cremation places on beach ridges in Eiderstedt and on the Pleistocene deposits in the hinterland of Dithmarschen, the area was abandoned. The hoard of *Katharinenheerd* with precious metals from the sixth century AD in Eiderstedt is the last sign of human habitation (Müller-Wille, 1986). Excavations in the Wiedingharde (f.e. Toftum) and single artefacts (ceramics and Roman coins) of the western part of the today North-Frisian Wadden Sea have demonstrated that these areas (Fig. 5.2) have been also settled during Roman period (Meier et al., 2013, 35–40). The end of this phase of settlement in the marshes along the German North-Sea coast may be the result of climatic deterioration. There is no doubt that during this same period groups of Angles and Saxons migrated to England, but the question is how large these groups were. Too little is known about the pattern of habitation between 400 and 700 AD in Schleswig-Holstein to answer this question.

III. From anonymity to history: settlements of the early medieval period

The early medieval settlement of wetlands in Schleswig-Holstein is connected with Saxonian and Frisian tribes. In the *Gesta Hammaburgensis* (II/15), written by Adam von Bremen in 1076 AD, the *Tedmarsgpoi* is described as one of three Saxonian tribes which lived north of the river Elbe. Written sources also state that Charlemagne conquered the area north of the Elbe as far as the river Eider in 798 AD. The area of Dithmarschen was protected by castles, *Bökelnburg* and *Kuden* in the south and *Stellerburg* (Meier, 2011: 93 pp.) in the north (Figure 5.3). The coastal area of Dithmarschen on the North-Sea coast of Schleswig-Holstein was a region with one of the highest population densities in the early medieval period and became part of the cultural maritime landscape, which spread with Frankish-Frisish trade over the Rhine and around the southern North-Sea.

Favourable natural conditions had permitted renewed settlement of the salt marshes of Dithmarschen since the end of the seventh century AD. During a period which saw a lowering of storm surge levels again, ground level settlements could be erected in the salt marshes on the elevated banks of tidal channels. These favourable conditions were only of short duration, however, because new *Wurten* had to be built soon after 800 AD. Botanical investigations of some of these settlements demonstrate that extremely halophytic conditions existed in the area at that time on both sides of the River Eider and in Dithmarschen.

The *Dorfwurten* (village *wurten*) of this period, located in a long line from the estuary of the Elbe to the Eider, reached more than 250 m in diameter and heights of up to NN +6.2 m.

Figure 5.3 North-Sea coast of Schleswig-Holstein. Reconstruction of the landscape *c.* AD 800 (source: Dirk Meier)

In the southern part of Dithmarschen, the large *Wurten* of this period occur in the same area as the settlements of the Roman period, which contrasts with the northern part where the early medieval *Wurten* were founded on new silted up marsh ridges westward of the old settlements from the Roman period. There, bogs had spread since the late Roman period over the inner part of the former salt marshes.

The best example with a series of occupation layers is *Wellinghusen* near Wöhrden (Meier, 2001a: 70–98). In 1994 extensive excavations (Figure 5.3 and 5.4) unearthed a ground level settlement erected at today's elevation of NN +1.8 m on a marsh ridge along a tidal channel. Relicts of *Phragmites* collected from the marsh surface below the settlement (KI-3936) yielded an age of 1440±740 BP (555–660 AD). The partly excavated houses of the ground settlement were built *c.* 691 AD on flat sod mounds. Thereafter, increasing storm surges between 764–820 AD apparently forced the settlers to raise their farmyards to NN +3 m using manure carefully covered with clay sods. The tidal channel was filled with dung in the course of the extension of the *Hofwurten* (dwelling mounds with a drainage system and long fields) and a bridge across this creek was rebuilt in 785 AD. The individual *Hofwurten* were consolidated to form one large *Dorfwurt* during a phase of further expansion of the settlement which attained an elevation of NN +3.8 to +4.0 m in the tenth century AD. In the twelfth century the *Wurt* was raised with clay to NN +5 m and in the fourteenth century to NN +6 m. The site was later abandoned. Four-row *Hordeum vulgare* (barley), *Linum usitatissimum* (flax) and some *Avena* (oats) were cultivated around the early medieval *Dorfwurt*. The economy of the settlement was rural, and more cows (over sixty-five percent) than sheep (just over fifteen percent), pigs (just over eleven percent) and horses (just over five percent) grazed in the salt marshes. The settlers used the coastal environment as well as the peat hinterland and the forests on the Pleistocene areas for fishing and hunting.

The tenth century saw the beginning of an increase in settlement which resulted in the founding of new *Wurten* like Hassenbüttel, 2 km north of Wellinghusen (Meier, 2001a: 101–114). The living conditions in the deep seated and often inundated saltmarsh there, with a height of NN +0.8 m, required the raising of the *Wurten* with layers of clay and dung. The earliest dwelling mounds, which reached a height of NN +2.0 m, were raised on a sediment layer that covered a pre-existing ploughed field. On the top of one of the later *Wurten*, which reach a height of NN +3.0 m, a large three-aisled stable-house and smaller buildings with walls made of clay sods were built in the eleventh century. The husbandry bone material indicates that comparable to Wellinghusen, relatively more sheep were kept (just over thirty-nine percent), although bone material of cows (just over forty-eight percent) still formed the majority of bone material in the more low lying salt marsh in Hassenbüttel as well.

The settlement of the salt marshes north of the River Eider during the eighth century AD was often attributed to a Frisian immigration. This cannot be proven from written sources and archaeological arguments for a specifically Frisian material culture are hard to define. A number of early medieval settlements and burials are known from Eiderstedt, including larger *Wurten* on the banks of the Eider, and smaller settlements or cemeteries on the sandy beach ridges (Meier, 2001b: 62–67). The landward parts of the peninusula of Eiderstedt were covered with bogs. The *Dorfwurten* on the northern side of the River Eider are normally related to the purported Frisian inhabitation in the eighth century AD (Figure 5.3). Palaeobotanical investigations in Elisenhof (Behre, 1976) demonstrate that extremely halophytic conditions existed during early medieval period in the area at the

Figure 5.4 Excavation of the Wurt Wellinghusen (Photograph: Dirk Meier)

mouth of the River Eider. A group of three-aisled stable-houses surrounded by fences were built as ground level settlements with yards which were raised only slowly. In contrast, the lower salt marsh area at Welt soon forced the settlers to raise their house sites (Meier, 1997). Animal husbandry was undoubtedly the economic base of these settlements as only modest farming was possible. However, there is no doubt that the farmers of the coastal area profited from the Frisian and Frankish trade which had developed around the North Sea since 700 AD (Meier, 2009: 29).

Unfavourable environmental conditions, due to the extended peat bogs and swampy areas in the inner part of the present-day North-Frisian tidal flat area, prevented a settlement before 1000 AD. Only areas with outcropping Pleistocene deposits and more elevated parts of the salt marsh were inhabited. A number of settlements and burial places are recorded on the Pleistocene deposits on the islands of Amrum, Föhr and Sylt (Kersten and La Baume, 1958) and the edges of the Pleistocene areas were densely populated too. Further archaeological artefacts are known from the western part of the tidal flat area in the vicinity of the *Hallig Hooge* and the island of Pellworm (Figure 5.3). These indicate that in early medieval times, salt marshes, protected by sand ridges, existed in the western part of the present-day North-Frisian tidal flats. Excavations have brought to light a ground level settlement of the ninth century which was inundated by salt water (Kühn and Müller-Wille, 1988: 188). From the twelfth century onwards, people could not live in their houses built on surface of the marshland in North-Frisia any longer, because the beach ridges in the west had been destroyed by the sea, and the salt marshes were too often inundated during storm tides. As a consequence, *Warften* (villages) had to be built.

IV. The constructed landscape: the salt marshes during high and late medieval time

Since the high medieval period the entire area of the salt and river marshes were intensively cultivated and were more densely populated than ever before. Therefore, investments in the cultivation of new land with embanking and the drainage of peaty areas were necessary. In places like Flanders and Holland dikes had been built along the German North Sea coast since the twelfth century AD (Meier, 2007: 100). The written sources describing this process in medieval Schleswig-Holstein are limited. A letter of the Pope Innocent III (1198–1216) from 1198 to the overseer of the Strand, a large marsh island in the southern part of North-Frisia, describes that the low lying land is often flooded (Meier, 2007: 100). The Danish historian *Saxo Grammaticus* (*c.* 1150–1220) wrote in his *Gesta Danorum* (XIV, VII) that the dike breaches in *Frisia Minor* (North-Frisia) had terrible effects. More and more salt marshes were embanked and inhabited, landward swampy areas were drained, and the local people generally took the initiative for the construction of dikes. Economic associations formed by people of high social standing on larger *Wurten* were organised as cooperatives, maintaining complete independence from outside nobles and landlords until late medieval times, in Dithmarschen, for example, until 1559 (Meier, 2001b: 127–133; 2007a, 91–93; 2012: 121–134). The wealth of these leading families was based on the systematic drainage and colonisation of the landward parts of the marshes, which transformed the natural environment into the present-day anthropogenic landscape.

The first dikes in Dithmarschen connected the larger village *Wurten* from the Elbe in the south to a beach ridge at the edge of the Pleistocene area near Meldorf, and from there on to the Eider and the sandy beach ridge of Lunden (Figure 5.5). *Geschlechter* (Cooperatives of farmers) decided about the location, design, construction, and maintenance of dikes, drainage systems, and sluices. In the newly drained areas the farms were built on single *Wurten* in long settlement lines to protect against flooding from inland waters. Their long and small fieldstrips penetrated farther and farther into the peat marsh. Moreover, in the salt marshes of Dithmarschen, new village *Wurten* constructed of clay such as Büsum (Meier, 2001a: 124–128), Schülp (Meier, 2001a: 124), Norderbusenwurth (Meier, 2003: 277–284) and Lütjenbüttel (Meier, 2003: 284–291) were established, reaching heights of around NN +3.0 m. Also, the old village *Wurten* founded in the seventh century AD were heightened again with clay between the twelfth and the fourteenth centuries (Meier, 2001a: 101).

Also, in the southern part of the peninsula of Eiderstedt, a dike protected a large part of the salt marshes along the Eider and the low lying peat areas in the hinterland, which were drained and colonised from the twelfth to the fourteenth centuries (Meier, 2001b: 84). This newly cultivated land was partly settled again by Frisian immigrants. For the most part, this embanked area belong to the district of Tönning-Harde (Meier, 2001b: 102), as it is described in the *Earth Book* of the Danish King Valdemar II (1170–1241). This marshland is bordered to the west by the tidal channel of the Süderhever, which divided Eiderstedt from the embanked island of Utholm. Utholm, as well as Westerhever, are labelled as islands (Holm and Haefrae), which were later connected with Eiderstedt and the northern part of the peninsula, called Everschop Harde. There, numerous tidal channels divided different marsh islands, which had been settled and embanked since the twelfth century. Because salt water then often inundated the low lying salt marshes, the

Figure 5.5 North-Sea coast of Schleswig-Holstein. Reconstruction of the landscape *c*. AD 1300 (source: Dirk Meier)

Warften were raised in height at the same time. Excavations in Hundorf have documented a *Warft* of the twelfth century with a height of NN +3.0 m, which was raised to NN +4.0 m in the fourteenth century (Meier, 2001a: 135). The cooperatives which settled on these larger *Warften* also built local dikes around their cultivated land. Later on, smaller house *Warften* were constructed inside of the polders with their irregular field strips (Meier, 2001a: 147 pp.). The surrounding low summer dike of the St. Johannis polder, with its single *Warften* and its church, reached a height of NN +1.5 m in the twelfth century and was raised again in the fourteenth century. Soon after their erection, these isolated ring dike systems were connected together and, since the middle of the fifteenth century, larger tidal channels like the Fallstief were dammed (Meier, 2012: 136–142).

To sum up, according to archaeological investigations, the outer salt marshes of North-Frisia were not colonised until the Roman period and the early Middle Ages (Figure 5.5), as archaeological artefacts from the tidal flats around Pellworm and Hallig Hooge demonstrate. As the twelfth century progressed increasing numbers of *Warften* were built in the recently silted up and colonised marshland. Similarly, in the north-part of Eiderstedt, the *Warften* of Pellworm reached heights of NN +3.0 m and were subsequently raised to +4.0 m in the fourteenth century. The first dikes had very gentle banked slopes, which became steeper when they were later raised (Meier, 2005). In the high medieval period, rural societies organised in parishes were responsible for the dikes and sluices, as well as the *Harden* (Hundreds) as larger districts (Meier, 2012: 147–161).

V. Natural powers: high and late medieval storm surges

The process of medieval colonisation was sometimes partly interrupted by heavy storm surges. After the medieval warm period (Lamb, 1982) the climate became colder after 1300 AD with heavy storms and rainfalls in the fourteenth century. This rapid climate change can be observed all over Europe as well as elsewhere in the world (Nunn, 2007). During the second half of the fourteenth century many disastrous storm surges occurred along the North-Sea coast, which partly shaped the coastline we know today. To be able to classify historical storm surges before the nineteenth century, it is necessary to know the historical development of (paleo)tidal levels because the sea level data are always referred to MHW. Unfortunately, detailed tidal measurements are available only after 1850. Historical chronicles, such as the *Gesta Danorum* of the Danish historian *Saxo Grammaticus* (1150–1220: Book XIV, VII) or the *Chronica Slavorum* of Helmold von Bosau (1120–1177: book II, 97) focused primarily on the effects and damage of the storm surges after the breaking of the dikes. Natural disasters were mostly interpreted as punishment from God. The chronicles are not very detailed and the pain is common, not personal. In bad times (poverty, epidemics, and wars), even smaller storm surges may have had catastrophic consequences due to the poor condition of the dikes. Finally, the numbers of people killed is not realistic (Meier, 2007: 119). Therefore, geo-archaeological investigations are necessary for any reconstruction of the coastal landscape and the older MHW levels.

As a consequence of dike building, the water level increase during storm surges was in part due to the reduction in size of flood plains. In particular, storms from the north-west forced water into the German Bight and the estuaries, and breaches of the small dikes were common. Therefore, in the fourteenth century the dikes were built higher and the *Warften* were raised. Unfortunately, natural powers were stronger than the human reactions to flooding. In Schleswig-Holstein, for example, smaller land losses of the late medieval period are known from Dithmarschen where the higher marsh banks near the Elbe estuary, together with the church village of Uthaven, were destroyed (Figure 5.5). Also, large parts of the island of Büsum were drowned. Along the conserved northern part of the island new salt marshes were silted up, which were embanked sometime between the fifteenth and sixteenth centuries. Shortly after the end of the independent farmer republic of Dithmarschen in 1559, the island was connected by a dam with the main land in 1585 (Meier, 2001b: 149–152). The effects of storm surges of the sixteenth century in this area are described in the *Dithmarschen chronicle* (I, 217) of Neocorus (1550–1630).

Large effects of storm surges could also be observed on the north part of the peninsula of Eiderstedt, where the bay of Offenbüll broke into the embanked marsh area between Osterhever and Uelvesbüll in the fourteenth century (Meier, 2007: 123–124; 2012: 136–142). At that time the land connection between Eiderstedt and the North-Frisian *praepositura Strand,* the Lundenbergharde, was also partly destroyed. The tidal channel of the Hever broke through the old sandy beach ridge of Witzwort-Lith and reached the edge of the Pleistocene hinterland near Husum, which later became a harbour town (Meier, 2007a: 130; Wohlenberg, 1989). The parishes in the northern part of the former Lundenbergharde, Morsum, Hamm, and Lith, were now divided from the southern parishes of Lundenberg, Ivelek, and Padelek (Meier, 2007a: 131). The rest of the Lundenbergharde, which became an island after 1362, was protected with dikes again and connected with Eiderstedt in 1525 before the last parts were destroyed by storm surges in 1634 and 1718 (Meier *et al.,* 2013: 67–73). Another tidal channel of the Hever, the Nordereider broke inside the cultivated marshland of Eiderstedt in the fourteenth century and reached the Eider, but was embanked during the course of the next 200 years by a lot of polders during an offensive phase of land reclamation under the duke of Schleswig (Meier, 2007b: 252).

The largest land losses in the fourteenth century along the North-Sea coast of Schleswig-Holstein were recorded in the North-Frisian *Uthlande* (Figure 5.5). The written sources, such as the *Earth Book* of the King Valdemar II, indicate that there had been embanked salt marshes in the parts of North-Frisia which changed after the *Second Marcellus Flood* of 1362 into tidal flats. This disaster happened shortly after a pestilence and wars have hit the Uthlande. In the registry of the chapter of the cathedral of Schleswig, which is known in a 1407 copy from Copenhagen, destroyed churches are also listed in the writing of Bishop Nicolaus Brun of Schleswig (1350/51–69). This information was partly reused in the *Catalogus vetustus* by Anton Heimreich (1626–1685), who also wrote a chronicle of North-Frisia (Heimreich, 1668; Hansen u. Jessen, 1904: 75). Heimreich argued that the storm surge was two *Ellen* (*c.* 1.38 m) higher than the dikes, which themselves mostly reached a height of NN +2.0 m. Other historians of the seventeenth century described this heavy storm surge. For example, Matthias Boetius (d.1624) called the flood the *Grote Mandrenke* (great drowning of men) in his book *De catalysmo Nordstrandico*. Peter Sax also gives a impression of the landscape and the disaster in his *Descriptio Insulae Nordstrandiae* (1636) and in his maps of *Frisia Minor* (1636).

From the sixteenth and seventeenth centuries, it becomes possible to combine maps with an analysis of written sources and geo-archaeological investigations. This allows a reconstruction of the coast lines of that time. Marcus Jordanus (1559) and Lucas Janszoon

Waghenaer (*Spieghel der zeevaerdt*, 1584) mapped the Strand (first labelled in 1198 in the letter of Pope Innocent III.) as a large island in the southern part of the North-Frisian Wadden-Sea. The map *Frisia Cimbria Antiqua* (*Landtcarte Von dem Alten Nordfrieslande Anno 1240,* Danckwerth 1652) *Clades Rungholtina* by Johannes Mejer, and the maps of Peter Sax of the seventeenth century, all of which recorded medieval situations before 1362, are not very realistic (Henningsen, 1998: 65; Meier *et al.*, 2013: 41–48).

It is not possible to reconstruct the medieval shore lines in detail, because the Strand with the Pellwormharde and Edomsharde was partly destroyed in 1362. The Pellwormharde probably bordered the Strand in the west at an old beach ridge which was destroyed in 1362. This storm surge also destroyed ten churches in the Pellwormharde. The central part of Pellworm, the *Großer Koog* (Large Polder), which outlasted the disaster, was protected by a medieval ring dike which reached a dike crest of NN +2.0 m (Kühn, 1989: 29). The old church of Pellworm, which is now situated at the west coast of the island, was located more in the centre of Pellworm before 1362, indicating just how much land was lost. Another result of the storm surge was that the tidal channel of the Norderhever broke inside the Edomsharde and destroyed the settlement of Rungholt (legends have greatly exaggerated its size and wealth). Matz Paysen, a rector of Oldesloe, was the first to write down the legend of Rungholt with its unholy people. Matthais Boetius (†1624) also described cultural traces near the Hallig Südfall in his chronicle *De Catalysmo Nordstrandico*. Impressed by the fate of this village Anton Heimreich (1626–1685) and Detlef von Liliencron (1844–1909) wrote poems about it.

In a testament from Hamburg the destruction of Rungholte is dated to 1345 (Meier, 2007a: 126). Other records note traffic connections in 1355 between Hamburg, Bremen, and Flanders with a *nostrum portum* (my port) (*Hansisches Urkundenbuch*, III: 320, 239) in the Edomsharde, which is noted again on 19 June 1361. A document of the *consules* declared on that day that the merchants of Hamburg would receive trade privileges until 1 May 1362 (*Hansisches Urkundenbuch*: IV, 20, 10). The habour of the Edomsharde might be Rungholt (Meier, 2012: 242–249; Meier et al., 2013: 82–85, 118–130). In another document the *consules* of the Edomsharde promised in 1358 that they would not hinder the aims of the counts Heinrich II and Adolf VII of Holstein and Stormarn, who wanted to attack the other *Harden* of North-Frisia (*Hansisches Urkundenbuch*: IV, 20, 10).

Following the plague which hit North-Frisia between 1347 and 1352, the Second Marcellus Flood was yet another disaster. The older town book of Schleswig and the list of Bishop Brun recorded that twenty-five churches on the Strand, and 28 churches on the North-Strand were lost in 1362. One of these might be Rungholt. Also, less certain lists from the seventeenth century, such as the *Catalogus Vetustus* and the *Designatio* by Johannes Mejer, recorded a lot of destroyed churches in this area (Hansen and Jessen, 1904; Meier, 2007a: 126; Meier *et al.*, 2013: 85–92). A heavy storm with a south-westerly wind is recorded in English written sources on 14 and 15 January 1362 (C. Pfister, M. Burri, U. Dietrich (Ed.), Datenbank Euro-Climhist. Oeschger Zentrum für Klimaforschung, Universität Bern. Version 2008) Taken together, these sources indicate that a massive south-westerly Atlantic gale swept across England. Thereafter, it possibly veered around in the Skagerrak before hitting Northern Germany around 16 January, 1362.

Cultural remnants of the Rungholt area around the younger Hallig Südfall, which date back to the twelfth to fourteenth centuries, were documented between 1921 and 1972 by the farmer Andreas Busch from Nordstrand (Busch, 1963a/b). He recorded at least twenty-nine *Warften* and ninety wells. Also he mapped the Niedam dike as well as two large wooden sluices, and 8 to 12 m wide field patterns with ditches, and two distinct groups of *Warften*. The older sluice (20.5 x 3.3 m) was possibly built *c.* 1200, the younger *c.* 1280 (25 x 5.36 m). One group with single *Warften* in a line was orientated south of the younger Hallig Südfall near the medieval dike (*Grote Rungholt*), the other with the church *Warft* westward of the Hallig (*Lütke Rungholt*). These two place names were first mentioned in the *Codex manuscriptus historiae* (Codex Manuscriptus Historiae, 102:107, Archive of Hamburg). The exact locatons of these names is not really clear, probably because they are situated north-east of the Hallig Südfall. In contrast, some of the cultural remnants may be traced on the map: *Clades Rungholtina* of Johannes Mejer from 1636, so the *Emißarius Rungholtinus* (the sluices), *Agger Ripanus* (a larger ditch), *Niedanum* and *Halgenessum* (dikes), the area with eight *Warften* (Rungholtum) and the area of the nine wells (Fedderingnam Capell velRip). Other recordings have been undertaken between 1981 and 2000 by Hans-Herrmann Henningsen (1998; 2000), a non-professional archaeologist. The high density of *Warften* in the small area suggests a population of at least 1500 to 2000 people, which is fairly large for that region and time. Also imported goods underlined the wealth of the social leading people (Meier *et al.,* 2013: 82–84, 118–131). It is likely that Rungholt was a major village that was destroyed by the 1362 flood.

Also, north of the Strand large areas of cultivated land changed into tidal flats during the late medieval period, as cultural remnants around there today indicate (Halligen of Langeness, Habel and Gröde) (Bantelmann, 1966; Meier, 2007a: 145–147, 149; Meier *et al.,* 2013: 157–177). The medieval settlers in this area sustained themselves on the basis of agriculture and on the production of salt. The raw material for salt production was peat from the upper organic deposits, systematically cut over large areas. Therefore, the surface of the cultivated land was partly lower than the MHW. The earliest report of salt production is recorded by Saxo Grammaticus (1180) and by 1230 AD taxes were paid on salt production. Remnants of peat cutting activities can be observed in the vicinity of the northern Halligen in the inner part of the North-Frisian tidal flats (Meier *et al.,* 2013: 157–177).

The magnitude of the land losses in the southern North-Frisian *Uthlande* was not only the result of anthropogenic activities like peat cutting, digging, and drainage, but also partly due to the geological conditions (Figure 5.6). A detailed evaluation has determined that the thickness of the region's marsh sediments is subject to considerable small-scale variation. The subsoil is also characterised by different compaction properties, and this was later compounded by the effects of peat-digging outside the dikes and by drainage within the enclosures. These human measures caused subsidence of the soils to under MHW level. The effects of soil subsidence were very high in the areas of the former Pleistocene valleys which were covered with yielded clay sediments.

Thus, although the exploitation of the coastal area by its inhabitants is responsible for the scale of the disaster, the coastal population could not know that the geological subsidence of the land, had not yet come to an end. Large tidal channels and gullies such as the Norderhever cut into the embanked salt marsh area in 1362. These tidal channels followed old, deep glacial melt water valleys of the pre-Holocene landscape which had been filled up with sediments after the North-Sea reached this area 6000 years ago (Meier, 2007a: 130; Meier *et al.,* 2013: 92, 133–135).

Figure 5.6 Model of the geology in the North-Frisian tidal flats with tidal channels and islands (source: Dirk Meier)

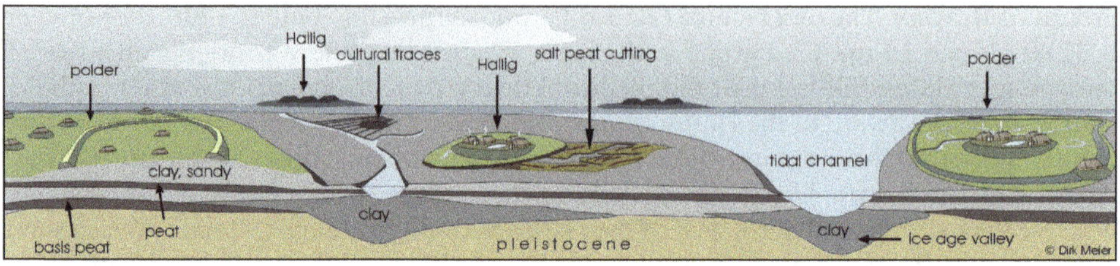

VI. Storm surges from the fifteenth to the seventeenth century

After the disasters of the fourteenth century other storm surges hit the North-Frisian North-Sea coastline. One of these catastrophes was the St. Elisabeth Flood on 19 November 1404 which caused breaches in the dikes along the coasts of Flanders, Zeeland, and Holland (Meier, 2007a: 133). The largest storm surge disaster of the sixteenth century was the All Saints Flood on 2 November 1532 which destroyed dikes along the whole North-Sea coast, including North-Frisia. But since the sixteenth and seventeenth centuries the embankment of new polders as well as drainage became more and more profitable for dukes as well as for contractors, who invested in new land for agriculture and taxes. Dike building and water management became more professional than ever (Knottnerus, 2005). At the North-Sea coast of Schleswig-Holstein the duke of Schleswig-Holstein-Gottorf, who ruled North-Frisia, Eiderstedt, and North-Dithmarschen (since 1559), as well as the Danish king in South Dithmarschen erected new polders (Figure 5.7) like the Gotteskoog (1566) and Old Christian-Albrechts-Koog (1662) in North-Frisia; the Old Augustenkoog (1611) and Dreilandenkoog (1625) in Eiderstedt, as well as the Wardammkoog (1609), Hedwigenkoog (1696), and Ammerswurth-Marne (1581) in Dithmarschen. In addition, bays (bay of Offenbüll in Eiderstedt – 1617) and tidal channels (Nordereider in Eiderstedt – 1579; Wardstrom between the island of Büsum and North-Dithmarschen, 1585) and rivers (Treene 1570) were dammed. Dike building (Meier, 2005b; 2007: 109) became the task of professional engineers – often from Holland – as employees of the dukes of Schleswig as well as of the Danish king. The newly erected dikes are characterised – such as in other coastal areas in this time – by straight lines, which no longer follow the natural landscape structures. The common section of the new dikes (Sieversflether Koog, Eiderstedt, 1610) is 25 m wide at their base and they reached a height of 4.2 m. The bank at the sea-side was rather gentle (1:4) but the dikes were very steep at the landward part (1:1.5). Only along some coastal areas without foreland were *Stackdeiche* (revetted dykes) with wooden front constructions were erected.

But still, these newly erected dikes were not high enough when the Buchardi Flood struck the North-Sea coast of Denmark and North-Frisia in the night between 11 and 12 October 1634 (Meier, 2007a: 134; Meier *et al.,* 2013: 104–113) during a period of cold climate (Fagan, 2000) and during the thirty years of war. The height of the storm surge was reconstructed by Prange (1965) to have reached a level between NN +4.3 and +5.08 m at the bay of Dagebüll, North-Frisia. The disastrous storm surge turned extended areas of coastal marshland into tidal flats, killed thousands of people (estimates range from

Figure 5.7 North-Sea coast of Schleswig-Holstein. Reconstruction of the landscape before AD 1634 (source: Dirk Meier)

8000 to 15,000) and caused catastrophic material damage. The 22,000 ha large island of Strand with its twenty-two parishes was divided into the islands of Pellworm and Nordstrand as well as a group of *Halligen*. On the Strand 6123 inhabitants were killed and 1336 houses as well as twenty-eight windmills were destroyed. Only three island churches remain today (Old and New church on Pellworm, Odenbüll on Nordstrand). The Dutch hydraulic engineer Jan Adriaanszoon Leeghwater (1575–1650), who worked on the embankment of the bay of Dagebüll in North-Frisia, described in a letter the effects of this disaster (Meier, 2007a: 136). The storm surge reached a height of around of 3 m above the MWL (NN +4.3 m) which was higher than the dike crests. Lobedantz, a priest from Nordstrand (Meier, 2007a: 134–138), wrote that shortly after a warm spell the weather suddenly became colder and rainy on 11 October. The wind became stronger in the evening at 9pm, and one hour later the first dike broke near Stintebüll. Most of the dikes were destroyed by 2am in the middle of the night. The dramatic dimensions of the landscape change associated with this event are exemplified by the deepening and widening of the tidal channel Norderhever, which evolved from the Fallstief channel, itself the result of previous storm surges that already had caused a land-loss of *c*. 4000 ha. By 1660 the floods had transformed the former large island into a smaller land mass with many ancillary islands which represented the former shoreline (Meier *et al.,* 2013: 104–113).

As a result of the storm surge of 1634 the island of Alt-Nordstrand broke into two larger islands (Pellworm and Nordstrand) and a lot of smaller islands, some of which partly followed the old shoreline of the former island. Also the *Wüstes Moor* (peatland), on to which humans had fled, outlasted the Burchardiflood. Considering the dramatic changes to the cultivated landscape following this disaster, it is not surprising that people attempted to visualise this destruction with topographical maps. These often artistic but inaccurate maps such as the *c*. 1597 *Nordstrand* of Johannes Petreus were supposed to illustrate the reports. On this map new *Halligen* are recorded on the former area of Rungholt. The detailed map *Lantcarte von Nordergoesharde, Amt Husum* (1641) by Mejer and *Frisor Minor* (1646) by Johann Wittemack show the part of the former island of Strand in which the dikes were completely destroyed during the storm surge of October 1634, following a series of previous floods. The situation *c*. 1659 is documented in a map of *Quirinius Indervelden*. The topographical elements in this map, such as positions of former dikes, roads, remnants of tidal channels, as well as the location of the *Warften* and churches, give an impression of the old island *c*. 1600 and after the storm surge of 1634 (Hoffmann *et al*., 1984: 180–184).

Subsequent attempts to reclaim the destroyed areas were unsuccessful. This disaster is witnessed by the cultural remnants in the tidal flats now found near the islands of Pellworm and Nordstrand. Traces of cultivation had already been found on the tidal flats in the sixteenth and seventeenth centuries. These consisted of old field forms as well as *Warften*, and wells; the dikes of Bupte, Bupsee, Oster- and Westerwohld (Figure 5.8), which were drowned in 1634, have been recorded near the tidal channel of the Rummelloch (Meier *et al.,* 2013: 135–156). In the tidal flats north of the Elisabeth-Sophien-polder of Nordstrand, parts of the Old Nordstrand´s village of Morsum (200 by 50 m in size) were also documented (Bantelmann, 1966; Kühn, 1998: 30; Meier *et al.,* 2013: 151–152). A road runs through this village on a north-south axis, bordering on at least ten rectangular plots separated by trenches. This form of typical medieval planned settlement, is called a *Marschhufendorf* and the single farmsteads were erected on small *Warften* to protect them from the inland water. Obviously, the people had relied exclusively on the protection from the sea afforded by the dikes, at least on the eastern part of Alt-Nordstrand.

Figure 5.8 Landscape development in the southern North-Frisian Wadden Sea with cultural remains and former coastlines (source: Dirk Meier)

As elsewhere in eastern Alt-Nordstrand, cultivation of the partly very boggy marshes did not begin in Morsum until the high Middle Ages. This apparently took place in the course of a well-organised campaign of land reclamation, as evidenced by the arrangement of the house sites. The original thickness of the peat layer is manifested by the peat layers conserved under the medieval *Warften* and the dikes of Nordstrand, which are up to 1 m high even when compressed (peat can be compressed to 10% of its original volume which gives an indication for the volume in the past). Though Morsum was first mentioned in documents of the fourteenth century the sources are too scarce to permit an assessment of the impact of the flood catastrophe of 1362 (Figure 5.8).

Subsequently, from the late Middle Ages onwards, Morsum developed into one of the principal villages of Alt-Nordstrand with a market as well as a ferry port, and it had one of the island's largest churches. According to a contemporary report, 396 people drowned during the Burchardi flood of 1634 and eighty-four houses plus three windmills were destroyed in the parish of Morsum. Only sixteen houses survived. Most of the parish of Morsum remained outside the new dike enclosures built in the seventeenth and eighteenth centuries. The ruins are now located in the north-western part of the newly reclaimed Betringarder polder (1987), covered by recent sediments.

After a long series of floods and damages, this last grave storm surge of 1634 resulted in the definite loss of the central section of Strand. The tides, which now had access to the greatly enlarged intertidal zone, led to a continuous expansion of the Norderhever tidal channel system. The islands of Pellworm and Nordstrand, as well as the Hallig areas remained separate, because dike construction was given up in this area. Emigration of a part of the population and the resulting decrease in economic capacity of the remaining islands prevented urgently needed dike improvements. Other inhabitants of the destroyed island of Strand escaped to the old high bog in the middle of the island, where they would live under very poor conditions during the next years (Meier, 2007a: 139). By 1637, a large part of the Harde Pellworm was again enclosed by dikes, despite new floods causing recurrent damage. These dikes withstood onslaught from the North Sea, and even the polders in the south of Pellworm were reclaimed in 1674. As a result, Pellworm attained a size about 3000 ha, half its size before the disaster. On the remaining area to the east, Nordstrand, the population had been decimated and impoverished to a greater extent. Therefore it took another two decades until a renewed enclosure of four polders with 2300 ha was achieved in 1691 upon the initiative of the duke of Schleswig-Holstein-Gottorf and Dutch investors (Meier et al., 2013: 113–117).

In spite of the fact that large parts of the cultivated land had been destroyed by heavy storm surges, more and more polders have been reclaimed since the sixteenth and seventeenth centuries. Although the storm surges of 1717/1718, 1721, and 1825 caused much damage, this did not result in large land losses in Schleswig-Holstein. Nevertheless, the loss of sediment from this part of the Wadden Sea (which began in the seventeenth century) have continued until today. The tidal volume has increased, as a result of both the ongoing relative rise in sea level, and because of the loss of land surface area to accommodate the tidal volume thanks to land reclamation. Therefore, more water flows in and out of the tidal basins through the channels with each tide, increasing erosion.

VII. Conclusions

Geo-archaeological investigations on the North-Sea coast of Schleswig-Holstein demonstrate that the settlements, which were founded in the un-embanked salt marshes, were totally depended from the environment from Roman times until 1000 AD. Since the twelfth century, dike building and drainage transformed the natural environment to a cultural, manmade landscape. The main push factor of this process in high and late Medieval times was an increase in population which forced new investment in agrarian areas and associated products. From the twelfth to the fifteenth century dike building and drainage was mainly done by groups of farmers who were organised in parishes. Since the sixteenth century land reclamation became more and more important for the dukes. As a result, dike building became more professional and new polders were reclaimed, but heavy storm surges still had terrible effects on the low lying settled marsh areas. Large sections of the cultivated land in North-Frisia were destroyed and became part of the tidal flats. The reasons for the land losses in North-Frisa were partly the effects of human measurements – like drainage and salt peat digging – but mostly the result of the geomorphology. Analyses of these changes in the landscape leads to the insight that greater attention must be paid to the geomorphological changes of the Wadden Sea, the reasons and effects of the large storm surges, as well as the human impacts in each coastal area of the North-Sea. The cultural remnants of the Wadden-Sea are an important witness to the past transformation of the landscape and a programme of historical environmental research helps to predict what might happen to these same coastal landscapes in the future as they become increasingly subjected to climate change.

Bibliography

Primary sources

Staatsbibliothek Hamburg, Hansisches Urkundenbuch, Volume 3 and Volume 4.

Edited sources

Aakjaer, S. (1926–1943) *Kong Valdemars Jordebog*, København.

von Bremen, A. (2000), *Gesta Hammaburgensis ecclesiae pontificum,* in W. Trillmich and R. Buchner (eds), *Quellen des 9. und 11. Jahrhunderts zur Geschichte der Hamburgischen Kirche und des Reiches*, FSGA 11, 7., gegenüber der 6. um einen Nachtrag von V. Scior erweiterte Auflage, Darmstadt.

Danckwerth, C. (1652) *Newe Landesbeschreibung der zwey Herzogthümer Schleßwich und Holstein,* Husum.

Helmold von Bosau (1986), *Chronik der Slawen.* J. M. Laurent u. W. Wattenbach (Translator), A. Heine (Ed.), Historiker des Deutschen Altertums, Stuttgart-Essen.

Anton Heimreich (1668) *Ernewerte Nordfresische Chronick.* Neu Ed. von N. Falck, Schleswig.

Neocorus (1978) *Johann Adolfi's, genannt Neocorus, Chronik des Landes Dithmarschen.* Unveränderter Nachdruck der Ausgabe Kiel 1827, Leer.

Saxo Grammaticus (1981) *Danorum Regum Heroumque Historia.* The text of the first edition with translation and commentary in three volumes, E. Chrisitansen (ed.), Oxford (BAR International Series; 118).

Secondary Sources

Austen, G. (1992) 'Sandwälle im südlichen Eiderstedt (Schleswig-Holstein)' *Meyniana,* 44, pp. 67–74.

Averdiek, F.-R. (1980) 'Geobotanik des Sylter Holozäns' in: G. Kossack, O. Harck, J. Newig, D. Hoffmann, H. Willkomm, F.-R. Averdieck and J. Reichstein (eds), *Archsum auf Sylt Teil I. Einführung in Forschungsverlauf und Landschafsgeschichte*, Archum, pp. 147–172 (Studien Küstenarchäologie Schleswig-Holstein, Ser. B).

Bantelmann, A. (1955) *Tofting, eine vorgeschichtliche Warft an der Eidermündung*, Neumünster (Offa-Bücher; 12).

Bantelmann, A. (1957/1958) 'Die kaiserzeitliche Marschensiedlung Ostermoor bei Brunsbüttelkoog', *Offa,* 16, pp. 53–79.

Bantelmann, A. (1966) 'Die Landschaftsentwicklung an der schleswig-holsteinischen Westküste – dargestellt am Beispiel Nordfriesland. Eine Funktionschronik durch fünf Jahrtausende', *Die Küste, Archiv für Forschung und Technik,* 2, pp. 5–99.

Bantelmann, A. (1975) *Die frühgeschichtliche Marschsiedlung beim Elisenhof in Eiderstedt. Landschaftsgeschichte und Baubefunde,* Bern-Frankfurt (Studien Küstenarchäologie Schleswig-Holstein, Ser. A, Elisenhof; 1).

Behre, K.-E. (1976) *Die Pflanzenreste aus der frühgschichtlichen Warft Elisenhof,* Bern-Frankfurt, (Studien Küstenarchäologie Schleswig-Holstein, Ser. A, Elisenhof; 2).

Brande, A. (1988) 'Zur frühsubatlantischen Vegetations- und Landschaftsentwicklung Westeiderstedts' in: M. Müller-Wille *et al.*, *Norderhever-Projekt 1. Landschaftsentwicklung und Siedlungsgeschichte im Einzugsgebiet der Norderhever (Nordfriesland)* 1988, pp. 139–147.

Busch, A. (1963a) 'Zur Rekonstruktion der Rungholter Schleusen', *Die Heimat,* Heft 70, pp. 11–16.

Busch, A. (1963b) 'Über die Kirchwarft im Rungholtwatt' *Die Heimat,* Heft 70, pp. 16–18.

Dittmer, E. (1952) 'Schichtenaufbau und Entwicklungsgeschichte des dithmarscher Alluviums', *Meyniana,* 1, pp. 138–168.

Fagan, B. (2000) *The Little Ice Age*, New York.

Haarnagel, W. (1937) 'Die frühgeschichtlichen Siedlungen in der schleswig-holsteinischen Elb- und Störmarsch, insbesondere die Siedlung Hodorf', *Offa,* 2, pp. 31–78.

Hansen, R. and Jessen, W. (1904/reprint1974) *Quellen zur Geschichte des Bistums Schleswig,* Kiel.

Henningsen, H.-H. (1998) *Rungholt. Der Weg in die Katastrophe. Aufstieg, Blütezeit und Untergang einesbedeutenden mittelalterlichen Ortes in Nordfriesland. Bd. I: Die Entstehungsgeschichte Rungholts, seine Ortslage, heutige Kulturspuren im Wattenmeer und die Geschichte und Bedeutung der Hallig Südfall,* Husum.

Henningsen, H.-H. (2000) *Rungholt. Der Weg in die Katastrophe. Aufstieg, Blütezeit und Untergang eines bedeutenden mittelalterlichen Ortes in Nordfriesland. Bd. II: Das Leben der Bewohner und ihre Einrichtungen, die Landschaft, der Aufstieg zu einem Handelsplatz, Rungholts Untergang, der heutige Zustand von Kulturspuren, der Mythos von Rungholt und ein Epilog: Die Geschichte im Zeitraffer,* Husum.

Hoffmann, D. (1980) 'Küstenholozän zwischen Sylt und Föhr', in: G. Kossack, O. Harck, J. Newig, D. Hoffmann, H. Willkomm, F.-R. Averdieck and J. Reichstein (eds), *Archsum auf Sylt. Teil 1. Einführung in Forschungsverlauf und Landschaftsgeschichte,* Mainz, pp. 85–130.

Hoffmann, D. (1988) 'Das Küstenholozän im Einzugsbereich der Norderhever, Nordfriesland' in: M. Müller-Wille *et al.*, *Norderhever-Projekt 1. Landschaftsentwicklung und Siedlungsgeschichte im Einzugsgebiet der Norderhever (Nordfriesland)*,1988, pp. 51–115.

Hoffmann, D. (2004) 'Holocene landscape development in the marshes of the West Coast of Schleswig-Holstein, Germany', *Quaternary International,* 112, pp. 29–36.

Hoffmann, D., Kühn, H.-J. and Müller-Wille, M. (1984) 'Landschafts- und Siedlungsgeschichte im Bereich der heutigen Marscheninseln und Watten Nordfrieslands. Ergebnisse archäologisch-geowissenschaftlicher Untersuchungen (1975–1981)', *Siedlungsforschung. Archäologie – Geschichte – Geographie*, 2, pp. 147–187.

Hoffmann, D., Meier, D. and Müller-Wille, M. (1997) 'Geologische und archäologische Untersuchungen zur Landschafts- und Siedlungsgeschichte des Küstengebietes von Norderdithmarschen', *Germania*, 1/1997, pp. 213–253.

Hummel, P. and Cordes, E. (1969) 'Holozäne Sedimentation und Faciesbildung beim Aufbau der Lundener Nehrung', *Meyniana,* 19, pp. 103–112.

Kersten, K. and La Baume, P. (1958) *Vorgeschichte der nordfriesischen Inseln,* Neumünster.

Kühn, H.-J. (1989) 'Deiche des frühen Mittelalters und der frühen Neuzeit', in H.J. Kühn and A. Panten, *Der frühe Deichbau in Nordfriesland. Archäologisch-historische Untersuchungen,* Bredstedt, pp. 11–62.

Kühn, H.-J. and Müller, M. (1988) 'Siedlungsarchäologische Untersuchungen im nordfriesischen Marschen- und Wattengebiet und in Eiderstedt', in: M. Müller-Wille *et al.*, *Norderhever-Projekt 1. Landschaftsentwicklung und Siedlungsgeschichte im Einzugsgebiet der Norderhever (Nordfriesland*), 1988, pp. 181–193.

Kühn, H.-J. (1998) 'Morsum – ein Marschhufendorf Alt-Nordstrands' in Landesamt für den Nationalpark Schleswig-Holsteinisches Wattenmeer und Umweltbundesamt (ed.), *Nordfriesisches und Dithmarscher Wattenmeer,* Stuttgart, pp. 30–31.

Knottnerus, O. (2005) 'Die Verbreitung neuer Deich- und Sielbautechniken entlang der südlichen Nordseeküste im 16. und 17. Jahrhundert', in: M. Fansa, (ed.), *Kulturlandschaft Marsch. Natur – Geschichte – Gegenwart.* Schriftenreihe des Landesmuseums für Natur und Mensch Oldenburg 33 (Oldenburg 2005) pp. 161–167.

Lamb, H. H. (1982) *Climate History and the Modern World*, London-New York.

Meier, D. (1997) 'Welt, eine frühmittelalterliche Dorfwurt im Mündungsgebiet der Eider', *Archäologisches Korrespondenzblatt,* 27, pp. 171–184.

Meier, D. (2001a/b) *Landschaftsentwicklung und Siedlungsgeschichte des Eiderstedter und Dithmarscher Küstengebietes als Teilregionen des Nordseeküstenraumes.* Untersuchungen der AG Küstenarchäologie des FTZ-Westküste, Universitätsforschungen zur Prähistorischen Archäologie 79, Teil 1: Die Siedlungen (a); Teil 2: Der Siedlungsraum (b), Bonn.

Meier, D. (2003) 'Ausgrabungen auf den mittelalterlichen Wurten Norderbusenwurth und Lütjenbüttel in Süderdithmarschen (Schleswig-Holstein)', *Probleme der Küstenforschung im südlichen Nordseegebiet,* 28, pp. 277–291.

Meier, D. (2004) 'Man and environment in the marsh area of Schleswig-Holstein from Roman until late Medieval times', *Quarternary International,* 112, pp. 55–70.

Meier, D. (2005a) 'Süderbusenwurth. Vorbericht der Ausgrabungen einer Marschensiedlung der Römischen Kaiserzeit in Dithmarschen', *Studien zur Sachsenforschung,* 15, pp. 343–363.

Meier, D. (2005b) 'Untersuchungen zum frühen Deichbau in Schleswig-Holstein und Dänemark. Vorträge Symposium Oldenburg 2004', in: M. Fansa (ed.), *Kulturlandschaft Marsch. Natur – Geschichte – Gegenwart.* Schriftenreihe des Landesmuseums für Natur und Mensch Oldenburg 33 (Oldenburg 2005) pp. 133–147.

Meier, D. (2007a) *Die Nordseeküste. Geschichte einer Landschaft,* Heide.

Meier, D. (2007b) 'De Dam geslagen wart twischen Eyderstede unde Husum...Die Bedeichung der "Nord-Eyder", Eiderstedt (Schleswig-Holstein)', in: J. J. J. M. Beenakker, F.H. Hortsen, A. de Kraker and H. Renes, *Landschap in ruimte en tijd. Liber amicorum G. Borger*, Amsterdam, pp. 252–262.

Meier, D. (2008) 'The Historical Geography of the German North Sea and the Baltic Sea' *Die Küste. Archiv für Forschung und Technik,* 74, 2008, pp. 18–30.

Meier, D. (2009) *Land in Sicht. Entwicklung der Seefahrt an Nord- und Ostsee,* Heide.

Meier, D. (2011) Schleswig-Holstein im Frühen Mittelalter. Lanschaft - Archäologie-Geschichte, Heide.

Meier, D. (2012) Schleswig-Holstein im Hohenm und Späten Mittelalter. Landesausbau - Dörfer - Städte, Heide.

Meier, D., Kühn, H. J. and Borger, G. (2013) *Der Küstenatlas. Das schleswig-holsteinische Wattenmeer zwischen Vergangenheit und Gegenwart*, Heide.

Menke, M. (1988) 'Die holozäne Nordseeküstentransgression im Küstenbereich der südöstlichen Deutschen Bucht', in M. Müller-Wille *et al., Norderhever-Projekt 1. Landschaftsentwicklung und Siedlungsgeschichte im Einzugsgebiet der Norderhever (Nordfriesland),* 1988, pp. 117–137.

Müller-Wille, M. (1986) 'Frühgeschichtliche Fundplätze in Eiderstedt', *Offa*, 43, pp. 175–179.

Müller-Wille, M., Higelke, B., Hoffmann, D., Menke, B., Brande, A., Bokelmann, K., Saggau, H.-E. and Kühn, H.-J. (1988) *Norderhever-Projekt 1. Landschaftsentwicklung und Siedlungsgeschichte im Einzugsgebiet der Norderhever (Nordfriesland),* Neumünster (Offa-Bücher; 66; Studien Küstenarchäologie Schleswig-Holsteins, Ser. C).

Nunn, P. D. (2007) *Climate, Environment and Society in the Pacific during the last Millennium,* London (Developments in Earth & Environmental Sciences; 6).

Prange, W. (1965) 'Die Höhe der Sturmflut vom 11. Oktober 1634 in Nordfriesland nach neuen Wasserstandsmarken. Zwischen Eider und Wiedau', *Heimatkalender Nordfriesland*, pp. 40–48.

Wohlenberg, E. (1989) 'Die Lundenbergharde. Eine historische küsten- und deichbaugeschichtliche Monographie aufgrund neuer Grabungen im nordfriesischen Wattenmeer (1962 bis 1977) nebst Freilegung eines doppelten Stackdeiches und Öffnung eines historischen Nüstersieles, beides vor Ort beim "Halbmond" im Seedeich Südermarsch – Lundenberg bei Husum' *Die Küste, Archiv für Forschung und Technik,* 48, pp. 1–119.

6 Natural and anthropogenic factors in the origin and evolution of the dune landscape on the islands of the south-west Netherlands

Anton M.M. van Haperen, Staatsbosbeheer Regio Zuid, Tilburg

Abstract

The landscape and vegetation of the coastal dunes of the south-west Netherlands show considerable variation. This is primarily caused by a geochemical and physical gradient in the dune sand that probably has its origin in the Late Pleistocene geology of the coastal zone. In addition to these natural factors, man acts in the dune landscape along a second axis, in which vegetation structure and the presence of herbivores are controlled. However, this influence was not always the same during the last 500 years. In the late Middle Ages and Early Modern period human dune use was rather similar in all the dune areas. In the mid-eighteenth century dune use changed, giving rise to a differentiation of landscapes. Ultimately, this led to the present dune landscape characterised by an appearance completely different from the dune landscape of two hundred years ago. The new types of dune use were in general much more sensitive to mobile sand than was the use of the late Middle Ages and the Early Modern period. This generated for more dune fixation. Consequently, the dune landscapes are now almost static. In this static state the dunes still show a significant differentiation along two axes: (1) a natural axis dominated by the carbonate content of the dune sand and (2) an axis of anthropogenic origin that has to do with the landscape structure. The first axis explains the natural differentiation between the different dune areas while the second axis explains many of the changes across time.

I. Introduction

From the northwest coast of France to Cape Skagen in northern Jutland a 0.1–10 km wide dune landscape borders the eastern coast of the Channel and the Southern North Sea. The dunes strongly differ from the inland landscape. Their sand originates from the nearby coastal zone and in fact the dunes form the outer boundaries of the coastal ecosystem. In the longitudinal direction the dune landscape is not a continuum; in certain areas it shows sharp contrasts and sudden changes within a relatively short distance. The dunes of the (former) islands in the south-west Netherlands (Figure 6.1) are an example of this phenomenon. The dune vegetation of Voorne is strikingly different from that on Schouwen and Walcheren. The vegetation of the dunes of Goeree is variable and has elements of both extremes. This paper summarises results of a recent study on the influence of natural and anthropogenic factors on the evolution of landscape and vegetation of these dunes (van Haperen, 2009). It illustrates the role of geology and coastal hydro-

Landscapes or seascapes? The history of the coastal environment in the North Sea area reconsidered, ed. by Erik Thoen, Guus J. Borger, Adriaan M.J. de Kraker, Tim Soens, Dries Tys, Lies Vervaet & Henk J.T. Weerts, Turnhout, 2013 (*CORN Publication Series*, 13), pp. 111–126.

BREPOLS PUBLISHERS DOI: 10.1484/M.CORN.1.101550

Figure 6.1 Topography of the coastal area of the south-west Netherlands. This map represents the present situation with partially or totally enclosed estuaries

dynamics as primary causes for landscape differentiation on the regional level. In addition, anthropogenic factors play a role as well. However, their influence is not constant in time and space. For the present landscape, differentiating anthropogenic activities, starting in the eighteenth century, and the fixation of the dunes (twentieth century) are of major importance.

Figure 6.2 Distribution patterns of four calcicole plant species in the coastal dunes of the south-west Netherlands.
Each square represents the presence of a species in a square of 0,5 x 0,5 km. Species names (from left to right): 1. Viola hirta (Hairy Violet), 2. Thymus pulegioides (Larger Wild Thyme), 3. Polygonatum odoratum, 4. Rhamnus cathartica (Buckthorn)

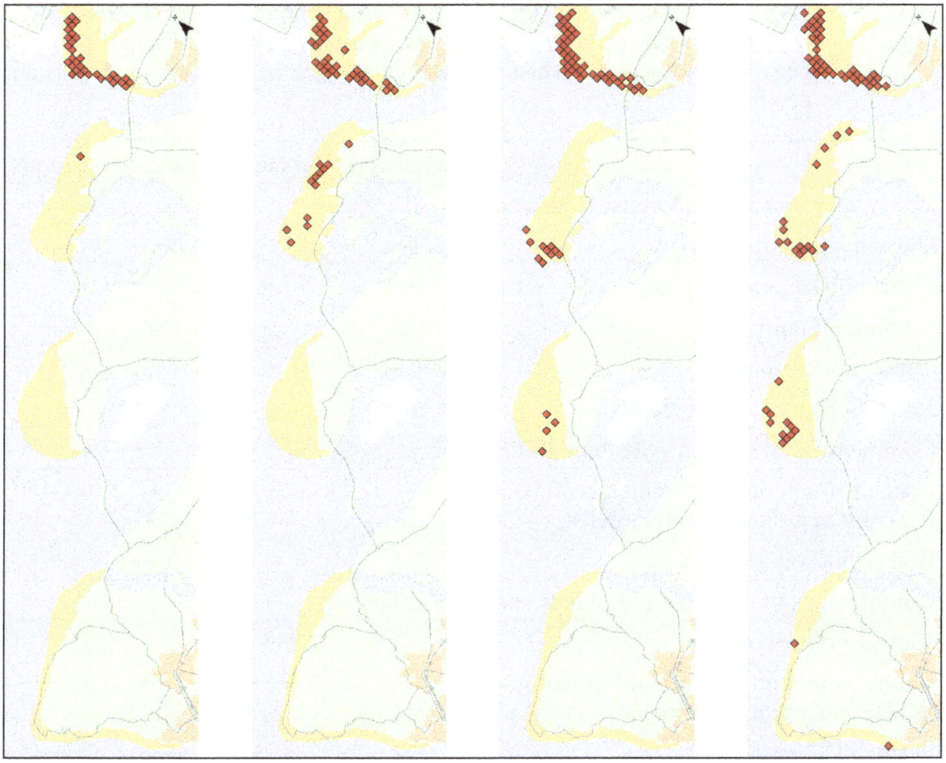

II. Variation in flora and vegetation

Both flora and vegetation of the dune areas in the south-west Netherlands show a distinct variation. The dune area of Voorne is relatively rich in species (Adriani and van der Maarel, 1968; van der Maarel, 1966). Approximately 130 species on the Dutch Red List of endangered plant species have been found here since the middle of the nineteenth century. In the dune areas of the other former islands, this number varies between eighty-seven and ninety-five species (van Haperen, 2009). Especially calcicole species are much more common in the dunes of Voorne. Some species are largely restricted to Voorne and have no or very few populations in the more southern dune areas. Others perform a distribution gradient: they are widely spread in the dunes of Voorne, have an intermediate distribution on Goeree and Schouwen, and are scarce or absent on Walcheren (Figure 6.2). Though less numerous, there are also species that generally occur frequently in the dunes of Walcheren and are scarce or absent in the more northern dunes.

Plant growth not only varies at the species level. Vegetation and landscape also show obvious differences. Table 6.1 summarises the distribution of some characteristic plant communities according to the French-Swiss School of Vegetation Science (Schaminée

Table 6.1: Spatial distribution of some selected plant communities in the coastal dunes of the south-west Netherlands. For explanation and a more complete survey see Van Haperen, 2009.
Legend: +++ common in different parts of the dune area ++ locally common + only very locally present and mostly fragmented - (almost) absent () symbols between brackets refer to historical situations since the middle of the nineteenth century

	Walcheren	Schouwen	Goeree	Voorne
Communities of wet and moist dune slacks:				
Cicendietum filiformis	+(+)	+++	+++	-
Caricetum trinervae-nigrae	++	++	+	-
Littorellion uniflorae	(+)	++	++	+++
Centaurio-Saginetum	+(+)	++	++	++(+)
Junco-Schoenetum nigricantis	+(+)	++	++	+++
Communities of mesotrophic humid grasslands:				
Frame community Danthonia decumbens & Hypnum jutlandicum [Nardetea]	+++	+++	+	-
Botrychio-Polygaletum	+	++	+++	+
Communities of dry dune grasslands and pioneer habitats:				
Violo-Corynephoretum	+	++	++	+
Frame community Dicranum scoparium [Koelerio-Corynephoretea]	+++	+++	++	+
Festuco-Galietum veri	++	+++	+++	++
Phleo-Tortuletum ruraliformis	++	+++	+++	+++
Frame community Hypnum cupressiforme [Koelerio-Corynephoretea]	++	+++	+++	+++
Taraxaco-Galietum veri	+	+	++	+++
Dune shrubland communities:				
Hippophao-Ligustretum rubetosum vigorosi	+++	+++	+	-
Frame community Crataegus monogyna [Rhamno-Prunetea]	+++	+++	-	+
Hippophao-Ligustretum asparagetosum	-	+	++	+++
Rhamno-Crataegetum	-	-	++	+++

et al.: 1995). In the dunes of Voorne typical calcicole communities like *Taraxaco-Galietum veri*, *Hippophao-Ligustretum asparagetosum*, and *Rhamno-Crataegetum* are very common. In contrast, they are almost absent or very fragmentally developed in the dunes of Walcheren and Schouwen. Here, they are replaced by communities that are characteristic for a lower carbonate content of the soil, like *Violo-Corynephoretum*, the frame community of *Dicranum scoparium*, *Festuco-Galietum veri*, *Hippophao-Ligustretum rubetosum vigorosi* (characterised by the presence of several species of brambles), and the frame community of *Crataegus monogyna*. In this respect, the dunes of Goeree have an intermediate position. Certain parts of this dune area are comparable to those of Voorne, while others have a closer similarity to Schouwen and Walcheren. Most strikingly, the vegetation differences are not restricted to individual communities. The dune landscapes as a whole show profound variations, in which mosaics of different plant communities are involved. The differences between the communities lie on a relatively high level in the hierarchical classification system of the French-Swiss School of Vegetation Science (assocation, alliance and even order), indicating important environmental differences between the dune areas. This differentiation in the distribution of plant species and communities must primarily be attributed to natural, abiotic factors in which the carbonate content of the soil plays an important role (see below).

Variation in flora and vegetation of the dunes in the south-west Netherlands is not only a matter of space. The dune landscape changed totally during the last two centuries. In the Middle Ages and Early Modern period an open landscape prevailed and shifting and mobile dunes were then the most dominant type of habitat. Pioneer thickets with *Hippophae rhamnoides* (Sea Buckthorn) and *Salix repens* (Creeping Willow) occurred locally, but mature bushes and woodland were very scarce. As a consequence of the planting of *Ammophila arenaria* (Marram Grass) and afforestation (especially with Pines, *Pinus spec.*) almost the entire dune landscape became fixed in the nineteenth and twentieth centuries. At present shifting dunes are scarce and, where still present (e.g. Schouwen), they are declining. In many dune areas thickets and woodland are the dominant type of habitat now. This fixation took place in all the dune areas of the south-west Netherlands, though there is some difference on the actual presence of dynamic and shifting dunes. The changes in the dune landscape of course had their consequences for flora and vegetation. Figure 6.3 shows for each of the four (former) islands the percentage of Red List-species that have been observed since the middle of the nineteenth century and subsequently disappeared. The graph illustrates that not only the total percentage of disappeared species differs between the various dune areas, but that there are also clear differences in the ecological spectra of the species that disappeared. In the dunes of Walcheren a large percentage of (ground)water dependent species also disappeared. In the dunes of Voorne species of thickets and woodland did relatively well. On Goeree species of dune grassland have been relatively preserved. This differentiated decline must primarily be attributed to anthropogenic factors (see below, chapter IV)

III. Geochemical and physical variation

Dune sands of Voorne, Goeree, Schouwen and Walcheren show clear physical and geochemical differences that run parallel to the botanical variation treated in the previous chapter. Figure 6.4 shows the results of a geochemical analysis of forty-six samples (depth

Figure 6.3 Percentage of extinct species of the Dutch Red List of endangered plant species

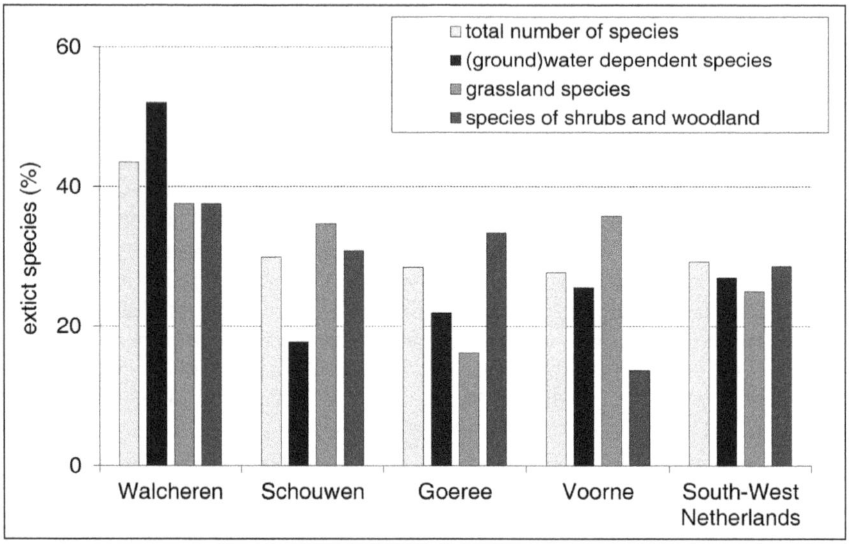

1–2 m) from the main dunes on these former islands. In order to interpret the geological origins of the dune sands some reference samples of a known geological origin were involved in the analysis. With the aid of multivariate cluster techniques the samples could be grouped in three main clusters with a distinct geographical distribution. The chemical differences between these clusters particularly show different levels of iron, calcium and magnesium, and carbonates in general (Van Haperen, 2009). At a depth of 1–2 m the carbonate content of the dunes of Voorne always is higher than 3.5 percent $CaCO_3$ (max. 5.9 percent). In the northern dunes of Walcheren the maximum is 0.7 percent $CaCO_3$. Elsewhere on this former island, on Schouwen and on Goeree the carbonate content may be somewhat higher (mostly 1–3.5 percent). Particularly on Goeree, in the dune sand belonging to cluster 3, a higher carbonate content has been found (3.5–5 percent).

The dune sands of the south-west Netherlands not only differ geochemically. They also show variation in grain size (Figure 6.5). The sands of Walcheren have very low amounts of particles smaller than 150 μm (average 0.2 percent, n=11). In the dunes of Voorne their share can amount 10–15 percent (average 9.3 percent, n=8). On the other hand, grains larger than 350 μm are much more dominant in the sands of Walcheren (29–56 percent, average 37 percent) than on Voorne (1.5–9 percent, average 4 percent). The grain size of the dune sands of Schouwen and Goeree is in between these extremes.

Research has made clear that carbonates and iron are crucial factors for the availability of phosphorus and nitrogen as plant nutrients in the dune landscape (Kooijman and Besse, 2002). Grain size is also of great importance for the plant nutrient status of the soil. Therefore, it is likely that the physical and geochemical variation in dune sand has great consequences for plant growth and vegetation and this explains the close relation-

Figure 6.4 Geographical distribution of the sand samples and their geochemical identity.
The symbols indicate the geochemical clustering: circles = cluster 1; squares = cluster 2 and triangles = cluster 3. The colour represents the carbonate content (% CaO3): white = < 1,5; grey = 1,5–3 and black = 3,5–6

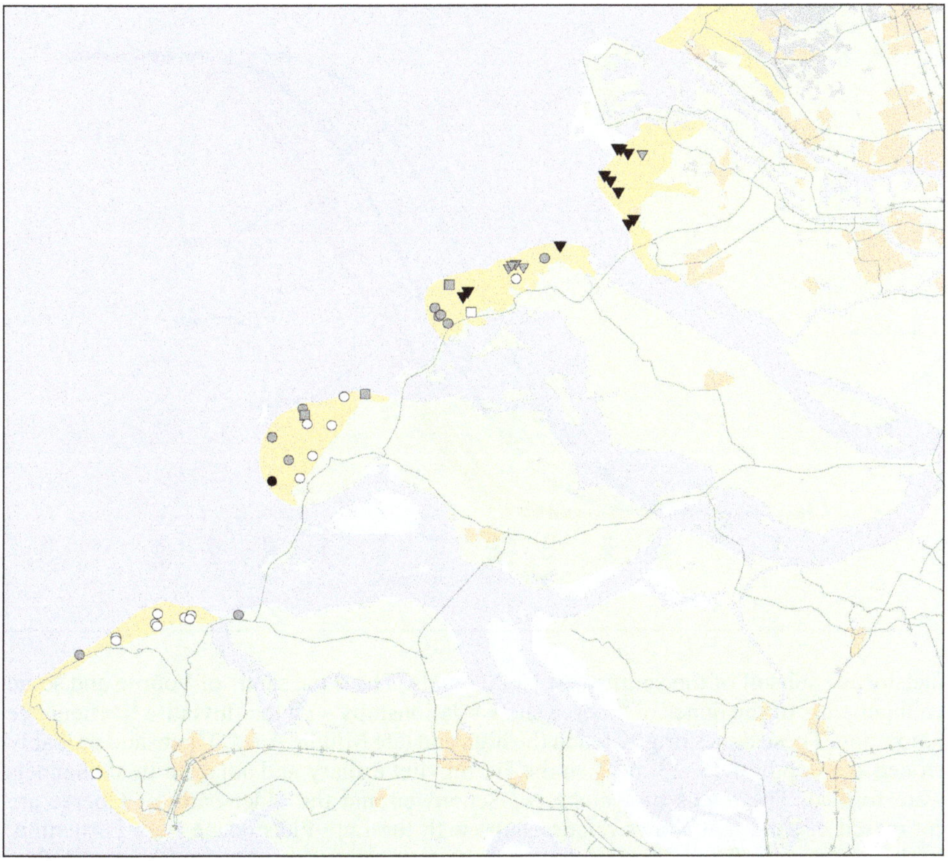

ship between the distribution patterns of plant species (Figure 6.2) and the results of the geochemical analyses (Figure 6.4).

The present dunes of the south-west Netherlands have been created since the early Middle Ages. The dune sand originates from the nearby coastal area and the ebb tidal deltas of the adjacent estuaries in particular. Therefore, there is a direct relationship between the development of the tidal inlets and estuaries and the history of the coastal dunes. The geochemical variation in the dune sands of the south-west Netherlands is probably caused by differences in their geological origin and by the presence of geological gradient in the

Figure 6.5 Cumulative distribution of the grain size in dune sands from the south-west Netherlands.
The samples from Voorne are indicated in green (n=8), Schouwen in blue (n = 11) and Walcheren in red (n = 11). Each thin line represents one sample. The thick lines indicate the average of all the samples from an island

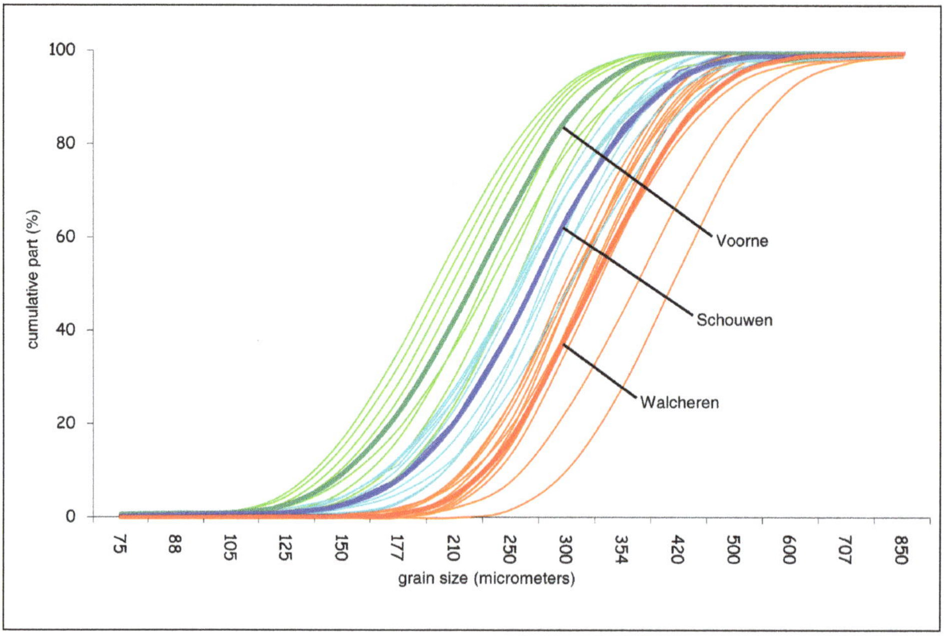

Pleistocene subsoil of the south-west Netherlands. The dune sands of Voorne and some younger parts of the dunes of Goeree show relationships with the fluviatile Kreftenheye Formation. These dunes originated in the high and late Middle Ages. Their sand probably eroded at a depth of 15–25 m when the Haringvliet estuary and adjacent tidal channels were formed. The dunes of Walcheren, Schouwen and the older parts of Goeree are composed of sand that shows relationships with the Late Pleistocene Eem Formation. It likely originates from ebb tidal deltas of the more southern estuaries (Westerschelde, Oosterschelde, Grevelingen), where this kind of formation is more important. The Eem Formation and the Kreftenheye Formation have a quite comparable position in the Pleistocene subsoil of the northern and southern estuaries of the south-west Netherlands (Laban and Rijsdijk, 2004; Ebbing and Laban, 1996).[1] In both cases tidal currents were

[1] The statements on the possible geological identity of the dune sand are based on reference samples of the Kreftenheye Formation and the Eem Formation, that were analysed in the same way as the dune sands. It cannot be excluded that also other geological formations are involved in the origin of the dune sand (e.g. Koewacht Formation and Waalre Formation). This possibility detracts however nothing from the hypothesis that there is a close relationship between the geochemical variation in the dune sand and the Pleistocene subsoil in the coastal area.

responsible for the erosion of large quantities of sand which was subsequently sedimented in the ebb tidal deltas. These were the depositories of sand for coastal dune formation. Accordingly, differences in the Late Pleistocene subsoil of the coastal area at a depth of 15–30 m likely account for differences in in the geochemistry and the present vegetation of the current dune landscape.

In addition to geology, water currents and tidal movement play an important role in the presence and the conservation of the abiotic gradients along the dune coast. The tidal current runs anti-clockwise along the coasts of the southern North Sea, resulting in a (small) net transport of sediment in northward direction (Terwindt, 1973; van den Berg, 1984; Beets and van der Spek, 2000). This explains why sediments related to the Kreftenheye Formation in the Haringvliet estuary never have reached the dune coast of Walcheren and Schouwen. The northward directed current is also responsible for a net transport of smaller sand particles from the ebb tidal deltas in the south (Westerschelde, Oosterschelde, Grevelingen) to the north (Haringvliet). This contributes to the existence of a gradient in particle size in the coastal waters and in the dunes. However, other factors like local wave-action, the discharge of river water in the Haringvliet, and the involvement of other sediments with a finer grain size might also play a role.[2]

IV. Anthropogenic factors

The present dune landscape of the south-west Netherlands is relatively young. It came into existence in the early Middle Ages. Human habitation of the coastal area was concentrated in these dunes then (Beekman, 2007; van Haperen, 2009). In the tenth and eleventh centuries human populations moved to new settlements on reclaimed salt marshes further inland, but humans probably continued to use dunes for grazing cattle, digging peat, and cutting wood etc. Therefore, in addition to natural factors, anthropogenic influences play an important role in the development of the dune landscape. However, this influence was not always the same, but acted in different ways during the last 500 years. In the late Middle Ages and Early Modern times dune use was rather similar in all the dune areas. In the mid-eighteenth century dune use changed in the various dune areas, giving rise to a differentiation of landscapes. Ultimately, this led to the present dune landscape characterised by an appearance completely different to the dune landscape of two hundred years ago.

IV.1. Late Middle Ages and Early Modern Time: rabbit warrens as the main form of dune use

Until recently rabbits were considered to be a pest for the dune landscape and their presumed overpopulation was generally regarded as an important cause for dune erosion and the presence of vast mobile dunes. Recent studies, however, suggest that the role of rabbits in dune erosion has been overestimated (van Haperen, 2009). As Beekman (2007)

[2] In general the Kreftenheye Formation has a relative coarse composition and this seems contradictory to the small average grain size of the dune sand of cluster 3 on Voorne and Goeree, that is related to this formation according to our geochemical analysis. The upper layers of this formation however have a relative fine grain size (Verbraeck, 1990). We suppose that these layers are the main source of the dune sand.

shows, the primary cause for large-scale shifting dunes on Schouwen in the Middle Ages and in modern times lies in erosion of the most seaward dunes by the tidal currents of the Oosterschelde. As a consequence, cliff dunes came into existence that were subject to wind erosion. This gave rise to mobile dunes that moved inland several kilometres and expanded enormously during the course of that movement. Indeed, rabbits were present in large numbers in that time and they were of great influence on the vegetation (van Haperen, 2009). Their contribution to large-scale dune erosion was, however, of minor importance. Rabbits were of great economic importance and were protected and culled by man. In areas with high rabbit densities men played a major role in safeguarding the fragile balance between the open pioneer landscape, profitable for rabbit management, and local erosion. When discussing the presence and causes of dune erosion one should distinguish between large-scale mobile dunes, with hardly any vegetation, and the open pioneer landscape where vegetation and mobile sand are more or less in balance.

From the late Middle Ages (fourteenth century) until well into the eighteenth century the dunes in the south-west Netherlands were mainly used as rabbit warrens. This use was very similar in different areas and on the different islands (van Haperen, 2009: 136–155; for rabbits and warrens in general: van Dam, 2007 and 2010). Central authorities (the Counts of Holland and Zeeland, Princes of Orange; after 1588 also the *Staten van Holland* and the *Staten van Zeeland*) subdivided the dunes in lots and granted leases of these lots for periods of either seven or ten years. Because of the economic importance of the warrens and the rabbits, lotted dune use was strictly regulated and supervised by local stewards. Authorities, stewards, as well as leaseholders were well aware of the functioning of the dune ecosystem and regulated rabbit density and the intensity of cattle grazing. When vegetation growth was too abundant for rabbits, additional grazing with cattle was applied to restore the short and young vegetation, rich in protein. From dune rents it can be concluded that during the Early Modern period cattle grazing was always of minor importance to the culling of rabbits (van Haperen, 2009: 153). Erosion of the dunes was prevented by local planting of Marram Grass. So in this period the dunes were intensively used by man, but this usage did not lead to overexploitation. Dune dynamics and incidental flooding were part of the ecosystem and kept it in a young stage, favouring high densities of rabbits (and biodiversity). This use continued at least several centuries and there are no indications that the dunes suffered from degradation in those days.

Figure 6.6 shows the indexes of dune rents on the former islands of Walcheren, Schouwen and Goeree. For further reference the rent of agricultural land in the Province of Zeeland and the average meat price in the Netherlands are also shown. These indexes can be considered representative for both rural and urban economies. In the first half of the seventeenth century all indexes trend upwards, with a maximum *c.* 1640–50. These trends correspond with the economic prosperity of the first half of the seventeenth century (the 'Golden Age'). In the middle of this century prices of agricultural land started to decrease in connection with the downturn in rural economy that continued until the middle of the eighteenth century. Meat prices were initially stable but in the late seventeenth century they also began to decline. Figure 6.6 shows that the rents of dune warrens run parallel to upward and downward movements of both the rents of agricultural land and meat prices, indicating that these rabbit warrens and their products (meat and fur) were probably connected to both the rural and urban economies of the seventeenth and eighteenth centuries. Figure 6.6 also demonstrates that the relative economic importance of rabbit warrens in the years 1640–1750 distinctly

Figure 6.6 Dune rents in comparison to the rent of agricultural land and the average meat price in the Netherlands.
The values represent average indexes per decade. The decade 1645–1655 is set on 100 (sources: Rents of agricultural land: Priester, 1998; Average meat prices: Institute for Social History, Amsterdam, http:/www.iisg.nl/hpw/)

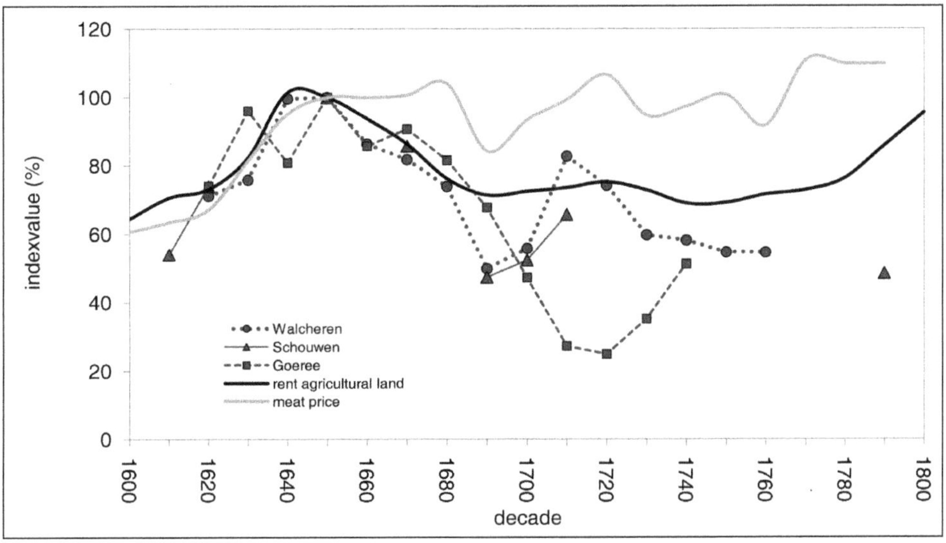

decreased in comparison to meat prices and the rent of agricultural land. The lower dune rents generated by the warrens in combination with the changing attitudes towards uncultivated land, made dune owners and authorities to put an end to the propagation and culling of rabbits in warrens in the middle of the eighteenth century. The motive and the moment of this change varied in the different dune areas and on different islands, but all over the south-west Netherlands the change of dune use was obvious. Rabbits were no longer considered as an economic commodity but as a pest that should be controlled. Other types of dune use like agriculture, cattle grazing, and forestry became more important.

IV.2. Eighteenth and nineteenth century: differentiation of dune use and dune landscape

The changing attitudes towards dune use in the eighteenth century found its cause not only in the weak economic position of the rabbit warrens but also in the philosophical renewal of the Enlightment. In particular the ideas of the physiocrats were very important. For them agriculture and forestry were the driving forces in the economy. *Terres incultes* (uncultivated areas) were considered to be a symptom of an uncivilised and backward society. As a consequence of this new philosophical movement the wealthy upper-class got involved in the cultivation of the dunes: they build estates, planted woodlands, and stimulated agricultural reclamations. Because of the divided and sometimes isolated geographical situation of the south-west Netherlands these initiatives were strongly

Figure 6.7.a **Three different types of dune landscape in relation to natural and anthropogenic factors.**
Large-scale dynamic dune landscape where geomorphologic processes dominate vegetation growth.
This type of landscape probably mostly originates in the most seaward part of the dunes and can move land inward later on. Natural processes (coastal erosion, wind action) play an important role in these developments (Large-scale dynamic dune landscape)

Figure 6.7.b **Three different types of dune landscape in relation to natural and anthropogenic factors:**
Small-scale dynamic dune landscape, in which geomorpholic processes and vegetation growth are in balance.
Pioneer vegetation and dune grassland play a dominant role.
This is an optimal habitat for rabbits. In Late Middle Ages and Early Modern Time humans highly contributed to keep the landscape in a pioneer stage like this (Small-scale dynamic dune landscape)

Figure 6.7.c Three different types of dune landscape in relation to natural and anthropogenic factors:
Fixated dune landscape, where vegetation growth is dominant and geomorpholic processes hardly play any role.
Woodland and dune scrubs are an important type of habitat.
Grazing by natural or domestic animals may contribute to a more open landscape. This landscape is the result of large-scale human activities, especially in the twentieth century (Zeepe- en Meeuwenduinen, Schouwen, autumn 2008)

dependent upon local socio-economic circumstances and the presence of motivated individuals with ample financial resources. Therefore, in the dunes we see different developments in different places from the middle of eighteenth century onwards. A good example is the fate of the rabbit warrens owned by the *Staten van Holland* in the inner dunes near Oostvoorne (Voorne) and Ouddorp (Goeree). These dune areas are largely comparable from an ecological point of view (calcicole inner dune grassland) and were used in a similar way as rabbit warrens up to the first decades of the eighteenth century.

The *Staten van Holland* decided to sell the inner dunes near Oostvoorne in 1725. In the following decades they were split up in separate lots and parcels of coppice wood were subsequently planted, agricultural fields were constructed, and houses were built. As a result of this development the former dune grassland area is now largely covered by houses, gardens, and woodland. Investors from urban environments, particularly from the nearby city of Rotterdam, played an important role in this transformation. On the island of Goeree the rabbit warrens were terminated in 1743 when the *Staten van Holland* decided to lease the dunes solely for cattle grazing. This type of dune use continued up to the beginning of the nineteenth century. Then, either the local farmers or authorities they dominated, purchased the inner dunes and continued the grazing until the middle of the twentieth century. In this case, where the dunes were more remote to urban areas, local farmers played a major role in conserving the inner dunes in their mediaeval state. This example also illustrates that anthropogenic factors contributed to the differentiation of the landscape and to the ecological differentiation of nearby dune areas. They explain why endangered grassland species are better conserved on Goeree than on Voorne (Figure 6.3). In the same way, the introduction of drinking water catchments in the northern dunes of Walcheren (1892) and the way how this has been carried out, explains the relative strong decline of (ground)water depended Red List-species on this island (see above, chapter II and figure 6.3). In other dune areas this type of dune use was less extensive or had a lower ecological impact.

VI.3. Twentieth century: fixation of the dune landscape

For centuries the dunes of the south-western Netherlands were in a relative dynamic stage during the late Middle Ages and in the Early Modern period. As a consequence of the grazing by rabbits and cattle, and the cutting of shrubs and trees, there was a fragile balance between vegetation growth and mobile sand. Man played an active role in keeping this balance by planting Marram Grass in parts of the dunes where sand mobility was going to dominate vegetation growth. Cattle grazing was only moderately applied, particularly when vegetation growth was too abundant for rabbits. In the second half of the eighteenth century attitudes towards dune use and dune habitats changed (see above): cattle grazing increased and, where conditions were favourable, woodlands were planted and agricultural fields were established. For the latter in particular, small-scale dune dynamics and sand mobility were a much greater nuisance than the rabbit warrens. In these places the perceived need for dune stabilisation and fixation increased. In some areas, like the Walcheren and Schouwen dunes which were inconvenient for forestry and agriculture, abandonment occurred in the nineteenth century, with a resulting strong increase of large scale dune dynamics (van Haperen, 2009; Beekman, 2007). In the second half of the nineteenth century large parts of the dunes on these islands consisted of shifting dunes with hardly any vegetation cover. Adjacent agricultural interests (Schouwen) and initiatives for drinking water catchment (Walcheren) also made it necessary to stabilise these dunes. As a result, in the course of the twentieth century all the dunes of the south-west Netherlands were largely fixed by the planting of Marram Grass and by afforestation. The immediate reasons for these measures vary: from drinking water catchment (Walcheren) to complaints by farmers about mobile sand (Schouwen), to private initiatives for afforestation (Voorne). Coastal defences played a minor role as primary motive for dune fixation. At present only on Schouwen can dynamic dunes still be found. From this perspective, wood and scrub will possibly be the most important type of habitat in the dunes in the foreseeable future. This will have a negative impact on the characteristic biodiversity of the dunes, that largely depends on the presence of open and dynamic habitats. It is probably also negative for coastal defences, because static dunes have little opportunity to adapt to new coastal conditions and a changing climate. Therefore, the challenge is to develop a way of management in which dynamics have their role in keeping the dunes in a vivid natural state.

V. Conclusions

Dunes are not an accidental sandy landscape between land and sea, but are part of the coastal ecosystem. Their origin and development must primarily be interpreted from the geology and hydrodynamics of the adjacent sea. In the south-west Netherlands the differentiation of vegetation in the dune landscape is primarily caused by a gradient in the Pleistocene subsoil. The origin of this gradient in its turn has to do with the tectonic and glacio-isostatic development of the North Sea area, with marked differences in subsidence between the Flemish and the northern part of the Dutch coast (Vos *et al.*, 2011: 14–17). The tidal currents in the North Sea highly contribute to the expression of this gradient, mainly because of a net transport of sediment in a northward direction. On a more local and regional scale the dunes of the individual islands are part of the ebb tidal deltas of the various estuaries. The tidal volumes of these estuaries are the driving forces behind the quantities of sand available for dune formation and the tidal currents are the main transport mechanisms.

In addition to these natural factors, man and his local communities play a minor role in the primary differentiation of the dune landscape. Mankind is responsible for major changes to this landscape across time. From the late Middle Ages until the eighteenth century his influence was very intensive, but similar on the various islands. In this period the dune ecosystems with their extensive rabbit warrens had a dynamic appearance with a relatively high resilience. This was due to negative feedback mechanisms, in which geomorphologic processes were kept in balance with vegetation growth. Man highly contributed to this balance by local dune fixation and by regulating grazing activities.

From the middle of the eighteenth century onwards a principal change and a differentiation in the anthropogenic influence took place. This was caused by economic, cultural, and political factors and had no relation with to the differentiating natural factors (geology and hydrodynamic processes). In fact, these anthropogenic factors act along a second axis in addition to natural processes. New forms of dune use were introduced, with a higher sensitivity to dune dynamics and rabbit grazing (agriculture, forestry; in the twentieth century also drinking water catchment). This asked for more fixation of dunes and increasing control over the rabbit population. In dune fixation, mechanisms with a positive feedback were dominant, ultimately leading to an almost total fixation of the dunes and excluding dynamic factors. The challenge for our generation is to develop a management in which coastal dynamics are restored to both safeguard coastal defences and to conserve dune biodiversity.

Bibliography

Adriani, M.J. and Maarel, E. van der (1968) *Voorne in de branding,* Oostvoorne.

Beekman, F. (2007) *De Kop van Schouwen onder het zand. Duizend jaar duinvorming en duingebruik op een Zeeuws eiland*, Utrecht.

Beets, D.J. and Spek, A.J.F., van der (2000) 'The Holocene evolution of the barrier and the back-barrier basins of Belgium and the Netherlands as a function of late Weichselian morphology, relative sea-level rise and sediment supply', *Geologie en Mijnbouw/ Netherlands Journal of Geosciences*, 79, pp. 3–16.

Berg, J.H. van den (1984) 'Morphological changes of the ebb-tidal delta of the Eastern Scheldt during recent decades', *Geologie en Mijnbouw*, 63, pp. 363–375.

Dam, P.J.E.M. van (2007) 'Ein Neubürger in Nordeuropa. Menschliche und natürliche Einflüsse af die Assimiliering des Kaninchens in den Niederländischen Dünen 1300–1700', in B. Hermann (ed.), *Beiträge zum Göttinger Umwelthistorischer Kolloquium 2004–2006*, Göttingen, pp. 163–176.

Dam, P.J.E.M. van (2010) 'Rabbits swimming across borders; micro-environmental infrastructures and macro-environmental change in Early Modern Holland', in: B. Scott (ed.) *Economies and Ecologies in Medieval and Early Modern Europe*, Leiden, pp. 63–92.

Ebbing, J.H.J. and Laban, C. (1996) 'Geological history of the area off Walcheren and Zeeuwsch-Vlaanderen (Southwestern Netherlands) since the start of the Eemian', in: D.J. Beets *et al.* (eds), *Coastal Studies on the Holocene of the Netherlands*, Haarlem, pp. 251–268 (Mededelingen Rijksgeologische Dienst; 57).

Haperen, A.M.M., van (2009) *Een wereld van verschil: landschap en plantengroei van de duinen op de Zeeuwse en Zuid-Hollandse Eilanden*, Zeist.

Kooijman, A.M. and Besse, M. (2002) 'The higher availability of N and P in lime-poor than in lime-rich coastal dunes in the Netherlands', *Journal of Ecology*, 90, pp. 394–403.

Laban, C. and Rijsdijk, K.F. (2004) 'De Rijn-Maasdelta's in de Noordzee', *Grondboor en Hamer*, 56, pp. 60–65.

Maarel, E. van der (1966) 'Dutch studies on coastal sand dune vegetation, especially in the Delta Region', *Wentia*, 12, pp. 1–61.

Priester, P. (1998) *Geschiedenis van de Zeeuwse landbouw circa 1600–1900, Wageningen-Houten-Middelburg* (A.A.G. Bijdragen; 37).

Schaminée, J.H.J., Stortelder, A.H.F. and Westhoff, V. (eds) (1995) *De Vegetatie van Nederland: deel 1. inleiding tot de plantensociologie: grondslagen, methoden en toepassingen*, Uppsala-Leiden.

Terwindt, J.H.J. (1973) 'Sand movement in the in- and offshore tidal area of the SW part of the Netherlands', *Geologie en Mijnbouw*, 52, pp. 69–77.

Verbraeck, A. (1990) 'De Rijn aan het einde van de laatste ijstijd, de vorming van de jongste afzettingen van de formatie van Kreftenheye', *Geografisch Tijdschrift*, 24, pp. 328–340.

Vos, P.C. and Heeringen, R.M. van (1997) 'Holocene geology and occupation history of the Province of Zeeland', in: M.M. Fischer (ed.), *Holocene evolution of Zeeland (SW Netherlands)*, Haarlem, pp. 5–109 (Mededelingen Nederlands Instituut voor Toegepaste Geowetenschappen TNO; 59).

Vos, P.C *et al.* (2011) *Atlas van Nederland in het Holoceen*, Amsterdam.

7 Living in a dynamic landscape: the dune area on the island of Schouwen, province of Zeeland, during the late prehistory and early historical period*

Bert van der Valk, Deltares, Delft
Frans Beekman, Den Haag

Abstract

The Schouwen dune coast and its human occupation over the last 3,000 years are the subject of this chapter. Human occupation of the narrow coastal dune area on the island of Schouwen, with a distinctively marshy and less accessible hinterland, seems to have been risky at best. Nevertheless, archaeology suggests almost continuous inhabitation, starting in the Late Bronze Age and continuing well into the Early Middle Ages, with a notable dip during the Roman period caused by unknown factors. It is argued here that the mouth of the River Scheldt was the dominant factor that facilitated and affected human settlement. Its coastal outlet has never been river-dominated with rapid aggradation of the beach barriers, such as in South-Holland. Instead, its coastlines prograded slowly, and must have shown periodical phases of coastal erosion and coastline retreat that were linked to dune formation.

The Zeeland coastline has been gradually eroding for thousands of years bearing witness to the continuous struggle of man against the sea, as well as human victory over the harsh environment. Van der Valk et al. *(1997) and Beekman (2007) have reconstructed the Younger Dune formation on Schouwen. The current chapter starts with a more detailed reconstruction of the Older Dune landscape.[1] We based our work on the data collected during two major observation periods, 1992–94 and 2007–11. For a report on the data of the latter period, see Van der Valk and Beekman (2011).*

---

* We sincerely thank Staatsbosbeheer (Dirk Fluijt and Camiel Beijersbergen) for access to the SBB management area. From SCEZ we thank Henk Hendrikse and Robert van Dierendonck, for support and access to the Hubregtse collection kept in SCEZ. Gemeente Schouwen-Duiveland (Michiel Verduijn) is thanked for permission to carry out the archaeological work. The Provincie Zeeland (Miek Geerts) is thanked for supporting the archaeological analyses in relation to this project by a provincial grant. Two reviewers provided useful suggestions that significantly improved the contents of this paper as well as the written English.

[1] Both geological units as described in this chapter, the Older Dune and Younger Dune Sands, now belong to the aeolian Schoorl sequence, Naaldwijk Formation (Weerts *et al.*, 2006). For convenience, however, the terms Younger and Older Dunes are used in this paper.

Landscapes or seascapes? The history of the coastal environment in the North Sea area reconsidered, ed. by Erik Thoen, Guus J. Borger, Adriaan M.J. de Kraker, Tim Soens, Dries Tys, Lies Vervaet & Henk J.T. Weerts, Turnhout, 2013 (*CORN Publication Series*, 13), pp. 127–143.

BREPOLS PUBLISHERS DOI: 10.1484/M.CORN.1.101551

I. Introduction

During and because of the formation of beach barriers, including the Older Dunes, the Netherlands' coast became vegetated and inhabitable since *c.* 5500 BP. The inhabitable area included the seaward part of the island of Schouwen, just to the north of the outlet of the River Scheldt. At that time the River Scheldt was a small rain-fed river, as it still is, that never carried large volumes of fresh water to the sea (less than *c.* 100 m^3/second). The sandy head of the Schouwen Island has long been connected with the coastal area further to the north that extended up to the River Maas outlet.

The island of Schouwen became disconnected from the north either sometime during late prehistory or post-Roman times. Though the occupational history of seasonal habitation (by e.g. performing transhumance) *versus* perennial inhabitation during prehistory, Roman times, and the early medieval age, is subject to ongoing research, fresh water must have been readily available. The sheer size of the area and the remains of palaeo-vegetation provide sufficient indications to accept this presumption. Another choice for living at the Scheldt River mouth was perhaps made from the point of exploitation of fish and/or marine mammal resources, although palaeo-zoological remains that should provide evidence for the latter have not yet been found.

During the formation of the beach barriers that area was probably safe enough for permanent inhabitation, but during the early medieval phase of coastal erosion, the area might not have been so safe, and the sandy area may have shrunk considerably in size, making it more of a refuge than anything else. At that point the landscape became a seascape.

The coastline prograded until the middle of the eighth century, thereby increasing the potential for arable land. However, the onset of the Western Scheldt formation greatly affected the Schouwen peninsula. Tapping the River Scheldt northwest of Antwerp after the eighth century, river water no longer flowed northwest towards the Eastern Scheldt area, but westwards into the rapidly developing Honte, as the Western Scheldt was called during the Middle Ages. As there was no longer a permanent fluviatile connection to the Eastern Scheldt that kept its estuarine inlet in place, coastal erosion began at Schouwen in earnest from the ninth to tenth centuries onwards. It may have had the consequence that the area became densely inhabited, and consequentially heavily exploited. As the sandy soil was already decalcified and leached it suffered badly from over-exploitation, which caused mobilisation of the dune sands people were living on. These two joint processes (coastal erosion and effects of overpopulation) spelled doom for the medieval use of the older dune area, covering it in the course of centuries by transgressing younger, secondary, coastal dunes. These dunes were large parabolic dune forms that left wide secondary dune valleys in their wake. The extensive coastline retreat (some 1,5 km in about 1000 years) caused widespread loss of arable, as well of small villages and some smaller semi-natural harbours. Secondary dune formation continued for centuries and could only be tamed with great difficulty and for the first time during the seventeenth and eighteenth centuries and once again in the early twentieth century. A period of fifty years neglect of dune plantations before the last event was enough to cause widespread extension of drifting dune sands.

Over time, repeated field surveys along the coast and in the few large natural blow-outs that still exist gave us a good view on such developments. The natural exposures show sedimentary structures that reflect the geological history. These exposures, in the form of 'peepholes', at the same time provided an opportunity to register stratigraphy and archeology. In the lower parts of these peepholes arable land with ditches acting as field partitions occasionally became visible after having been buried for *c.*1000 years. In many cases we found pottery shards and other archeological materials distributed over the lower parts of these valleys in the form of a desert pavement, so documenting the older, and sometimes various phases, of land-use. Analysis and integrated interpretation of these various types of data is a specific element of non-destructive archeological research in this area, as well as a challenge (Van der Valk and Beekman, 2011). In an earlier paper we studied the formation of the Younger Dunes (Van der Valk *et al.*, 1997), while we now focus on both the Older as well as the Younger Dunes. Both geological units, as described in this paper, the Older Dune and Younger Dune Sands, now belong to the Schoorl sequence, Naaldwijk Formation (Weerts *et al.*, 2006).

The field surveys which form the basis of this paper were carried out during 1992–94 and 2007–11. We also used notes made by schoolmaster J.A. Hubregtse who compiled massive archeological data (including photographic pictures of the landscape) during the period 1910–40, when the dune area was much more mobile than today. Since only a few scanty references survive from the early Middle Ages we have employed a varied array of resources, perhaps making this contribution to the book a bit different from the others.

To understand this dynamic and complex landscape that has to a large extent been fragmented, archaeologists and historians must team up with geologists. Only then the vicissitudes of its human occupation can be put in perspective. Mapping of peat layers, of directions of coastal progradation as measured from the sedimentary beds being registered in Older Dune- and beach sands, and archaeological and/or radiocarbon datings of organic material are all helping to reconstruct palaeo-environments. The medieval human occupation on a retreating coastline with an increasingly flooded hinterland must have been a risky matter in contrast to the human occupation of the prehistoric coast which was still accreting. For researchers it is a challenge to demonstrate the continuity of occupation – or not – in such dynamic landscapes.

The natural settings of the sandy head of the Schouwen Island are presented in this paper. Lack of data (as so far, scientific excavations were not carried out in the area) prevents the current authors of drafting more conclusive answers to the actual configuration of the island during the late prehistory. In addition, there are some solid research questions still remaining: how and when did the different late prehistoric and proto-historic communities flourish and how did they exploit the natural resources of the area? We are convinced that further study of the occasional (natural and/or man-made) exposures in the area will reveal some of the necessary data, perhaps along with some intentionally placed smaller research locations.

II. Geology of the Schouwen dune area

The island of Schouwen-Duiveland once consisted of two separate islands that were joined by salt marsh reclamation in the seventeenth century. This paper refers to the most

Figure 7.1.a Simplified geological map of the Netherlands with the position of Schouwen relative (square) to other beach barrier areas.
Legend: 1. Walcheren Member (previously Dunkirk Deposits); 2. Den Haag Member (previoulsy Younger Dunes); 3. landward boundary of Wormer Member (previously Calais Deposits) 4. Voorburg and Ypenburg Members (previously Beach barriers and Older Dunes); 5. Pleistocene and older formations at surface; 6. Holland Peat Member and river deposits (modified after Van der Valk *et al.*, 1997)

Figure 7.1.b The topographical map 1: 25.000 shows the transgressive coastal dunes (Younger Dunes-yellow) covering the 'Binnenduinen' landscape. The Binnenduinen belong to the Older Dunes (light green), as is indicated by surface finds of pre-medieval Iron Age shards amongst many hundreds of medieval pottery shards (source: Van der Valk and Beekman, 2011; Map: Topografische Dienst, Emmen, 2004)

western part called Schouwen. The Schouwen Island consists of a sandy barrier fronting the shallow fore-delta along the southern North Sea. The coastal barrier widened and expanded during *c.* 2000 years. The barrier never took on the wide, predominantly river-dominated forms with wide beach plains in between the older barriers, as it did in central Holland (Van der Valk, 1996), but instead developed a wave-dominated form, much like the younger barriers of Holland where, in between barriers, primary dune slacks prevail instead of beach plains (Van der Valk and Beekman, 2011).

Obviously, the reason for this different barrier architecture was that the River Scheldt has always been much smaller than the River Rhine, indicating that the River Scheldt probably never provided sand in any volume to the coast. The widening of the barrier caused a regional rise in the groundwater table. This, in turn, caused Holland peat to start to accumulate along the primary dune slacks occasionally visible at low tide along the coast of Schouwen. Human activity at a later stage likely caused the barrier dune sands to become mobile, possibly starting in the Middle Iron Age, which in phases subsequently covered the accumulated peat and caused the dune areas south and north of the peat-filled dune slack to join. Peat layers were seen in earlier days when the coastline was still eroding, but this is less common today under the prevailing coastline management system. In some places, due to groundwater stagnation, a second peat layer grew at some 1–2 m above the first peat layer (unpublished data). In restricted areas such as west of Renesse and around Nieuw-Haamstede, a third peat layer possibly formed at an even higher level; it is unknown when this local peat growth started and ended. It could previously be seen in a dune cliff at RSP 11–12 (Rijks Strand Paal = beach pole) west of the Nieuw-Haamstede light house after storm surge erosion. Unfortunately, due to a general lack of boring data and radiocarbon dated peat samples in the dune area we cannot further resolve the dune and peat stratigraphy.

Behind and in the protection of this barrier is the remainder of a previously vast coastal lagoon/tidal embayment filled in by fine sand and clay that subsequently became covered by Holland Peat starting from *c.* 4500 BP, but now eroded on all sides. Over wide areas the peat was excavated for fuel and saltmaking during medieval times.

The object of our study, West-Schouwen, now consists of an extensive younger (transgressive) dune lobe overlying the remnant of an older coastal barrier, much eroded over time (Figure 7.1). The Schouwen Island has been subjected to intense tidal scouring from the south by the channels such as the very deep (30 to 50 m) Hammen channel forming part of the tidal inlet, the Eastern Scheldt. This has been a true estuary until well into the sixteenth century when the River Scheldt drainage was tapped by the Western Scheldt as the latter's drainage area expanded eastward and tidal volume increased.

The Verklikkerduinen area, being the northern fringe of the barrier area is 400 years younger than the southern part of the Younger Dune area. Based on stray archaeological finds, a Roman military outpost is presumed to have been located some 2 km southeast of the current southernmost tip of the Schouwen Island, while the early medieval settlement, called *villa Scaltheim*, was situated some 500–1000 m to the immediate west of the study area, completely reworked by the earlier version of the Krabbengat, now in the Eastern Scheldt outer delta (Beekman, 2007: 112–13).

III. Archaeology of the dune area (based on field campaign 2007–2011 and older data)

The sandy head of the former island of Schouwen forms a small, but important remnant of the Subboreal Zeeland aggradational coast, documenting an almost continuous inhabitation since the Late Bronze Age (Van Heeringen, 1992; Van der Valk *et al.*, 1997).

In that, it strongly resembles the much larger and wider fossil beach barrier area of the Western Netherlands. The immediate vicinity of a permanent tidal inlet-cum-river mouth, however, provides a varied geological setting. Continuous as well as episodic changes occur: examples of the first type of change are tidal channel shifts and sea-level rise; the second type of change includes storm-surge flooding of the back barrier area and periods with intense aeolian erosional activity in the dune area. Archaeological research is necessarily taking place based on surface finds from secondary blow-outs, as targeted excavations have never taken place in the area (and cannot take place under the prevailing legislation). Use is made of a large collection of finds from deflation areas in the archaeological window picked up by J.A. Hubregtse during the first half of the twentieth century, as well of our own observations (1992–1994 and 2007–2011). There is an important limitation on the nature of such archaeological finds in the study area: assumptions must be made on their reliability. Essentially, the spot where we find them is their provenance in terms of (x,y) coordinates. It can safely be assumed that the archaeological finds were not shifted from place to place: during deflation of dune sands they only drop down the stratigraphical sequence of sand layers i.e. their (z) coordinate is nearing the groundwater table, being bottom line during the formation of wind-scour hollows (or blow-outs). From our own experience it showed that traces of postholes and other man-made structures are almost non-existent (but we found a few, see below); these traces are easily obliterated by wind action in the blowouts.

Regular inspection of these blow-outs, chiefly Vlierendal and Nieuwe Westerse Laagte (field names for two blow-outs just north of the *Pinus* plantation on West-Schouwen; see Vd and Nwl in Figure 7.3), is therefore of importance to preserve the archaeological heritage. We have actually seen archaeological finds in a private collection either deteriorate in quality or disintegrate (and get disposed of).

Almost continuous habitation has taken place during the Neolithic, Late Bronze/Early Iron Age, Middle Iron Age, Late Iron Age/Roman Age, over time in different areas over the changeable and periodically dynamic Older Dune surface. The Late Neolithic presence is limited to Brabers (actually behind the larger Older Dune area, a sandy rise of uncertain geological origin in the lagoonal back-barrier area), and to stray finds of smaller flint implements in the extreme west of the Meeuwenduinen – if these are indeed of Neolithic age. So far, Late Neolithic and Early to Middle Bronze Age pottery seems to be lacking (some stray metal finds were made by Mr. Hubregtse), and some early medieval finds from the period 400–800 AD were recognised (again some metal (coin) finds are known, but not much pottery). All other inhabitation phases have been suitably documented, and their locations can now be understood in respect to their position in the (palaeo-) landscape. These positions relate to erstwhile dune landscape dynamics occurring before and after inhabitation phases, as well as to renewed and also recent deflation.

III.1. Late Neolithic and Early to Middle Bronze Age inhabitation missing

Based on Van Heeringen (1988a, 1988b) and from the finds made in the area we conclude that on West-Schouwen, Late Neolithic finds are lacking, even though this seems not to be so at first sight.Neither Hubregtse nor we could find beaker culture pottery shards or typical implements, the beaker culture being the next successive archaeological period following the Vlaardingen group remains that were found on Brabers, a few kilometres southeast of our study area (For a recent publication on the Vlaardingen cultural group including the Brabers site see Verhart, 2010).

Equally, Early- and Middle Bronze Age pottery finds also seem to be lacking: the observation- and collection activities should have brought these to light, as they did with other cultural periods. The status of the Middle Bronze Age metal finds from the area, published by Van der Feen (1952) seems ambiguous due to the lack of culturally associated beaker and *Hilversum* pottery fragments – the latter were never found.

Recently we checked the Hubregtse collection for these types of finds and found none. The artefacts that could indicate Neolithic inhabitation most typically, being also the most resistant against weathering in the light sandy soils, would be flint artefacts. Hubregtse found some typical Bell Beaker (3) and Single Grave culture (1) flint artefacts (Van der Feen, 1952: his figure 7.2). Equally, some flint artefacts were found during our two field campaigns, of which the most typical is a so-called 'transverse' arrowhead usually associated with the Vlaardingen group inhabitation. Remarkably, these artefacts cannot be matched with any of the pottery types found so far in West-Schouwen (which does not incorporate the Brabers site). The only reason we can think of to explain this anomaly is severe erosion of the Schouwen island head due to large-scale re-arrangement of the Scheldt River mouth during the Middle Bronze age (about 1300–1100 BC). This lead to the uprooting of the palaeo-landscape, the destruction of pottery shards, and the re-distribution of flint implements across younger landscapes. We suppose the Late Neolithic to Middle Bronze Age barrier/river mouth to have been eroded as also has been supposed for Walcheren Island south of the River Scheldt estuary mouth after the Neolithic period (Vos and Van Heeringen, 1997). The only artefacts surviving such an erosional event would be flint (and perhaps metal) objects. This would then imply that Late-Neolithic and Early and Middle Bronze Age flint and metal artefacts would have been found in secondary deposition positions. To prove this a renewed study of these artefacts is required. It means that we cannot use their find spots for landscape reconstruction other than in a very general way.

III.2. Late Bronze Age (LBA)/Early Iron Age (EIA) start continuous inhabitation

Pottery finds dating from this period are frequent and widespread in Schouwen-Meeuwenduinen, indicative that the landscape was suitable for inhabitation (Van Heeringen, 1988b). The finds of the LBA and EIA pottery cannot be separated well enough on the basis of available material. Supplementing the map made by Van Heeringen (1988b) using unpublished data form the Hubregste collection in the Middelburg, Archaeological Depot (SCEZ), we conclude that the LBA/EIA material occurs further south in the dunes than previously thought. There is a marked 'empty' area between the northern concentration as mapped by Van Heeringen and the recent additions made to the data. We presume that the area was inhabitable due either to the presence of a higher dune ridge, or due to the presence of a low lying wet (secondary) dune slack in between higher barrier and dune ridges.

III.3. Middle to Late Iron Age and Roman Age inhabitation: highest number of find spots

Observing the occurrence of the MIA to LIA pottery finds we find their distribution mirrors the LBA/EIA pottery finds. Exactly the same division is present, which we interpret as the general landscape character remaining stable from one cultural period to another. During the last field survey campaign, we noticed for the first time the majority of shards of a LIA pottery vessel in a contemporary man-made pit. Close by, in the same general location, a larger part of another similar vessel was found in pieces lying closely together at the surface, freshly exposed by wind action (but also bearing traces of older sand blasting). These finds are particularly indicative of the repeated cycle of aeolian sedimentation and denudation, typical for coastal dune landscapes. Radiocarbon datings are now available starting from 2210 ± 30 to 2030 ± 30 BP on bone and soot samples from MIA and LIA archaeological context (Van der Valk and Beekman, 2011).

During the following cultural period, however, we presume a change. Roman Age artefacts are spread thinly, but evenly, amongst the entire western segment of Schouwen. It is generally assumed that a Roman military outpost (or perhaps the fleet station Roompot) was situated where now the Hammen channel spreads out and changes into the Eastern Scheldt ebb-tidal delta area (Beekman, 2007: 41).

There could be two reasons for this thin spread of Roman finds in the western Schouwen area:

1. It could be a true thin spread, reflecting a sparse areal use by the Roman age population.
2. The Roman finds could be later pick-ups from erstwhile Roman sites (such as the above mentioned military outpost/fleet station) by Merovingian and/or Carolingian settlers. Until excavations are made in the area, we cannot decide which of these two reasons would be valid.

III.4. Early medieval inhabitation

From the profuse presence of medieval pottery, near-stoneware, and stoneware shards, sometimes 20–30/m^2, as well as other artefacts, it can readily be assumed that during the Middle Ages widespread inhabitation in an agricultural landscape on the Older Dune landscape must have been present. Nothing is known about the names for localities other than there had been an early medieval *villa Scaltheim,* mentioned in 828 AD, located at a creek with access for ships and in the protection of a sand dune barrier that presumably was located west of the current coastline: [...] *villa Scaltheim iuxta ostium Scaldis fluminis in maritime Frisionum regione* (the *villa Scaltheim* situated at the mouth of the River Scheldt in the Frisian coastal area, see Beekman, 2007: 40). This site may have been located some 500–1000 m immediately west of *Oude Eendenkooi*, see Figure 7.3). This *villa Scaltheim* must have been the northern counterpart of *villa Walacria*, the latter being situated on the southern bank of the River Scheldt. Not a single excavation has yet been carried out on medieval dwellings or land parcelling systems except for some measurements on the site of *het Akkertje* (half a silver penny dating from 1040 was found here; see Ak in Figure 7.3) during the 1992–94 field campaign (Beekman, 2007: 48 and 177). We can, however, safely presume that a correlation exists between the finds of pottery and coins (and other metal finds) from the area. In the following sections some notes on the separate medieval periods are made.

III.4.1. Merovingian period

As mentioned above we have recently performed another quick scan of the material that Mr. Hubregtse collected and which we did not survey during the previous campaign. A number of find spots in the south-western part of the Schouwen dune area surveyed by

Figure 7.2 West-Schouwen geomorphological map (fragment from: Wolfert and Buitenhuis, 1991).
On this map, the Younger Dune landscape has mainly deep yellow and brown hues, while the Older Dune landscape has greenish yellow and yellow hues. The map shows the predominant area where archaeological finds were made, namely west of the high dune ridge (dense brown colours) in the Meeuwenduinen, in a composite archaeological window; for explanation, see text and Figure 7.3. (source: Alterra, 2011)

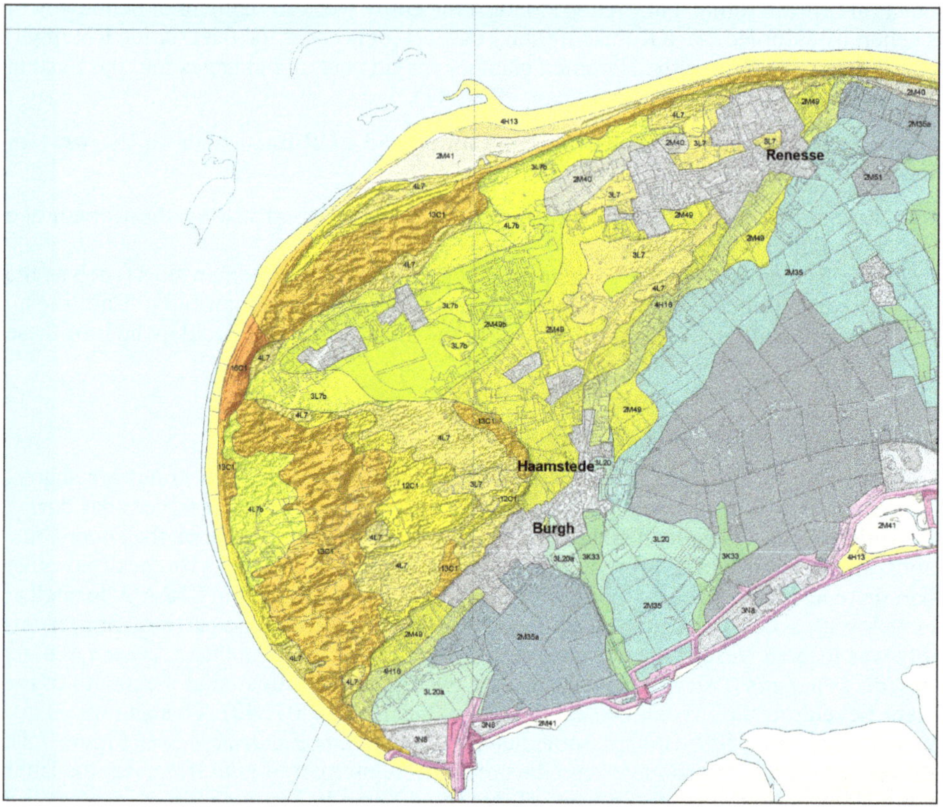

him, demonstrated large quantities of typical Merovingian pottery (*Wölbwandtöpfe*: red earthenware with typical prominent turning marks on the inside). It was not previously known that these pottery wares are also present on Schouwen.

In 1999, on the Westenschouwen beach a shard of unmistakably seventh-century Merovingian age and Eifelian/Rhineland make pottery was found (Anonymous, 1999). Not much is known about native pottery of this period and we do not know how it looks like in this area. Presumably some forms of *Kugeltopf* local wares were either made or imported from elsewhere.

Figure 7.3 Field map of West-Schouwen
Legend: 1. polder area; 2. Older Dunes (inner dunes); 3. Younger Dunes (outer dunes); 4. Blow-outs with archaeological finds, with recent additions (based on data from Van der Valk and Beekman, 2011): locations Vlierendal (Vd), location Nieuwe Westerse Laagte (Nwl) and the estimated location of the Scaltheim settlement; distribution of: LBA/MIA occupation (grey oval), LIA occupation (terracotta oval), RA stray finds (oval, red outline) (source: Van der Valk *et al.*, 1997, Figure 7.2b).

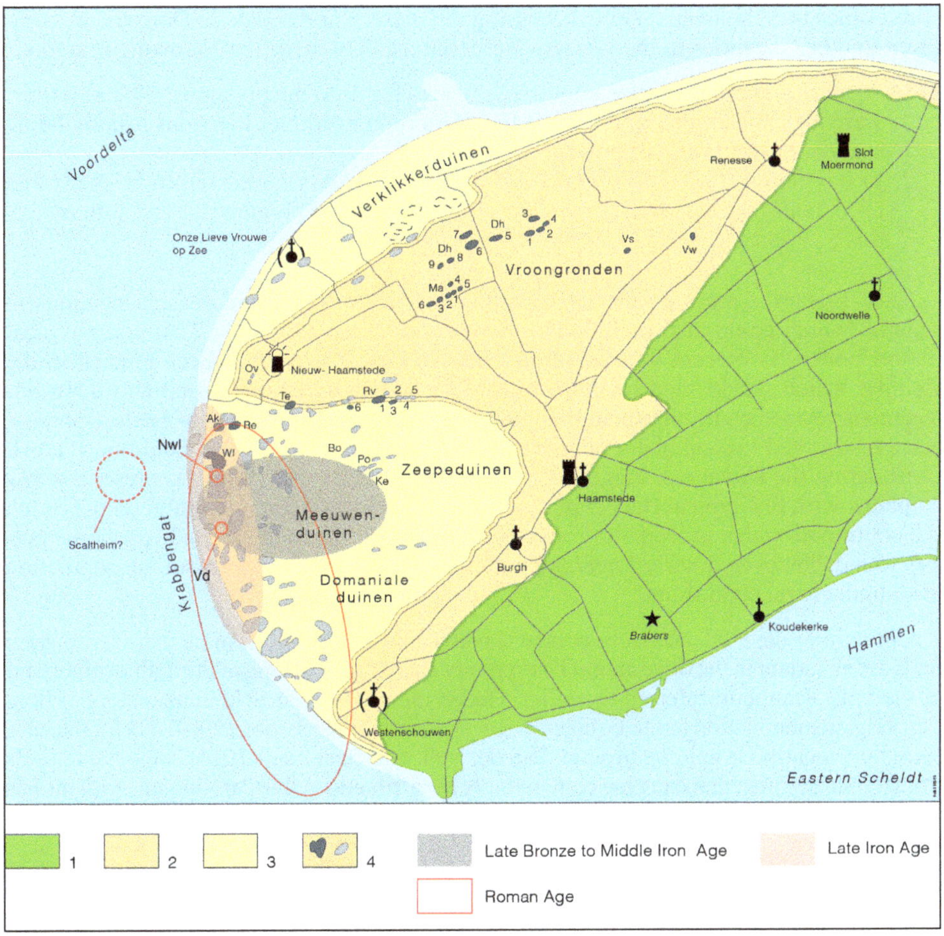

In fact the finds of Merovingian import pottery tally with the finds of Merovingian coins: sparse finds in some of the secondary dune blow-outs in our study area (two trienses from the sixth to seventh centuries) and a sizable number on the beach (seventy-nine sceattas from 710–760 AD). Also, a very restricted number of early medieval fibulas were found by Hubregtse (cf. Capelle, 1978: plate 7, nrs. 1 and 2).

III.4.2. Carolingian period

Carolingian coins were found on the beach, directly north of the *Pinus* plantation. These amounted to forty-six silver coins dating from 768–875 AD and, in a dune blow-out, a coin struck during the reign of Louis the Pious (822–840 AD). The type of import pottery commonly found during this period is Badorf pottery. A complete Badorf amphora dating from the period 775–850 AD was found on the beach after storm wave erosion west of Nieuw-Haamstede (Van Dierendonck, 2001, the pot is currently kept in the Burgh Museum).

Recently some statements were made by Van Es and Verwers (2010) that seem to be of relevance to Schouwen. While discussing the pottery finds made in Dorestad as compared to other localities in the Western Netherlands, they made the following remarks:

1. Agricultural communities import less 'wares for drinking purposes' such as Badorf ware jugs and amphora; these are found more in wealthier locations usually being seats of business and/or power.
2. In the more rural, agricultural areas *Kugeltopf* pottery continued to be made and used, being the standard pottery wares available to its population. The authors call this a 'Frisian' feature.

This 'Frisian' feature covers the mouth of the River Rhine and coastal locations in general, which could also include the mouth of the River Scheldt. That some typical Badorf pottery is present indicates that a seat of power may have been present in the vicinity, which would then be the former *villa Scaltheim*, as discussed above. In our recent survey we could not unambiguously identify 'typical' shards of Badorf pottery, but we observed low numbers of 'white pottery with fine sand', that could be Badorf. Considering the above our suggestion is therefore that Badorf pottery is present in low numbers, with few typical Badorf pottery vessels in the area (the pot found on the beach is a confirmation that indeed these vessels were imported into the area). Most of the pottery found, however, is early *Kugeltopf* pottery, indicating that majority of the population lived under rural conditions.

Combining the evidence from coin and pottery finds we now assume that Carolingian finds are abundant in the dune area. These pottery shards may belong to the arable hinterland of the villa Scaltheim referred to earlier. Hence the notion that Schouwen does not hold Carolingian inhabitation has to be abandoned. Beekman's suggestion (2007: 41) for the *villa Scaltheim* location is gaining ground. Van der Valk and Beekman (2011) show metal finds from the survey area that can now convincingly be attributed to the Carolingian era. This has also implications for the age attributions of Capelle (1978) for metal finds from Schouwen.

IV. History of the dune landscape after medieval times

IV.1. Older Dunes

The older barrier and dune sands have been subject to aeolian reworking from the start of their existence. From Vlierendal there is evidence for repeated soil formation on the older dunes even before Middle Iron Age inhabitation. Only in two locations (Van Heeringen,

Figure 7.4 West-Schouwen, Westerse Laagte blowout (Wl in Figure 7.3; period 1992–1994).
The lower part of the profile shows a prograding sequence of Older Dune and Beach Sands (the North Sea is to the right; radiocarbon date on shell pair of *Macoma balthica* ('1' in the figure, c. 2900 y. BP) covered by transgressive Younger Dunes. In between shallow soils in which both infrequent LIA pottery shards as well as frequent medieval pottery shards were found (both at '2' in the figure). The top soil is a desert pavement indicating severe aeolian erosion, as in so many locations in West-Schouwen (source: Van der Valk *et al.*, 1997; Van der Valk and Beekman, 2011)

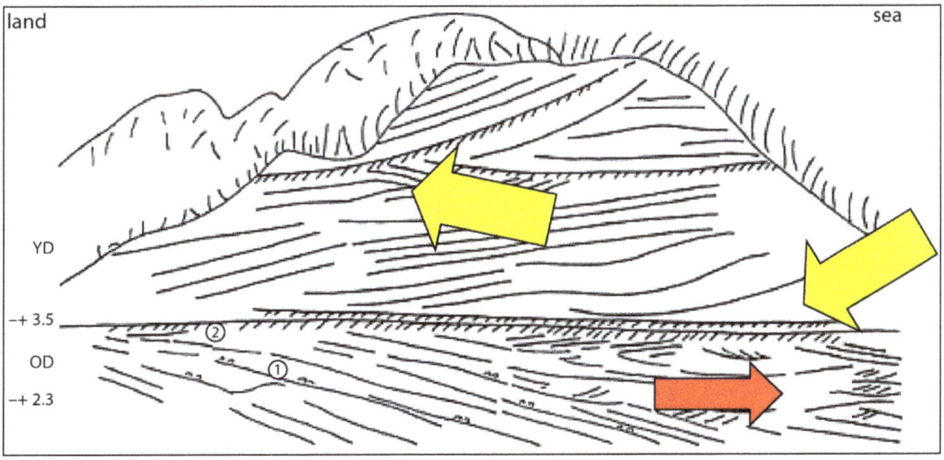

1988b and Van der Valk and Beekman, 2011) are found two separate archaeological period materials in the same dip. This is a strong indication for 'shifting cultivation' in a dynamic Older Dune landscape. It is largely unkown if this is because of depletion of local soil quality, or due to moving dune sands. The Roman and medieval finds in the Hubregtse collection of Schouwen need to be collected, re-evaluated, and contextualised against other categories of archaeological materials and with the landscape history. In doing so, better insights into continuity of habitation and settlement could be demonstrated.

IV.2. Younger Dunes

Transgressive dune development started around 900–950 AD, and probably earlier in the area that has since been eroded off the west coast of Schouwen. The older dune surface was (partly) covered by wide mobile dune field transgressing the Older Dunes until stabilising *c.* 1250 AD, forming a 5 km wide and equally deep dune field (Meeuwenduinen, Domaniale duinen and Zeepeduinen) currently with (younger) large parabolic forms superimposed. Due to further coastline erosion a next phase of dune formation occurred along the north side of the island (Verklikkerduinen, sixteenth to nineteenth centuries) while human neglect reshaped the Meeuwenduinen and Domaniale duinen

Figure 7.5 The actual dune blow-out Nieuwe Westerse Laagte with members of the volunteer archaeological working group AWN systematically field walking the area in 2008 observing a dense spread of medieval pottery shards. On top of the multiple LIA-medieval soil horizons (the dark couloured zone at the level of the persons' heads) a phased build-up of the Younger Dunes can be seen, separated by faint soil horizons (Photograph: L. van der Valk, The Hague)

in the south, which started in the sixteenth and lasted well into the nineteenth century. Beyond the younger dune mobile dune phases that froze *c.* 1250 AD (the central area) and again during the early twentieth century (the northern area), the older beach barrier/older dune area (Westeren Ban and Oosteren Ban) remained uncovered by Younger Dunes and was in agricultural use until the 1950s. In the latter area intermittent small-scale dune mobilisation took place well into historical and recent times (Van der Valk and Beekman, 2011).

IV.3. Subrecent dune landscape development

Neglected coastal defence and dune stabilisation in the period from the end of the nineteenth century and into the twentieth century caused extensive remobilisation of the Meeuwenduinen and Domaniale dunes, as well as the formation of numerous large, deeply scoured dune valleys with extensive sheets of blown sand in between these scour pits. The southern area was finally planted with *Pinus* trees, while the Meeuwenduinen

Figure 7.6 Aerial photograph of the West-Schouwen dune area, 1930.
Compare with Figure 7.1b for development of the forest vegetation and other indications of substantial changes in the dune landscape since 1930 (Photograph: anonymous)

remained somewhat mobile until the late twentieth century when the area became stabilised by vegetation development under the influence of rain depositing NOx which acted as a fertiliser. At present only a narrow area behind the foredunes is subject to erosion by wind and then only locally.

From the mid-1990s beach nourishments and dune reinforcements have been laid down (1991, 1994, 1997, 2003 and 2007), but these seem to be of little importance to the long term coastal development. The latter is characterised by the eastward shift of the Krabbengat tidal channel in front of the coastline, and erosion of the island head directed towards prevailing north-westerly gales, and sedimentation to the north, probably into the Hammen channel mouth. New experimental breaching of the coastal dune is permitted in areas where the safety situation allows such experiments (e.g. near RSP 13–14 directly north of the *Pinus* plantation). Allowing these experiments to take place encourages deflation of dune areas immediately behind the current foredune, and it is particularly in these areas where we have now documented archaeological finds (Van der Valk and Beekman, 2011). If and when dune remobilisation projects are planned for further inland in the Meeuwenduinen area, the preservation and documentation of any archaeological material should be of crucial importance.

Table 7.1 Schouwen: occurrence of features in relation to inhabitation per archaeological period (research status 2010; based on literature).
Legend: + : present; - : not present; -- : will not occur; ? : possibly present but not found/uncertain age
(source: mostly Van Heeringen (1988a; -b) and Van der Valk and Beekman, 2011).

	pottery	metals	animal bones	coins	worked flints	post holes
Roman Age	+	?	?	+	-	?
Late Iron Age	+	very rare	+	?	?	+
Middle Iron Age	+	very rare	+	-	?	?
Early Iron Age/ Late Bronze Age	+	very rare	?	-	?	?
Middle and Early Bronze Age	?	?	?	-	?	?
Late Neolithic Age	?	-	?	-	?	?

V. Conclusions and recommendations

Late Neolithic, Early and Middle Bronze Age inhabitation on the sandy western Schouwen Island seems doubtful (the presence of Vlaardingen group cultural remains on the Brabers sand ridge exempted). Extensive Late Bronze Age, all through the Iron Age, and possibly up to and including Roman Age, inhabitation and use of the landscape is proven. Medieval archaeological remains are profusely present on the Older Dune landscape. Based on renewed study of the pottery and metal finds dating from the sixth to ninth centuries from the Hubregtse collection, we now suggest that Merovingian as well as Carolingian inhabitation was present on the Older Dune landscape, particularly southwest of the study area.

The current state of archeological and geological research is scant at best (cf. Table 7.1), allowing only general conclusions on coastline progradation, retreat as well as dune landscape formation and the associated geomorphological processes, and on human inhabitation. For the sake of nature management and landscape restoration in the area, the land owners are recommended to coordinate their research efforts. For landscape reconstruction and dune restoration efforts a comprehensive geological survey of dune deposits using modern observation techniques and age-determinations is needed. There is also an urgent need for the systematic documentation of archaeological data that becomes available through exposure by the coastal winds, otherwise the data is lost through wind and weather. Regular area field surveys are to be organised documenting the archaeological evidence that is being exposed by the wind. Unless this information is saved it will be destroyed. Future dune system reactivation projects will need a proper archaeological assessment before the work on this type of projects is started.

Bibliography

Anonymous (1999) 'Vondstmelding strand Westenschouwen', *Nieuwsbrief Archeologie Provincie Zeeland*, 7, p. 4.

Beekman, F. (2007) *De Kop van Schouwen onder het zand. Duizend jaren duinvorming en duingebruik op een Zeeuws eiland*, Utrecht.

Capelle, T. (1978) 'Die karolingische Funde von Schouwen 1 and 2', *Nederlandse Oudheden*, 7, pp. 1–31.

Dierendonck, R. van (2001) 'Een gave strandvondst uit Haamstede', *Scaldis, Nieuwsbrief archeologische werkgroep Schouwen-Duiveland*, 9.

Es, W.A., van and Verwers, W.J.H. (2010) 'Early medieval settlements along the Rhine: precursors and contemporaries of Dorestad', *Journal Archaeology of the Low Countries*, 2, pp. 5–39.

Heeringen, R.M. van (1988a) 'De bewoning van Zeeland in de IJzertijd', *Archief, Mededelingen van het Koninklijk Zeeuwsch Genootschap der Wetenschappen*, pp. 1–43.

Heeringen, R.M. van (1988b) 'Iron Age occupation of the dunes near Haamstede on the island of Schouwen-Duiveland, Province of Zeeland, The Netherlands', *Helinium*, 28, pp. 63–80.

Heeringen, R.M. van (1992) *The Iron Age in the Western Netherlands*, Amersfoort.

Feen, P.J., van der (1952) 'Early populations in the delta of the river Scheldt', *Beaufortia*, 9, pp. 1–5.

Valk, L. van der (1996) 'Coastal barrier deposits in the central Dutch coastal plain', in: D.J. Beets, M.M. Fisher and W. de Gans (eds), Coastal studies on the Holocene of the Netherlands, *Mededelingen Rijks Geologische Dienst*, 57, pp. 133–199.

Valk, L. van der and Beekman, F. (2011) *Geologie en archeologie van de Kop van Schouwen, Gemeente Schouwen-Duiveland. Verslag over de periode 2007–2011. Rapport AWN Den Haag en Omstreken, Werkgroep 's-Gravenhage*, Den Haag.

Valk, L. van der, Beekman, F. and Heeringen, R.M. van (1997) 'Stuivend zand over boerenland. Geologie, archeologie en historische geografie van het duingebied van Schouwen in de Provincie Zeeland', in: D.P. Hallewas and P.J. Woltering (eds), *Dynamisch landschap, archeologie en geologie van het Nederlandse kustgebied*, Amersfoort, pp. 131–152.

Verhart, L. (2010) *De geur van veen; Vlaardingen en de ontdekking van de Vlaardingencultuur*, Utrecht.

Vos, P.C. and Heeringen, R.M. van (1996) 'Holocene geology and occupation history of the Province of Zeeland', *Mededelingen NITG*, 59, pp. 5–110.

Weerts, H., Cleveringa, P., Westerhoff, W. and Vos, P. (2006) 'Nooit meer: Afzettingen van Duinkerke en Calais', *Archeobrief*, 10, pp. 28–34.

Wolfert, H.P. and Buitenhuis, A. (1991) *Geomorfologische kaart van Nederland, 1:50.000, Toelichting op kaartblad 42 Zierikzee*, Wageningen-Haarlem.

Zagwijn, W.H. and Staalduinen, C.J. van (1975) *Toelichtingen bij de geologische overzichtskaarten van Nederland door de Rijksgeologische Dienst*, Haarlem.

8 Holocene sea-level change, sedimentation, coastal change and palaeogeography in the southern North Sea lowlands. A 2012 geological literature overview*

Henk J.T. Weerts, Cultural Heritage Agency, the Netherlands

Abstract

In the past decades, few regions have seen as many detailed regional Holocene sea-level reconstructions and palaeogeographical reconstructions as the southern North Sea coastal lowlands. Several curves for regions of the UK, France, Belgium, the Netherlands, Germany, and southern Denmark have become available and several series of palaeogeographical maps have been published. However, most of these reconstructions are thought to be of local or regional interest only. Since the second congress on the evolution of the coastal area of the Low Countries (held at Ghent in 1978), our knowledge of the factors that control relative sea-level rise (RSL) has increased significantly. Furthermore, we now better understand the influence of changes in the rate of sea-level rise, sediment budget and human interference on coastal change. Finally, the idea of dealing with uncertainties in ^{14}*C data when using them as sea-level index indicators has gained ground. It is now generally accepted that Holocene sea-level reconstructions for the southern North Sea have regional validity only. This is caused by regional differences in land movement and palaeogeographical evolution. For the latter half of the Holocene, it has become clear that regional coastal changes have also been caused by human interference rather than by sudden fluctuations in RSL. Finally, the uncertainty in exact age and height of RSL makes it very problematic to speculate on small real RSL fluctuations, as we shall see.*

* I would like to thank my co-editors of this volume for their invitation to write this paper. Especially Guus Borger, Adrie de Kraker and Erik Thoen made me realise that there is a need for such a paper. Over the past years, I also had many discussions with colleagues from several institutes working on coastal development and sea-level history of the southern North Sea. I especially thank Cecile Baeteman, Friederike Bungenstock, Roland Gehrels, Patrik Kiden, Bart Makaske and Peter Vos on this matter. They were always willing to share their ideas and provide their contributions in reports with difficult accessibility. I thank Cecile Baeteman for her critical review and valuable comments that have improved the manuscript. Marjolein Haars (re)drew Figs. 8.3–8.5, 8.9 and edited the other figures except for Fig. 8.14. Klaas van der Veen drew the block-diagrams of Fig. 8.13 Friederike Bungenstock drew Figs. 8.8, 8.10 and 8.15 the latter based on a slide used in a presentation by Prof. dr. Paolo Orru (University of Cagliari) in 2009. Finally, I thank the Cultural Heritage Agency of the Netherlands for giving me permission to visit the 2010 Ghent Conference, to be a member of the editorial committee of this volume and for granting me time to write this paper.

Landscapes or seascapes? The history of the coastal environment in the North Sea area reconsidered, ed. by Erik Thoen, Guus J. Borger, Adriaan M.J. de Kraker, Tim Soens, Dries Tys, Lies Vervaet & Henk J.T. Weerts, Turnhout, 2013 (*CORN Publication Series*, 13), pp. 145–173.

BREPOLS PUBLISHERS DOI: 10.1484/M.CORN.1.101552

I. Introduction

In the past decades, a wealth of Holocene sea-level reconstructions from all over the world has been published in peer-reviewed journals. If any general conclusion can be drawn from this vast amount of literature it is that RSL history of a given region is influenced by regional to local components that make worldwide correlation impossible and useless. Regional components have to be considered for every RSL curve. If the authors of a RSL curve are not fully aware of these influences, the sea-level curve they publish has to be rejected (see e.g. Baeteman *et al.*, 2011: 557–572).

Nowadays most of the sea-level literature is well known among earth-scientists and it has become part of university BA and MSc programs in (quaternary) geology or physical geography, at least in some countries. It is common knowledge within the sea-level community and amongst quaternary geologists and physical geographers worldwide and it has made its way into their textbooks. However, as already stated in Baeteman *et al.* (2011: 557) and Baeteman (this volume) when reading papers in journals dealing with archaeology, historical geography or history, it becomes clear that this 'common knowledge' is much less common than most quaternary geologists and physical geographers realise. Apparently, these research communities do not read the geological literature. Hence this paper. It is not meant to give an extensive overview of all the sea-level related literature of the past decades, nor to give a complete overview of all palaeogeographical reconstructions of the southern North Sea coastal areas (stretching from Sangatte in northern France to Esbjerg in Danmark). Its aim is merely to summarise the present state of the art with respect to factors that control RSL movement in general and their effects on coastal change, with an emphasis on the coastal lowlands of the southern North Sea region.

A short summary of relevant papers for the continental southern North Sea region will be followed by a discussion of the factors controlling RSL movement. Thereafter, the paper will explain how to deal with the uncertainties of RSL indicators for RSL history before evaluating the possible effects of the changing sediment budget and anthropogenic activities on coastal change.

II. A concise southern North Sea literature review

II.1. Regional sea-level curves

Several regional Holocene sea-level curves for the southern North Sea coastal zones have been published over the past decades. Several of these curves are summarised in Vink *et al.* (2007). They also present a map of the data-locations (Vink *et al.*, 2007: 3253) and a figure with the reviewed regional curves in one diagram (Vink *et al.*, 2007: 3263). For more details about these curves, the data and their location the reader is referred to Vink *et al.* (2007: 3251–3263, particularly 3253, Figure 1, 3263, Figure 2). For Belgium, Denys and Baeteman (1995) presented a reliable curve that spans the period between 9.5–3 cal. ka BP[1]. Kiden (1995) did the same for Zeeland (southwest Netherlands) from 8–4 cal. ka BP.

[1] Ages are presented here as they were given by the original authors. The ages are always inferred from ^{14}C analyses. Sometimes BP is used, meaning before 1950, uncalibrated. Cal. BP means

Figure 8.1 Research areas and locations mentioned in the text

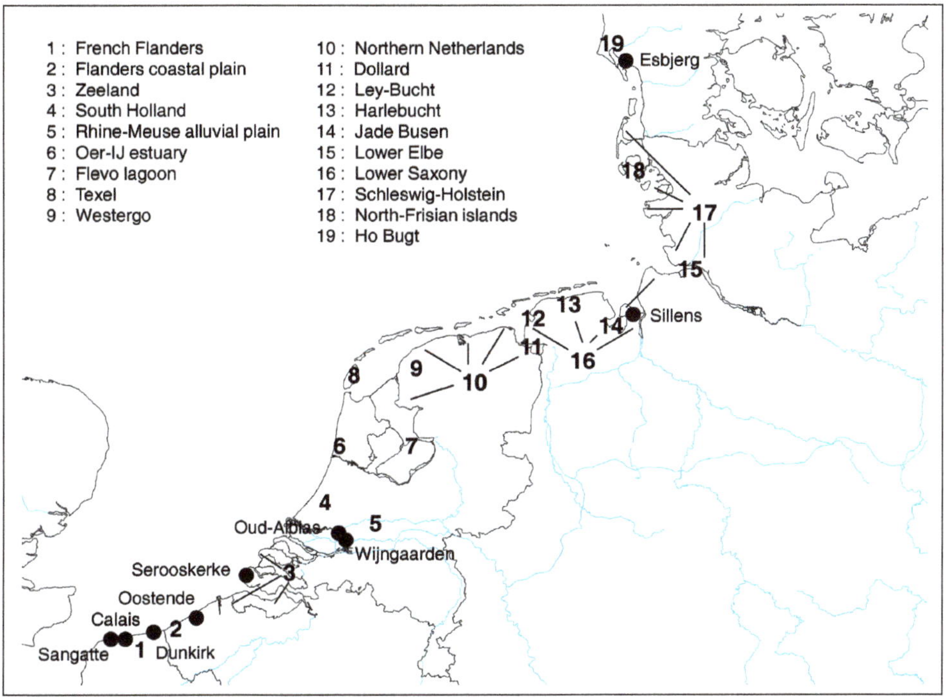

For the central Dutch coast, curves by Van de Plassche (1982; 1995) can be used for the period 8–2 cal. ka BP. For the Flevo-lagoon area in the central Netherlands, Van de Plassche (1982) and Van de Plassche *et al.* (2005) presented curves for 5300–1500 BC. For these curves data was collected from the present-day coastal zone. Behre (2003, 2007) published a sea-level curve with data from the entire German North Sea coast for 7500 BC to 2000 AD. This curve, however, should not be used. Baeteman (2007: 2–4) already argued that the curve is not reliable and discussed the data points and Baeteman *et al.* (2011) showed that many data points of his curve should not be used for sea-level reconstructions at all. Bungenstock and Weerts (2010) used Behre's data points in a 'first step' (Bungenstock and Weerts, 2010: 1687) to demonstrate that several regional curves for the German North Sea coast referring to smaller regions are necessary. To make sure that these regional curves are reliable, many new data points are necessary (Bungenstock and Weerts, 2012: 1087). They agree with Baeteman *et al.* (2011) that many of the Behre curve data points have to be rejected and cannot be used to construct the curves (Bungenstock and Weerts, 2012: 1088–1089). For the French Flanders coastal plain, Mrani-Alaoui and Anthony (2011:

that the ^{14}C analyses have been calibrated by one or other calibration program. These ages present 'true' years before 1950. BC and AD refer to Before Christ and Anno Domini. In these cases, the authors also used calibrated ^{14}C analyses. Ka: 1000 years.

Figure 8.2 Structural tectonic movement in the Netherlands in the last 2.5 million years
Note the striking differences in regional subsidence and rise in the southeastern part of the Netherlands along the active Peel Boundary Fault
(source: Westerhoff *et al.* 2003b: 172; modified after Brand *et al.*, 2002; Kooi *et al.*, 1998).

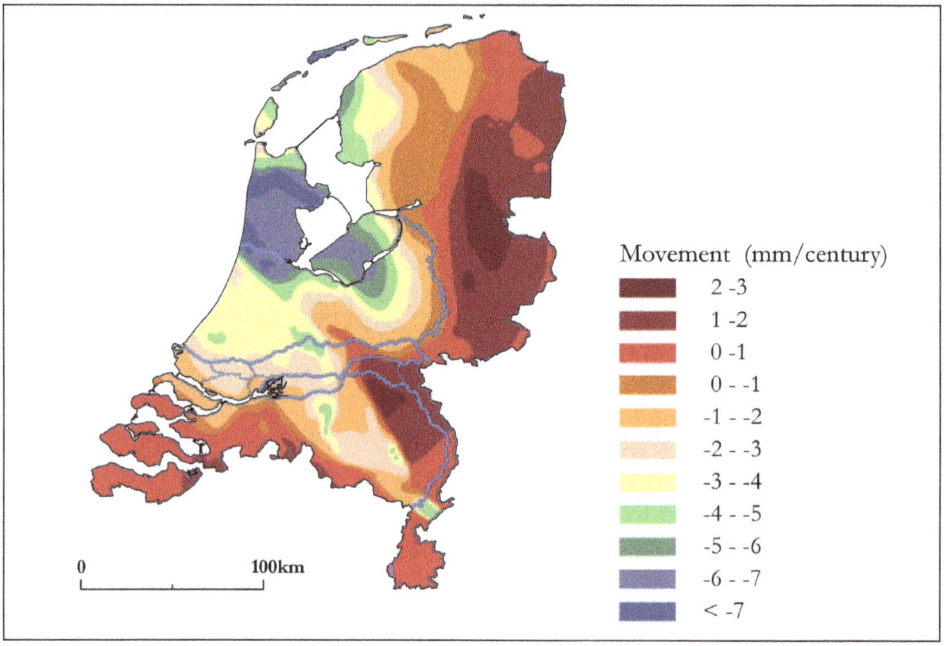

450) present an age-depth graph with radiocarbon dates between 5000–1000 BP of a palaeosoil, intercalated peat beds and shells, which, however, have no direct relation with sea-level. For the Danish Wadden Sea coast, two sea-level reconstructions have recently been published for the Ho Bugt Bay area near Esbjerg. Gehrels *et al.* (2006: 298) present a curve between 2500 BC and 2000 AD. That curve has been extended by Pedersen *et al.* (2009: 76) to 8500 Cal. BP. For the southern North Sea itself, data is more scarce and scattered. When comparing the existing curves, they all differ to a certain degree. The explanation for the differences will be explained in the following sections.

II.2. Regional palaeogeographical reconstructions of the southern North-Sea coastal zone

Palaeogeographical reconstructions of the southern North Sea coastal lowlands have been made over the past decades at di'fferent time scales and spatial scales using borehole information, temporary outcrops, and historical sources. In the sediment succession the alternation of marine clastic deposits and peat beds is striking. This led scientists to identify transgressions, when clastic marine sediments were deposited, and regressions when peat formation occurred (see e.g. Hageman, 1969: 377–379; Roeleveld, 1974: 15; Zagwijn, 1986: 14–23). The availability of ^{14}C-dating led them to identify periods of trans- and regressions (see Baeteman, 1981; 1999; Baeteman, this volume). A better understanding of the factors controlling coastal change since the late 1970s resulted in the abandonment of the use of the concept of trans- and regressions and their interpretation in the sense of sea-level fluctuations (at least by most of the coastal geologists). For this reason, the terms trans- and regression are no longer in use among coastal geologists. Where these terms have been used in the past as synonyms for land gain (regression) or land loss (transgression) it is better to replace them by the terms 'progradation' and 'retrogradation' respectively.

No detailed recent palaeogeographical reconstruction is available for French Flanders. Mrani-Alaoui and Anthony (2011: 451), however, hint at the erroneous interpretation of the palaegeographic evolution of a particular area and they stress the important role that changing sedimentary circumstances and the auto compaction of peat (caused by drainage) have played. For the western part of the Belgian coastal plain, a detailed palaeogeographical reconstruction was made by Baeteman (1981) and Baeteman and Declercq (2002) for the first half of the Holocene. In 1999 a preliminary map was produced for the periods 1500BP and the seventh to eighth centuries AD (Ervynck *et al.*, 1999: 103–106).

Such a detailed reconstruction is still lacking for the Belgian coastal plain between Oostende and the Dutch border. A nation-wide palaeogeographical map series for the entire Holocene for the Netherlands was published in Vos *et al.* (2011). This map-series is in part based on more detailed regional reconstructions. Examples of such reconstructions for the south-west Netherlands are Vos and van Heeringen (1997) for Zealand (entire Holocene), and Bos *et al.* (2011) for a part of Walcheren (500 BC-1250 AD). Vos and Eijskoot (2011) made a detailed reconstruction for a part of the coastal area of south Holland (3850 BC-1500 AD) and Vos (2008) did the same for the Oer-IJ estuary in North Holland (west Netherlands, 1500 BC-1000 AD). Detailed maps are also presented for the Rhine-Meuse alluvial plain by Berendsen and Stouthamer (2001) (central Netherlands, entire Holocene), extended by Hijma (2009) to the south Holland coast for 9100–6300 Cal. BP (Hijma, 2009: 184–189). Lenselink and Koopstra (1994) made reconstructions for the Zuiderzee – North-Holland Region (northwest and central Netherlands, 7000 BP-1350 AD), Woltering (1997) for Texel (northwest Netherlands, 1250 BC-1200 AD), Vos and Gerrets (2005: 71) for Westergo (Friesland, northern Netherlands, 600 BC, 300 BC, 0 AD, 500 AD) and Vos and Knol (2005: 130–135) for northern Netherlands (500 BC, 100 AD, 800 AD). The only widely published palaeogeographical reconstructions for the German North Sea coast are those of Behre (1999: 13) for the entire Lower Saxony coast at low resolution (0 AD, 800 AD, 1500 AD), Behre (1999: 15, 17, 19, 29) for the Dollard, Ley-Bucht, Harlebucht and Lower Elbe respectively (historical times, detailed), Behre (1999: 22–25) for the Jade Busen (0 AD – historical times, detailed), Meier (2004) for several small regions at the Schleswig-Holstein coast from Roman until late Medieval times and Meier (this volume) for the entire Schleswig-Holstein coast from 5000BC to modern times. No maps for the Danish North Sea coastal zones currently exist.

For the Netherlands in particular, many recent mostly local reconstructions for specific small timescales have been made during archaeological research. The number of these archaeological research projects has exploded since the European La Valetta treaty was ratified by the Netherlands in 1997 and was incorporated into Dutch legislation in 2007 (Anonymous, 2009: 107–109). Noticeably, this increase in projects has not been so large in the Belgian and German North Sea coastal zones. Although the enormous increase in the local reconstructions in itself is very positive, most of these local reconstructions are not readily accessible because their publication is limited to research reports. In the Netherlands, for instance, these reports can only be physically accessed through the library of the Cultural Heritage Agency of the Netherlands and, at best, through the DANS e-depot by query. The latter is not freely available to all.

II.3. Lithostratigraphy

The alternation of clastic marine deposits with peat beds led twentieth-century geologists to identify these beds as separate lithostratigraphical units. The first to publish on the stratigraphy of the Holocene deposits in the continental southern North Sea coastal zone were Dubois (1924) and Tavernier (1947: 332–343, 1948: 609–641). They distinguished a classical lower and an upper marine clastic complex in the Flemish French Flanders and Belgian coastal zone. These were the Calais and the Dunkirk Members, separated by a peat bed. Hageman (1969: 373–388) followed and extended their observations into a lithostratigraphical proposal in the Netherlands that was subsequently formally adopted by Doppert *et al.* (1975: 47–48). Unfortunately, the clear lithostratigraphical position of the two coastal marine complexes, separated by an often thick peat bed, was thought to be chronostratigraphically synchronous along the entire continental southern North Sea coast as well. When increasingly more ^{14}C data became available along the continental southern North Sea coast, this appeared not to be the case and substantial amounts of criticism arose about this combined litho- chronostratigraphical classification (summarised by Weerts *et al.*, 2005: 162–166). In the Netherlands, new lithostratigraphical terminology has been introduced to prevent any confusion and the classical Calais-Dunkirk subdivisions are no longer formal stratigraphical terms (Ebbing *et al.*, 2003: 1584–1586; Westerhoff *et al.*, 2003a: 313–317; Weerts *et al.*, 2005: 165–167). Although some authors still adhere to the Calais-Dunkirk scheme in Germany and Belgium (Behre, 2003: 21–40; Behre, 2007: 82–95; Gullentops and de Moor, 2001: 157), most scientists have stopped using it.

III. Factors controlling RSL movement

It is interesting to note that on the palaeogeographical maps mentioned in the previous section, the timing of coastal change with respect to shifting shorelines and infill of the tidal basins or back-barrier areas does not coincide for the different regions. These differences arise from the fact that the development of a coastal plain is a function of the following controlling factors: changing rate of RSL rise, sediment budget, and initially depending on the morphology of the pre-Holocene surface that reflects the original topography at the beginning of the post-glacial sea-level rise (Beets and van der Spek, 2000; Baeteman, 1999, Baeteman this volume). The relative impact of the controlling factors changed in the course of the Holocene and can be different for each region. In the latter half of the Holocene, also human activities are to be considered as a significant factor. The factors controlling RSL movements will be explained below. The changing impact of the controlling factors on coastal change will be illustrated with some examples from the Netherlands.

It has long been known that global and regional sea-level changes over geological timescales have been driven by several factors. These factors are not constant over space and time. Continental drift, growing and melting of ice caps, expansion or contraction of ocean water, regional glacio- and hydro-isostatic rebound, regional differences in structural uplift or subsidence, and compaction of soft sediments have all played their roles (see e.g. Berendsen, 2004: 229–237; Bowen, 1985: 158–168; de Bakker and Van den Berg (eds), 1982, and publications therein; Wilgus *et al.* (eds), 1988: and publications therein, Baeteman, Waller and Kiden, 2011: 557). This section summarises this knowledge without further references except where regional examples are used.

For every coastal location on earth, one can express RSL as follows:

$$\mathbf{RSL} = f(\mathbf{ESL} + \mathbf{LSM}) \tag{1}$$

Where: **ESL** = Eustatic sea-level

LSM = Land-surface movement

ESL and **LSM**, on their turn, are resultants from the effect of several factors:

$$\mathbf{ESL} = f(\mathbf{CD} + \mathbf{IC} + \mathbf{Ex}) \tag{2}$$

Where: **CD** = sea-level change caused by continental drift

IC = sea-level change caused by the amount of water that is trapped in ice caps

Ex = sea-level change caused by thermal expansion of ocean water

$$\mathbf{LSM} = f(\mathbf{T} + \mathbf{I} + \mathbf{C}) \tag{3}$$

Where: **T** = land-surface movement due to structural tectonics

I = land-surface movement caused by glacio- and hydro-isostasy

C = land-surface movement by compaction of soft sediments and oxidation of peat

The eustatic sea-level factors all operate on a global scale. Their time scales, however, are quite different. Sea-level change caused by continental drift acts on millions of years (see e.g. Wilgus *et al.* (eds.), 1988, and publications therein). Although it is irrelevant at Holocene time-scales, it is important to realise that it exists. In times when the continents were all contracted to one super-continent, the area of continental flat was small. This leads to a lower world-wide sea-level. When the continents drifted over the globe after the super-continent broke apart, the area of continental flat increased – and global sea-level rose.

The time-scale of sea-level change caused by the amount of water that is trapped in ice caps is shorter, ranging from hundreds to tens of thousands of years. We are living in a geological period of Ice Ages. In such periods, cold glacials alternate with warm

Figure 8.3 The concept of glacio-isostasy
(source: Vos *et al.*, 2011b: 17; modified after Ruddiman, 2001)

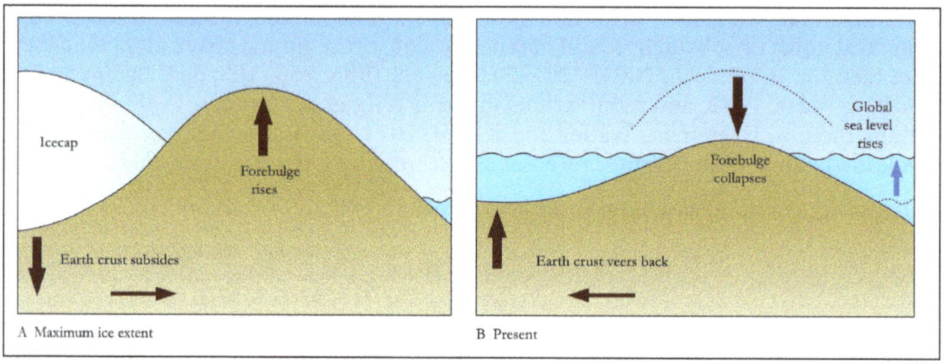

interglacials. In a glacial, more ice is trapped within continental ice caps than during an interglacial. A good example of this is the Laurentide ice cap that sits above northern America during an Ice Age. It is estimated that during the pleniglacial of the last Ice Age, some 20,000 years ago, world-wide sea-level was about 120 m lower than today because of the amount of water that was trapped in the ice caps.

Sea-level change caused by thermal expansion of ocean water acts faster, on time-scales between thousands to less than 100 years. The global sea-level rise we are currently experiencing is largely caused by the thermal expansion of ocean water. Ocean water gets warmer due to the global temperature rise and it consequently expands.

The land-surface movement factors all act on a regional to local scale, at different time-scales. Structural tectonic movement of the earth's crust acts on timescales from millions to hundreds of years. This also plays a role in the North Sea basin which has been tectonically subsiding since the Early Cretaceous (Ziegler, 1990). The North Sea basin itself is slowly subsiding whereas adjacent regions are tectonically stable or slowly rising. Along the present North Sea coast, there are large regional differences of land-surface movement due to structural tectonics. Figure 8.2 shows that the regional differences in subsidence / uplift for the Netherlands are in the order of about 1 cm/century. On a Holocene time-scale, this results in regional differences in land surface movements of over 1 m, a figure also observed by Cohen *et al.* (2002: 389–405) for areas on both sides of the Peel Boundary Fault in the Rhine-Meuse delta. Thus, for the southern North Sea coast, this factor is not negligible over Holocene time-scales.

Isostatical crustal adjustment is caused by the growing and subsequent melting of continental ice caps (glacio-isostasy) and by the variations in sea-level (hydro-isostasy). For the southern North Sea regions, the effects of glacio-isostasy are much larger than those of hydro-isostasy. As soon as an ice cap starts to form, its weight pushes the underlying crust down. This results in a flow of molten mantle material to the surrounding ice free areas, where the earth's crust is pushed up by this process (Figure 8.3). After the ice cap has melted the earth's crust that was pushed down by the ice cap slowly veers back up, molten mantle material flows back and the surrounding areas slowly and gradually subside. For an

Figure 8.4 Glacio-isostatic subsidence in the southern North Sea since the last glacial maximum
Highest values indicate strongest subsidence
(source: Vos *et al.*, 2011b: 15; modified after Steffen, 2006)

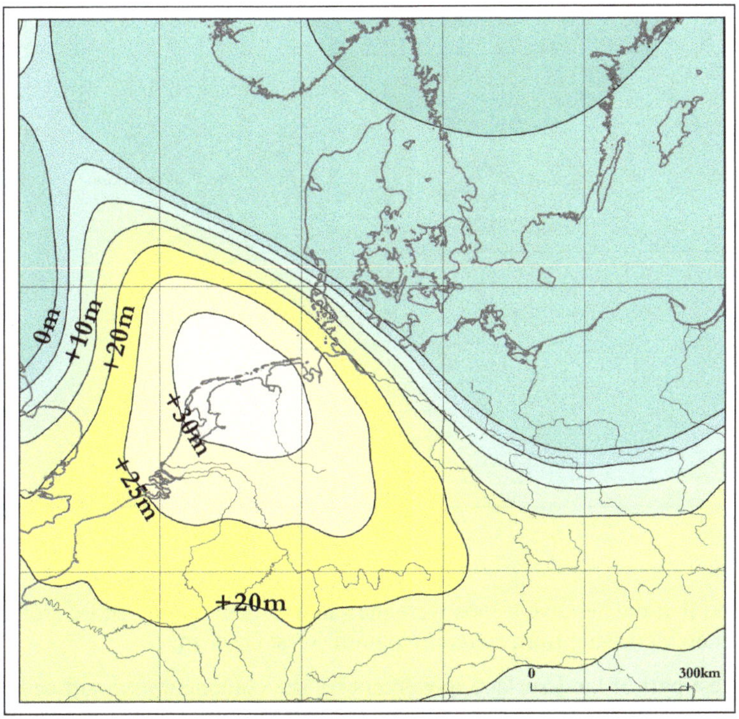

extensive review of the impact of glacio- and hydro-isostasy on southern North Sea sea-level reconstructions, the reader is referred to Kiden *et al.* (2002) and Vink *et al.* (2007). Glacio- and hydro-isostasy act over thousands to hundreds of years with, for the southern North Sea, regional differences of over 30 m since the last glacial maximum about 20,000 years ago (Figure 8.4). This reconstruction by Steffen (2006; reprinted in Busschers *et al.*, 2007: 3243) fits well with the depths of ^{14}C dated basal peat remnants in the southern North Sea and is thus likely to reflect the 'real' glacio- and hydro-isostasy. His model results are far more conservative for this region than the more well-known results of Peltier (2004: 111–149) who modelled 100 m of rebound. The differences in their results are caused by different assumptions concerning the viscosity of the earth's mantle in their geophysical models. Hydro-isostasy acts in the same way as gloacio-isostasy. An exposed continental shelf that is submerged by rising sea-level after deglaciation, will subside somewhat more by the weight of the water column upon it. When it gets exposed again during the next glaciation, it will veer back. For the southern North Sea, both glacio- and hydro-isostasy act in the same direction during glacials and interglacials. Vink *et al.* (2007: 3263) summarised all the known regional RSL curves for the continental southern North Sea coast in one diagram (Figure 8.5).

Figure 8.5 Regional Holocene relative sea-level reconstructions along the southern North Sea coast
(source: Vos *et al.*, 201b1: 17; modified after Vink *et al.*, 2007: 3263)

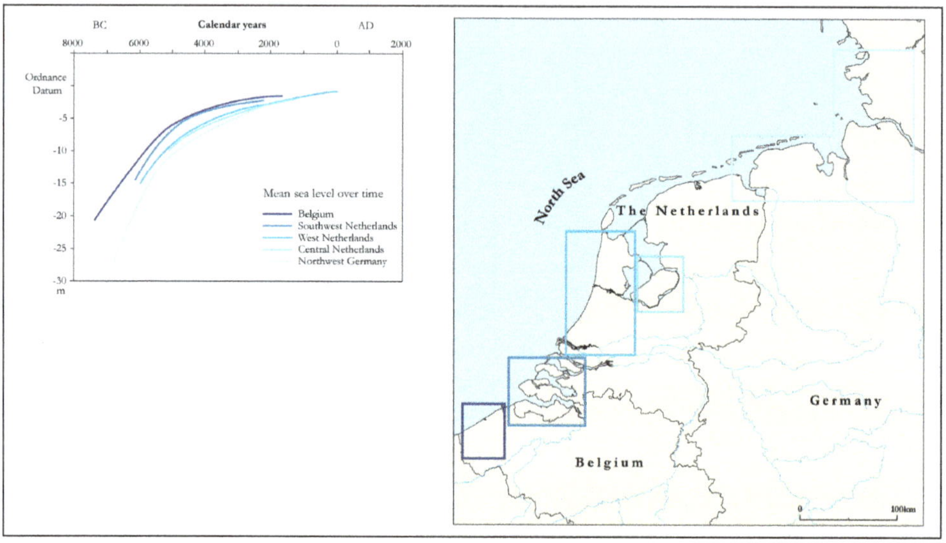

They showed that the regional trends were the same throughout the entire Holocene, with subsidence rate increasing from Belgium to north-west Germany.

The largest part of land-surface movement by the compaction of soft sediments and oxidation of peat along the southern North Sea coast is caused by the compaction of soft Holocene clays and the compaction and oxidation of Holocene peat, acting over thousands to tens of years. The effects of peat oxidation in particular may have catastrophic consequences. Peat consists of organic matter and water. As long as the organic matter remains under water, it is preserved. If the peat is drained, e.g. by digging ditches to make the peat surface suitable for agriculture, the organic matter is exposed to the air. This leads to oxidation of the peat; it simply evaporates into the air. This process is irreversible and it can lead to a lowering of the surface of several meters over centuries. An often forgotten component of land-surface movement in the southern North Sea region is that of compaction of deep buried Tertiary marine clays and older deposits. Figure 8.6 shows the land-surface movement caused by the compaction of Tertiary clays for The Netherlands. Figure 8.7 shows the total expected surface-land movement in the Netherlands between 2000 and 2050. The huge local differences in total land-surface movement are merely due to local differences of land-surface movement by compaction and oxidation of Holocene deposits.

The main conclusion of this section is that RSL rise always has a local and a regional component. In particular, this holds for the southern North Sea continental coast that is

Figure 8.6 Present-day surficial movement due to compaction of deep buried Tertiary clays for the Netherlands
(source: Westerhoff *et al.*, 2003b: 172; modified after Kooi *et al.*, 1998)

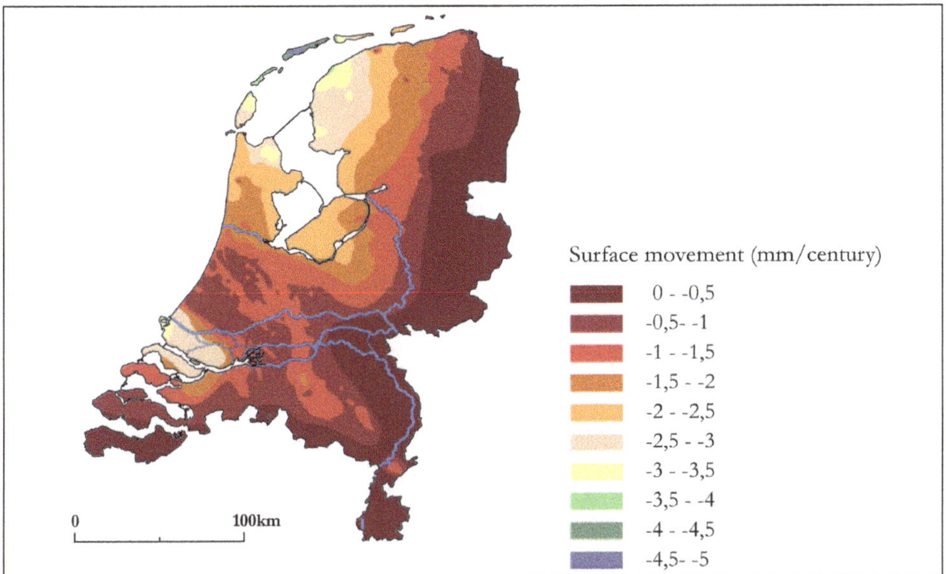

located in an active tectonic zone with huge differences in isostatic rebound and local compaction. This means that RSL reconstructions for the southern North Sea coasts should always be carried out in relatively small areas that account for these regional to local differences.

IV. Factors influencing local Mean High Water

When reconstructing sea-level history, it is essential to know which water level is reflecting/representing the dated sample, particularly in tidally influenced coastal areas. This is particularly true for the use of basal peat that is an excellent sea-level indicator. Holocene basal peat started to accumulate when the groundwater in a coastal region had risen to the Pleistocene surface, caused by sea-level rise. The Pleistocene surface is usually considered to be free of compaction. However, as discussed above, their present-day absolute depth may be different from the formation depth due to isostastic and tectonic movement and by the ongoing compaction of underlying deposits. Apart from this, it should be realised that the basal-peat level in tidally influenced coasts is not equivalent to local mean RSL, but to local mean high water (MHW) (Jelgersma, 1961: 17–22; Roeleveld, 1974: 29; Van de Plassche, 1982: 17–28; Vink *et al.*, 2007: 3251, Baeteman *et al.*, 2011).

Figure 8.7 Total expected LSM of the Netherlands between 2000 and 2050 (source: De Mulder and Ritsema, 2003: 67; modified after Brand *et al.*, 2002)

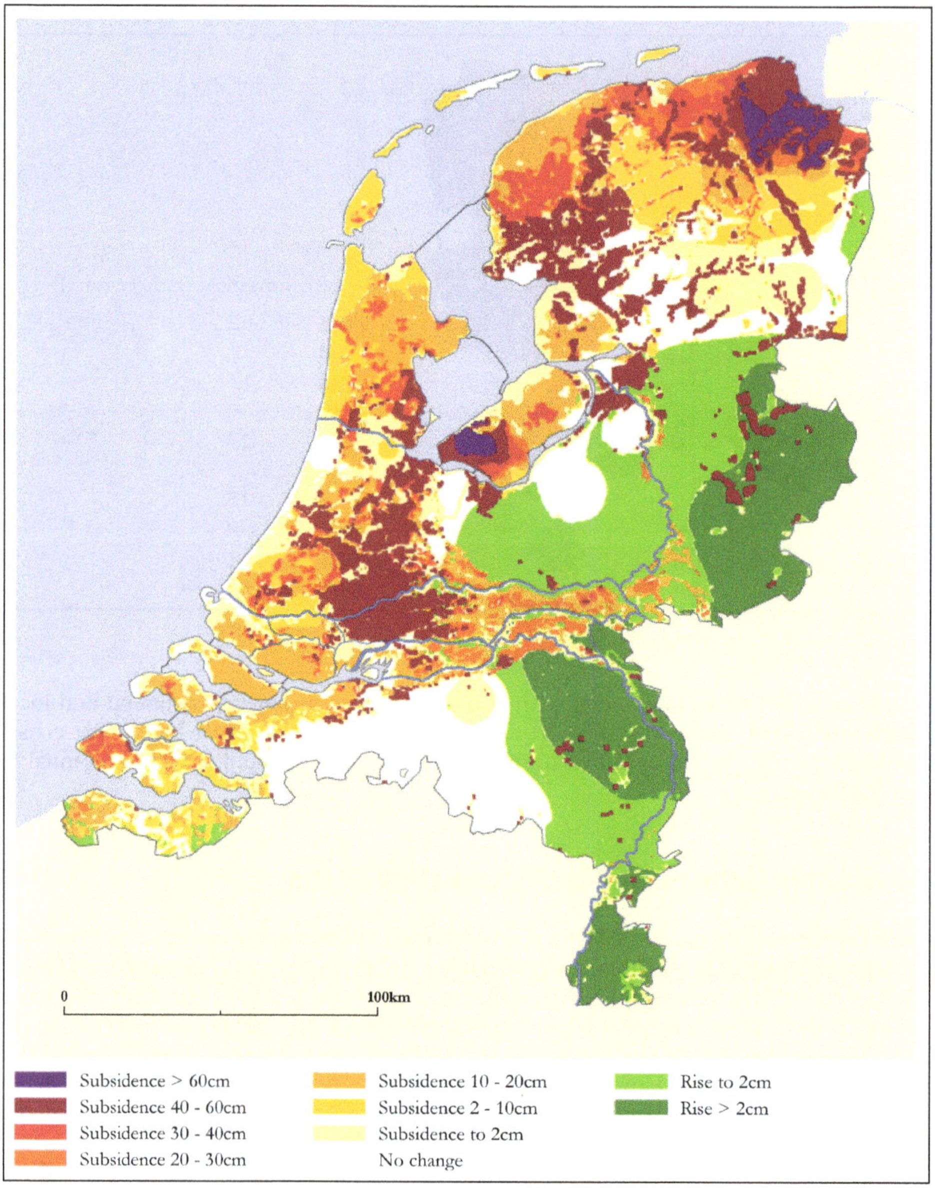

Figure 8.8 Present-day variability in storm surge heights in the North-Frisian islands, Germany
Along the coast of Langeness, the difference in surge height is 40 cm. Difference in surge height between Nordstrandischmoor and the westernmost tip of Langeness is 75 cm, over a distance of only 20 km
(adapted from Bungenstock and Weerts, 2012 from data kindly provided by the Schleswig-Holstein Agency for Coastal Defence, National Park and Marine Conservation)

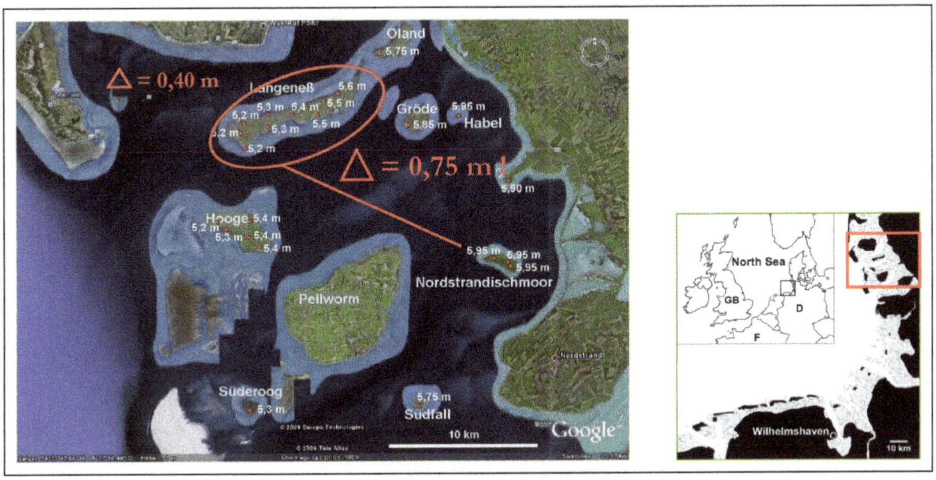

Furthermore, this local MHW can be either higher or lower than real MHW at the coast. This has several causes that vary over space and time:

- Tidal amplification, or the *estuary effect*. This is often seen in estuaries (Fairbridge, 1961; Vink *et al.*, 2007: 3269–3270) and it can also play a role in tidal basins. It is caused by confinement of the tidal wave in a funnel-shaped embayment. This leads to higher daily mean high water where the estuary or tidal basin gets narrower. Figure 8.8 illustrates this effect for present storm surge heights in the North-Frisian islands, Germany. Note that this figure presents storm surge heights and not daily MHW, so the order of magnitude of the differences is higher than for daily MHW. Due to the exposure of the marshland islands to the sea and the wave surge caused by the local coastline morphology, differences of up to 0.75 m in height can be observed over a distance of only 20 km (Bungenstock and Weerts, 2012: 1084–1085). To complicate matters further, this effect changes across time with the changing morphology of the embayment over time. The present estuary effect (funnelling) in an estuary or tidal basin does not necessarily reflect the effect in the past; it is therefore impossible to extrapolate this effect over time – a problem that has not been solved yet and requires further research.

- The *floodbasin effect (kombergingseffect)* is the 'upstream' decrease in the tidal wave (and so, local MHW) by frictional dissipation of tidal energy in large tidal basins and/or by changing morphology of tidal inlets between barrier islands (Zonneveld,

Figure 8.9 Apparent changes in RSL caused by regional landscape changes that influenced MHW.
(A): floodbasin effect; (B): avulsion effect (modified from Makaske *et al.*, 2003: 128 (A), Kiden *et al.*, 2008: 58 (A), 60(B))

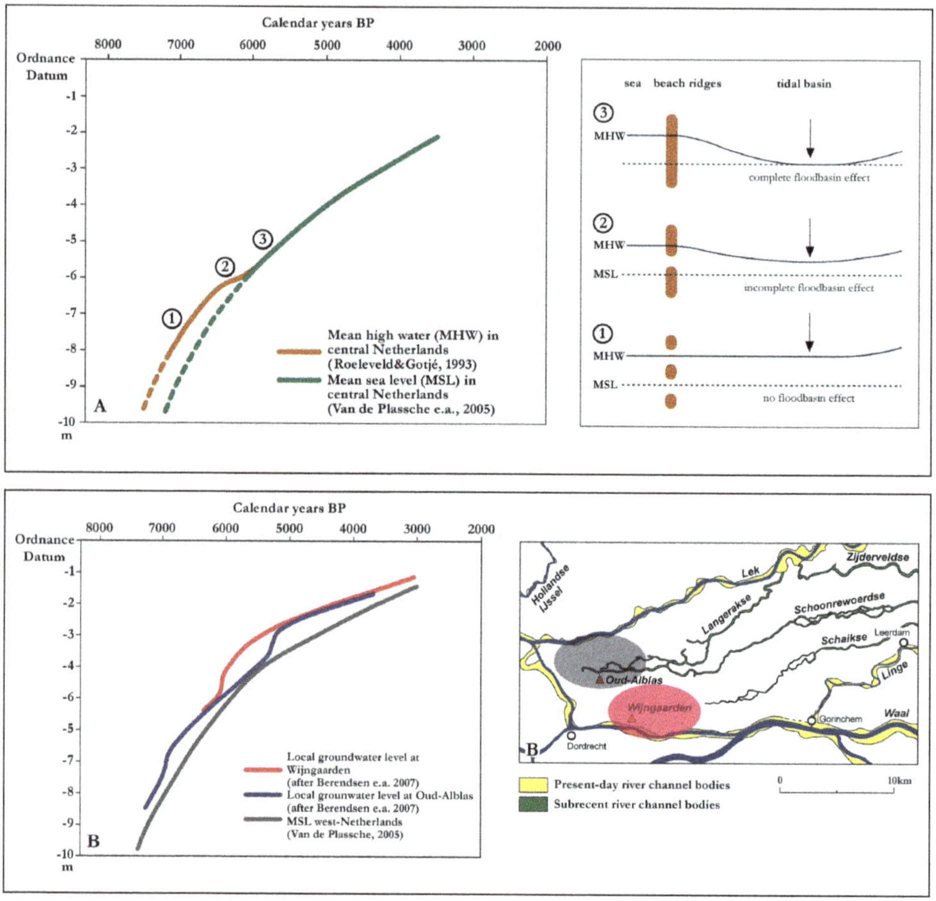

1960: 160, 373–377; Van de Plassche, 1982: 47–48; 1995: 113–128; Makaske *et al.*, 2003: 128; Kiden *et al.*, 2008: 58–59). Figure 8.9 (A) schematically shows this effect for the former Flevo tidal basin in the Netherlands. At t=1, the barrier islands at the coast are small and the tidal inlets wide. The entire volume created by the tidal wave can enter the tidal basin behind the barriers. In this situation, local Flevo MHW equals coastal MHW. At t=2, the barriers have started to grow. The inlets have become too small to permit the entire tidal volume to pass through. This leads to a 'sudden' fall in local Flevo MHW, that is now lower than coastal MHW. At t=3, the barrier islands have joined together to a closed barrier coast. The tidal wave no longer enters the Flevo basin. This process results in a dip in the Flevo MHW curve. This dip might be misinterpreted as a temporary decrease in relative sea-level rise while it is actually caused by sedimentary processes at the coast many kilometres to the west of the Flevo basin. Comparable situations might have occurred along the southern North Sea coast elsewhere. So, one has to know the palaeogeographical development of a given region in order to explain features in curves such as in Fig 8.9 (A). Also note that in closing tidal basins with a river outlet the end-situation of Figure 8.9 (A) might reverse into an estuary effect with funnelling, leading to even higher local MHW than at t=1 in Figure 8.9 (A). In such a case, local MHW first drops due to the floodbasin effect and then rises again due to the estuary effect.

- The *river gradient effect*, (e.g. Louwe Kooijmans, 1974: 57–61; Cohen, 2005: 341–364) refers to a gently sloping groundwater surface in river deltas and estuaries. It leads to an increase in local MHW altitude in a longitudinal upstream direction along the river. This is shown in Figure 8.9 (B). MHW in the western Netherlands at the coast is lower then local MHW at Oud-Alblas and Wijngaarden. These two locations are situated slightly upstream in the Rhine-Meuse delta.

- Figure 8.9 (B) shows yet another effect that influences MHW curves in river deltas, not mentioned by Vink *et al.* (2007): the *avulsion effect* (Kiden *et al.*, 2008: 59). The MHW curves at Oud-Alblas and Wijngaarden show sudden jumps that might be (mis) interpreted as sudden relative sea-level jumps. What actually happened is that many kilometres upstream from the areas, river avulsions of the Rhine formed new river courses in the delta – in this case at 6200, 5300 and 4500 calendar years BP. These new rivers did not have well developed natural levees yet, causing permanent flooding of their downstream floodbasins. This led to a sudden MHW rise in these areas. The 'jump' at Wijngaarden and the two 'jumps' at Oud-Alblas correspond well with the onset of the 'Schaikse' Rhine-branch and the 'Langerakse' and 'Schoonrewoerdse' Rhine-branches, respectively (Berendsen and Stouthamer, 2001: 213, 232–234).

Whenever interpreting (or constructing) MHW curves, it is important to realise the role these effects can play.

Archaeological and historical data is also often used for the reconstruction of regional RSL curves of the southern North Sea coasts. Because a wealth of this data is available for along the continental southern North Sea coast, this is understandable. However, the context of the artefacts and/or the historical sources in relation to sea-level must be considered carefully.

In the first place, we tend to think that prehistoric people reacted the same way we do to the threats from the sea. This assumption may not be right. To cite Eichfeld (2010: 273): '…there is no such thing as 'environmental determinism' governing people's reactions to environmental change.' He compared the height of neighbouring first- and second-century AD dwelling mounds and found that there was no strict causality between environmental change and the height of the dwelling mounds (Eichfeld, 2010: 278–281); Figure 8.10. This led him to the conclusion that the heights and frequencies of prehistoric storm surges have not necessarily led to adequate adaptations in dwelling mound heights. In Lower Saxony in the first and second centuries AD, life expectancy was rather short and knowledge of the past was passed orally between generations (Kossack, 1995: 37). Also, 'To remember a storm surge does not imply knowledge about the height of the storm surge' (Eichfeld,

Figure 8.10 Different heights of "Wurten" (dwelling mounds) in Lower Saxony. The Sillens dwelling mounds are located within 1 km from each other on a salt marsh open to the North Sea (modified after Eichfeld, 2010)

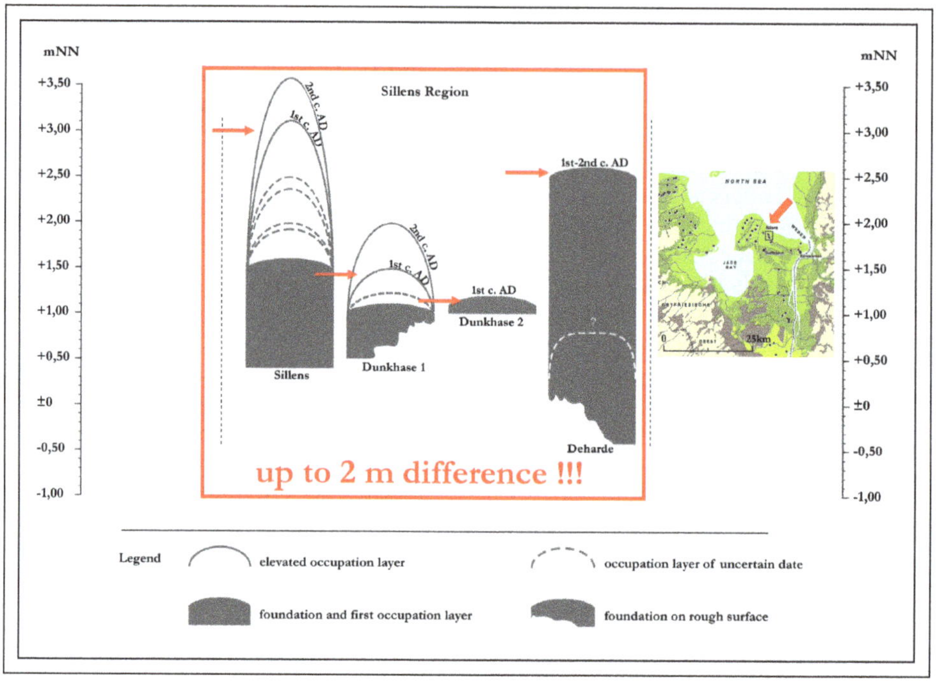

2010: 281). Furthermore, prehistoric disaster precautions might have been taken in the religious sphere rather than in the material world (Eichfeld, 2010: 282). So, one farmer might immediately heighten his dwelling mound above the surge flood level, while his neighbour offered his youngest daughter to the marsh to please the Gods, trusting that no future floods would occur (Hauke Jöns, personal communication, 2011). On the other hand, a local chief might have viewed the height of his dwelling mound as a symbol of power (cf. Medieval moated site). Its height symbolised his power – and had nothing to do with storm surge heights. This all illustrates that the height of prehistoric dwelling mounds as storm surge height indicators in RSL curves can only be justified if large error bands are used. Their heights can not be used in an absolute way.

Some authors argue that the height of *Flachsiedlungen* (settlements constructed directly upon salt marshes) and the height of old soil horizons in salt marsh deposits can be used as RSL indicators (Behre, 2003: 12, 17, 50–55; Behre, 2007: 83, 89). This too is wrong. These *Flachsiedlungen* and soil horizons are all situated on (or in) salt-marsh deposits prone to compaction that is enhanced by compaction of the underlying soft Holocene clay and peat. The actual site compaction can only be estimated if detailed cores reaching to the top of the Pleistocene sub-surface are available. Even if this is the case, the uncertainty in the rate of compaction is large. Large error bands correcting for the estimated compaction

Figure 8.11 From RSL index points to RSL index areas.
Horizontal axis: age, oldest to the left. Vertical axis: depth, deepest at the bottom. See text for further explanation

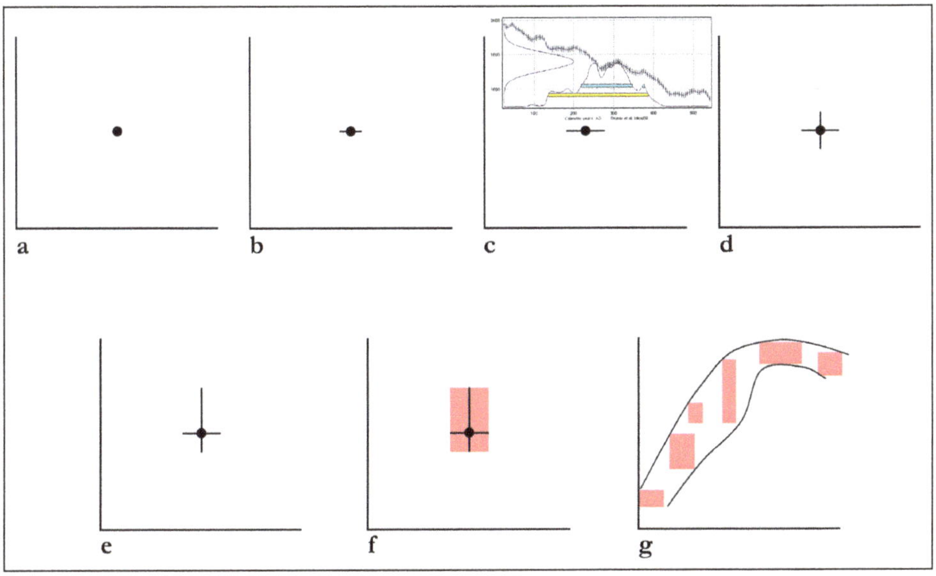

are necessary if one uses these *Flachsiedlung* heights in relative sea-level reconstructions (Baeteman *et al.*, 2011: 565; Bungenstock and Weerts, 2012: 1084–1085). Even then, this data has to be handled with care as compaction is a source of error that cannot yet be quantified, despite attempts to use decompaction models that are calibrated against field data (Baeteman *et al.*, 2011: 565 and references therein).

V. Uncertainties related to age determination

Another uncertainty that occurs in a sea-level curve is that of the age determination itself. The most common measurements are radiocarbon (^{14}C) datings, but archaeological 'datings' (e.g. pottery sherds) or OSL (optically stimulated luminescence) datings are also used. All these datings suffer from the same kind of uncertainties, albeit to a different degree.

A radiocarbon date is always given in years Before Present (BP), i.e. before 1950. Figure 8.11a shows a plot of an imaginary ^{14}C dating in a graph with age versus depth. Every date has an uncertainty around it, usually given in 1 sigma (66% certainty) or 2 sigma (95% certainty). That is the uncertainty of the dating itself (Figure 8.11b). Such a measurement uncertainty also is given for laboratory OSL datings. The age uncertainty of pottery differs from region to region. The ^{14}C content in the earth's atmosphere slightly varied in the past due to variations in solar activity. Every ^{14}C dating gives a 'solar' age that has to be translated in a real age. This can be achieved by tree-ring calendars but

Figure 8.12 Progradation along the central Dutch coast from 3850 BC to 1500 BC.
RSL was slowly rising during this entire period
Blue: North Sea, tidal inlets and channels. Light green: tidal flats and salt marshes. Yellow: beach ridges, beach plains and dunes. Brown: peat (source: maps from Vos *et al.,* 2011a: 47, 51, 55)

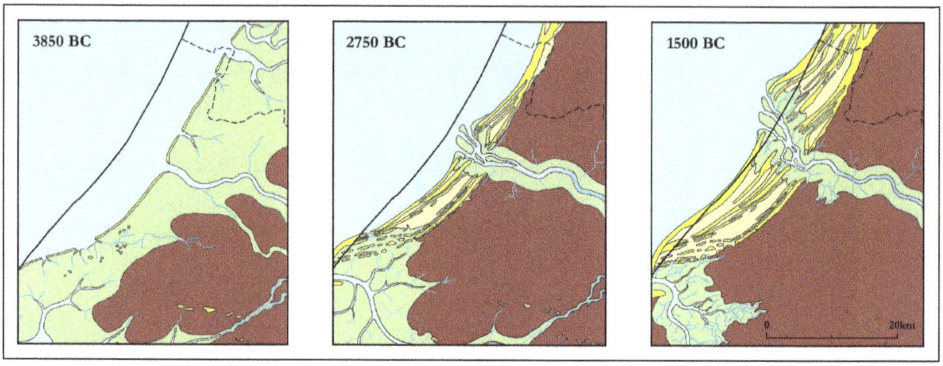

such calendars only have regional validity. Many tree-ring calendars exist and several programs are available to calibrate a ^{14}C solar dating into a calendar dating. For north-west Europe, Intcal 09 (Reimer *et al*., 2009). is often used. Usually, calibration leads to another uncertainty range that is often asymmetrical (Figure 8.11c). OSL datings also suffer from this kind of uncertainty, because assumptions have to made regarding the degree of bleaching and burial history (groundwater saturation, clay and feldspar content of neighbouring deposits, see Wallinga *et al.* (2007: 179–196)) and references therein.

We now have to add the vertical uncertainty to the plot of the dating location. This uncertainty stems from uncertainties in the absolute height of the sample due to coring methods (Figure 8.11d) and from possible post-depositional compaction (Figure 8.11e). If we quantify the uncertainties in height and age of a single dating, we get an error box like that seen in Figure 8.11f. Several of these boxes in a graph might result in a curve like Figure 8.11g. It is clear that it is very hard to prove decimetre to meter scale RSL (or MHW) fluctuations from such a curve, let alone relating such fluctuations to global sea-level fluctuations. Centennial-scale sea-level fluctuations related to climate fluctuations have a maximum amplitude of the order of decimetres. They are certainly not greater than about 0.55 m which is within the resolving limit of existing methodology (Baeteman *et al.*, 2011: 569 and references therein). Finally, the relation of the sample height with respect to RSL or MHW has to be considered. What does the position of, for example, a ^{14}C dated fish weir tell us about local RSL at that time? Such considerations have to be made for each individual point of the curve and this often leads to additional vertical uncertainty.

VI. The relative impact of RSL, sediment budget, and human interference in coastal change

It is still not sufficiently acknowledged amongst non-coastal geologists that a rising RSL in a region does not automatically lead to a retrograding coastline and a falling RSL to a prograding coastline. A coastal area can experience a progradation (seaward shift of the coastline) while RSL is rising slowly. The palaeogeographical maps of the Netherlands produced by Vos *et al.* (2011a: 39–79) illustrate this clearly (Figure 8.12) (see also Baeteman, this volume). What are the possible causes of this intuitive contradiction? This can be expressed in a simple formula:

$$\mathbf{Pr} = f(\mathbf{vRSL} - \mathbf{vS}) \qquad (4)$$

Where: **Pr** = progradation (seaward shift of the coastline) or retrogradation (landward shift of the coastline) in a given region
vRSL = the volume in a region over a given amount of time created by RSL rise (positive or negative). This volume is often referred to as the accommodation space
Note that this includes volume created by all the factors mentioned in Section II, including in situ compaction of soft deposits and oxidation of peat
vS = the volume of sediment delivered to that region or removed from that region by sediment transport in the same amount of time
v may be expressed in km^3 and it can have either positive or negative values

If the volume of sediment delivered to a given region over a given amount of time is larger than the volume created by RSL rise in that region over the same amount of time, the region will experience a seaward shift of the coastline. If it is smaller than the volume created by RSL rise, the region experiences a landward shift of the coastline. Both the volume of sediment delivered to a given region over a given amount of time and the volume created by RSL rise in that region over the same amount of time differ over time; the volume created by RSL rise is due to differences in RSL rise, the volume of sediment delivered to a given region over a given amount of time is due different sediment budgets over time.

In the time slices shown for the region in Figure 8.12 (a part of the central Dutch coast), the volume of sediment delivered to that region surpassed the volume created by RSL rise in that region over the same amount of time. This led to a closing of the coast followed by progradation in spite of the slowly rising RSL for this region at that time. Most of the sediment delivered to the coast of Figure 8.12 came from the North Sea floor, transported there by currents and wave action. A smaller amount of sediment was delivered by the Rhine, whereas peat bog formation filled the accommodation space behind the coastal barriers (Beets and Van der Spek, 2000: 8–14). Due to the closing of the coast and the following progradation, the landscape of the coastal zone changed significantly (Figure 8.13). The examples of Figures 8.12 and 8.13 are typical for shallow seas with a gently sloping sea-floor. When the sea-floor slope is steep, more sediment is needed to move the coastline in a seaward direction because most of the volume of sediment is stored under water. With the same amount of sediment delivered to a region, the net coastal progradation area will be much smaller in that case.

Figure 8.13 The changing landscape of the Holland coast between 3850 (left) and 1500 BC (right).
Note that RSL was continuously rising in that time (source: Vos *et al.*, 2011b (a): 15; Weerts *et al.*, 2011 (b): 20)

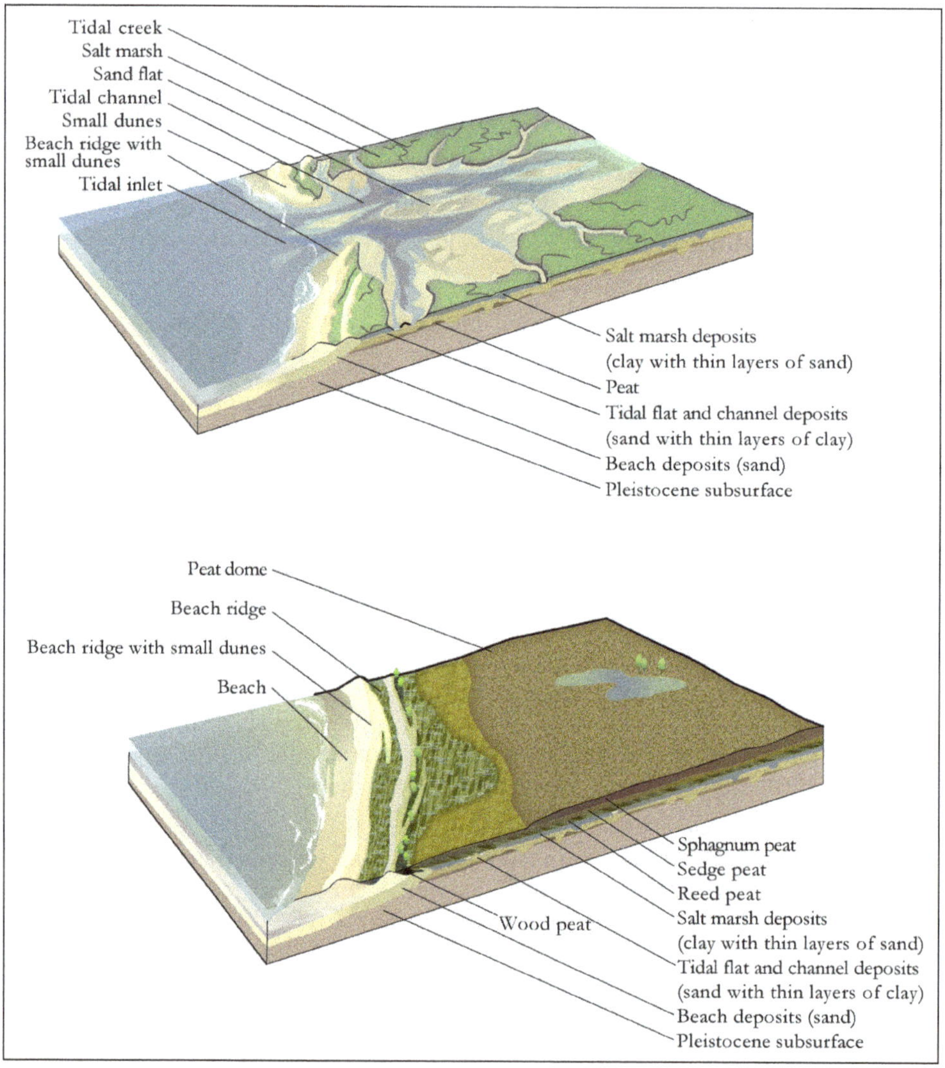

Once humans actively start enhancing drainage in the coastal zone, compaction of soft clay and peat and oxidation of peat increases. Traces of prehistoric ditches are often found in the Dutch coastal zone. The oldest ditches date from the Bronze Age. In the Iron Age and the Roman Age, ditches were common features in the coastal zone and in the

Figure 8.14 Intensified land-use in Zeeland led to unintended land subsidence with catastrophic results from an anthropogenetic point of view. A-C: drowning of the rural landscape at Serooskerke (Walcheren, Zeeland, source: Bos *et al.,* 2011: 48). Between 100 and 800 AD the entire province of Zeeland changed from a peat area behind a closed dune coast to an area of intertidal flats disected by tidal creeks, while by 1500 AD most of the area had silted up again (D, E, F) (source: Vos *et al.,* 2011a: 63, 67, 71). Legend D, E, F, see figure caption of Figure 8.12. Additional units: Green: embanked coastal plain. Red dots: towns with city rights (1500 AD only). The yellow areas in the southeastern part of the map represent higher Pleistocene grounds.

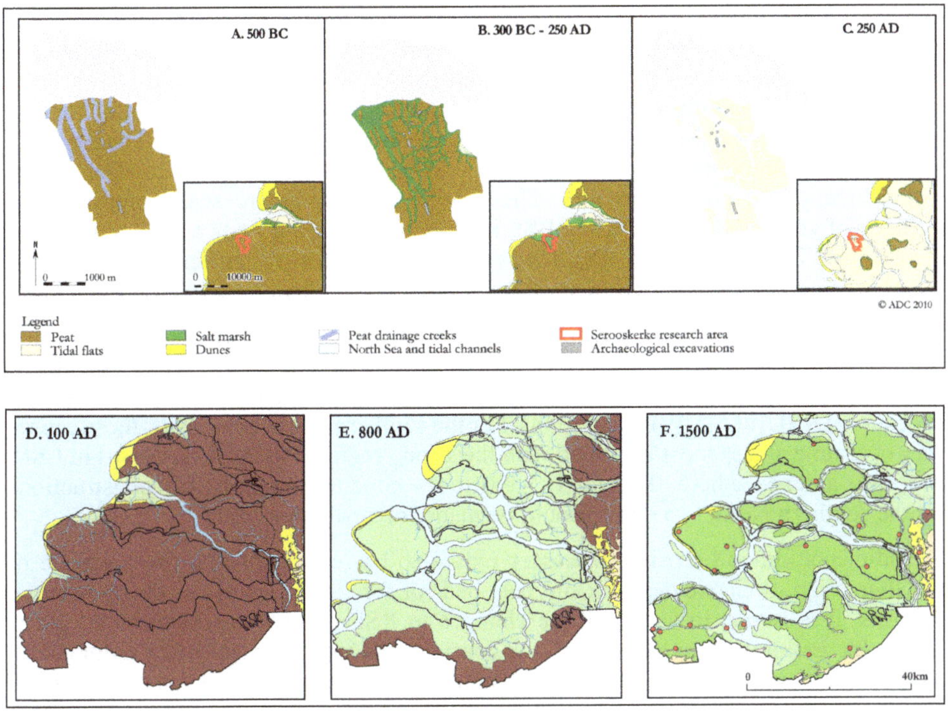

Rhine-Meuse alluvial plain (Van Londen, 2006: 25–41; 59–71; 75–91; 100–102; 132–138; 183–186; Vos, 2009: 8; 92–116; Lascaris and de Kraker, this volume). Sometimes, even remnants of Iron Age or Roman Age dikes are found in excavations (see Lascaris and de Kraker, this volume). All these traces have been preserved under younger marine deposits that cover them. A fine example is given by Bos *et al.* (2011: 35–58). The authors describe how an Iron Age rural landscape with many ditches in a peat area at Serooskerke (province of Zeeland) gradually drowned and changed into a landscape with tidal channels and sand flats (Figure 8.14A-C). They attribute the drowning to drainage by ditches and the subsequent compaction and oxidation of peat. The entire province of Zeeland was lost to the sea by this process between 100 AD and 800 AD (Figure 8.14 D, E) as the compaction and oxidation of peat due to human activities led to significant subsidence. Hence, the tidal system could enter the area because it became lower than the MHW level. The area was then changed into an intertidal area with sand flats dissected by tidal channels and creeks. With time, sufficient sediment was supplied, and the area silted up and eventually changed again into a salt marsh above mean high water level that was subsequently embanked. This seems to have happened elsewhere along the North Sea coast as well (see also Rippon, this volume, for a general description of this process).

VII. Conclusions and possible future research directions

This article has shown that RSL at a given coastal location is a function of eustatic sea-level (ESL) and land-surface movement (LSM). ESL is a function of sea-level changes caused by continental drift, by the changing amount of water that is trapped in ice caps and by the thermal expansion or contraction of sea-water. LSM depends on structural tectonical movement, on glacio- and hydro-isostasy and on compaction of soft sediments and oxidation of peat. These factors operate on timescales ranging from millions of years to decades. While ESL is global, LSM always has a local to regional component. Therefore, every RSL reconstruction has local to regional validity only.

Fluctuations in regional RSL curves may be caused by real global sea-level fluctuations, although recent research has shown that on a centennial scale such fluctuations are in the decimetre scale at most. Regional causes such as the estuary effect, the floodbasin effect, the river gradient effect, and the avulsion effect probably have a much larger contribution to observed RSL (or MHW) fluctuations. For the southern North Sea coasts, these causes vary over time and space. Combined with the strong regional to local variations in LSM over time and space here, this clearly leads to the conclusion that RSL reconstructions for the southern North Sea coast should be made for small areas only.

Sedimentation and erosion processes may also be responsible for the presence of the alternation of clastic deposits and peat beds or for significant facies changes in the deposits. It is wrong to ascribe these changes to RSL fluctuations (Baeteman, 2008: 1226). Increasing human land-use in the coastal zone has led to an (often unintended) anthropogenic impact on these sedimentation and erosion processes (Vos *et al.*, 2011a; Knottnerus, this volume; Meier, this volume; Rippon, this volume; Waller and Long, this volume). This should be borne in mind when reading texts that deal with Holocene coastal zone sediment successions and processes and mechanisms that caused them.

The use of archaeological proxy-data as RSL indicators should also be handled with extreme care because most are not compaction free. Others, such as the height of dwelling mounds, cannot be interpreted safely as indicative for palaeo storm-surge heights. Cultural aspects may also have played a role in adjusting the height of the dwelling mounds to storm surge levels, as may the egos of our prehistoric ancestors.

Finally, the sea-level proxy data itself always has intrinsic uncertainties in age and height that should be quantified when plotting this data in a graph. Sea-level data points without their uncertainties cannot be interpreted unambiguously.

Figure 8.15 Which of these gentlemen is right?
(drawing by Friederike Bungenstock, based on a slide used in a presentation by Prof. dr. Paolo Orru (University of Cagliari) in 2009. Unpublished, used with permission of the authors)

So, if we look at Figure 8.15 which of these gentlemen is right? From the cartoon alone, we cannot tell. We do not know how their curves were constructed. The location and size of their areas is also unknown. What do their curves represent? Local RSL or regional RSL? Is the gentleman to the right aware of the fact that the fluctuations in his curve are likely to be caused by local effects or does he think that they represent global sea-level fluctuations? Has the gentleman to the left eliminated index-points that do not fit to the continuously rising curve, just because he knows that there is no mechanism for recent sudden global sea-level changes – without being aware of changing local circumstances that may explain them? For every sea-level curve that is presented in literature, we have to make sure that we understand what it represents. And we have to make sure that all influencing factors acting on different timescales and spatial scales have been interpreted correctly. Only if this is the case, we can safely use a sea-level curve for further interpretations.

Because of our better understanding of coastal change and sea-level change, the curves published before about the mid-1990s have to be considered critically. Not all of the knowledge of how the controlling factors discussed above may influence RSL dating from the past decades may have been included in the final representation of the older curves. Archaeologists and historians often have used these older curves to explain changes in palaeogeography. Also the role of human impact on coastal change has been neglected far too much in the older literature.

For a better understanding of the palaeogegraphical development of the southern North Sea coast, several future research directions should be followed:

1. True interdisciplinary research. With palaeogeography of the coastal zone being the resultant of natural causes and human interference, cooperation between earth scientists, archaeologists and for the youngest periods historians and historical geographers is necessary. In Belgium and The Netherlands, this cooperation already exists in the many – albeit small-scale- 'La Valetta' excavations (see for example Dijkstra and Zuidhoff

(eds, 2011, the original source of Figure 8.14a-c). In Germany, the recent Jadebusen Projekt of the Karl von Ossietzky Universität Oldenburg and the Institut für Chemie und Biologie des Meeres (ICBM) (http://www.jade.icbm.de) serves as an example.

2. For those areas that still lack good Holocene palaeogeographical reconstructions, these should be made (e.g. French Flanders, the eastern part of the Belgian coastal plain, Lower Saxonian coast, Danish North Sea coast, Schleswig-Holsteinian North Sea coast for the first half of the Holocene). The same holds for RSL reconstructions. Only some small-scale RSL reconstructions exist for the entire Wadden Sea area.
3. Regional and local effects of Holocene compaction, floodbasin effect, estuary effect and avulsion effect should be assessed.
4. Results of the many small-scale 'La Valetta' research projects should be systematically used to refine or adjust existing larger-scale regional palaeogeographical reconstructions. Unfortunately, much of the 'La Valetta' reports are hard to access. These smaller research projects often contain data that can lead to new insights in coastal zone development. This alone would justify a systematic integration of these data into existing reconstructions on a periodic basis, say every five years.

Bibliography

Anonymous, (2009) *Erfgoedbalans 2009. Archeologie, monumenten en cultuurlandschap in Nederland*, Amersfoort.

Baeteman, C. (1981) *The Holocene Development of the Western Coastal plain of Belgium*. Unpublished dissertation Brussels Free University.

Baeteman, C. (1999) 'The holocene depositional history of the IJzer paleovalley (western Belgian coastal plain) with reference to the factors controlling the formation of intercalated peat beds', *Geologica Belgica*, 2, pp. 39–72.

Baeteman, C. (2007) 'Roman peat-extraction pits as possible evidence for the timing of coastal changes. An example from the Belgian coastal plain', in: J.J.J.M. Beenakker, F. H. Horsten, A.M.J. de Kraker, H. Renes (eds), *Landschap in ruimte en tijd.* Amsterdam, pp. 16–25.

Baeteman, C. (2008) 'Radiocarbon dated sediment sequences from the Belgian coastal plain: testing the hypothesis of fluctuating or smooth late Holocene relative sea-level rise', *The Holocene*, 18, pp. 1219–1228.

Baeteman, C. (this volume) 'History of research and state of the art of the Holocene depositional history of the Belgian coastal plain'.

Baeteman, C. and Declercq, P-Y. (2002) 'A synthesis of early and middle Holocene coastal changes in the western Belgian lowlands', *Belgeo,* 2, pp. 77–107.

Baeteman, C., Waller, J. and Kiden, P. (2011) 'Reconstructing middle to late Holocene sea-level change: A methodological review with particular reference to "A new Holocene sea-level curve for the southern North-Sea" presented by K.-E. Behre', *Boreas,* 40, pp. 557–572.

Bakker, H. de and Berg, M.W. van den (eds) (1982) *Proceedings of the symposium on peat lands below sea level*, Wageningen (ILRI publication; 30).

Beets, D.J. and Spek, A.J.F. van der (2000) 'The Holocene evolution of the barrier and the back-barrier basins of Belgium and the Netherlands as a function of late Weichselian morphology, relative sea-level rise and sediment supply', *Geologie en Mijnbouw/Netherlands Journal of Geosciences*, 79, pp. 3–16.

Berendsen, H.J.A. ([4]2004) *De vorming van het land. Inleiding in de geologie en de geomorfologie,* Assen.

Berendsen, H.J.A. and Stouthamer, E. (2001) *Palaeogeographic development of the Rhine-Meuse delta, The Netherlands*. Koninklijke van Gorcum, Assen.

Behre, K.E. (1999) 'Die Veränderungen der niedersächsischen Küstenlinien in den letzten 3000 Jahren und ihre Ursachen', *Probleme der Küstenforschung im südlichen Nordseegebiet*, 26, pp. 9–33.

Behre, K.E. (2003) 'Eine neue Meeresspiegelkurve für die südliche Nordsee', *Probleme der Küstenforschung im südlichen Nordseegebiet*, 28, pp. 9–63.

Behre, K.E. (2007) 'A new Holocene sea-level curve for the southern North Sea', *Boreas*, 36, pp. 82–102.

Bos, J.A.A., Zijverden, W.K. van and Zuidhoff, F.S. (2011) 'De ontwikkeling van het landschap op Walcheren met de nadruk op het onderzoeksgebied rondom Serooskerke', in: J. Dijkstra and F.S. Zuidhoff (eds) *Kansen op de kwelder,* Amersfoort, pp. 35–58 (ADC Monografie; 10).

Bowen, D.Q. (1985) *Quaternary Geology. A Stratigraphical Framework for Multidisciplinary Work. Reprint with additions*, Oxford.

Brand, G.B.M., Bruijne, A.J.Th. de and Molendijk, R.E. (2002) *Oorzaken van de bewegingen van de meetpunten van het NAP,* Delft (Ministerie van Verkeer en Waterstaat, Directoraat-Generaal Rijkswaterstaat, Meetkundige Dienst; Rapport MDGAP-2002.26).

Bungenstock, F. and Weerts, H.J.T. (2010) 'The high-resolution Holocene sea-level curve for Northwest Germany: global signals, local effects or data-artefacts?', *International Journal of Earth Sciences (Geologische Rundschau)*, 99, pp. 1687–1706.

Bungenstock, F. and Weerts, H.J.T. (2012) 'Holocene relative sea-level curves for the German North Sea coast. Reply on K.-E. Behre', *International Journal of Earth Sciences (Geologische Rundschau)*, 101, pp. 1083–1090.

Busschers, F.S., Kasse, C., Balen, R.T. van, Vandenberghe, J., Cohen, K.M., Weerts, H.J.T., Wallinga, J., Johns, C., Cleveringa, P. and Bunnik, F.P.M. (2007) 'Late Pleistocene evolution of the Rhine-Meuse system in the southern North Sea basin: imprints of climate change, sea-level oscillation and glacio-isostacy', *Quaternary Science Reviews*, 26, 25–28, pp. 3216–3248.

Cohen, K.M. (2005) '3D geostatistical interpolation and geological interpolation of palaeo-groundwater rise within the coastal prism in the Netherlands', in: L. Giosan, J.P. Bhattacharaya (eds), *River Deltas: Concepts, Models, and Examples,* Oklahoma, pp. 341–364 (SEPM Special Publication; vol. 83).

Cohen, K.M., Stouthamer, E. and Berendsen, H.J.A. (2002) 'Fluvial deposits as a record for Late Quaternary neotectonic activity in the Rhine-Meuse delta, The Netherlands', *Geologie en Mijnbouw/ Netherlands Journal of Geosciences*, 81, pp. 389–405.

Denys, L. and Baeteman, C. (1995) 'Holocene evolution of relative sea level and local mean high water spring tides in Belgium – a first assessment', *Marine Geology*, 124, pp. 1–19.

Dijkstra, J. and Zuidhoff, F.S. (eds) (2011) *Kansen op de kwelder. Archeologisch onderzoek op Walcheren langs de N57*, Amersfoort (ADC Monografie; 10).

Doppert, J.W.Chr., Ruegg, G.H.J., Staalduinen, C.J. Van, Zagwijn, W.H. and Zandstra, J.G. (1975) 'Formaties van het Kwartair en Boven-Tertiair in Nederland', in: W.H. Zagwijn and C.J. Van Staalduinen (eds), *Toelichtingen bij geologische overzichtskaarten van Nederland*, Haarlem, pp. 11–56.

Dubois, G. (1924) 'Recherches sur les terrains quaternaires du Nord de la France', *Mémoires de la Société Géologique du Nord*, 8, pp. 3–355.

Ebbing, J.H.J., Weerts, H.J.T. and Westerhoff, W.E. (2003) 'Towards an integrated land-sea stratigraphy of the Netherlands', *Quaternary Science Reviews*, 22, pp. 1579–1587.

Edwards, R.J. (2006) 'Mid- to late-Holocene relative sea-level change in southwest Britain and the influence on sediment compaction', *The Holocene,* 16, pp. 575–587.

Eichfeld, I. (2010) 'Diluted Causalities: The Variability of Human Response to Environmental Change: Examples from the North Sea Coast, Lower Saxony, Germany', in: *Landscapes and Human Development: The Contribution of European Archaeology. Proceedings of the International Workshop 'Socio-Environmental Dynamics over the Last 12,000 Years: The Creation of Landscapes (1st-4th* April, 2009)'. Universitätsforschungen zur prähistorischen Archäologie, Band 191 aus der Graduiertenschule 'Human Development in Landscapes' der Universität Kiel, pp. 273–284.

Ervynck, A., Baetemen, C., Demiddele, H., Hollevoet, Y., Pieters, M., Schelvis, J., Tys, D., Strydonck, M. van and Verhaeghe, F. (1999) 'Human occupation because of a regression, or the cause of a transgression? A critical review of the interaction between geological events and human occupation in the Belgian coastal plain during the first millennium AD', *Probleme der Küstenforschung im südlichen Nordseegebied*, 26, pp. 97–121.

Fairbridge, R.W. (1961) 'Eustatic changes in sea level', *Physics and Chemistry of the Earth*, 4, pp. 100–185.

Gehrels, W.R., Szkornik, K., Bartholdy, J., Kirby, J.R., Bradley, S.L., Marshall, W.A., Heinemeier, J. and Pedersen, J.B.T. (2006) 'Late Holocene sea-level changes and isostasy in western Denmark', *Quaternary Research,* 66, pp. 288–302.

Gullentops, F. and Moor, G. de (2001) '2.2 Remaining marine-estuarine deposits', in: P. Bultynck and L. Dejonghe (eds), *Guide to a Revised Lithostratigraphic Scale of Belgium, Geologica Belgica*, 4, pp. 153–164.

Hageman, B.P. (1969) 'Development of the western part of The Netherlands', *Geologie en Mijnbouw,* 48, pp. 373–388.

Hijma, M. (2009) 'From river valley to estuary. The early-mid Holocene transgression of the Rhine-Meuse valley, The Netherlands', *Netherlands Geographical Studies,* 389, pp. 1–192.

Jelgersma, S. (1961) 'Holocene Sea Level Changes in the Netherlands', *Mededelingen van de Geologische Stichting*, Serie C-VI-7, pp. 1–100.

Kiden, P. (1995) 'Holocene sea-level change and crustal movement in the southwestern Netherlands', *Marine Geology*, 124, pp. 21–41.

Kiden, P., Denys, L. and Johnston, P. (2002) 'Late Quaternary sea-level change and isostatic and tectonic land movements along the Belgian-Dutch North Sea coast: geological data and model results', *Journal of Quaternary Science,* 17, pp. 535–546.

Kiden, P., Makaske, B. and Plassche, O. van de (2008) 'Waarom verschillen de zeespiegelreconstructies voor Nederland?', *Grondboor en hamer,* 62, pp. 54–61.

Knottnerus, O.S. (this volume) 'Reclamations and submerged lands in the Ems River estuary (900–1500)'.

Kooi, H., Johnston, P., Lambeck, L., Smither, C. and Molendijk, R. (1998) 'Geological causes of recent (-100 years) vertical land movement in the Netherlands', *Tectonophysics*, 299, pp. 297–316.

Kossack, G. (1995) 'Kulturlandschaft und Naturkatastrophe. Bäuerliche Verhaltungsweisen in prähistorischer Vergangenheit', *Siedlungsforschung*, 13, pp. 31–50.

Lascaris, M.A. and Kraker, A.M.J. de (this volume) 'Dikes and other hydraulic engineering works from the Late Iron Age and Roman periods on the coastal area between Dunkirk and the Danish Bight'.

Lenselink, G., Koopstra, R. (1994) 'Ontwikkelingen in het Zuiderzeegebied; van meer Flevo, via de Almere-Lagune, tot Zuiderzee', in: M. Rappol, C.M. Soonius (eds), *In de bodem van Noord-Holland, geologie en archeologie*, Amsterdam, pp. 129–140.

Londen, H. van (2006) *Midden-Delfland. The Roman native Landscape. Past and Present,* Unpublished Dissertation, Universiteit van Amsterdam.

Louwe Kooijmans, L.P. (1974) 'The Rhine/Meuse delta. Four studies on its prehistoric occupation and Holocene geology', *Analecta Praehistoria Leidensia,* 7, pp. 1–421.

Makaske, B., Smeerdijk, D.G. van, Peeters, H., Mulder, J.R., Spek, T. (2003) 'Relative water-level rise in the Flevo lagoon (The Netherlands), 5300–2300 cal. yr BC: an evaluation of new and existing basal peat time-depth data', *Netherlands Journal of Geosciences*, 82, pp. 115–131.

Meier, D. (2004) 'Man and environment in the marsh area of Schleswig-Holstein from Roman until late Medieval times', *Quaternary International*, 112, pp. 55–69.

Meier, D. (this volume) 'From Nature to Culture: Landscape and Settlement History of the North-Sea coast of Schleswig-Holstein, Germany'.

Mrani-Alaoui, M. and Anthony, E.J. (2011) 'New data and a morphodynamic perspective on mid- to late-Holocene palaeoenvironmental changes in the French Flanders coastal plain, southern North Sea', *The Holocene*, 21, pp. 445–453.

Mulder, E.F.J. de and Ritsema, I. (2003) 'Duurzaam gebruik en beheer van de ondergrond', in: E.F.J. de Mulder, M. Geluk, I. Ritsema, W.E. Westerhoff, Th.E. Wong (eds), *De ondergrond van Nederland. Geologie van Nederland, deel 7*, Utrecht, pp. 9–118.

Pedersen, J.B.T., Svinth, S. and Bartholdy, J. (2009) 'Holocene evolution of a drowned melt-water valley in the Danish Wadden Sea', *Quaternary Research*, 72, pp. 68–79.

Peltier, W.R. (2004) 'Global glacial isostasy and the surface of the ice-age Earth: the ICE-5G (VM2) model and GRACE', *Annual Review of Earth and Planetary Sciences*, 32, pp. 111–149.

Plassche, O. van de (1982) 'Sea-level change and water-level movements in the Netherlands during the Holocene', *Mededelingen Rijks Geologische Dienst*, 36, pp. 1–93.

Plassche, O. van de (1995) 'Evolution of the intra-coastal tidal range in the Rhine-Meuse Delta and Flevo Lagoon, 5700–3000 yrs cal B.C.', *Marine Geology*, 124, pp. 113–128.

Plassche, O. van de, Bohncke, S.J.P., Makaske, B., Plicht, van der J., (2005) 'Water-level changes in the Flevo area, central Netherlands (5300–1500 BC): implications for relative mean sea-level rise in the Western Netherlands', *Quaternary International*, 133–134, pp. 77–93.

Reimer, P.J., Baillie, M.G.L., Bard, E., Bayliss, A., Beck, J.W., Blackwell, P.G., Bronk Ramsey, C., Buck, C.E., Burr, G.S., Edwards, R.L., Friedrich, M., Grootes, P.M., Guilderson, T.P., Hajdas, I., Heaton, T.J., Hogg, A.G., Hughen, K.A., Kaiser, K.F., Kromer, B., McCormac, F.G., Manning, S.W., Reimer, R.W., Richards, D.A., Southon, J.R., Talamo, S., Turney, C.S.M., van der Plicht, J., Weyhenmeyer, C.E. (2009) 'INTCAL 09 and MARINE09 radiocarbon age calibration curves, 0–50,000 years Cal BP', *Radiocarbon*, 51, pp. 1111–1150.

Rippon, S. (this volume) 'Human impact on the coastal wetlands of Britain in the medieval period'.

Roeleveld, W. (1974) 'The Holocene evolution of the Groningen Marine-Clay District', *Berichten van de Rijksdienst voor het Oudheidkundig Bodemonderzoek*, 24, pp. 1–132.

Ruddiman, W.F. (2001) *Earth's Climate. Past and Future*, New York.

Steffen, H. (2006) *Determination of a consistent viscosity distribution in the Earth's mantle beneath Northern and Central Europe*, Published Dissertation, Institut für Geologische Wissenschaften der Freie Universität Berlin.

Tavernier, R. (1947) 'L'évolution de la plaine maritime belge', *Bulletin de la Société belge de Géologie,* 56, pp. 332–343.

Tavernier, R. (1948) 'Les formations quaternaires de la Belgique en rapport avec l'évolution morphologique du pays', *Bulletin de la Société belge de Géologie*, 57, pp. 609–641.

Vink, A., Steffen, H., Reinhardt, L. and Kaufmann, G. (2007) 'Holocene relative sea-level change, isostatic subsidence and the radial viscosity structure of the mantle of northwest Europe (Belgium, the Netherlands, Germany, southern North Sea)', *Quaternary Science Reviews*, 26, 25–28, pp. 3249–3275.

Vos, P. (2008) 'The geological development of the Oer-IJ area', in: M.S.M. Kok, *The homecoming of religious practice: an analysis of offering sites in the wet low-lying parts of the landscape in the Oer-IJ area*, Dissertation, University of Amsterdam, pp. 81–95.

Vos, P.C. and Heeringen, R.M., van (1997) 'Holocene geology and occupation history of the Province of Zeeland', in: M.M. Fischer (ed.), *Holocene evolution of Zeeland (SW Netherlands), Mededelingen Nederlands Instituut voor Toegepaste Geowetenschappen TNO*, 59, pp. 5–109.

Vos, P.C. and Gerrets, D.A. (2005) 'Archaeology: a major tool in the reconstruction of the coastal evolution of Westergo (northern Netherlands)', *Quaternary International,* 133–134, pp. 61–75.

Vos, P.C. and Knol, E. (2005) 'Wierden ontstaan in een dynamisch getijdelandschap', in: E. Knol, A.C. Bardet and W. Prummel (eds), *Professor Van Giffen en het geheim van de wierden. Boek bij de gelijknamige tentoonstelling,* Groninger Museum, pp. 119–135.

Vos, P.C and Eijskoot, Y. (2011) '3 Geologie', in: Y. Eijskoot, O. Brinkkemper, T. de Ridder (eds), *Vlaardingen-De Vergulde Hand-West,* Amersfoort, pp. 69–146 (Rapportage Archeologische Monumentenzorg; 200).

Vos, P.C., Bazelmans, J., Weerts, H.J.T. and Meulen, M.J., van der (eds) (2011a) *Atlas van Nederland in het Holoceen. Landschap en bewoning vanaf de laatste ijstijd tot nu*, Amsterdam.

Vos, P., Weerts, H.J.T., Hoogendoorn B. and Meulen, M., van der (2011b) 'De zeespiegel stijgt', in: P.C. Vos, J. Bazelmans, H.J.T. Weerts and M.J. van der Meulen (eds), *Atlas van Nederland in het Holoceen. Landschap en bewoning vanaf de laatste ijstijd tot nu*, Amsterdam, pp. 14–17.

Vos, W.K. (2009) *Bataafs platteland. Het Romeinse nederzettingslandschap in het Nederlandse Kromme-Rijngebied*, Dissertation Vrije Universiteit Amsterdam/NAR Nederlandse Archeologische Rapporten 35. Rijksdienst voor het Cultureel Erfgoed/Vrije Universiteit/Hazenberg Archeologie.

Weerts, H.J.T., Westerhoff, W.E., Cleveringa, P., Bierkens, M.F.P., Veldkamp, J.G. and Rijsdijk, K.F. (2005) 'Quaternary geological mapping of the lowlands of The Netherlands, a 21st century perspective', *Quaternary International*, 133–134, pp. 159–178.

Waller, M. and Long, A. (this volume) 'The Holocene Coastal Deposits of Sussex: a Re-evaluation'.

Weerts, H.J.T., Vos, P., Hoogendoorn, B. and Meulen, M., van der (2011) 'Veen bedekt het land', in: P.C. Vos, J. Bazelmans, H.J.T. Weerts and M.J. van der Meulen (eds), *Atlas van Nederland in het Holoceen. Landschap en bewoning vanaf de laatste ijstijd tot nu*, Amsterdam, pp. 20–23.

Westerhoff, W.E., Wong, T.E. and Mulder, E.F.J., de (2003a) 'Opbouw van de ondergrond', *Geologie van Nederland, deel 7*, Utrecht, pp. 247–352.

Westerhoff, W.E., Geluk, M. and Mulder, E.F.J., de (2003b) 'Geschiedenis van de ondergrond', in: *Geologie van Nederland, deel 7*, Utrecht, pp. 119–246.

Wilgus, C.K., Hastings, B.S., Kendall, C.G.St.C., Posamentier, H.W., Ross, C.A. and Wagoner, J.C., Van (eds) (1988) 'Sea-level Changes: An Integrated Approach', *SEPM Special Publication*, 42, pp. 1–407.

Wallinga, J., Davids, F. and Dijkmans, J.W.A. (2007) 'Luminescence dating of Netherlands' sediments', *Geologie en Mijnbouw/ Netherlands Journal of Geosciences*, 86, pp. 179–196.

Woltering, P.J. (1997) 'Occupation History of Texel. III The Archaeological Survey, Palaeogeography and Settlement Patterns', *Berichten van de Rijksdienst voor het Oudheidkundig Bodemonderzoek*, 42, pp 209–363.

Zagwijn, W.H. (1986) 'Nederland in het Holoceen', in: *Geologie van Nederland, deel 1*. Haarlem.

Ziegler, P.A. (²1990) *Geological atlas of western and central Europe*, Den Haag.

Zonneveld, I.S. (1960) 'De Brabantse Biesbosch. Een studie van bodem en vegetatie van een zoetwatergetijdendelta', *Mededelingen van de Stichting voor Bodemkartering/ Bodemkundige Studies*, 4, pp. 1–396.

Part II

Landscapes versus seascapes? Social organisation and changing coastal landscapes

9 Dikes and other hydraulic engineering works from the Late Iron Age and Roman period on the coastal area between Dunkirk and the Danish Bight

Michel A. LASCARIS, Cultural Heritage Agency, the Netherlands
Adriaan M.J. DE KRAKER, VU University Amsterdam

Abstract

This article examines the building of the oldest known dikes, dams, sluices and culverts in the southeastern North Sea coastal area. Giving an outline of the archaeological excavations carried out since the 1950s, the following key themes are addressed. The building of these dams and dikes goes back to the end of the Late Iron Age, which also applies to the oldest sluices and culverts. In the northern Wadden area, coastal Flanders, and Zealand, dams and especially dikes were dominant whereas in the Rhine-Meuse area sluices and culverts were most prevalent. The top levels of dikes and dams barely reached one meter in height, while their width could vary much more. Sluices and culverts were wooden constructions often made of hollowed tree trunks. The distribution of the earliest hydraulic engineering works seems very uneven and large areas on the map are still blank. This is likely to be caused by differences in the number of archaeological excavations, the recognisability of these features and the archaeological methods employed at any particular time. Finally, a striking resemblance between the oldest dike and the medieval dikes at Kampereiland suggests a land use which was not exclusively devoted to pasture.

I. Introduction

At the time of the second congress on the evolution of the coastal area of the Low Countries, held at Ghent in 1979, hardly anything was known about the earliest hydraulic engineering works which already had been dated to the Late Iron Age and Roman periods. Some excavations had already taken place in Schleswig Holstein and Lower Saxony since the mid-1950s, but archaeologists at that time were very reluctant to interpret these remains as primitive dikes and dams. It took some decades and several other excavations in the coastal area before we began to learn about the Roman occupation and that some vague remnants from this period could be interpreted as dikes and dams. Since the 1990s much new evidence all along the North Sea coast has been unearthed, ranging from settlement patterns to farming systems, and from reclaiming the fringe of peat areas to what must be considered to be the first hydraulic engineering works in the area. This article gives an overview of the most important discoveries by archaeologists in the coastal area of

Landscapes or seascapes? The history of the coastal environment in the North Sea area reconsidered, ed. by Erik THOEN, Guus J. BORGER, Adriaan M.J. DE KRAKER, Tim SOENS, Dries TYS, Lies VERVAET & Henk J.T. WEERTS, Turnhout, 2013 (*CORN Publication Series*, 13), pp. 177–198.

BREPOLS PUBLISHERS DOI: 10.1484/M.CORN.1.101553

Figure 9.1 Late Iron Age and Roman dikes, culverts and sluices in North Western Europe (source: de Ridder, 1999)

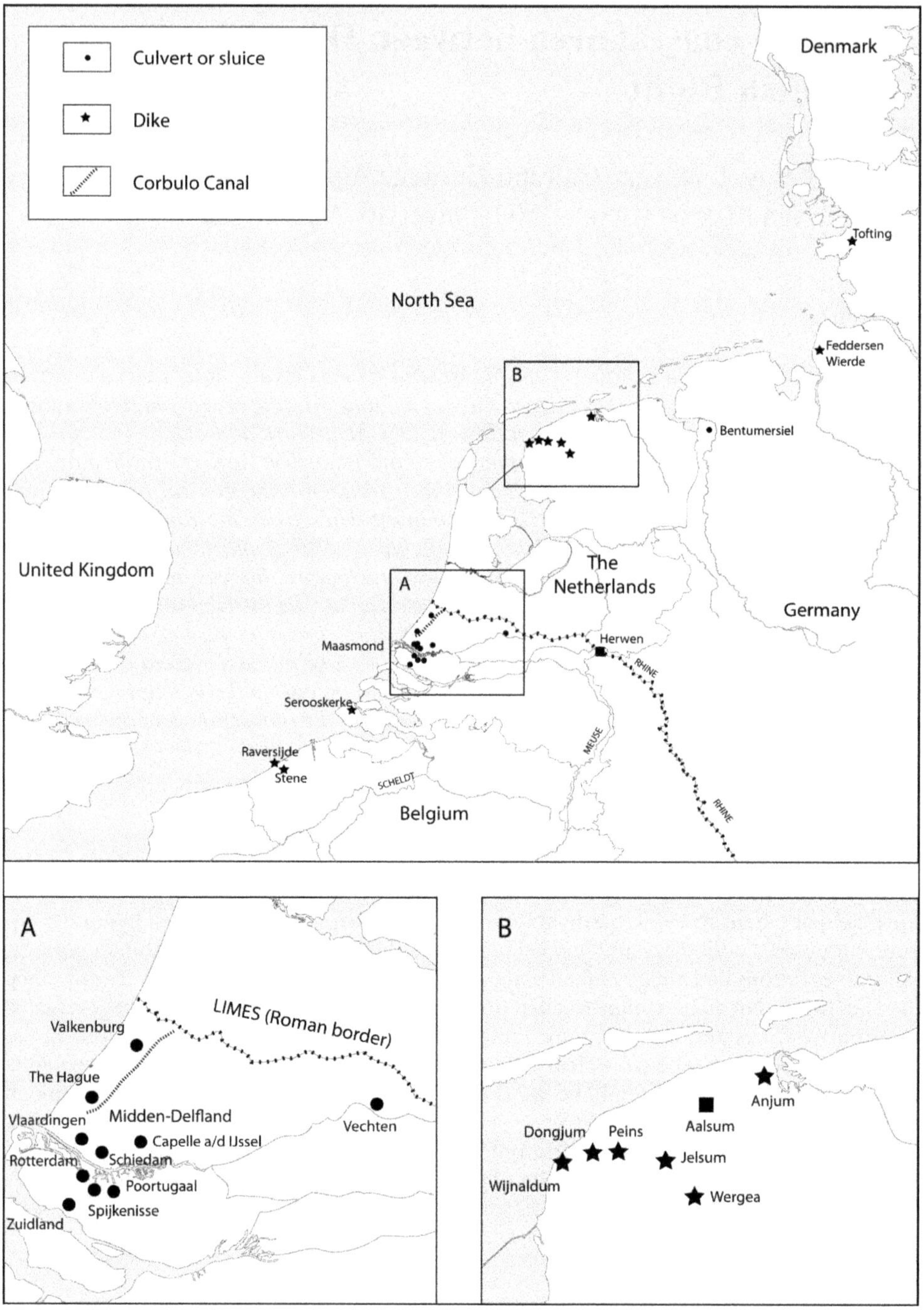

North West Germany, the Netherlands and Belgium of the past few decades. This last coastal area is divided into several main estuarine areas. After this overview the evidence will be analysed whereby the following research questions need to be addressed. When were the first hydraulic engineering works built and what purpose did they serve? Is it possible to distinguish a specific Roman influence in this respect? When and why did it end? Furthermore, this article deals with the gap in our knowledge of coastal development in the late Iron Age and Roman periods, so connecting to the main themes of this publication. It not only refers to the complex interaction between human and physical evolution of the coastal areas but it also addresses the issue of how different or similar local coastal communities have developed using the best of their technology in order to become more resilient to risks that might threaten their environment.

II. Remains of small scale embankments in the dwelling mound area

When Pigorini visited the excavation of the Aalsum dwelling mound near Dokkum in 1881 he recognised an earthen ring dike (Bazelmans *et al.*, 1999: 7–8). At that time he was an archaeologist who had become famous for his many publications on the Bronze Age *terramares* (dwelling mounds) in the Po area. Pigorini believed these circular structures with traces of posts in the centre to be settlements with pile dwellings alongside a water basin surrounded by a ring dike. These dikes did not serve as a defence against the sea but were meant to retain fresh water. Pigorinial so believed he had recognised a similar ring dike several metres high at Aalsum, which he therefore interpreted as a historic variant of *terramares*. This intriguing pile village theory was influential for some time in the Netherlands. As no further ring dikes were uncovered during the big excavations of dwelling mounds in the first half of the twentieth century, one of the most famous archaeologists of that time – Albert Egges van Griffen – could only conclude that older ring dikes in the dwelling mound area have never existed prior to the early medieval embankments.

Did this close the discussion about dikes hidden under dwelling mounds? This depends largely on the definition of *dike*. This word has its roots in the verb digging and is therefore related both to underground and above ground structures which can be traced backwards in different European languages.[1] The word does not imply limitations on dimensions or the exact use but most common in modern Dutch is the definition of a large man-made and meters high embankment as protection against high water levels in rivers and (spring) tides in coastal areas. However, if dikes are defined as a modest kind of structure protecting fields and lots rather than as a high sea wall, this does make a difference. Frequent flooding and the low salt tolerance of most crops make arable use of salt meadows very risky. Therefore it has been presumed for a long time that prehistoric and protohistoric settlers of open salt marshes could have lived from livestock farming alone. While this interpretation did not agree with the finding of parts of a plough, plough traces and grains of barley underneath the dwelling mounds, the 'livestock only' theory was questioned by Van der Poel in the early 1960s (Van der Poel, 1960/1961: 176–177; Pals, 1999: 140).

[1] In some modern languages this still can be seen, i.e. Deich, Teich (German), dike, dyke, ditch (English). In modern Dutch only the overground definition of dike (dijk) still exists (Philippa *et.al.*, 2003–2009).

Moreover, by 1955 Bantelmann had published the findings from a number of drainage ditches and beds of fields or plots protected by a wall underneath the Tofting dwelling mound (Schleswig Holstein) (Bantelmann, 1955: 21–22). Carefully avoiding the term 'dike' in his publication, he suggested that the construction was made to protect the small field or garden from flooding. In that same year also Halbertsma mentioned the idea of low dikes surrounding grain fields (Halbertsma, 1955: 60–61).

In addition, excavations of the famous Feddersen Wierde dwelling mound at Cuxhaven in Lower Saxony, 1954–63, uncovered the remains of a Roman Iron Age dike (Haarnagel, 1979; Ey, 2007: 92–99). This dike underneath the dwelling mound had been reinforced twice and was about 5.5 m wide. The most recent change occurred during the second century A.D. The results of the paleobotanic research of the FeddersenWierde were published by Körber-Grohnein 1967. One of the findings was that some crops have a relatively high salt water tolerance (i.e. barley, flax and field beans) and could therefore be grown on the higher parts of the salt marshes. However, the results of later experiments were less optimistic about growing crops on these places and state that this is certainly possible but difficult and rather risky (Van Zeist *et al*,. 1976: 139; Bottema e*t al.* 1980: 139).

As these results were published in 1967 and 1979 it is unknown if Halbertsma could have known of this theory back in 1955. In any event, neither the excavations or the published material immediately led to the abandonment of the generally accepted idea of arable becoming possible only as late as the medieval process of embankment. It was the 1990s before there was a growing acceptance among archaeologists that the early settlers of the salt meadows did not only live from livestock farming but also from cultivating small arable fields, the latter made possible by securing fields with small-scale embankments. Excavations of the Wijnaldum-Tjitsma dwelling mound (1991–93) led to the unearthing of two dikes (Gerrets/ de Koning, 1999: 98–99; Gerrets, 2010: 58–59). The archaeologists were very prudent in their interpretation, referring to them as 'dike-like' structures. The first dike was about 0.75 m high, 10 m wide and was built of sods that had been piled one on top of the other in a regular pattern. Sandy layers and repairs showed that the dike had been eroded by floods twice. The dike was constructed before the building of the dwelling mound (before 175 AD). A second dike, that was contemporary with the first occupation phase (175–250 AD), was raised on top of the first one.

A second remnant of a dike found at Dongjum-Heringa in 1998 was built of carefully piled up sods (Bazelmans *et al.*, 1999: 36–38). The first building phase of the dike consisted of sods and the structure was about 2 m wide and 0.70 m high. Thereafter the dike was extended both landward and seaward giving it a wider and flattened top level. The top level of the dike reached about 1.20 m while its width ranged from 14 to 18 m. The dike must have been flooded several times because underneath one of the later extensions was a layer containing material deposited from the salt meadow. On the landward side a layer of brown earth was interpreted as a field deposit. The building of the dike must be dated in the first century AD.

Although it is often impossible to reconstruct the length of such dikes, as the size of the dwelling mound area excavations are usually too small, the excavation of the dwelling mound remnant of Peins-Oost was an exception (Bazelmans *et al.*, 1999: 39–48). From the large size of the excavated terrain it was shown that the Peinser dike must have been over 54 m long. Unfortunately only a few decimeters of its height had been preserved, while the remaining top levels must have been used as fertiliser during the nineteenth century. However, using the information obtained from the steep slope of the remnant of the dwelling mound, the initial top level of the dike must have been about 1 m high while its width was over 14 m. In the transformation process of the dike in order to make it more solid four distinct phases can be distinguished. Most probably the dike was built in the first century BC. More recently another dike feature was found during the excavation of the dwelling mound of Anjum, Friesland (Nicolay, 2010: 116–117). This dike had a base line width of 2.40 m and must have had a top level of 1.20 m. The first phase was constructed of plaggen which had been piled one on top of the other, cluttered in the centre, with heavy and much more carefully piled up sods at the outside of the dike. In a second stage the dike was strengthened with a gently sloping flat layer of sods on the seaward side.

Concerning all the locations in the Dutch and German Wadden area so far, it needs to be noticed that the dikes were not strong enough to protect the embanked area against seasonal storms. They simply had too low a top level for that purpose but they could protect fields and gardens against the sea during summertime, thereby enabling the continuous deposition of a new thin fertile layer each year during winter time (Bazelmans *et al.*, 1999: 60–61; Gerrets, 2010: 39; Vos, 1999: 63). This layer could be used for the natural fertilising of fields and lots, making it possible to make hay.

III. Other dikes dating in pre- or proto-history

Outside the Dutch and German Wadden area remnants of prehistoric dikes have been found elsewhere. Because of the building of a new motorway (the A59) on the former island of Walcheren, archaeological research carried out near Serooskerke revealed a dike which had been built during two phases (Figure 9.2) (Dijkstra, 2011: 67–68; Campenhout/Dijkstra, 2011: 272–275). In spite of its location so far to the south, the dike has much in common with those found in the Wadden area. Again, this is a summer dike having a low top level on a salt meadow of which the main purpose must have been to protect the vulnerable fields against flooding by the sea. Due to medieval peat cutting and erosion the oldest phase has largely disappeared so only about 70 m of its length could be investigated. The sections that could still be distinguished have a top level of barely 0.16 m. As high-quality dating material could not be found, only a rough dating of the dike was possible. Using C14 dating and its location underneath a section of the younger dike the dike most probably dates between 25 BC and 200 AD (Campenhout/Dijkstra, 2011: 272–281). The younger building phase, during which the dike must have been at least 5 m wide, 0.8 m high and over 82 m long, connects to the dwelling mound and partly follows the same track as the older phase. Using pottery dating, the dike of the younger phase must have been in use between 2 and 250 AD. Similar to the Wadden area the Serooskerke dike was built from locally-dug plaggen, using both clay and peat sods. Furthermore, the dike must have been flooded regularly because there was a thick layer of maritime sediment deposited behind it. Apparently temporary loss of land did not force settlers of the dwelling mounds to abandon the area permanently. Somewhere between 200 and 250 AD they built a new dike at precisely the same place.

Figure 9.2 Dike built of locally dug plaggen (Photograph ADC-Archeoprojecten)

Another Roman dike was found underneath the medieval settlement of Raversijde in Flanders in 2005 (Priester, 2006). This dike must have been 11 m wide and 1.10 m high. It was built perpendicular to the present coast line and was later buried by marine deposits of several decimeters thickness. Some 110 m of this dike have been preserved. The centre of the dike was built of carefully heaped up sods, while the western slope was covered with a layer of peat for extra protection against the sea. Because of its perpendicular location it is assumed that it must have been built on a tidal channel levee protecting the hinterland. Pottery found in the top layer of the soil and on top of the dike clearly dates from Roman times but a precise dating is not possible yet. After it lost its function the dike was covered by marine deposits. Also at Stene, near Oostende, remnants of a dike have been excavated (Demey, Vanhoutte and Pieters, 2010). However, further details on this excavation are not available yet.

IV. Dams and sluices

No doubt the most-mentioned and speculated-about Roman dam in the Netherlands is the Drusus dam (Moles) mentioned by Tacitus (Tacitus Historia V, 19; Vollgraf, 1939). This dam, actually a groyne, is supposed to have been built about c.12BC at the junction of rivers Waal and Rhine to ensure enough water would flow through the River Rhine which that at that time formed the Roman Limes.

The discovery of an inscription on a tombstone of a Roman soldier has made us believe this *moles* must have been located near the village of Herwen. However, despite references to this impressive hydraulic engineering work, so far no remnants of it have been uncovered. This comes as no surprise because later fluvial erosion around this site has been massive (Berendsen and Stouthamer, 2001).

Of more modest dimensions are the Roman dams and sluices further downstream in the Rhine-Meuse estuary which have been excavated since the 1990s. Between 1992 and 1997 seven dams were found on the premises of Hoogstad at Vlaardingen, three of them having a sluice (De Ridder, 1999: 10–15; De Ridder, 2007; 18). A fourth sluice was found in an area eastward of this location. The former location is shown in Figure 9.3. The oldest dam lay crosswise over an east-west running channel and had been built of posts with an earth filling. The wood of this dam dates from the Late Iron Age (170/180 cal BC) which makes it one of the oldest dams unearthed in Western Europe to date. After some time the channel of the old dam must have silted up because in Roman times it was intersected by a younger creek that had a totally different shape. The other dams and sluices were built in this younger network of tidal channels. Four of them were aligned crosswise over a creek. These structures did not function at the same time: each of them succeeded the other, gradually shifting towards the river Meuse. This was most probably due to gradual sedimentation in the channel caused by the dams. According to the pottery evidence, the first dam was built between 70 and 125AD, while the three southern dams were built in the period 120–175AD. In the northwest section there was a branch of the channel in which there may also have been a dam, but this remains yet to be proven. Close to the Vlaardingen-Hoogstad location Roman dams and sluices have also been found at Cappelle aan den IJssel-Middenwatering West and Barendrecht-Zuidpolder (Carmiggelt, Trierum and Wesselingh, 2010: 127–132).

Like the dikes from the Wadden area the dams dating from the Roman period were built of plaggen. But this is the only similarity because the dams found at Vlaardingen were built of layers of sand and clay sods alternating with thin layers of plant remnants which were laid down in order to enhance drainage. This made the dams more solid and provided better water-proofing. All the dams found in the Rotterdam region consist of lumps of clay with thin layers of material that look like reeds. In order to stabilise the dam, piles had been driven into the dam on both sides. In some of the dams mentioned earlier, remnants of sluices have been discovered that were used for draining the closed-off creeks. Culverts have also been found which were used to drain ditches on a creek and which also served to connect two ditches. The construction of the sluices and culverts varies. Some were made of tree trunks hollowed out by auger. Some sluices and culverts consisted of tree trunks that had been first split to make it easier to hollow the separate parts out. Others consisted of tree trunks covered with thick planks and one sluice was even made out of a broken canoe (De Ridder, 2007: 20). The technically most advanced culverts had a valve which closed at high tide outside the drainage area (Figure 9.4), thereby preventing the flow of water in to the drainage creek. In a creek such a sluice functions exactly like sluices in medieval sea walls with slamming doors: fresh water could be discharged and salt water kept out. Thanks to the drainage the lower and wetter parts of the coastal zone could also be used as arable although not for a very long period. The drainage inevitably was followed by land subsidence due to clay and peat

Figure 9.3 Dams, culverts and sluices at Vaardingen Hoogstad (source: de Ridder 1999, 12 Figure 3)

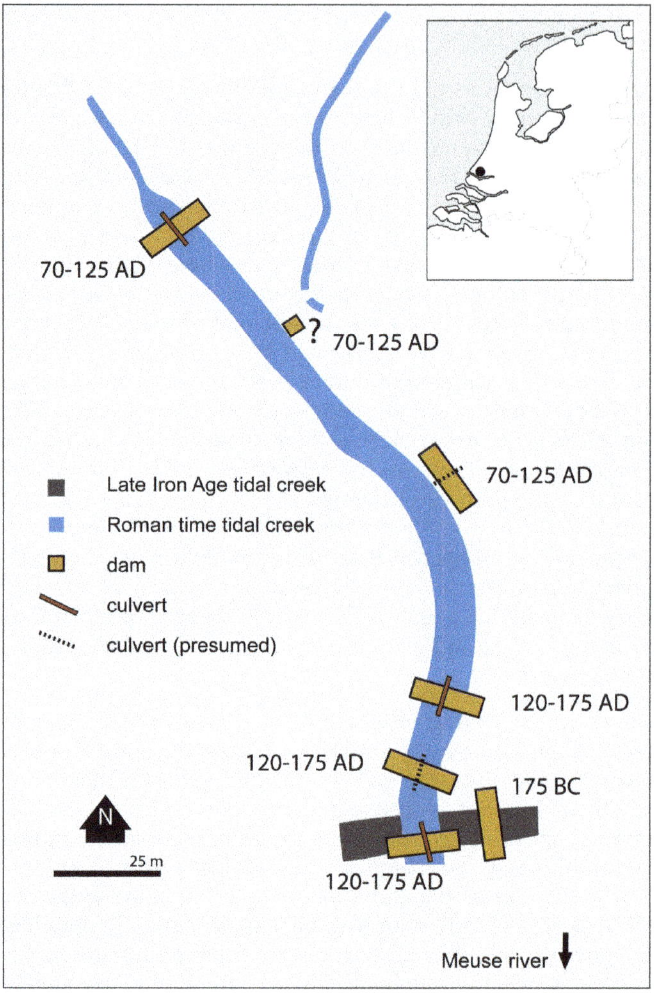

compaction (Van Asselen, Stouthamer and Van Asch, 2010: 36–43). This made the land sensitive to high waters from sea and rivers, just like in the Middle Ages when extended areas of drained land were again lost to the sea.

V. Drainage ditches and canals

The culverts and sluices discussed so far did not function as isolated elements but must have functioned as part of larger drainage systems. The oldest ditches which were used

Figure 9.4 The Valkenburg culvert with valve and a reconstruction of a dam with culvert (source: Bult/Hallewas, 1990)

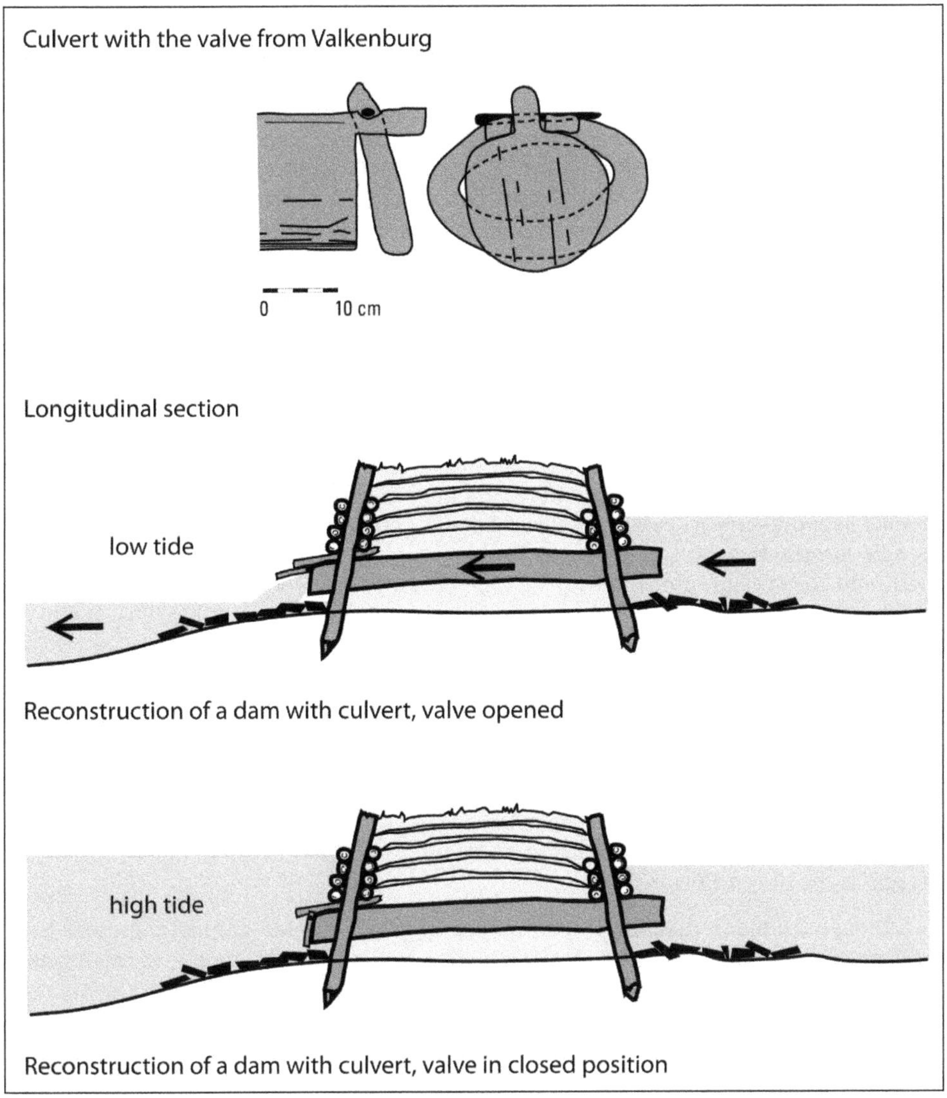

for drainage of parts of the coastal marshes have been found on excavation sites of Bronze Age settlements in the northwest of the Netherlands (West-Friesland) (IJzereef, Van Regteren and Altena, 1991: 76–77). Contemporary with the culverts and sluices found at the *Maasmond* (mouth of the river Meuse) are the regular patterns of Roman ditches which have been found in *Midden Delfland* and in the central Dutch river region (Van Londen, 2006: 186–221; Vos, 2009: 109–116). While ditches are common features on archaeological sites canals are definitely not (See forinstance Boersma, 2005; Heeren, 2009; Van Heeringen, 2005;Hiddink, 2005; Van Renswoud and Kerckhove, 2009; Van der Velde, 2011). Besides the already mentioned Drusus dam there must have been at least two other Roman canals in the Netherlands. The Roman authors Publius Cornelius Tacitus and Suetonius described a fossa Drusiana (Drusus canal) which served as a means of transport for Roman troops to the northern parts of the Netherlands (Tacitus Annales, II, 8; Suetonis Claudius, 2–4). Although there is much speculation about its location, the archaeological remains of this canal never have been found. Another canal mentioned by Roman authors is the fossa Corbulonis (Corbulo canal) (Tacitus Annales, XI, 20; Dio Hist. Rom., LXI, 61, 30, 4–6). This was a canal between the River Rhine and the *Maasmond* which was dug in 47AD. Remains of this canal have been found at several places which made it possible to reconstruct its course (Figure 9.1) (Hessing, 1990, 1991, 1992, 1993; De Kort, 2008: 24–26; Vos, 2007; De Jonge *et al.*, in prep).

VI. Discussion

Studying the oldest hydraulic engineering works, questions about their construction can now be raised. Judging from the size of the dams and the dikes found in general and looking at the technical dimensions and the many construction problems in active channel networks in particular, it is clear that their creation went beyond the capacity of a single farmer. Pieters, Demerre and Zeebroek estimate the sheer volume of the dike uncovered at Ravesijde alone was about 1000 m^3 and clearly indicate this dike must have been built on a supra-local scale (Pieters, Demerre and Zeebroek, 2006: 94). It should be noted that 1000 m^3 is equal to 12,500 modern wheel barrow loads. This went beyond the capacity of a single farmer, but if five wheel barrows could be working for at least some weeks, this could do the job. Besides, the building of these dikes was carried out in several phases, starting from a small dike and ending with a larger one several years later, as has been described. Also most of the building material – mostly clay sods – came from the immediate vicinity. Therefore it may be assumed that the building of a dike or dam could very easily have been carried out by a group of people of the size of a local peasant community (See Bazelmans *et al.*, 1999: 66–67 for the same line of reasoning as regards the oldest Frisian dykes).

Although the oldest dikes and dams are pre-Roman it was not until after the arrival of the Romans in the area that hydraulic construction works seem to have gained momentum. This raises the question whether the new technology was introduced by the Romans or already existing hydraulic engineering technology became more efficiently used on a much wider scale. No doubt the Romans, particularly the Roman army, had enough knowledge of hydraulic engineering, the right equipment and the logistics to build complex water management structures. A good example of what they were capable of is probably the northwestern section of *the Limes*: the Roman border. In the western parts of the Netherlands the Limes was in fact a small corridor along the rivers with extended moors on both sides. More than a physical border, the northwestern Limes was an intricate infrastructure for transport and communication over land and water, connecting the higher parts of the Netherlands with the coast. Its location in this wet and highly dynamic fluvial landscape made an extended system of bridges, revetments, dams, ditches and culverts necessary which had to be regularly maintained (Graafstal, 2002: 12–14, 19–20). The investment in manpower, technology, and material must have been tremendous and therefore far beyond the scope of the local communities. On the other hand, dams, dikes, and drainage ditches are basic structures which could have been made by anybody as can be seen by children playing on the beach. It seems all a matter of complexity and scale: the building and maintenance of the more complex and larger hydraulic engineering works require detailed logistical organisation.

An even more essential aspect of these investigations is how representative the present data base is. So far, in the mouth of the River Meuse about twenty dams and sluices from the Roman era have been uncovered but there is no evidence of dikes yet. In contrast, until now only dikes have been uncovered in the Wadden area and in the province of Zealand, but no dams or sluices have yet been found there. In vast areas of the Flemish-Dutch-German coastal area no remnants of early hydraulic engineering have been found at all. The same goes for the greater part of the Dutch Limes area. Therefore, a map plotting the distribution of the earliest dams, dikes and sluices of this period show very large areas without finds (Figure 9.1). Since almost all Roman dams and sluices have been found in the *Maasmond*, De Ridder suggests that in this territory large scale reclamations took place, comparable with the late medieval reclamations of the peat bogs in Utrecht and elsewhere in Holland (De Ridder, 2007: 21–24). In his opinion, the Roman dams and culverts were part of a large system made to improve indigenous agricultural production in the area. However, it still does not explain why no Roman dams and culverts have been found in other parts of the Rhine-Meuse delta, a territory that was also used as agricultural hinterland both for the local and Roman military and for an urban population (Vos, 2009: 251–259; Heeren, 2009: 241).

While, on the one hand, at many locations all along the coast vast amounts of building material (plaggen/sods) must have been available to people with the requisite construction skills, on the other, it is striking that only so little pre- and proto-historic hydraulic engineering structures have been found during the explosive growth of archaeological research from 2005 onwards. We believe that the blank areas on the map are caused by the limited recognisability of the archaeological features, combined with the archaeological methods used and the locations of archaeological research. As the earliest dikes, dams, and sluices have functioned in a landscape that does not exist any more, so they are no longer part of the present hydraulic engineering structure, which makes provenancing their location very difficult. In the case of the dikes this is even more difficult because of erosion of the top layer, long term agricultural use, and other kinds of natural and anthropogenetic processes which combine to influence the preservation of soil features (For an overview on post-deposition processes on soil features: Huisman and Deeben, 2009: 147–176). Only the base line of the dike is preserved in a situation where the original surface level has been completely covered. There is also a methodological problem. The culverts and sluices found at the *Maasmond* lie in a very intensively used part of the country where many excavations have taken place in the last twenty-five years. This can explain the relative large number of these elements found there. Furthermore, these Prehistoric and Roman hydraulic engineering structures are typically sited away from areas of primary settlement. Hence they are often neglected by archaeologists focusing on settlements and burial sites and they are rarely recognised when the preliminary archaeological investigations are limited to coring only. Large scale, off site, landscape-oriented research is necessary to find these structures. In fact the research into the oldest dikes and other pre-medieval hydraulic engineering structures has only just begun and it is still in its initial phase. To date, we are likely only dealing with the tip of the iceberg.

Finally, there is a very intriguing resemblance between those earliest dikes and dams and those built in late-medieval times, both in terms of technical aspects as well as land use of the protected area. This was already noticed by Bazelmans (Bazelmans *et al.*, 1999: 60–61; Bazelmans, 2005: 78). But there is another striking example. When the area of Kampereiland was sold to the town of Kampen in 1364 it embarked on a campaign of land reclamation. Starting from mud flats, salt marshes, and salt meadows the town first built low dams (0.5 m) that were used to enhance the process of deposition with reeds also being grown to accelerate this process (de Kraker, 2010; de Kraker, 2011). As soon as the area did not flood anymore during the summer it was embanked by building a higher dike (1.00 m) while the lower dams were partly removed. Within a year this latter dike was raised to 1.50 m, which was high enough to protect the area during summertime but simultaneously low enough to allow it to be flooded during the winter. This way, the area could still benefit from winter flooding along with the deposition of fertile sediment as a natural fertiliser. These conditions allowed local communities to use the land as arable during summertime and, in spite of the impact of salt water during winter flooding, it remained possible to grow most grain crops through the year (Bazelmans, 2005: 69–70). Given the same physical conditions and given the same technology, why would local communities living in different time periods not think of the same solutions to deal with their coastal environment? Therefore, it would be interesting to study this kind of dike building and look for aspects of continuation from the late Iron Age onwards. The focus here needs to be put on both the hydraulic engineering and the kind of land use that results from this. It implies a re-assessment of the idea that ring dikes should have served as high rise dikes keeping out all floods all year round and instead beginning to think in terms of a variety of summer dikes depending on local conditions and land use (Bazelmans, 2005: 68–69; Tys, 2005: 110–116).

Hopefully we shall begin to learn and understand more about the reasons why these earliest dikes and dams were built by using, for example palaeo-geographical reconstructions that enable us to better provenance the precise location of these artifacts. Theoretically speaking, provenancing their location could significantly be enhanced by looking at the settlement system, the patterns of contemporary drainage ditches, and the complex network of natural creeks.

VII. Conclusion

In this article it has become clear that although the earliest dikes, dams, and sluices date from the pre-Roman period (first century BC), the majority discovered so far were built and later regularly altered during the Roman era. While in the Wadden area generally small dikes were raised, in the *Maasmond* dams and sluices were also built. These dikes were by no means comparable to the several meters high and hundreds of kilometers long modern embankments. Most of them were built of plaggen and clay sods and show traces of flooding and repairs. Since the 1990s a lot of new evidence has come from the coastal area, but it remains very difficult to provenance other possible sites where more dams, dikes and sluices could once have been located. The main reason for this lies in the dramatic transformation of the coastal landscape, which has caused the disappearance of vast parts of the (post)-Roman landscape. Getting back to the key question of the 2010 *Landscape or Seascapes?* Congress, it has become clear that local communities have demonstrated similar resilience in developing a basically similar technology. However, local differences remain. It also has become clear that late Iron Age and Germanic/Roman settlers in various parts of the coastal landscape have had a far greater impact on their environment than was assumed in the late 1970s. This has brought to light complex hydraulic engineering works such as dams, dikes, ditches, and sluices in an area from southwest coastal Flanders as far as Schleswig-Holstein. Nevertheless, the representativeness of the present data base can be questioned. The distribution of the hydraulic engineering works shows very large blank areas, largely due to archaeologists not recognising such features, the archaeological methods employed while excavating, and/or lack of excavations. These kinds of off-site structures will only be found by extended trenching around settlements and burial places. This means that the desk-based archaeologists who are responsible for formulating project outlines need be convinced of the importance of these still somehow neglected archaeological features.

These earliest dikes and dams have a large similarity with dikes and dams from, for example, Kampereiland of the late medieval period. To enable better predictions about the precise locations of these hydraulic engineering features, the next step in the research is most likely going to be to look for similarities between both dike building of the first century BC and locations where summer dikes were built during the medieval period.

Table 9.1 Late Iron Age and Roman dikes, dams, culverts and sluices in northwestern Europe

Location	region/country	type	observed length (l), width (w) and height (h)	material	date	reference
Tofting	Schleswig-Holstein, Germany	dike	l ? h? w 3 m		*c.* 100 AD	Bantelmann, 1955: 22
Feddersen Wierde	Ost-Friesland, Germany	dike	h 1,2 m	clayey sods: gradually sloped at the seaside	2nd century AD	Haarnagel, 1979: 187–188; Tafel, 159–160
Bentumersiel	Ost-Friesland, Germany	dam with culvert	culvert: l 4,5 m		Roman	Prison, 2009: 127–129
Anjum	Friesland, the Netherlands	dike	l not published, w 2,4 m, h 1,2 m			Nicolai (ed.), 2010
Dongjum-Heringa	Friesland, the Netherlands	dike	l ? w 14 m, h 1 m		1st century AD	Bazelmans et al., 1999: 33–39
Jelsum	Friesland, the Netherlands	dike	unpublished (in press)		unpublished	
Peins-East	Friesland, the Netherlands	dike	5 phases, l 54 m, w 14 m, h 1 m		1st century BC	Bazelmans et al., 1999: 39–48
Wergea terp groot Palma	Friesland, the Netherlands	dike	2 phases, h 0,14 m and h 0,28 m. Length and width unknown.		1st century AD	Zandboer, 2010: 41–42
Wijnaldum-Tjitsma	Friesland, the Netherlands	dike	w 6,5 m (eventually 11,5 m), h 0,95 m		2nd century AD	Gerrets, 2010: 58–59; Gerrets/de Koning, 1999: 98–99
Wijnaldum-Tjitsma	Friesland, the Netherlands	dike	l ? w 3,5 m, h 1 m		2nd century AD	Gerrets, 2010: 58–59

Barendrecht-Zuidpolder	South-Holland, the Netherlands	dam with sluice	sluice l 7,5 m; dam l 14 m, w 8 m	dam: sods with horizontal layers of vegetal material, protected by vertical poles; functioning as an revetment. The sluice was made of oak.	150–250 AD (ceramic)	Carmiggelt/Trierum/Wesselingh, 2010: 129–132
Capelle aan de Ijssel	South-Holland, the Netherlands	dam with culvert	culvert l 5 m, inner bore 0,15 m	culvert made from a cloven and hollowed trunk (alder) bond with willow twigs; dam protected with vertical poles	Roman	Van Trierum, 1988: 49–50
Capelle aan de Ijssel-Middelwatering West	South-Holland, the Netherlands	dam with three sluices	dam l 9 m, w 5–8 m, h 2,2 m; sluice 1. l 4 m, w 0,4 m; sluice 2. not conserved; sluice 3. fragment l 0,2 m, w 0,4 m	dam: sods with horizontal layers of vegetal material, protected by vertical poles; functioning as an revetment. The sluices are made of black or gray alder wood	*c.* 300 AD (ceramic, coin)	Jacobs/Pavlovic, 2001: 20–27, 33
Den Haag Oosteinde	South-Holland, the Netherlands	culvert, part of an extended Roman drainage system		wood, very badly conservated	125–250 AD (estimated)	Waasdorp, 2001: 11; Siemons/Lanzing, 2009: 352, 375–376
Poortugaal	South-Holland, the Netherlands	culvert	l 6,6 m	hollowed trunk	100–225 AD (ceramic)	De Roo/Brouwer, 1972: 212–215

(Continued)

Table 9.1 (Continued)

Location	region/country	type	observed length (l), width (w) and height (h)	material	date	reference
Poortugaal	South-Holland, the Netherlands	culvert	l 5,7 m	hollowed trunk	100–225 AD (ceramic)	De Roo/Brouwer, 1972: 212–215
Rotterdam-Hartelkanaal	South-Holland, the Netherlands	dam with culvert or sluice	culvert l over 4 m	the culvert or sluice is made from a trunk of black or gray alder	cal 100 BC-75 AD	Van Trierum, 1988: 45–46
Schiedam Westabtspolder	South-Holland, the Netherlands	sluice	l 3,6 m, w 0,4–0,6 m, bore 0,2–0,37 m	culvert made from a cloven and hollowed trunk bond with willow twigs	1st/2nd century AD (ceramic)	De Roo, 1972: 209–216; Bult, 1983: 28
Spijkenisse Hartel West	South-Holland, the Netherlands	dam with culvert	culvert l > 3,3 m	culvert made from a trunk (alder)	probably 175–225 AD	Döbken, 1989: 328
Valkenburg, Marktveld	South-Holland, the Netherlands	dam with sluice	sluice l 2,3 m, w 0,24 m	sluice made from the trunk of an ash tree. The valve was made of oak.	end 2nd century AD	Bult/Hallewas, 1990: 14–15
Vechten	Utrecht	culvert(?)	unpublished		Roman?	Byvanck, 1943: 407
Vlaardingen-Hoogstad	South-Holland, the Netherlands	dam with sluice	dam l 9 m, w 6,3 m; sluice l 6,3 m	dam: several decimetres of clay- or sand sods separated by thin layers of vegetal material with wooden poles functioning as revetment and over 20,000 sherds functioning as an foundation layer.	70–125 AD	De Ridder, 1999: 11–12 (Figure 3. D1)

Vlaardingen-Hoogstad	South-Holland, the Netherlands	dam with sluice	unpublished	dam: several decimetres of clay- or sand sods separated by thin layers of vegetal material with wooden poles functioning as revetment	70–125 AD	De Ridder, 1999: 11–12 (Figure 3. D2)
Vlaardingen-Hoogstad	South-Holland, the Netherlands	dam (?) with culvert (?)	unpublished		70–125 AD	De Ridder, 1999: 11–12 (Figure 3. D3)
Vlaardingen-Hoogstad	South-Holland, the Netherlands	culvert	unpublished	two Alder three trunks. One of them cleaved down its lenght and hollowed	70–125 AD	De Ridder, 1999: 11–12 (Figure 3. D4)
Vlaardingen-Hoogstad	South-Holland, the Netherlands	dam (with sluice?)	unpublished	dam: several decimetres of clay- or sand sods separated by thin layers of vegetal material with wooden poles functioning as revetment	120–175 AD	De Ridder, 1999: 11–12 (Figure 3. D5)
Vlaardingen-Hoogstad	South-Holland, the Netherlands	dam	unpublished	dam: several decimetres of clay- or sand sods separated by thin layers of vegetal material with wooden poles functioning as revetment	120–175 AD	De Ridder, 1999: 11–12 (Figure 3. D6)

(Continued)

Table 9.1 (Continued)

Location	region/country	type	observed length (l), width (w) and height (h)	material	date	reference
Vlaardingen-Hoogstad	South-Holland, the Netherlands	dam (with sluice?)	unpublished	dam: several decimetres of clay- or sand sods separated by thin layers of vegetal material with wooden poles functioning as revetment	120–175 AD	De Ridder, 1999: 11–12 (Figure 3. D7)
Vlaardingen-Hoogstad	South-Holland, the Netherlands	dam with sluice	unpublished		between 180 and 175 cal BC	De Ridder, 1999: 11–12 (Figure 3. D8)
Vlaardingen-Hoogstad	South-Holland, the Netherlands	dam with culvert or sluice	culvert/sluice: l *c*. 2 m, w *c*. 0,3 m		between 91 cal BC and 57 cal AD	De Ridder, 1999: 16
Zuidland	South-Holland, the Netherlands	culvert	unpublished	culvert made from a cloven and hollowed trunk (alder) bond with wooden pins (Buckthorn)	150–275 AD	Trierum, 1986: 49–75
Seroorskerke	Zealand, the Netherlands	dike	phase 1: l 70 m, w 8 m, h 0,16 m; phase 2: l 82 m, w 5 m, h 0,8 m		phase 1. 25–200 AD (14C, 2sigma), phase 2. 200–250 AD	Dijkstra/Zuidhoff, 2011: 272–275
Raversijde	Flanders, Belgium	dike	l 110 m, w 11 m, h 1,1 m		Roman	Pieters et al., 2006
Stene	Flanders, Belgium	dike	unpublished		Roman	Demey et al., 2010

Bibliography

Asselen, S., van, Stouthamer E. and Asch, Th.W.J., van (2009) 'Effects of peat compaction on delta evolution: a review on processes, responses, measuring and modelling', *Earth Science Reviews*, 92, pp. 35–51.

Bantelmann, A. (1955) *Tofting, eine vorgeschichtliche Warft an der Eidermündung*, Neumünster.

Bazelmans, J. (2005) 'Die Wurten von Dongjum-Heringa, Peins-Oost und Wijnaldum-Tjitsma: kleinmaßstäblicher Deichbau in ur- und frühgeschichtlicher Zeit des nördlichen Westergo', in: *Küste und Marsch*, Oldenburg, pp. 68–84.

Bazelmans, J., Gerrets, D., Koning, J. de, Vos, P. (1999) 'Zoden aan de dijk. Kleinschalige dijkbouw in de late prehistorie en protohistorie van noordelijk Westergo', *De Vrije Fries*, 79, pp. 9–76.

Berendsen, H.J.A., and Stouthamer, E. (2001) *Palaeogeographic development of the Rhine-Meuse delta, The Netherlands*, Assen.

Besteman, J.M., Bos, J.M., Gerrets, D.A., Heidinga, H.A., Koning J., de (eds) (1999) *The excavations at Wijnaldum*, Rotterdam (Reports on Frisia in Roman and medieval Times; 1).

Boersma, J. (2005) 'De trek naar de klei', in: L.P. Louwe Kooijmans, P.W. van den Broeke, H. Fokkens and A. van Gijn (eds), *Nederland in de Prehistorie,* Amsterdam, pp. 561–576.

Bosman, A., De Clercq W., Hoeveberg, J. (2006) *Romeinendag/ Journéed'archéologie Romaine*, Ghent.

Bult, E.P., Hallewas, D.P. (1990) *Graven bij Valkenburg III, het archeologisch onderzoek in 1987 en 1988*, pp. 1–36.

Campenhout, K. van, Dijkstra, J. (2011) 'Vindplaats 4 Wattelwegsweg, sporen en structuren', in: J. Dijkstra, F.S. Zuidhoff (eds), *Kansen op de kwelder,* Amersfoort, Appendix 4, pp. 272–281.

Carmiggelt, A., Trierum M.C., van, Wesselingh D.A. (eds) (2010), *Bijdragen aan de bewoningsgeschiedenis van het Maasmondgebied*, Rotterdam, (BOORbalans; 6).

Demey, D., Vanhoutte, S., Pieters, M. (2010) 'Een Romeinse dijk met woonplatform te Stene bij Oostende (West-Vlaanderen)', *Romeinendag/Journée d'ArchéologieRomaine – 2010*, pp. 95–97.

Dijkstra, J. (2011) 'De ontwikkeling van de bewoning vanaf de Midden-IJzertijd tot en met de Late Middeleeuwen', in: J. Dijkstra and F.S. Zuidhoff (eds), *Kansen op de kwelder,* Amersfoort, pp. 59–78.

Dijkstra, J., Zuidhoff, F.S. (eds) (2011) *Kansen op de kwelder*, Amersfoort (ADC Monografie; 10).

Döbken, A.B. (1989) 'Spijkenisse Hartel West', *Archeologische kroniek Zuid-Holland. Holland*, 21, pp. 328–329.

Ey, J. (2007) 'Early dike construction in the coastal area of Lower Saxony', in: J.J.J.M. Beenakker, F.H. Horsten, A.M.J. de Kraker and H. Renes, *Landschap in ruimte en tijd,* Amsterdam, pp. 92–99.

Gerrets, D. A. (2010) *Op de grens van land en water. Dynamiek van landschap en samenleving in Frisia gedurende de Romeinse tijd en de Volksverhuizingstijd*, Groningen.

Gerrets, D.A., Koning, J. de (1999) 'Settlement development on the Wijnaldum-Tjitsma terp', in: J.C. Besteman, J.M. Bos, D.A. Gerrets, H.A. Heidinga and J. de Koning, *The excavations at Wijnaldum*, Rotterdam, pp. 73–123 (Reports on Frisia in Roman and medieval times; 1).

Graafstal, E. (2002) 'Logistiek, communicatie en watermanagement. Over de uitrusting van de Romeinse rijksgrens in Nederland', *Westerheem*, 51, pp. 2–27.

Haarnagel, W. (1979) *Die Grabung Feddersen Wierde: Methode, Hausbau, Siedlungs- und Wirtschaftsformen sowie Sozialstruktur*, Wiesbaden.

Halbertsma, H. (1955) 'Barradeels oudste historie terpen', in: W. Dijkstra, M. Wiegersma (eds), *Barradeel. Rapport betreffende het onderzoek van het Lânskip-genetysk Wurkforbân fan de FryskeAkademy*, Drachten, pp. 58–141.

Heeren, S. (2009) *Romanisering van rurale gemeenschappen in de civitas Batavorum*, Amersfoort-Amsterdam (Nederlandse Archeologische Rapporten; 36).

Heeringen, R. van (2005) 'Op zompig veen en stuivend zand', in: L.P. Louwe Kooijmans, P.W. van den Broeke, H. Fokkens and A. van Gijn (eds), *Nederland in de Prehistorie*, Amsterdam, pp. 581–596.

Hessing, W.A.M. (1990) 'Leidschendam: Rietvink', in: W.A.M. Hessing (ed.), *Archeologische kroniek van Holland over 1989, II Zuid-Holland. Holland 22/6,* pp. 342–343.

Hessing, W.A.M. (1991) 'Leidschendam: Rietvink', in: W.A.M. Hessing (ed.), *Archeologische kroniek van Holland over 1990, II Zuid-Holland. Holland 23/6,* pp. 344–345.

Hessing, W.A.M. (1992) 'Leidschendam: Rietvinkpolder', in: W.A.M. Hessing (ed.), *Archeologische kroniek van Holland over 1991, II Zuid-Holland. Holland 24/6*, pp. 366–367.

Hessing, W.A.M. (1993) 'Leidschendam: Rietvink', in: W.A.M. Hessing (ed.), *Archeologische kroniek van Holland over 1992, II Zuid-Holland. Holland 25/6*, pp. 336–338.

Hiddink, H.A. (2005) *Archeologisch onderzoek aan de Beekseweg te Lieshout*, Amsterdam (Zuid-Nederlandse Archeologische Rapporten; 18/1).

Huisman, D.J. (ed.) (2009) *Degradation of archaeological remains*, Den Haag.

Huisman, D.J. and Deeben J. (2009) 'Soil features', in: D.J. Huisman (ed.), *Degradation of archaeological remains*, Den Haag, pp. 147–176.

Hoek, C. (1971) 'Oudheidkundig Bodemonderzoek te Rotterdam en omgeving in 1970', *Rotterdams Jaarboekje 1971*, pp. 131–143.

IJzereef, G.F., Regteren Altena J.F. van (1991) 'Nederzettingen uit de midden- en late bronstijd bij Andijk en Bovenkarspel', in: H. Fokkens, N. Roymans (eds), *Nederzettingen uit de bronstijd en de vroege ijzertijd in de lage landen*, Amersfoort, pp. 61–82 (Nederlandse Archeologische Rapporten; 13).

Jacobs, E., Pavlovic, A. (2001) *Capelle aan den IJssel, een archeologisch onderzoek van een dam met duikers uit de Romeinse Tijd*, Rotterdam (BOORrapport; 77).

Jonge, W., Bakels, C.C., Bazelmans, J., Deurnhouwer, P., Kort, J.W. de, Veen, M. van, Vos, P. (in preparation) *Het kanaal van Corbulo. Nieuw inzicht, nieuwe vragen. De resultaten van multidisciplinair onderzoek in en na 2004. Rapporten Archeologische Monumentenzorg*, Amersfoort.

Körber-Grohne, U. (1967) *Geobotanische Untersuchungen auf der Feddersen Wierde,* Wiesbaden.

Kort, J.W. de (2008) 'Het kanaal van Corbulo', in: *Forum Hadriani. Romeinse stad achter de Limes*, Voorburg, pp. 24–28.

Kraker, A.M.J. de (2010) 'Lage dijken overstromingen en bemesting. Het einde van de bemesting van de polders op en rond het Kampereiland tijdens de negentiende en twintigste eeuw', *Tijdschrift voor Waterstaatsgeschiedenis*, 18, pp. 25–40.

Kraker, A.M.J. de (2011) 'Sustainable coastal management, past, present and future or how to deal with the tides', *Water History*, 3, pp. 145–162.

Londen, H. van (2006) *Midden-Delfland. The Roman Native Landscape Past and Present*, Amsterdam.

Nicolay, J.A.W. (ed.) (2010) *Terpbewoning in oostelijk Friesland. Twee opgravingen in het voormalige kweldergebied van Oostergo*, Groningen.

Pals, J.P. (1999) 'Preliminary notes on cropplants and the natural and anthropogeneous vegetation', in: J.M. Besteman, J.M. Bos, D.A. Gerrets, H.A. Heidinga, J. de Koning (eds), *The excavations at Wijnaldum*. Rotterdam, pp. 139–149 (Reports on Frisia in Roman and medieval Times; 1).

Philippa, M., Debrabandere, F., Quak, A., Schoonheim, T., Sijs N. van der (2003–2009) *Etymologisch woordenboek van het Nederlands*, Amsterdam.

Pieters, M., Demerre, I., Zeebroek, I. (2006) 'Dijk uit de Romeinse tijd aangesneden onder het Middeleeuwse vissersdorp Walraversijde', in: A. Bosman, W. De Clercq and J. Hoeveberg; *Romeinendag/Journée d'archéologie Romaine*, Ghent, pp. 93–97.

Poel, J.M.G. van der (1960/1961) 'De landbouw in het verste verleden', *Berichten Rijksdienst voor het Oudheidkundig Bodemonderzoek*, 10/11, pp. 125–194.

Prison, H. (2009) 'Von Prielen und Sielen', *Archäologie in Niedersachsen*, 12, pp. 127–129.

Renswoude, J. van, Kerckhove J. van (2009) *Opgravingen in Geldermalsen-Hondsgemet*, Amsterdam (Zuid-Nederlandse Archeologische Rapporten; 35).

Ridder, T. de (1999) 'De oudste deltawerken van West-Europa', *Tijdschrift voor Waterstaatsgeschiedenis*, 8, pp. 10–22.

Ridder, T. de (2007) 'Een verloren strijd. Watermanagement in de Romeinse Tijd', *Holland, Historisch Tijdschrift*, 39, pp. 17–24.

Siemons, H., Lanzing, J.J. (eds) (2009) *Bewoningssporen uit de Romeinse tijd in het Wateringse Veld*, Den Haag (Haagse Oudheidkundige Publicaties; 11).

Trierum, M.C. van, Henkes, H.E. (1986) 'Landschap en bewoning rond de mondingen van Rijn, Maas en Schelde', in: *A contribution to prehistoric, Roman and medieval archaeology. Teksten van lezingen gehouden tijdens het symposium te Rotterdam van 5–6 oktober 1984*, Rotterdam (Rotterdam Papers; 5).

Trierum, M.C. van, Döbken, A., Guiran, T. (1988) *Bijdragen aan de bewoningsgeschiedenis van het Maasmondgebied*, Rotterdam (BOORbalans; 1).

Tys, D. (2005) 'Landscape, settlement and dike building in coastal Flanders in relation to the political strategy of the counts of Flanders, 900–1200', in: M. Fansa (ed.), *Kulturlandschaft March. Natur – Geschichte – Gegenwart. Vorträge anlässlich des Symposiums in Oldenburg vom 3. Bis 5. Juni 2004*, Oldenburg, pp. 106–126.

Velde, H.M. van der (2011) *Wonen in een grensgebied*. Amersfoort.

Vollgraff, C.W. (1938) 'De dijk van Drusus'. *Mededelingen der Koninklijke Nederlandsche Akademie van Wetenschappen, Afdeling Letterkunde, Nieuwe Reeks*, 1, pp. 555–577.

Vollgraff, C.W. (1939) 'De Moles van Drusus', *Mededeelingen der Koninklijke Nederlandsche Akademie van Wetenschappen. Afdeling Letterkunde. Nieuwe reeks,* 2, pp. 141–143.

Vos, P.C. (1999) 'The subatlantic evolution of the coastal area aroud the Wijnaldum-Tjitsmaterp', in: J.M. Besteman, J.M. Bos, D.A. Gerrets, H.A. Heidinga and J. de Koning (eds), *The excavations at Wijnaldum,* Rotterdam, pp. 33–72 (Reports on Frisia in Roman and medieval Times; 1).

Vos, W.K. (2009) 'Bataafs platteland. Het Romeinse nederzettingenlandschap in het Nederlandse Kromme-Rijngebied', *Nederlandse Archeologische Rapporten*, 35, Amersfoort-Amsterdam.

Waasdorp, J.A. (2003) 'III M.P. naar M.A.C. Romeinse mijlpalen en wegen', *Haagse Oudheidkundige Publicaties*, 8.

Zandboer, S. (2010) *Wergea Terp Groot Palma, Twee terpen in de vaart*, Amersfoort (ADC-rapport 2053).

Zeist, W. van, Hoorn T.C. van, Bottema S. and Woldring H. (1976) 'An agricultural experiment in the unprotected salt marsh', *Palaeohistoria*, 18, pp. 111–153.

10 The medieval embankment of coastal Flanders in context*

Dries Tys, Brussels Free University (VUB)

Abstract

In North-West Europe, most coastal plains were embanked and reclaimed between the eleventh and thirteenth centuries. This step was, however, not a necessary one for societies to develop means of existence and settlements in coastal tidal environments. In early medieval Flanders, people and coastal society lived and existed in a situation of equilibrium with the natural resources at their disposal. Archaeological research shows that certain zones of the salt marsh environment of Flanders were already inhabited from the seventh century onwards (if not earlier). From the tenth century the choice was made to move from equilibrium to modification and transformation of the coastal environment. The use of the natural resources of the salt marshes was replaced by the organisation of embanked wetlands, with a growing potential for a diverse range of economic activities. We see how princely initiatives initiated this large scale and systematic embankment of this maritime region. The choice of these environmental solutions led to the particular material development of the landscape of the coastal plain. Another choice of a different concept under the same ecological conditions (for instance to leave the salt marshes to the free farmers) would have led to a different landscape. Landscapes are indeed contextual social products and we must avoid deterministic approaches and methods in order to understand the relationship between humanity and the environment.

I. Human response to the environment: questions and methods

In North-West Europe, most coastal plains were embanked and reclaimed between the eleventh and thirteenth centuries (Allen, 1997; Rippon, 2001). The ecological consequences of this control of tidal and inland waters were important, as described by Cook and Williamson (1999: 2): '*Within the embanked areas, the natural patterns of creeks and levees and the peat and the mineral alluvium were manipulated and ultimely transformed by various forms of agricultural use; new patterns of soil and topography were created, new landscapes and new habitats formed*'. This step was, however, not a necessary one for societies to develop means of existence and settlements in coastal tidal environments. This is shown by many examples, amongst others in the British Fenlands and early medieval Frisia (see A.O. Darby, 1983; De Langen, 1992; Heidinga, 1997; Allen, 2000: 1211; TeBrake, 2000). In both areas, as well as early medieval Flanders, settlements and economies developed in an open salt marsh environment without large-scale embankments in order to protect people from floods.

* Acknowledgments: Cecile Baeteman, Pieterjan Deckers, Henk Weerts, Barbora Wouters, Guus Borgers, Gilles de Langen, Adrie de Kraker, Erik Thoen, Tim Soens, Steve Rippon, Lies Vervaet.

Landscapes or seascapes? The history of the coastal environment in the North Sea area reconsidered, ed. by Erik Thoen, Guus J. Borger, Adriaan M.J. de Kraker, Tim Soens, Dries Tys, Lies Vervaet & Henk J.T. Weerts, Turnhout, 2013 (*CORN Publication Series*, 13), pp. 199–239.

BREPOLS PUBLISHERS DOI: 10.1484/M.CORN.1.101554

The step to convert a landscape and a way of life without dikes and water control systems into an embanked landscape and lifestyle was therefore a deliberate choice. It raises the question why people in a coastal region took this step to control the impact of flooding, to develop technological solutions to organise the landscape, and to invest in a more complicated relation with the environment. The embankment of a tidal area contained so much more than extensive land reclamation and enlargement of the occupied and worked rural area and living space. It involved views on land produce, perceptions of the environment, scales of impact, means of investment, and the knowledge and distribution of technology. In other words, the choice to embank was and is a social practice, based on the social and political perspectives, choices, and strategies. Landscapes can indeed be seen as material structures that are '*re-worked, interpreted and understood in relation to differing social and political agendas*' as Tilley and Bender state (Tilley, 2006: 8; Bender, 2006).

In coastal Flanders, we see how princely initiatives initiated the large scale and systematic embankment of this maritime region. The Counts of Flanders and their administration seem to have had the power to support and initiate the transformation of the early-medieval salt marsh landscape to the well-known Flemish polder area. In order to understand this process and why and how social and political powers had a thorough impact on the formation of maritime landscapes (and lifestyles), we will try to approach the landscape, or seascape, from coastal Flanders from the perspective of the impact of social groups and actors on their environment. Or put otherwise: in order to understand the coastal landscape, we not only have to look to the sea, but also to the land, power(s), and to society.

This is done via the study of the spatial and temporal contextual social property relations as indicators of power in the landscape (Hoppenbrouwers and Van Zanden, 2001: 35–36; Tys, 2004). The reconstruction of the dynamics and role of social property relations in medieval coastal Flanders is both confronted with and imbedded in the reconstruction of the material formation processes of the medieval coastal landscape and to the consequent environmental behaviour and perception behind these processes. In order to understand landscape transformation processes, we will have to look at landscape morphology, morphogenesis, and life trajectories of the material elements related to the embankment of the coastal plain. This does not only imply that we have to look at the contextual manner in which people and groups have dealt with these elements, but also that we have to be careful not to work with too strict, static, or formal typological classifications. One morphogenetic type of landscape can hide very different historical phenomena and static typologies can result in teleological understandings of landscapes. Every landscape is also always contemporary and its subsequent landscape forms and elements have a contextual position in it, as a result of the transformation of the same landscape following the spatial conduct and agency of its inhabitants (Chouquer, 2000).

We will apply this integral approach on a specific case study: namely, the central part of the Flemish Coastal Plain, the so-called Kamerlings Ambacht at the east bank of the present River Yzer, between the present towns of Ostend and Nieuwpoort (Figure 10.1). This 8000 ha large area was carefully studied in relation to field structures, property relations, earthworks, place names, general landscape morphology, settlement history, and more. The tool that was used to study these aspects was the retrogressive landscape reconstruction, starting from the early nineteenth century land registry plans of the different

Figure 10.1 Situation of Kamerlings Ambacht and places mentioned in the text in the Belgian coastal plain

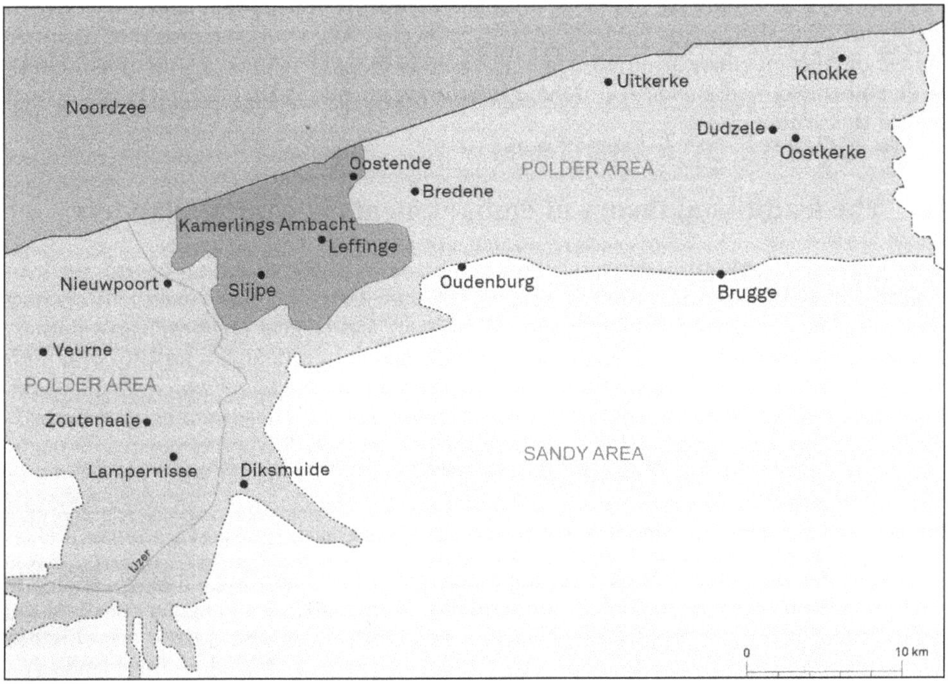

parishes in the area. This reconstruction was done through a wide range of sources. These include maps, aerial photographs, geological and geomorphologic data, earthwork evidence and archaeological data, written sources that include land registry books from between the fourteenth and the seventeenth centuries, and fragments of charters and other early medieval documents. A source of critical importance to this are the sixteenth- and early-seventeenth-century land books from the local medieval hydrology institutions called *wateringen* (water boards). These books cover each individual field in every parish, including the exact position of field-names, moated sites and farmsteads, and other property[1].

Through this reconstruction it was possible to study the evolution of the landscape and its materialisation, as well as its morphogenetic development, and this in the historical

[1] The combination of nineteenth-century land-registry-plans and historical data is not a novel. Already in 1950 M. Gysseling succeeded in the reconstruction of the fifteenth-century field pattern of a part of the parish of Oudenburg. However, Gysseling did not go further than the descriptive reconstruction of the field pattern and neglected the interpretation of the morphology and horizontal stratigraphy of this fifteenth-century field structure and pattern (see Gysseling, 1950). After Gysseling this method of reconstructing in detail the late-Medieval and/or early-Modern field structure of a parish was not used until the 90s (see for instance Tys, 1997).

context of the micro-regional property relations. In this way we might hypothetically deduce which decisions and choices were made concerning the development of a landscape at any given time. Partly, this has been done to try and understand why powerful agencies, like the Counts of Flanders, invested in transforming coastal landscapes within their lands. Was there a link between strategy, expected outcomes, and projected income? Where did they get their inspiration to embark upon the systematic construction of large-scale embanked landscapes? All of these interesting questions are returned to in the final part of this article.

II. The traditional theory of embankments in coastal Flanders

The question of how to understand 'human response' to the maritime environment in coastal Flanders was addressed by several scholars from the early twentieth century onwards. The concept of 'human response' in itself was typical for these early studies, because they started from a perception of the coastal environment as a hostile condition. To a certain extent these views fitted into the physiocratic tendency which favoured the importance of land development over environmental values. These studies were embedded in the 'classic' opposition between 'natural' and 'cultural', whether positive or negative (Barry, 1999: 76–77). Accordingly, human behaviour in coastal areas was bound to 'struggle against the forces of nature' and determined to transform the environment into a humanised landscape (for Flanders: see for instance Blanchard, 1906 and Beekman, 1932).

One of the characteristics of these approaches was that maritime behaviour and perception in itself were not looked at, so no role for human choice and agency towards the environment was considered. This is very clear in the way scholars dealt with the old Transgression Model for the development of the coastal area in which the human landscape was determined by the forces of destructive phases of sea-level rise. Therefore, in early studies like the writings of Verhulst, the environment, including human environmental policy, was not the object of research but only the context for the actual research topic: the occupation and settlement history of the coastal landscape in medieval Flanders (Verhulst, 1959; 1964). People were looking for a framework for the explanation for occupation and they found them in the model of cyclical marine transgression (sea-level rise) and regression, when coastal plains and lowland estuaries became permanently inundated by water during periods of sea-level rise (Tavernier, 1938; Ameryckx, 1958; Behre *et al.*, 1979, see Baeteman in this volume). Within these transgression and regression models, the period between 300 and 800 AD has been characterised as an era of marine transgression (the Dunkirk II transgression), and exploitation and occupation of the North Sea coastal plains has been suggested as seasonal and limited in nature, mainly due to the inhospitable conditions.

The written and material sources from the period between the Roman area and the eleventh century have also tended to be used in an inconsistent way, in Flanders as well as in Britain[2]. For example, the first references to settlements or individuals associated

[2] Rippon, 2001: 32: '*Inconclusive or poorly dated evidence was being forced into a preconceived straight-jacket;*'.

with settlements in the Flanders coastal plain date from the end of the eighth century. These were land grants to the abbey of St. Peter's Ghent, comprising seasonal grazing, and an estate associated with at least one settlement in the coastal plain, near Oostburg (Gottschalk, 1955: 16). Verhulst and others saw these grants as an indication of the end of the Dunkirk II transgression, and an onset of land reclamation, primarily sponsored by major monasteries from the Scheldt valley (1959, see also Boterberge, 1962). The general absence of textual evidence relating to activities in the coastal plain prior to the late eighth century was then used as evidence of an absence of permanent activity and settlement in this area, i.e. textual visibility governed all interpretation of early medieval activity in the Flemish, northern French and Zeeland coastal clay lands[3].

Yet, in Frisia (from the Rhine delta eastwards to Friesland), the oldest textual references to settlement and activities in the coastal plain also date from the eighth century, but the archaeological research on terp settlements has long demonstrated the permanent habitation of this region, in the early medieval period, several centuries before activity became visible in the historical records (see above; De Langen, 1992). In Flanders and northern France, however, the archaeological record was forced into the textually dated chronological framework and '*the transgression model and the gaps in the archaeological record were continuously used to reinforce each other*' (Ervynck e.a., 1999: 103).

On another level, the first appearance of dikes in written sources from the early eleventh century (*Tubindic* and *Isendycke*, Gottschalk, 1955: 20) were considered to be the start of of local embankment construction, supposedly as a result of the combination of the end of so called transgression of the sea and a rise in population. These first historical attestations were considered as a *terminus post quem*, as a start date for the embanking of the coastal area while, as Blok said, it should be a *terminus ante quem*, meaning that these dikes had already been present in the landscape for some time before they were first mentioned in historical sources (Blok 1984).

Soil scientists and historians also reconstructed three long rectangular dikes (the *Oude Zeedijk* east of Veurne, the *Silinghe* between Oudenburg and Bredene and the *Gentele* between Brugge and Uitkerke, Figure 10.2) in relation to the soils attributed to the so-called transgressions of the model (Moormann, 1951; Ameryckx and Verhulst, 1958). According to these views, these three dikes were built as defensive reactions against flooding of the estuaries of the Zwin and the IJzer caused by the so called Dunkerque III transgression (most probably exceptional storm surges) in the course of the eleventh and early twelfth centuries in Flanders. We now know that at least two of these dikes (the *Silinghe* and the *Gentele*) are more recent (late twelfth century or even thirteenth century) and were raised laterally alongside late twelfth century canals which led drainage waters from the inland to the sea, so protecting the poldar area from flooding by rainwater (Tys, 2001).

[3] In the same line of reasoning, the toponymist Gysseling used the transgression model as a chronological occupation argument to state that in coastal Flanders early medieval place names were used until the 10th century (Gysseling, 1960: 1126). See also Warlop, who uses the transgression model as a chronological framework to date institutional developments (Warlop, 1968).

Figure 10.2 In the fifties of the twentieth century, three rectangular dikes were reconstructed in the landscape without any historical, archaeological and other research.
These dikes were: 1: Oude Zeedijk – 2: Silinghe (also called Dijk van de Blankenbergse watering) – 3: Gentele or Blankenbergse dijk
(source: Verhulst, 1959, Figure 3)

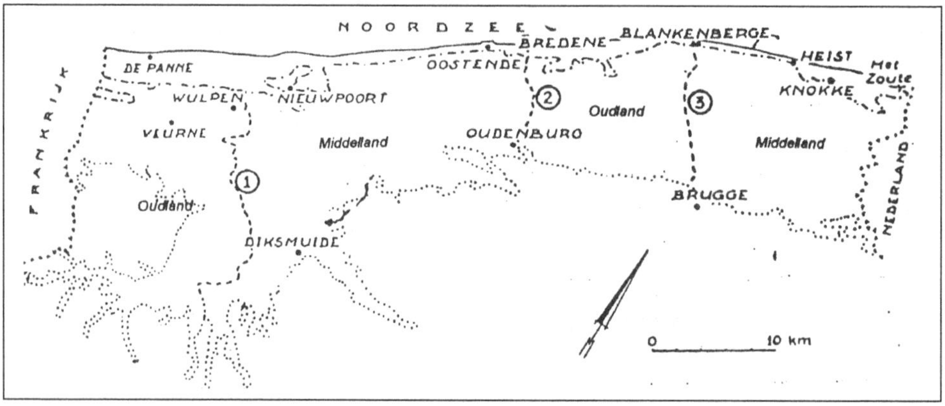

The methodology they used to reconstruct these first dikes was in fact nothing more than the static and formalistic projection of twentieth-century landscape features in an eleventh- and twelfth-century context, without proper retrogressive landscape analysis, without attention for the spatial-temporal context of these hypothetical embankments, and without any archaeological research (Tys, 2001). At the same time, the occupation zones demarcated by these linear embankments were entirely dependent upon the Dunkerque chronological framework and physical determinism used by its designers (Verhulst, 1980: 15). The embankment model was as artificially established as the entire Dunkerque Model and its chronology themselves (Ervynck e.a., 1999: 117). Scholars including Dekker, Henderickx, and Besteman criticised the way in which settlement history was made entirely dependent on the theories of the soil scientists and argued for more research, both archaeological and historical, together with a consideration of wider causal factors (Dekker, 1980; Henderikx, 1987 and Besteman, 1990).

Unfortunately, the Dunkerque Model is deeply rooted in public history and its simplicity is precisely what makes it so attractive[4]. However, the vast majority of coastal sedimentologists in Belgium, Germany, Britain, and The Netherlands have now realised that this model is obsolete. Instead, it is now understood that trans- and regressions in a coastal setting are the result of an interplay of several factors, including human action (see Weerts, this volume, for an overview).

[4] See for instance Monteyne and Vandaele (2008: 12–16), where the authors bring a late thirteenth-century dike in relation to the so-called Dunkerque III Transgression of the early eleventh century, see also Boterberge, 2000: 13.

III. Environmental dynamic and its controlling factors

Detailed and systematic geomorphological research in Flanders, based on sediments and their radiocarbon dating, by Cecile Baeteman and her team from the Belgian Geological Survey and the Brussels Free University (VUB), has shown that in the early medieval period the Flemish coastal plain was a stable and not significantly inundated area of salt-marsh, interspersed with tidal channels which were actively silting up and evolving into channel ridges (Baeteman 1999; Baeteman *et al.* 2002 and see Baeteman in this volume). These findings render the Dunkirk transgression model redundant as a significant barrier to permanent human activity and settlement in this coastal plain. They may not only have localised implications for coastal Flanders, they may also have considerable relevance to other areas of the Channel and North Sea littoral.

The complex of tidal influences leading to the deposit 'history' of the Belgian Coastal Plain is well described in Cecile Baetemans' publications (Baeteman, 1999; Baeteman *et al.*, 2002 and Baeteman in this volume). In order to understand coastal dynamics, one has to take into account the interaction between a variety of controlling factors, such as the relative sea-level rise, tectonics, accommodation space, the morphology of the pre-existing landscape, the sediment budget, compaction, the own natural dynamics of tidal environments, and human action (Weerts, this volume). By the end of the Roman period, between about 1600 and 1500 cal BP (*c.* 350–450 AD), the coastal plain had changed again into a vast tidal environment. This was mainly a result of the penetration of peat bogs by tidal channels and their associated systems, returning to a landscape that had previously existed between 5500 and 2100 cal BP (Baeteman, this volume). This was a very gradual process during which the tidal systems developed, peat was covered with sediments, and the coastal shoreface eroded. The intertidal flat transformed rather rapidly into a supratidal flat as is shown by the data from Cecile Baeteman. Already from 550 AD, channels started to fill and supratidal salt marshes became suitable for human activities. Cecile Baeteman's research highlights the diversity in the potential for human activity in low-lying coastal regions during the early medieval period, rendering allegiance to generalising models dangerous. From the sixth century onwards, the Belgian Coastal Plain was suitable for pastoral activities and human action, as is testified by archaeological excavations of pastoral settlements in this region (infra). The research area was from at least 650 AD onwards a largely silted up natural tidal environment under low-energy conditions. It is remarkable that exactly this specific period of equilibrium between the controlling factors of the coastal environment was regarded as a period of environmental imbalances and catastrophes in the former Dunkerque transgression model (Baeteman, this volume). Thereafter, between the seventh and the tenth centuries many remaining tidal channels silted up and the salt marshes became even more suitable for human activities and settlement. Only the largest tidal channels, like the one that separated *Testerep* from the other saltmarshes, remained open watercourses up to their embanking in the first half of the twelfth century.

IV. The unembanked landscape mirrored in place-names

Scarce written testimonies testify to this supratidal landscape. For example, they describe how in 944 AD Count Arnulf was stopped in the unembanked salt marshes by the rise of the tide at a place called *Drievlit* (the three channels) (Verhulst, 1995: 20). This suggests that the count traveled in the salt marshes, where people lived, that he trusted the environment, but that at the same time, some (large) channels remained active as inlets of tidal waters. Written sources also provide us the names of these tidal channels, as they were present in the landscape before the embankment. Water-names ending on *-aa* are related to the earlier Germanic *ahwo* (Gysseling, 1981 and 1983; Gelling and Cole, 2000: 14). For instance, the place-name *Salthan ahwo* refers to a salty channel and is preserved in the modern name Zoetenaaie in Flanders (Gysseling, 1960: 1104–1105). The *Holt auua* (Houtave), mentioned in charters in 1003 AD, refers to the channel that runs into peaty marsh bushes, possibly at Meetkerke (Gysseling, 1960). The *Gersta* running east from present Dunkerque could refer to a relation between the coastal plain and the production or trade of barley, while the tidal channel that formed the south edge of *Kamerlings Ambacht* was referred to as the *Suthana* (southern channel) in 1116 AD, indicating that the medieval boundaries of the research area were determined by features of the natural supratidal environment (De Coussemaker, 1882–1891; No. 22)[5]. Water-names ending on *-fliet*, *-flit*, *-vlit*, *-kreec*, *-swen*, and *-daerm* seem to refer to smaller channels and branches of the channel network. The muddy tidal flats and lower salt marshes next to channels seem to have been named and referred to as *sclipis* (*slib* in Dutch) or *waes* (muddy soil) (Tys, 2003: 201). *Waes* is certainly related to the Anglo-Saxon *waesse*, which was used for lower alluvial lands next to watercourses that were flooded regularly (Gelling and Cole, 2000: 93). Its presence in the toponymical vocabularly is amongst others testified in the place-name *Uuas-kapelle* (present Westkapelle near Knokke) (Gysseling, 1961)[6]. The place-name element *nesse* is used for local higher grounds in wetlands, such as higher saltmarshes or salty meadows, or even for eroded saltmarshes next to migrated channels. As a topographical noun it was known and used on both sides of the Channel in the early Middle Ages (Gelling 1998 and De Langhe, 1967). The oldest testified in Flanders is the *Lampanesse* (actual village of Lampernisse), or the higher ground where *lampri* or lambs were herded (Gysseling, 1960). Other Flemish examples are *Fronenes* (testified in 976 AD), *Gaternesse* (near Oostburg), *Scarpenisse*, *Merlodonisse* (near the abbey of Ter Doest), the *Vercanesse* or pigs nesse, and the *Vaghenesse* (in Middelkerke) (Tys, 2003: 199). There is also the place-name element *wold* which refers to local marshy bushes and peat bogs in the tidal landscape (Halbertsma, 1963: 190–193).

These place names and water-names describe elements and aspects of the natural environment and landscape that existed in coastal Flanders before embankment. As such, they are difficult to date. What we can say is that they show that the unembanked landscape was named and perceived with a certain degree of complexity familiar to any tidal zone. Margareth Gelling considered the typical and specialised character of this vocabulary and these place-names as an indication of the orienting importance of the topographical features of natural landscapes as these (Gelling and Cole, 2000: xiii). Place-name forma-

---

5 Other familiar waternames are the 'old *ahwo*' or '*Veteri A*' near Diksmuide, the '*Diapanha* or steep channel east of Bruges (Henderickx, 1995: 77) and the 'royal *ahwo*' or '*Cueninckx Ee*' near Gistel (Tys, 2003; 2005).

6 On the level of micro place-names in the research area, we found names like *Coppers Waese*, *Hof ten Waese* (a saltmarsh island in a tidal channel!) and *Nederwazeken* near a former tidal channel (Tys, 2003: 201).

tion is closely related to a sense of place and to what is significant and of major concern in the landscape to those who shared the names. In the unembanked tidal landscape of early medieval Flanders, the natural topographical structuring elements had clearly an orienting value. Tidal channels, for example, were the main axes in the landscapes which had to be crossed when travelling, with tidal waters and waves posing some danger to the traveller. But more than this, what was named and shared were those landscape features which were of economic and social significance. They indicate which wild and other resources could be found and exploited. Reeds, peat, salt, wild fowl, saltmarshes, and grazing grounds for sheep seem to have been of importance in these shared perceptions of the unembanked coastal landscape. The *nesses* and *ahwo*s surely testify to people living in equilibrium with the tidal environment; moreover, these people clearly recognised the central role of tidal waters in their salt marsh landscapes. To name places is also to bring them in cultural circulation, and through them the wild natural landscape was occupied and exploited by people

V. Settlements in an unembanked landscape, written and material sources (sixth/ seventh – tenth centuries)

The evidence for the development of settlement in the coastal plain of Flanders was for a long period restricted to the interpretation of attestations of *marisci* (sheep farm units) in written sources (Verhulst, 1964 and 1995). In the light of the re-assessment of the geomorphological evidence in coastal Flanders, and the demonstrable danger of following only textual sources as indicators of settlement, recent and current archaeological research in the coastal plain, integrating a wide range of approaches and sources, has sought to evaluate the evidence for the development of settlement in this area from an inter-disciplinary perspective (Tys, 2003; 2010; Deckers, 2010; 2011). This research has suggested the existence of a settlement hierarchy of farmsteads on the contemporary ground surfaces of silted-up tidal channel ridges, focused on nucleated settlements of multiple households located on terp mounds like Leffinge and Bredene (Tys, 2003: 588–598; 2005; Loveluck and Tys, 2006). There are also indications for the existence of larger trade centers in or nearby the coastal plain, such as the supposed trade center near Veurne and/or De Panne that is often referred to as a possible smaller *emporium* (Deckers, 2010).

The physical nature of the settlements surrounding the terps in coastal Flanders is gradually becoming apparent through new excavations on the basis of surface concentrations, as well as the reconsideration of previously excavated material and current surveys. It is not unlikely that the farms on the channel ridges in the coastal plain had a semi-permanent or permanent character. Recent excavations have shown that human presence and settlement in the coastal plain was already underway by the seventh century. The oldest activities in the small terp-like site of *Oude Werf* in Leffinge were dated by ^{14}C to the seventh century, the same age the oldest traces of the settlement at Uitkerke (Deckers, 2011; Pype 2002). Other similar sites have not yielded clear evidence of settlement structures, but indicate a similar origin in the seventh or eighth centuries, such as the site of Uitkerke-Groenwake (Hermans, 2009). At Lampernisse, in West Flanders, a sequence of settlement deposits dating from the seventh to ninth centuries was excavated on the sand ridge of a tidal channel, demonstrating the presence of a settlement which was the predecessor of a later manor (Ervynck *et al.*, 1999). Another with material from the ninth to twelfth centuries was excavated at Plassendale (Vanhoute and Pieters, 2003). Further east, in the area around Oostburg, in the Dutch part of the former County of Flanders, Van Heeringen and others have also excavated a series of settlement sites, dating from the ninth to tenth centuries (Vos and Van Heeringen, 1997: 96–97). Recent excavations in both the area around Dunkerque in present day France and on the former isle of Walcheren have shown that this pattern seems to be present across the entire coastal plain of the southern North Sea area (Lançon, 2005; Dijkstra and Zuidhoff, 2011). The apparent terp-like morphology of the so-called 'churchring villages' on Walcheren and Schouwen (for instance Noordwelle) suggests that also the phenomenon of wierdes and terpen in this region was not isolated to Leffinge and Bredene alone.

The presence of some sixth to seventh century *-inga* (the clan of) place-names like *Latfinga/Leffinge* or *Greveninge* and several seventh to ninth century *-ingaheim* (the settlement of the clan of) and *-heim* (the settlement of) place-names, like *Cavergheem, Ouenghem, Eyerghem, Heyteghem, Michem, Suethem* and many others, also indicate the role of individual farms in the early medieval coastal settlement structure (Henderikx, 1995: 79; Tys 2004).

We can assume that these individual settlements are encountered in written sources as the so-called *marisci*, which appear in written sources from 707 AD. In that particular year, twelve *marisci* in the saltmarshes around Aardenburg were donated to the abbey of Saint Peters near Ghent (Verhulst, 1959; Henderikx, 1995: 76). Before we had archaeological evidence for permanent farmsteads in the saltmarshes, historians considered these *marisci* to be seasonal grazing meadows for sheep (Ibid.). The new archaeological evidence, however, sheds new light on their significance and role in the landscape, being not only sheep meadows but also true sheep farms in the salt marshes and salt meadows. Their presence suggests calm environmental conditions in the tidal area, which is not contradicted by the geological and environmental data. A particular donation of a *mariscus* in 794 AD states that it could feed 190 sheep in summer and 130 sheep in the winter (Henderikx, 1995: 77)[7]. Wool and textile production was undoubtedly the main activity of these early medieval sheep farms, of which several hundred must have been scattered around in the large tidal landscape of Flanders and Zeeland. In 830 AD, for example, ten sheepfarmers in *Frisia* (which extended up to Flanders at that moment) were to deliver each year 855 *pallia* (woollen coats) to the abbey of Fulda who possessed those farms (Lebecq, 1983: vol. II, 382–383). If the production of at least eighty or more woolen coats per sheep farm per year in the salty meadows and marshes of Frisia and Flanders can be extrapolated outwards, the quantity of coats produced by the coastal inhabitants must have been impressive. Next to specialised textile production, these farmers were also involved in meat production, as is also shown by the existence of tithes on meat lambs

[7] The difference between the number of sheep in winter and summer is probably to be explained by seasonal slaughter for the winter, and the breeding of lambs in spring. However, also the type of vegetation available, and it's availability, might have been a factor in the winters before embankment. They might also just have been moving a proportion of the flock (a specific age group) to another feeding ground to over-winter. Nevertheless, seasonal slaughter was present in later ages, as testified by tithe specifications.

in Kamerlings Ambacht, that seem to have gone back to at least the eleventh century (Tys, 2003: 494)[8]. Excepting the exploitation of wild resources like reeds, fowl, and fish, another activity at these sites likely involved the production of salt in at least some of the farms or centres. Evidence for this can be found in an Act where seventeen *culinas ad sal faciendum* (saltpans) belonging to one farm in Zeeland were transferred to the abbey of Lorsch in 776 AD (Dekker, 1971: 34)[9].

The settlement structure of these individual farms with their specialised production was apparently organised around communal centres (villages) on dwelling mounds or terpen, comparable to the situation in early medieval Friesland. An example of these is Leffinge in Kamerlings Ambacht (Figures 10.1 and 10.3, Tys, 2004). This village is one of the oldest on the Belgian coastal plain. Its name ends with the element *-ingas*, which indicates that the origin of this settlement goes back to at least the seventh century (Gysseling, 1969: 22; 28–30.). The oldest historical attestation of the name, *Latfinga*, dates from 988 AD, and mentions Leffinge as a mother parish (Van Mingroot 1988). Small-scale test excavations there have shown the presence of several super-imposed layers in the centre of the presumed dwelling mound[10]. Around the more or less hexagonal village centre of Leffinge (*c.* 100 m wide), a radial field system stretches outwards by up to 400 m across the surrounding landscape. It resembles the radial patterns that are found so often in Friesland at sites like Feddersen Wierde, Foudgum, Ezinge, and many others. This indicates that we are dealing with the same type of settlement, as this particular field system is related to the proper nature of communal terp settlements. The radial system of Leffinge shows that different farms were set on the platform in a radial way. Behind each farm, a more or less equal property stretched out into the salty meadows and salt marshes, resulting in a radial organisation of property relations. Modern historical data also demonstrates that property relations in and around the village of Leffinge were still organised in a radial way in the sixteenth century, as was also the case in seventeenth century Foudgum (De Langen, 1992; Tys, 2003). We can assume that nucleated settlements like Leffinge became the central foci for their micro-regions of approximately 15 km^2, at the administrative, religious and economic levels, from at least the tenth century. Leffinge is the center of a large 'mother parish', following a minster-like parochial organisation. It is the center of local government from the eleventh century and there are clear indications that Leffinge was a micro-regional market center (Tys, 2003). This has parallels with the suggested role of 'trade terps' in Frisia, suggested by Halbertsma (1963) and others. Leffinge was probably not the only terp settlement in the actual Belgian coastal plain since Bredene and Uitkerke also seem to have performed a similar role in the landscape (Loveluck and Tys, 2006).

---

8 Archive Nationales de Paris, S5350: 16r°: '*Primo cest assavoir que la Maison de Sclipes a Rentes de Aigneaulx Mors par an selon que les vieux livres et registeres remonstrent a la somme de 80 lb 7 sol 10 den par, mais de ceste somme ylia en obscuer becop et demoura en obscuer jusques autant quil soit par acort messures les terres sus quoy la rente se doit estre assignes et souchy doit estre par le consent de Prinche Monseigneur le Conte de Flandres, Mon seigneur le Prevost de Saint Donas et de autres seigneurs a qui yl tient.*'

9 Early medieval salt production leaves few visible features in the landscape and in the archaeological record, contrarily to Roman salt burning (Fenwick, 2001).

10 Personal communication Inge Zeebroek, Agentschap Onroerend Erfgoed.

Figure 10.3 Image of the terp village of Leffinge in World War I and reconstruction of the radial field system in the field structure around Leffinge (Copyright Community of Middelkerke)

Retrogressive analysis of property relations in the coastal plain also shows that the populations of these terp-focussed settlement hierarchies were free proprietors (Tys, 2003: 266–273; Tys, 2004). They may have owed some dues to respective regional lords, whether counts or kings, but with the exception of these obligations there is no evidence that they came under any other significant socio-political control. The populations of this coastal landscape, largely located on tidal channels up until the tenth/eleventh centuries, possessed a maritime focus within a landscape, which is unlikely to have fulfilled all their subsistence needs. The primary activities for the nutritional support of their households would have focussed on sheep husbandry and fishing (Verhulst, 1995; 2002). The population, like others involved in specialist production of limited products, must also have been involved in a significant degree of exchange for the provision of the cereal component of their diet, and commodities such as wood. The presence of significant quantities of imported pottery, such as black burnished wares (seventh to eighth centuries AD), Mayen (sixth to eighth centuries AD), Badorf (eighth to ninth centuries AD), buff coloured wares (ninth century AD), and others, in almost every single farmstead in the unembanked coastal area shows that the costal sheep farmers had access to networks of trade and exchange with other parts of the southern north sea world (Loveluck and Tys, 2006; Tys, 2010; Deckers, 2011). While the exact nature of the exchange of such commodities is not known, the maritime inhabitants seem to have profited from a combination of specialist husbandry in relation to coastal resources and access to exchange networks (Loveluck and Tys, 2006). The likely major commodities for exchange would have been wool or finished wool garments (the so-called *pallia fresonica*), fish, and salt (Ervynck *et al.*, 2004; Tys, 2005). As such the sheep farmers with their specialist produce in the maritime regions were both involved in and contributed to forms of exchange and circulation of trade goods. Although the social context and framework in present-day Friesland were not altogether the same as in Flanders, it seems that in both regions (free) farmers lived during the early medieval period (the sixth to ninth centuries AD) in unembanked supratidal salt marshes, trading specialised products of animal husbandry to other commodities. In both regions, terp-settlements apparently formed the centres of communities and their landscape. In Flanders, the peasants and herders in the saltmarshes were confronted by new neighbours with a different attitude towards the coastal landscape, namely the Count of Flanders and those in his network.

VI. New agents in the landscape: the counts of Flanders

As a result of political events in North West Europe at the end of the ninth century, princes gained political control and developed strong (nearly royal) principalities on a regional scale. One of the most important of these new princely families were the counts of Flanders, who established a powerful new county in the coastal area of the North Sea *c.* 900 AD. The strategy developed by the earliest counts like Baldwin II and Arnulf I shows some interesting parallels to the developments of the late-Saxon kingdom of Alfred the Great, Baldwin's father in law. One of the remarkable parallels between Wessex and Flanders at the end of the ninth century was the decision to develop a network of fortified strongholds as centres of power with collegial churches and centres of administration. In Wessex this was the system of *burhs*; in Flanders the system of *ringwalburgen* (Hall, 2011)[11].

In Flanders, the development and acquisition of power was strongly connected with the acquisition and employment of the rights on tolls, on minting coins, and on the development of landed property. It was also used as a means to invest in in the construction of an influential and powerful court and administration, to develop a feudal military force, and to found rich collegiate churches to support the lord's position (Dhondt, 1941 and 1942; Voet, 1942; Thoen, 1993; Bisson, 1995; Davis and Fouracre, 1995; and Tys, 2004). The means and wealth that the early Flemish counts needed in order to undertake all of the above was based on their private territorial possessions. They took large surfaces of lower salt marshes in the estuaries of the open tidal channels and rivers (the Zwin, the Yzer, the Gersta, and others) into possession using a regal right on coastal and other waste grounds. This occurred sometime between 885 and 918 AD, during the lifetime of Count Baldwin II (Tys, 2003 and 2004). There is a distinctive concentration of comital properties and estates in the coastal *pagi* that were integrated into the medieval county of Flanders by Count Baldwin II before the end of the ninth century, like the estates on the east bank of the River Yzer (1600 hectares carrying 10,000 sheep), the huge comital

[11] It is not surprising that there existed strong links between both courts since Baldwin II was also married to one of Alfreds daughters, Aelfrith (Koch, 1978).

estates around Veurne which specialised in animal husbandry and sheep herding but also contained several fisheries, and the estates of the *lardarium* (larder) and *spicarium* (granary) of Bergues on both side of the Gersta estuary (Tys, 2003 and 2004). Comital property was much less extensive in the regions that were integrated into the county after 918 AD, which indicates that the actual property rights on the large coastal comital estates seem to go back to Baldwin's reign in the chaotic last decades of the ninth century AD[12].

Earlier interpretations of the coastal estates in the salt marshes regarded these properties as waste grounds that were transferred to ecclesiastical institutions in turn for rents (Verhulst, 1995; Thoen, 1993 and 2001; Augustyn, 1992; Brenner, 2001). The assumptions were that the counts and their administration could not have had the capacity to invest in embankments and technology, and that the counts would not have been able to develop feudal surplus extraction methods in the coastal salt marshes since it was supposed that peasants were absent from the area (ibid.). As such, comital property would only have had an indirect influence on the salt marshes.

Thanks to archaeology we know now that the comital estates as well as the comital fortesses (Veurne, Bourbourg, and Oostburg) did not emerge in a vacuum (Loveluck and Tys, 2006). Research has confirmed that the construction of the fortress of Furnes goes back to the end of the ninth century AD (De Meulemeester, 1979), and the lands inside and outside the fortress were part of the comital domain (Meijns, 2007). Through the establishment of these fortresses, comital power exercised direct military and political control over the economically important coastal fringe of the young county. The resemblance of these fortifications to the role of the *burhs* in Wessex is striking. The donation of important relics to the collegiate church of Furnes before 918 AD by Count Baldwin II supported this ambition (Meijns, 2007). In the eastern part of the Flemish coastal plain, Bruges seems to have played a similar role.

VII. The formation of comital estates on the east bank of the Yzer (Kamerlings Ambacht) (tenth to twelfth centuries)

Detailed analysis of the medieval landscape developments (with attention to spatial structures, field systems, property relations, and site catchment analysis) of the area on the east bank of the River Yzer resulted in the reconstruction of 1600 hectares of *Oud domein*, old comital demesne in coastal Flanders which goes back to Count Baldwin II (878–918 AD). In fifteenth and sixteenth century land books of the parishes in that region, these lands were described either as *Camerland* (demesnes of the privy chamber) or as *Proostland* (demesnes of the Lord Chancelor), being in fact the lands of the *Brevia Camera* and *Magna Brevia* of Bruges (money rents that were part of old comital demesne accounts that were mananged from the Comital Palace in Bruges). The *Brevia Camera* and *Magna Brevia* were two of the more important estate accounts of the comital treasury (Verhulst and Gysseling, 1962; Lyon and Verhulst, 1967). Both accounts went back to an older united

[12] These estates can be found in the archives of the administration of the so-called 'old estates' or *Oud Domein* of the comital financial administration (*Rekenkamer*) (Verhulst and Gysseling, 1962; Bautier and Sornay, 1984 and Tys, 2003; 2004).

estate account in the castellany of Bruges, which would have originated around the end of the eleventh century AD[13]. This account collected all the revenues in money of the comital *brevia*-estates in the castellany of Bruges, which were mainly estates leased out *in census* (rents in money) since the end of the eleventh century. Before the end of the eleventh century, these domains were part of the estate management called the *spicaria* of Bruges, which refers to the comital granary in the comital fortress of Bruges. This storehouse acted as rent-collecting centre of the comital estates in the region of Bruges. Spatial and historical analysis of the estates in relation to the descriptions of the revenues and incomes from the comital treasury in later sources, in particular the *Grote Brief* of 1187 AD – the oldest known general account of the revenues of the comital territorial estates, show that these incomes could be brought back to domains with specialised rural products such as wool from sheep, dairy products, meat, fish, peat, and corn (see also Verhulst and Gysseling, 1962). According to Verhulst and Lyon, the castral *spicaria*-system came into existence *c.*1000, together with the division of Flanders in castellanies (Lyon and Verhulst, 1967).

The estates of the *Brevia Camera* and *Magna Brevia* of Bruges on the east bank of the River Yzer were all situated on the salt mashes on the isle of Testerep and in the area called *Slijpe* (Figure 10.4). The estates themselves can be divided in a group of estates situated on *oudland* (old lands), lands that were secured by dike construction probably before the end of the eleventh century AD and a group of estates situated on *nieuwland* (new lands) that were gained after active reclamation during the twelfth century AD. These old and the new lands are delineated and dived by a sharp line of linear embankments along the last open tidal channels.

According to the fifteenth-century landbooks of the *Brevia Camera* and *Magna Brevia* of Bruges, the comital estates on the old lands on Testerep (the medieval parishes of Westende, Middelkerke, Mariakerke, and Oostende) comprised:

- an estate of the *Brevia Camere* in Westende with a surface of 197.64 ha
- an estate of the *Brevia Camere* in Middelkerke with a surface of 87.5 ha
- an estate of the *Brevia Camere* in Mariakerke with a surface of 32.5 ha
- an estate of the *Magna Brevia* in Middelkerke with a surface of 78.61 ha (*Vranckx Ambacht* or *Ministerio Lamberti,* Tys 1998)
- a comital *vaccaria* (specialist cow pasture) in the dunes near the village of Westende (12.48 ha, called the *vakerie* in 1459)[14]
- feudal lands and goods from the feudal courts of Bruges, Veurne and Oudenburg (82 ha)
- several donations of comital lands to ecclesiastical institutions, amongst others the sheep estate of the abbey of St Peter's near Ghent with a surface of 175.5 ha that was donated to the abbey in 992 and 995 AD[15].

---

[13] Around 1170, the treasury was reorganised in the context of the development of a new estate account administration, the so called '*reddeninge*', see Declercq, 1997 and Soens and Tys, 2001.
[14] Archives départementales du Nord, B 3979 131v°.
[15] Fayen, 1906 (*Liber Traditionum)*, nr. 100 ('*Item anno incarnati verbi 992 … quidam Eilbode*

Figure 10.4 Reconstruction of the comital estates in Kamerlings Ambacht (coloured fields), situated on the geomorphological map of the region with the last open tidal channels (source: Tys, 2003)

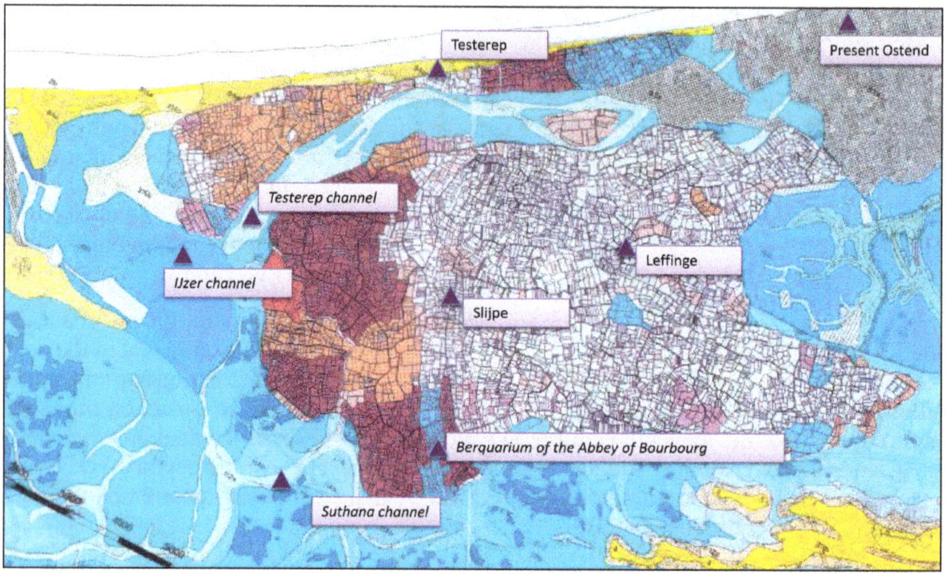

At least eighty-five percent of the old lands in Westende and eighty-seven percent of the old lands in Middelkerke clearly originated as comital estate lands. Since these figures are based on fifteenth-century land books and represent a situation that dates after significant loss of land during the thirteenth, fourteenth, and fifteenth centuries, the lad surface of these estates must have been much larger in the tenth and eleventh centuries. In the oldest known estate account, the *Grote Brief* of 1187 AD, these revenues were booked as, [...] *ex Testerep Siger*: *77lb 17s* [...] *ex Testerep Westenda pro lana*: *13lb½* [...] *ex Testrep 210lb 9s* [...] *ex agnis Testerep 24lb preter 15d* (Verhulst & Gysseling 1962, 170 and 177).

Wool was without any doubt the main produce from these estates, but it is interesting that the core of the estates in the old dunebelt of Westende was a *vaccaria* or well delined specialist cow pasture.

According to the fifteenth-century landbooks of the *Brevia Camera* and *Magna Brevia* of Bruges, the comital estates on the old lands in Slijpe (the medieval parishes of Slijpe and Mannekensvere) comprised:

- an estate of the *Brevia Camere* in Slijpe with a surface of 202.6 ha
- an estate of the *Brevia Camere* in mannekensvere with a surface of 144.53 ha

tradidit Deo et Sancto Petro alodem suum, id est ... terram quoque in Testereph ad 600 oves ... ') en nr. 102 ('*Susanna Regina ... tradidit alodem suum, id est ... et in Testerep terram ad 600 oves ... Actum est ... anno verbi incarnati 995 ...*'.).

- an estate of the *Magna Brevia* in Slijpe with a surface of 743.88 ha
- an estate of the *Magna Brevia* in Mannekensvere with a surface of 143.3 ha
- lands from the receiver of the *Spicaria* of Bruges with a surface of 7.55 ha
- feudal lands and goods from the feudal courts of Bruges, Veurne and Oudenburg (108 ha.) and a *berquaria* (leased-out sheep farm) of 77.89 ha which was donated by the comital family to the Abbey of Notre-Dame-de-Bourbourg in 1116 AD (Figure 10.4)[16].

Ninety percent of the old lands of Slijpe and more than ninety-nine percent of the lands in Mannekensvere originated as comital estate of the *oud domein*. The estates in the Slijpe-area formed a topographical unity that was separated into different *ministeria* (organisations), as was the case on Testerep (Tys, 1998; 2003; 2004).

Thus we can see how the salt marshes on the east bank of the (present) River Yzer were transformed into comital estates and sheepfarms for thousands of sheep from the end of the ninth century onwards. We can assume that the house of the Counts of Flanders possessed almost the entire isle of Testerep and area of Slijpe. Only at the end of the eleventh century were large parts of the estate leased out against money rents (Verhulst and Gysseling, 1962; Lyon and Verhulst, 1967).

VIII. Comital estates and the start of the embankment process

The actual sheep domains were not only organised as *ministeria* (administrative units) under the responsibility of a *preco* (estate administrator), but they also seem to have had distinctive material forms. A number of oval enclosures or nuclear embankments in the centres of the comital coastal estates we described earlier demonstrate that the comital estates also formed distinctive units in the landscape. In the parish of Middelkerke, for example, an oval enclosure of *c.*12.5 ha was located near the Testerep tidal channel (Figure 10.5). Its presence in the landscape was still traceable in the early nineteenth-century parcel structure and by the place name *cromme diken* (curved or bent dikes). It lay in the centre of a large comital sheep estate of 96 ha. Around the initial enclosure a pattern of protruding land-claims and drainage ditches was laid out. Although we were not able to undertake excavations inside the enclosure, the (later) farms related to comital estate activities are located on the edge of the enclosure, not inside. Other similar enclosures were found in Westende, Stene, and Slype. These oval enclosures in areas of old comital estate closely resemble the patterns of nucleated embanking that Allen considers to be the earliest systematic embankments in the coastal plains of north-western Europe (Allen, 1997: 16; 2000: 1212). They have also many similarities with the oval enclosures Steve Rippon discovered in the Somerset Levels in Wessex, and have some resemblance to oval enclosures discovered in Holland (for example in Assendelft, Rippon

[16] Vercauteren, 1938, nr. 78: '*berquarian unam apud Slipan, singulis annis 52 libras persolventem, que iuxta Suthana iacet*'. In a land survey from 1454, this *berquaria* is described as'*abdesseland ... ghemeten ende ghescreven uten bouke van mevrauwen der abdessen van Brugburch.*' (Archives d'Etat à Mons), C103, 23v°). The estate can be situated in the field structure exactly (Figure 10.4).

Figure 10.5 Reconstruction of the oval enclosure in Middelkerke (1).
This enclosure is situated in the centre of a comital domain. Before the end of the 11th century, a longitudinal embankment alongside the Testerep channel replaced the oval embankments of the enclosure (4). Nearby is a possible terp site (3) (source: Tys, 2003)

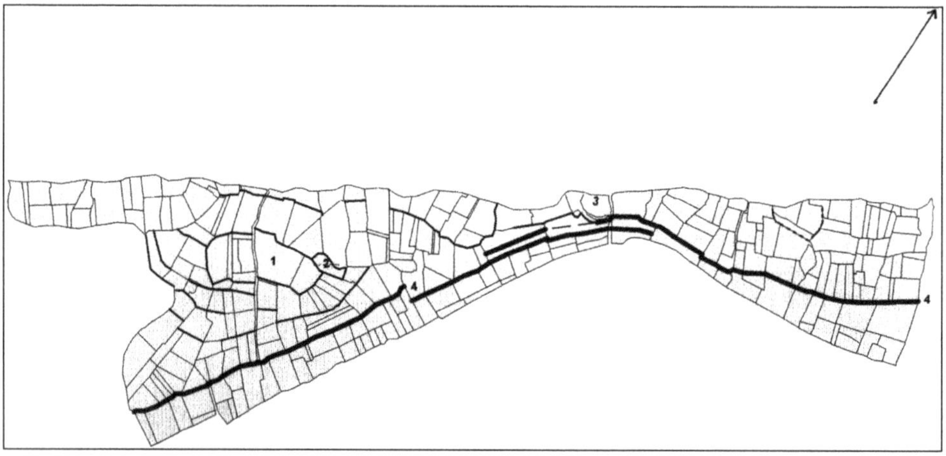

2002 and this volume; Besteman and Guiran, 1986). In both areas the enclosures can be dated to the end of the first millennium, which is not contradicted by the evidence from Flanders. In or until the nineteenth century similar systems also existed on the isle of Ameland in the Dutch Wadden-area, protecting hay lands (Bazelmans, 1999). The use of such *ringdijken* (ring-dykes) in Zeeland seems to be related to these nuclear embankments. What they have in common is that they seem to be the first step in the embankment process, notwithstanding the fact that they surround much larger areas than the examples from the Severn estuary and the north of Holland. Unfortunately, the ring dikes in Zeeland were until now only dated in relation to the Duinkerke-system so it is difficult to assess their value in the broader framework. (Hallewas and van Regteren Altena, 1980). It is also possible that the first written attestations of dikes in Flanders, like *Obekinskinderdyc* (the dike of the children of *Obekin*) and *Tubinsdyc*, refer to similar local nuclear embankments.

In the Somerset Levels colonists would have erected the enclosures, while in Flanders they were erected on the initiative of those who administered and managed the comital estates[17]. However, their function and intention seems to have been the same: protecting a small area of agricultural land against exceptional floods in spring or summer (Rippon,

[17] It is not possible to explain this in detail in this paper, but the connection of the cores of the old comital estates to honorary positions of *ammans* (*ministeriales*) confirms that the comital estates were administered by (unfree?) members of the comital services (see also Koch, 1951 and Tys, 2003).

2002). It is not unlikely that the enclosures, like those in Ameland, were designed in order to develop seasonally dry hay lands to assure the provision of hay for the comital sheep after the summer season. As such, these enclosures could be regarded as investments in the intensification of the sheep breeding in the comital coastal estates, and the enclosures as a kind of infield in the outfield salt marshes (as Rippon suggests, 2002). Therefore, it would not be surprising if the examples in the Somerset Levels were also the creation of an unknown power with a capacity to invest in the intensification of its activities in the coastal plain, just like the Flemish counts.

The problem is that we cannot date the nuclear embankments by archaeology. No excavations have enlightened us about their date of construction and the start of occupation. Accordingly, it is dangerous to try and date these structures too strictly. The *terminus post-quem* seems to be the formation of the comital *Oud Domein* under Baldwin II (878–918 AD). The *terminus ante quem* is more difficult. Some data for the region of Testerep suggests that the secondary embankments and reclamations had already expanded beyond the nuclear embankments before the end of the tenth century, when a large estate of *terrae ad oves* (land and sheep) was donated from comital property to the abbey of Saint-Peters near Ghent in 994/996 AD (Tys, 1997). We need, however, increasing amounts of data of a more specific nature into the appearance, nature, and character of these early forms of embankments in the southern North Sea area.

Notwithstanding the difficulties we have with establishing the exact chronology of the actual phases in the embankment process, we can see that shortly after the construction of the ring dikes of the oval enclosures the decision was made to construct defensive dike systems alongside the remaining tidal channels in order to embank an even larger area. In tenth- and eleventh-century Flanders, only the main tidal channels in the coastal plain seem to have remained active because large parts of the tidal channel network had previously silted up. One of the most striking of these dynamic processes was the shift in the course of the River Yzer from the Avekapelle Channel near Veurne to the more eastward channel that currently hosts the river. But shifts of channel like that were the exception rather than the norm and the coastal plain generally had a rather calm environment (Baeteman, this volume). The Sincfala channel and its side-channels between Kadzand and the area around Damme in the west, together with the Gersta-channel and its side-channels between Bergues and the later town of Duinkerke, are well-known examples of the last large tidal channels that remain present in the coastal plain. In the central coastal plain, the (new) Yzer channel with its several side channels east and west of the river also remained open. On the east-bank of the river we find the Testerep Channel and to the south the mouth of the *Suthana* or *Spermalie* channel (Figure 10.4).

The longitudinal dikes at each side of these channels enclosed and protected the comital sheep estates of the *oud domein*. The names of these dikes are the *Kadedijk* (shore dike) and *Hoge Dijk* (high dike). The element *kade* or *quay* is also to be found in the name of the isle of Cadzand, probably also in the context of the lay-out of the embankments before 1111 AD (Verhulst 1995). Due to lack of archaeological research we know nothing of their initial dimensions, although landscape and iconographic evidence suggests that they resembled the dimensions of the oldest known sea dikes in Flanders, with a general width of approximately 6 m and a height of approximately 3 m (Gottschalk,

Figure 10.6 Reconstruction of the medieval field structure in Mannekensvere. We can clearly see the longitudinal defensive dike (Hoge Dijk), separating older estates (oud domein) with irregular feeld patterns (tenth-eleventh century) on the eastside, and new estates in the embanked salt marshes of the river IJzer (twelfth century) on the westside (source: Tys, 2003)

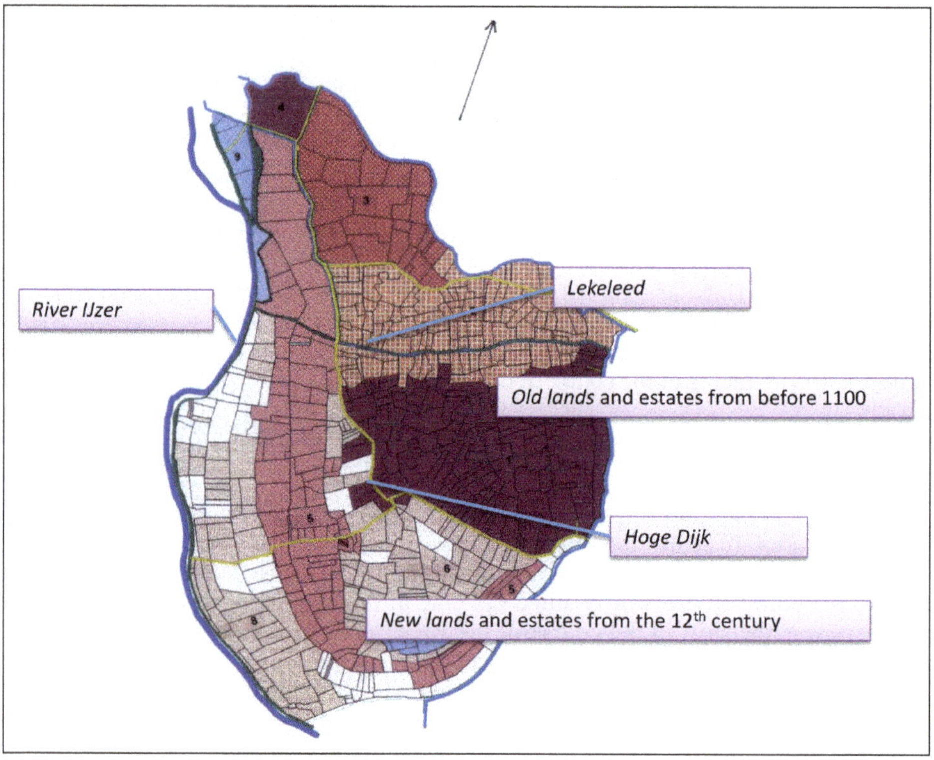

1955, 116–119)[18]. These dikes enclosed large surfaces of several square kilometres, which the historical sources call *oudland* (old lands). Similar situations are encountered in most of the northwest European coastal plains that were reclaimed in the medieval period (Gottschalk, 1955; Hallewas, 1984; Eddison, 1995; Meier, 1997 and this volume, Silvester, 1999; Rippon, 2001, 2002 and this volume and many others). These defensive dikes demarcated two different field patterns. First, the irregular, sometimes curved, field pattern of the older embanked lands on the inner side. Second, the more systematic and rectangular field pattern of the reclaimed new lands on the outer side (Figure 10.6 shows the situation in the parish of Mannekensvere). The first written testimonies of the pres-

[18] The Kaaidijk in Middelkerke was leveled in the nineteenth century.

ence of new lands dates from 1115 AD[19]. The demarcation between old and new lands in the field structure provides a *terminus ante quem* for the construction of the longitudinal channel dikes *c.*1100 AD. The history of the organisation of the old domains behind the dikes suggests strongly that these dikes and the estates behind them are much older, even possibly older than the end of the tenth century AD as the testimony about the donation in 992/995 AD of the large *terrae ad oves* estate on Testerep suggests. Without archaeological data, however, we have to be careful to date the construction of these longitudinal dikes too precisely. Therefore, it seems acceptable to say that the process of the definitive embankment of the old lands by these defensive channel dikes dates from sometime between the tenth and the end of the eleventh century.

IX. The embanked landscape

This act of embanking had several consequences and implications. One of them was that comital power and organisation must have had the ability to organise manpower on a large scale, and that the counts must have had the financial capacity to carry out these expensive works. These resources were required to apply the necessary hydrological technological knowledge concerning flood control and drainage on such a large scale. The embanking meant that the natural drainage of the former tidal environment, which was based upon the ready-made system of gravitational drainage of the salt marsh creeks, had to be modified and improved by a large-scale artificial drainage network of new *leed* (drainage ditches). The drainage of the embanked area also required the construction of sluices and outlets. Such works are much earlier than authors like William TeBrake supposed (2000). The sluices were used to lead the drainage waters at low tide through the dikes in the open tidal channels, but they also helped to irrigate the fields in the summer by opening the sluices at high tide and closing them at low tide. Another practical use was that they allowed people to regulate the water level in the trade canals that ran through the domains.

At first the embanked comital estates seem to have been directly exploited by *ministeriales* (comital officials). Evidence for this is found in the late medieval (but fossilised) feudal positions of *ministerial* services connected to the organisation of the old estates. This is what Koch essentially suggested in 1951, though he did not see the spatial context for the functions, nor the estates (Koch, 1951). In other words, he analysed social organisation without its spatial context. More evidence relating to this is found in the form and situation of the *berquariae* (leased out sheep farm) that was given to the abbey of Bourbourg in 1116 AD. The *berquaria* of Bourbourg was very clearly taken out of a larger and older organised landscape (Figure 10.4). Essentially, the position and presence of the *berquaria* itself indicates that there was an older phase in the management of such estates. The best indication of this is the link between investments in dikes and the oldest reconstructable forms of the estates themselves, in the form of nucleated embankments. The scale and

[19] Platelle, 1962: 178, in 1115 Count Baldwin VII exchanged or leased out a *berquarium* or leased out sheep farm on 'new alluvial lands' situated on Testerep with the Abbey of Saint-Amand to the villa *Herlengova*.

organisation of the old estates, both spatially and administratively, strongly suggest the presence of a central power. Any argument in favour of several individual leaseholders, each organising a part of the comital salt marshes for themselves, must now be rejected.

Some smaller parts of the estates during the eleventh or early twelfth centuries given in fief to comital knights and servants. The count's knights could well have been responsible for the protection and organisation of the revenues of the estates as they seem to have been the representatives and executors of comital power and prestige on the local level. This position gave them a certain amount of influence on the landscape and its organisation. For example, they were responsible for the development of manors and new villages on the comital estates, often named after the knights (like *Arleboudscapelle*)[20]. A detailed study of the duties that were connected to the fiefs and of the positions and functions held in the comital administration, reveals that the group of comital estate knights were also involved in the organisation of the control of waterworks and hydraulic systems that were created by the defensive embanking of the area (Tys, 2004 and 2010). The main family of estate knights in the area, the knights of Slype, held also in fief the toll that was raised at the point where the canal between Ieper and Bruges had crossed the defensive dike between the comital estates and the tidal channels since *c.*1100. A sluice must once have existed at this crossing since it would be strange if these knights controlled the toll but not the economically important sluice and water control system at the same spot. Second, some knights also held the position of *scabini* of the coastal area (Kasselrij Brugge). This title, probably from the twelfth century AD onwards, held juridical power over water household matters. It seems clear that through his vassals the count tried to control the economic, strategic, and also the social importance of the systems for controlling water. Until the middle of the twelfth century the knights who were located on the comital estates also kept large and important portions of the tithes of the old mother parishes in the area, such as the parish of Leffinge or the parish of Testerep, after which those fiefs were resumed by the count so that he might transfer their ownership to ecclesiastical institutions.

X. Consequences of the embankment and the construction of dams and new waterworks

An environmental consequence of the embanking was that since the water had less space to occupy it caused rising storm surge levels in tidal seawater channels (Vos and Van Heeringen, 1997 and Baeteman, 1998). Higher storm surge levels thus placed greater pressure on the dikes and an increase of the risk of dikes bursting. Indeed, historical sources describe an increase of disastrous floodings in the eleventh and early twelfth centuries, accompanied by many casualties (Verhulst, 1959). These could have been the result of dike bursts although we have to remember that these embankments must have made people feel more secure about colonising the areas immediately behind the dikes.

---

[20] *Erlebaldi Capella*, 1141 (de Hemptinne and Verhulst, 1988, No. 58). Similar new, eleventh- and twelfth-century estate villages in which we suppose the hand of comital estate knights are *Snelghers Kerke* (or Snaaskerke), Eggewaertskapelle, Ramskapelle, *Volcravenskinderkerke* (Oostkerke), *Reinilini Capella* (Oudekapelle), *Simonis Capella* (Koksijde), *Eustacii Capella* (Zoutenaaie), etc. (Tys, 2003)

So when such dams burst the casualty rates would naturally be higher. In the defensive dikes we also find *wielen*, which are semi-circular restorations of dike bursts (Allen, 1997: 12–16). At the same time, the areas between the dikes belonged to the last remaining tidal zones that remained open, meaning that the natural sedimentation of sands by tidal waters resulted – paradoxically to the problem of the increased danger of inundations by storm surges – in the silting up of the lands between the defensive dikes alongside the channels. This is a process we can follow in the written sources, since the new lands meant new landed capital could be distributed to members of the comital elite network as well as to ecclesiastic institutions. In 1115 AD the Count of Flanders donated the right on lands between a *berquarium* (leased out sheepfarm) on the east side of Testerep and the sea, lands that people called *sant* (sand), to the abbey of Voormezele[21]. This attestation indicates the presence of newly silted-up sandy lands (or even dunes) in the Testerep channel or at the beachside. In 1133 AD, a conflict over ownership of new alluvial lands in the Testerep channel, that had appeared next to the old lands of the estates of the abbey of Saint-Peter's near Ghent: [...] *in Testerep praefacta ecclesia Sancti Petri berquarias super litus maris possedit, quibus per maris alluvionem nova quaedam terra conglutinate est et concreta* (Tys, 2001)[22].

So, according to written sources, the sedimentation of the Testerep channel mostly occurred in the first half of the twelfth century. In order to secure the new lands and to defend the old lands from inundations by storm surges, *c*.1150 AD a decision was made to block the mouths of the partially silted up channels by the construction of large stone dams, like the *novum dam* (new dam) on the mouth of the Testerep channel and the *Steendam* (stone dam) and *Streeckxdamme* (stretched dam) on the mouth of the *Suthana* (southern channel), the last remaining tidal channels in the research area (Tys, 2003; Figure 10.7)[23]. The *Steendam* was constructed near the hamlet of *Houckevliete* (curve in the channel), the later Sint-Pieterskapelle, to consolidate silted up lands in the channel (Tys, 2003: 325). The *Streeckxdamme* was constructed at the mouth of the channel in the River Yzer, which means that it was part of the same embankment operation as the the *novum dam* at the mouth of the Testerep channel. The latter was first mentioned in 1167 AD when a small parcel of land behind the dam was transferred to the abbey of Watten[24]. This dam can still be seen in the landscape as a

---

21 Vercauteren, 1938: 172–174, No. 76: '... *quandam terram, latine Harenam, vulgo autem sant nuncupatum, que interjacet berquarie ipsius Isaac et mari in orientali fine in Testrep, Sancte Marie de Formesela et fratribus ibidem deo servientibus ... et pro anima Adam, patris Isaac de Formosela qui ipsam terram a me et ab antecessoribus meis in pheodum tenuerunt, me libere concessisse, et omnia secularia jura ad dominium meum pertinentia ab ea prorcus removisse*'.

22 de Hemptinne and Verhulst 1988: No. 25.

23 The *novum dam* was first mentioned in a charter of 1167 (de Hemptinne and Verhulst, 1988: No. 275).

24 de Hemptinne and Verhulst 1988: No. 275: '*In nomine patris et filii et spiritus sancti amen. Ego Philippus dei gratia Flandrensium comes notum facio tam futuris quam presentibus quod ecclesie Watinensi dedi in Sclipen XVII mensuras terre perpetuo possidendas iuxta Nouum Dam et extra fossatum maris pro anima fratris mei Willelmi cognomine Broon rogatu ipsius et ob salutem anime mee meorum que tam succedentium quam precedentium de terra noua illa uidelicet quam primitus sibi in feodum donaui et quam propriis sumptibus contra marinos fluctus ac impetus circumfodi fecerat...*'.

Figure 10.7 The *novum dam* on the mouth of the Testerep channel, constructed around 1150 is still present in the landscape (Photograph: Tys, 2003)

clear elevation on the low-lying lands of the former Testerep-channel (Figure 10.7)[25]. Together with the construction of the dam, a complex system of drainage channels was designed and laid-out in order to organise both the commercial traffic that used the former Testerep channel (as part of the connection between Bruges and Ieper, via the *Ierperleed*), and for the drainage of rain flowing from the embanked lands behind the dam. These and similar dams (such as the dam on the Zwin in the town of Damme and the *Hooghen Steendam* on the *Gersta* channel between Bergues and the sea) were clearly very impressive and complex constructions. The place-names indicate that they were made in stone from well-designed masonry, as the description of the dam near Sint-Pieterskapelle suggests.

The construction of these dams *c.*1150 AD reduced the tidal zone in the central part of the coastal plain to the mouth of the River Yzer. The Testerep channel and the Suthana channel disappeared as active natural features from the landscape (Figure 10.8). The dams on both channels were connected with each other via the newly developed *Havendijk* (harbour or mooring dike) (Figure 10.6). The name of this dike refers to

[25] The position of the dam is not to be confused with the place called *Nieuwendamme* which was a port and later a fort 100 m to the east of the dam itself.

Figure 10.8 Remains of a large tidal channel that separated the former island of Testerep from the inland salt marshes.
In the background, we see the impact of modern tourism on the coastal landscape (Photograph: Tys, 2003)

the use as the river as mooring place for Nieuwpoort, the port that was simultaneously developed on comital estate grounds as the dam-constructions. The *Havendijk* in itself secured new polders and reclaimed lands on the east-bank of the Yzer, outside of the older longitudinal defensive dikes in Mannekensvere. The new lands in the former channels and in the new Yzer-polders were leased out as *Eggewaertsleen* (*census*-lands) around 223 ha in area, organised as feudal property belonging to the feudal court of Bruges (76 ha) or donated to ecclesiastical institutions (220 ha) (Tys 2003, Tys 2004). No less than 140 ha of embanked new lands in the former Testerep-channel were donated to the Cistercians of Beauprés-su-la-Lys in France (Tys, 2003: 397). Perhaps more remarkable was the foundation of a new religious house on the new lands of the former *Suthana* channel in Slype. The *Monialia de Nova Terris in Slipis* originated out of a small religious congregation in Sint-Pieterskapelle *c.*1200 AD and received 75 ha of lands (Strubbe, 1942; Tys, 2003: 441–442). An abbey was duly constructed on the newly donated lands but in 1241 the congregation moved to the hamlet called Spermalie, east of Bruges, after which the original site became a dependant farm of the congregation[26].

[26] The place-name *Oude Hof van Spermalie* still refers to the original site of the abbey.

XI. The new organisation of drainage and waterflood control

These new lands and polders developed as a result of a planned and well-organised policy of systematic land reclamation, so the field structure of these new lands was much more rational and regular compared to the older irregular field structure of the old lands, based upon the old network of salt marsh creeks. The rational and regular field structure was also an integral part of the water management. The drainage function of the former tidal channels had to be replaced by artificial drainage systems. New drainage canals were necessary in order to lead the inland waters safely to the sea from the second half of the twelfth century onwards. One of these was the so-called *Lekeleed*, a broad and straight canal that ran directly from the hinterland around Leke to conduct river floods and inland waters safely to the sea (Figure 10.6). The *Lekeleed* was first mentioned in written sources in 1258, which acts as a *terminus ante quem* for this particular canal (Tys, 2003: 541). It was part of systematic operation to re-organise the drainage of inland waters in the coastal plain from the second half of the twelfth century onwards. There were many similar canals, like the one between Oudenburg and Bredene that formed the western edge of the *Blankenbergse watering*. To prevent the coastal area from being inundated by inland rain floods carried by these canals a new type of dike was invented, the *silinghe* or *zijdelinghe*. These dikes were built alongside the new drainage canals, to protect the lands next to these canals from flooding. These 'side dikes' appear in the historical sources from the end of the twelfth century. The *Gentele* and the *silinga*-dike of the *Blankenbergse Watering* (first mentioned in 1201) are nice examples of such dikes next to the new drainage canals (Gysseling, 1950). In the 1950s it was thought these dykes formed the first phase of the embankment process on the coastal plain. Now, it is clear that they represent a specific reaction in the landscape that relates to the growing complexity of the drainage of the coastal plain after the actual process of embankment.

The old and new drainage canals led (and still lead) vast amounts of drainage waters from the interior through the polders to the sea. This implied the construction of larger sluices, leading these drainage waters through the dams into the estuary of the River Yzer or through the dunebelt straight into the sea. One of the sluices in the dunebelt was the *Kamerlingssluis* in Middelkerke. We can assume that the exercise to drain these canals through the dunebelt must have been rather difficult. The most important sluices were situated around the estuary of the River Yzer. Today, a complex of eight sluices in Nieuwpoort still drains the western and central coastal plains. Excepting one sluice on the river itself, the remaining sluices regulated the flood regime of the drainage canals from the late twelfth century. The sluice that drained the *Lekeleed* was the *Lekesluis*. Another important sluice that drained the area around Gistel and the central coastal plain was the *Westsluis* or *'s Kamerlins slus* (Tys, 2003: 541). The *Magna Sclusa* (great sluice) of 1184 that drained the area west of the River Yzer was another important feature in the complex of sluices at Nieuwpoort (Degryse, 1987; Soens, 2009).

From the thirteenth century the inhabitants of the maritime region had to pay contributions to maintain this hydraulic system and for the drainage, protection, and irrigation of their lands. These contributions were called *geschot* and were raised by particular micro-regional institutions called *watering* (Fockema-Andreae, 1960; Soens, 2001 and 2009). The *wateringen* were organised on the level of the micro-regional public institutions and circumscriptions called *ambachten* (most often in accordance with the tenth to eleventh century mother parishes). The head of each *ambacht* (the *ambachtsheer* or *amman*) was an agent of central juridical power and often belonged to the group of comital knights (Koch 1951: 125–129). The count's main coastal knights also held a position as *scabini* (*aldermen*) of the coastal area (burggraafschap Brugge), which had at least in the thirteenth century, but probably earlier, juridical power in water management matters. These functionaries also represented the central juridical power in the field of the organisation of the waterhousehold, meaning that they mediated in juridical conflicts inside or between the *wateringen*[27]. Jurisdiction and organisation of these institutions could have been of comital origin (Fockema Andreae, 1960: 190–191). It seems clear that through his knights and vassals the count tried to control the economic, strategic, and also the social importance of the water control systems.

The *wateringen* themselves were, although local public authorities, organised as an association and assembly of the landowners in the territory of the *ambacht* and *watering*. The assembly was called *the meentucht*, and decided on water management, the appointment of the executive officers like the *sluismeesters* (sluice masters) (who belonged to the families of the comital estate knights and *ambachtsheren*) and the height of the contribution (*geschot*). As such the origin of the *wateringen* probably goes back to the twelfth century, after which comital estates were gradually leased out to individual landholders, who subsequently became the landowners in their own rights.

Soens has shown that the democratic appearance of these institutions was merely an idealistic anachronism (Soens, 2001). As he argues, decisions were controlled and steered by the small group of larger landowners and executive officials who belonged to the network of the regional elite, with antecedents in the group of comital knights involved in many (feudal and other) offices and functions in the region (supra). Their policy did not necessarily benefit the smaller landowners, and the financial contributions were apparently not always well spent, as many sources indicate that the hydraulic system was often neglected, particularly during the fourteenth century (Augustyn, 1992).

XII. Coastal erosion and seadikes

From (at least) the thirteenth century the coastal plain of Flanders was confronted with heavy coastal erosion. Because the coastal plain was now embanked and tidal influence was restricted to the estuary of the River Yzer (embanked between *c.*1200 and *c.*1300, see Boterberge, 1962 and Tys, 2003), the beach and dunebelt were now confronted with the full power of the high-energy conditions of the tidal waves. Without human influence, the rising sea level and a presumed shortage in the sediment budget would have caused the re-activation of the natural tidal environment and a new cycle of inundations and sedimentation (Baeteman, 1998: 268–269). Now the coastal plain was 'closed' to tidal influence, tidal forces started to erode both the beach barrier and the submersed shore of the coastal plain (idem). Simultaneously, intensive and early use of the dunebelt as pasture (for instance the presence of a comital *vaccaria* or cow farm in the dunehead

[27] This function was later declutched and assigned to the so called *waesschout* (watersherrif), the forerunner of the *dijkgraaf*.

of Westende) had weakened the dunebelt considerably. An interesting testimony to this process of coastal erosion is found in a charter of 1230 in which the ownership of the peat on the beach of Testerep was discussed (Warlop, 1968). The significance of this charter is that already by 1230 the sea had eroded the original Holocene sand barrier and dune system of Testerep. Already by that date the dune belt was blown inland over clay and peat deposits that became visible on the surface of the beach at the shoreline. The dunes at the seashore of Testerep, and probably also elsewhere in Flanders, were already considerably weakened by the first decades of the thirteenth century.

Sand drifts and land losses threatened valuable agricultural lands and new dike systems were applied in order to confront the environmental imbalance. In the course of the thirteenth century we see the first 'seadikes' appear in the landscape. In 1277 a charter mentions the *dike nord jeghen die zee omtrent die Middelkerke oest wart ende west wart*[28]. These seadikes were meant to keep the weakened dunes in position and to prevent further sand drift and coastal erosion. The impressive *Boondijk* inland from the dunes of Westende is a very well preserved example (Figure 10.9). Also, the *Evendijken* in the eastern coastal plain (in Wenduine and Knokke) that appear in written sources of 1281 seem to have been similar thirteenth-century sea dikes (and not tenth-century embankment dikes as local historians claim) (Coornaert, 1974 and Monteyne and Vandaele, 2008)[29].

Notwithstanding these attempts, the erosion and sand drift continued and worsened considerably between the end of the thirteenth and the start of the fifteenth century. The stories of the land losses and catastrophic storm surges like the Saint-Clemens flood of 23 November 1334, the Saint-Vincentius flood of 22 January 1394 and the Saint-Elisabeth flood of 18 November 1404 are well known (Buntinckx, 1968; Gottschalk, 1977 and 1980; Augustyn 1992, and others). The thirteenth- and fourteenth-century town of Ostend, as well as several villages like Scarphout, Arendycke, Walraversyde, and the entire island of Wulpen (including four villages on the isle) fell victim to the tidal surges and coastal erosion. Several thousands of hectares, the actual number is impossible to reconstruct, were lost between the early thirteenth and the early fifteenth centuries. The Saint-Elisabeth flood alone was responsible for the loss of no less than 3,025 hectares. In the central coastal plain, Ostend and Walraversyde had to be rebuilt further inland after the flood of 1394.

The only technical solution was to keep up the attempts to reinforce the dune belt by new inner dikes behind the dunes. In the area of Ostende, the historical sources mention the construction of new inner dikes in 1399, 1404, 1427, and 1540 (Tys, 1997)[30]. These dikes were part of a large-scale and systematic program of the construction of new sea dikes on the coast of Flanders between the end of the fourteenth and the first half of the fifteenth centuries, particularly in those parts of coastal Flanders that seemed to have been most vulnerable to land loss and dune drifts (Augustyn, 1992). After the sixteenth century these dikes that were renamed *Gravejansdijk*, after Duke John the Fearless.

---

[28] State Archives of Bruges, Blauwe Nummers, 6752.

[29] State Archives of Bruges, Blauwe Nummers, 4399.

[30] For instance State Archives of Ghent, Abbey of St.Peter's of Ghent, Account 832c: f°23r°: '*zyn gheghaen in den nieuwen dyc ghemaect int jaer 1404 metgaders pitlande daer huut ghenomen also 't blyct 5 ghemten 1 line 37 roeden lands*'

Figure 10.9 The thirteenth-century Boondijk in Westende, as visible in the present landscape (Photograph: Tys 2003)

The construction of these dikes was organised and financed by the *wateringen*. For the construction of these new dikes, each *watering* asked for a significant additional *geschot* from each landowner, which must have been a high extra burden for the large group of impoverished small landowners on the coastal plain (Soens, 2001 and 2009)[31]. The accounts of estate of the abbey of Saint-Peter's demonstrate that between 1394 and 1417 many owners of the surrounding area of Walraversyde could not pay this relatively low rent, *mets der aermoede van den volke* (seen the poverty of this people)[32]. In 1401 the arrear in rents had risen to seventy-five percent of the rents that had to be paid[33]. The financial demands of the extra costs of the dike constructions following on the environmental catastrophes impoverished most of the landowners on the coastal plain. These insolvency problems would doubtless have resulted in expropriations, desertions, and land sales, all of which could have triggered

[31] Unfortunately, the available sources do not inform us on the actual sum of this extra *geschot*, nor on the amount of the normal sum of money the landowners had to pay annually for their property. Information for the year 1682 throws however some light in this matters. In that year the waterboard asked for the repair of seadike an extra *geschot* of 1 schelling and 8 penningen for each *gemet* (0,4 ha) of every landowner on top of the normal *geschot* of 3 schellingen 1 penning for each *gemet* (State Archives of Bruges, Registers of the Brugse Vrije, 4039: church accounts of Middelkerke 1670–1684: 72r°). The price for an entirely new dike undoubtedly must have been much higher!

[32] State Archives of Ghent, Abbey of St.Peter's of Ghent, Account 832a: f°8v°.

[33] State Archives of Ghent, Abbey of St.Peter's of Ghent, Account 832b.

Figure 10.10 The sixteenth-century dike in Raversyde that was constructed to reinforce the dunes (*inlaagdijk*), as visible in the present landscape (Photograph: Tys, 2003)

the amalgamation of coastal properties in the hands of the small group of capitalist urban landowners during the fifteenth century (Verhulst, 1990; Soens, 2001; Tys 1997). As such, the water management institutions acted as surplus extraction systems, with an increased social and economical control of the rural community, with important consequences for the development of a rural capitalist landscape from the late fifteenth century onwards, a development that still determines to a large extent the actual landscape of coastal Flanders.

The newly built inner dikes from the fifteenth century did not stop the dunes from moving. In 1540, for example, new inner dikes were built to secure the rebuild village of Walraversyde from sand drift (Figure 10.10). The shifting dunes covered the dike that had been built between 1404 and 1427, as excavations have shown (Pieters, 1995). During the entire sixteenth century the coastal society suffered from shifting dune sands, threatening even the drainage capacity of the coastal hydrology as sluices were filled with sands[34]. The water boards and landowners were forced to continue investments in new innerdikes and breakwaters (Soens, 2001). Landlosses continued, albeit not in the same severity as

[34] In the 1580s, accounts of the waterboard describe how the important *Kamerlings Sluis* in Middelkerke in the dunes had fallen into decay and how the damage seemed to be unreparable. The sluice was given which was a severe blow to the medieval drainage network and capacity in the region (State Archives of Bruges, Registers of the Brugse Vrije, 14324, Account of the water board Camerlynkx Ambacht, 1582).

in the late fourteenth and early fifteenth centuries. Still, on the night of 31 October 1570 the old village of Wenduine almost entirely disappeared into the sea.

Remarkably, since the seventeenth century there have been no new catastrophic inundations and sand drifts with the same impact as those of previous centuries except, of course, the flood of 1953; this despite the fact that the dunes of coastal Flanders have remained as narrow and confined as they were back in the sixteenth century. At the same time, sea level rise continued and even accelerated during the last century. This raises the question of what caused this apparent stability from the seventeenth century onwards. Was there a change in the environmental parameters that we do not fully understand? Was there an increase of human investments in the enforcement of the dunes and dikes? Were there important technological and engineering innovations and advancements towards the stabilisation of sands, dunes, and dikes? Unfortunately, our knowledge of the environmental policy towards the dunes and the water management is very limited. We need a better historical understanding of the environmental dynamics, as well as the human choices and technical advances between the seventeenth and nineteenth centuries, to understand the link between the present situation and the landscape dynamics since the Roman period.

XIII. Conclusions

In the early medieval period, people and coastal society lived and existed in a situation of equilibrium with the natural resources at their disposal. Archaeological research shows that certain zones of the salt marsh environment of Flanders were already inhabited from the seventh century onwards (if not earlier). Silted up tidal channels were particularly favorable for these earlier settlements. Also, artificial dwelling mounds were raised in order to construct collective settlements in the coastal plain. People raised sheep in the salt marshes, collected natural resources, and produced and traded textiles in a North Sea network. This stage of living and dwelling in the coastal saltmarshes can be regarded as a stage in which people went from exploitation to modification, largely in peace with the environmental rhythm. The society of the coastal sheep farmers was apparently a peasant society without either official power or hierarchy present in the salt marshes itself. It is only when official power started to interfere with the ecological conditions after the tenth century, thus becoming 'maritime' itself, that the North Sea lifestyle and social identity was transformed.

From the tenth century the choice was made to move from equilibrium to modification and transformation of the coastal environment. The use of the natural resources of the salt marshes was replaced by the organisation of embanked wetlands, with a growing potential for a diverse range of economic activities. The embankments were first implemented in the estates of the Counts of Flanders, who invested in dikes, sluices, and drainage channels in their territories in order to raise profits that they could use for political purposes. The counts used hydraulic technology to create and expand the framework of conceptual possibilities of environment and space.

The landscape was thoroughly transformed; polders replaced the salt marshes after a progressive process to protect more and more lands against tidal influence and inundations.

Longitudinal dikes were constructed alongside the tidal channels, which in their turn were blocked in the twelfth century by large dams. In these transformed landscapes, feudal agents dependant upon the Counts of Flanders controlled resources and the organisation of hydraulic technology, aimed at the management and manipulation of the environment of the comital landed possessions. They integrated advanced technical skills, a knowledge of natural history, and the ambitious quasi-royal policy of the counts in the period between the tenth and twelfth centuries.

The eventual choice of new solutions to facilitate control over the landscape and environment was deliberate. Yet, it is impossible to state that these developments were purely planned steps in a well-designed master plan for the coastal region. It is possible that the first nuclear embankments in the comital estates were trial-and-error attempts to transform small parts of the supratidal flats into reserves for crops and animals. We also have no idea of where the knowledge or inspiration came from in these initiatives. We also have no clear evidence for the presence of similar oval embankments in the older early-medieval terp settlement landscape. Or, at least, they have not left any traces either in the landscape or in the place-names. Only archaeological fieldwork could prove the presence of nuclear embankments outside the comital estates.

Yet, what is striking is the systematic character of the transformation of the coastal landscape. Probably well before the end of the twelfth century, the largest part of the former supratidal environment of coastal Flanders was embanked, leaving only the larger tidal channels in their natural state. This scale is the best prove that, although the whole process was probably not preconceived in a single economic master plan, we still have to consider the transformation of coastal Flanders into an embanked landscape as the result of deliberate choices. What triggered those choices will be discussed here, but at a certain moment the Count of Flanders and his administration(s) followed an entirely different economic policy than the peasants of the early medieval period. This policy was large-scale investment on a in technological solutions to raise the profits of the maritime region.

These developments had a lasting influence on the relationship between humans and ecology. On one hand, it made the relation of men towards the environment more complex with consequences that were not always foreseen. Defensive embankments alongside the channels resulted in dike bursts and inundations, later followed by the construction of dams to block and embank the last tidal channels. The reclamation of the last tidal channels influenced coastal erosion and caused both the construction of defensive seadikes and new drainage canals through the polder area. Coastal erosion triggered new investments and an on-going search for environmental solutions in order to restore the balance between humans and maritime forces. The Dutch archaeologist Van Der Leeuw declared twenty-two years ago that cognitive practices dominated the relation between men and their environment: *La cognition de l'homme domine la relation entre lui et son environnement. Plus l'homme transforme ce qui l'entoure, moins il peut comprendre ce qui s'y passé* (*Human cognition dominates his relationship with the environment. How more people transform their environment, the less they understand it*) (1990: 305). The choice to intensify the profits from the environment by applying technological enhancements alienated people from that same environment, consequently leading to more complex solutions and more alienation.

But as we said, the choice was deliberate and the expectations high. The transformation of the landscape allowed the counts to develop a valuable power base for their new principality and to organise a landscape and society in the coastal plain that depended almost completely on the comital initiatives, with longstanding consequences in the late medieval period. Through this the counts succeeded in creating a centrally organised, intensive, specialised and technical hydraulic landscape, like Brenner calls it (2001), where the environment was used to secure and increase power and control over economy and society.

Through their comital estates and through their intervention in the management of the ecology and hydrology of the coastal area, the counts created a dominant market position in the supply and trade of bulk commodities for themselves. During the tenth and eleventh centuries this was organised via castral storehouses and marketplaces, including the ringfort castles of Veurne, Bergues, and Bruges, the main comital stronghold at the time, namely. This hydraulic policy of the counts was also clearly part of a larger spatial program, including the foundation of ports under direct comital control at the mouth of the main watercourses in the coastal plain, which in turn were the main axes of the trade with the inland centres and the main axes of the drainage of the coastal area.

Market values and market developments, the knowledge that rural produce could be transformed in commercial profits, must have been a strong stimulus for the strategy for systematic embankment. In this respect, the embankment of the coastal plain is directly connected to the issue of the rise of the large Flemish market towns such as Bruges, Ieper and Ghent in the period between the tenth and twelfth centuries (Verhulst, 1999). Further north, the position of the landlords and princes was not that strong as those in Flanders so the choice of systematic embankments as well as the rise of market towns happened considerably later. Accordingly, market developments seem to have been a considerable force in how the landscape of coastal Flanders was transformed. It would also be interesting to take a closer look at the triangular relationship between landscape-markets and towns in the period from the tenth to the twelfth century elsewhere in the North Sea basin.

What we must not forget is that whatever the economic importance of the tenth- to eleventh-century comital territorial possessions in the coastal plain, it was probably not the count's purpose to develop an economic position of his own. Instead we must assume that he wanted to use the generated means to purchase a princely ideology with a powerful territorial position. The collected rents in species or natura could well have been used to support the organisation of a centralised public authority, military campaigns and territorial conquests in the neighbouring counties, socio-political networking through gifts out of the territorial properties or the construction of collegiate churches in Bruges, Furnes and elsewhere.

The hydraulic technology itself (dams, sluices, dikes, ports, canals) formed a distinct material culture of power in the coastal landscape. The counts effectively created a landscape in which the ordering principles of the ruler were organised through the organisation of the relation with the environmental conditions. This was done in such a way that these principles had to be accepted *'without any conscious thought or consideration as to the way things might otherwise be'* (Miller, 1996: 404). The coastal landscape virtually became a geographic dialogue of power with the people and other power groups of the county materialised in the waterways, dikes, and sluices (Allen, 2003: 2–3). One question remains regarding the origin of the technological knowledge behind this comital spatial policy.

The choice of these environmental solutions led to a particular material development of the landscape of the coastal plain. Another choice of a different concept under the same ecological conditions (for instance to leave the salt marshes to the free farmers, or either to follow a more 'classical' feudal policy) would have led to a different landscape. Landscapes are indeed contextual social products and we must avoid deterministic approaches and methods in order to understand the relationship between humanity and the environment. Every coastal landscape represents a chosen developmental trajectory and is a result of social strategies. Different social and environmental strategies of power lead to different landscapes.

This type of research also demonstrates how important it is to use an integrated interdisciplinary approach using all kinds of data. Many questions also need to be worked out in more detail. What was the human presence in the landscape before the final infill of the tidal channels in the sixth and seventh centuries AD? What was the importance of the early medieval free peasants' society for the economic development of coastal Flanders? What were the exact technological features of the dikes and dams that were constructed and what was the extent of the workforce needed to construct them? How was technological knowledge on environmental issues transferred? And what were the social, political, and environmental strategies and measures that were taken by the coastal societies of the seventeenth, eighteenth, and nineteenth centuries in order to keep the danger of coastal erosion under control and the coastal environment in balance?

Meanwhile, the rise in sea level has accelerated and it is not certain whether the medieval drainage systems or the twentieth-century versions of sea dikes are sufficient to keep the landscape and environment in balance. Nonetheless, we are compelled to maintain the actual coastline and beach because of tourism, while the 1000-year-old medieval polder landscape now lies several meters below high tide level. The technological relationship between human society and tidal forces is complex and fragile. It takes social and political choices to decide which environmental strategy we consider fit for organising the landscape and its hydrology. In the near future, we will have to invest in new sustainable solutions for an integrated and balanced relationship with the rising tides, as is shown by the historical development of the environmental agency in coastal Flanders.

Bibliography

Allen, J. (2003) *Lost Geographies of Power*, London.

Allen, J.R.L. (1997) 'The geoarchaeology of land-claim in coastal wetlands: a sketch from Britain and the north-west European Atlantic-North Sea coasts', *Archaeological Journal*, 154, pp. 1–54.

Allen, J.R.L. (2000) 'Morphodynamics of Holocene salt marshes: a review sketch from the Atlantic and Southern North Sea Coasts of Europe', *Quaternary Science Reviews*, 19, pp. 1155–1231.

Ameryckx, J. (1958) 'Bodem en bewoning in de zeepolders', *Natuurwetenschappelijk tijdschrift*, 40, pp. 176–193.

Ameryckx, J. and Verhulst, A. (1958) 'Enkele historisch-geografische problemen in verband met de oudste geschiedenis van de Vlaamse Kustvlakte. Een colloquium van bodemkundigen en historici. Verslag', *Handelingen van de Maatschappij voor Geschiedenis en Oudheidkunde Gent,* 12, pp. 1–24.

Augustyn, B. (1992) *Zeespiegelrijsing, transgressiefasen en stormvloeden in maritiem Vlaanderen tot het einde van de 16e eeuw. Een landschappelijke, ecologische en klimatologische studie in historisch perspectief,* Brussels.

Baeteman, C. (1998) 'Is het de natuur of de mens die een bedreiging vormt voor kusten en polders?', *Mededelingen van de Koninklijke Academie voor Overzeese Wetenschappen*, 44, pp. 257–273.

Baeteman, C. (1999) 'The Holocene depositional history of the IJzer palaeovalley (western Belgian coastal plain) with reference to the factors controlling the formation of intercalated peat beds', *Geologica Belgica*, 2-1-2, pp. 39–72.

Baeteman, C., Scott, D.B. and Strydonck, M. Van (2002) 'Changes in coastal zone processes at high-sea-level stand: a late Holocene example from Belgium', *Journal of Quaternary Sciences*, 17, pp. 547–559.

Barry, J. (1999) *Environment and Social Theory*, London.

Bautier, R.H. and Sornay, J. (1984) *Les sources de l'histoire économique et sociale du Moyen Âge. Les États de la maison de Bourgogne. Vol. I. Archives des principautés territoriales. 2. Les principautés du Nord,* Paris.

Bazelmans, J. (1999) 'Zoden aan de dijk: kleinschalige dijkbouw in de late prehistorie en protohistorie van noordelijk Westergo', *De Vrije Fries. Jaarboek uitgegeven door het Fries Genootschap voor Geschied-, Oudheid- en Taalkunde*, 1999, pp. 7–74.

Beekman, A.A. (1932) *Nederland als Polderland*, Zutphen.

Behre, K.E., Menke, B. and Streiff, H. (1979) 'The Quaternary geological development of the German part of the North Sea', in: E. Oele, R. Schuttenham and A.J. Wiggers (eds), *The Quaternary History of the North Sea*, Uppsala, pp. 85–113.

Bender, B., (2006) 'Place and Landscape', in: C. Tilley, W. Keane, S. Küchler, M. Rowlands and P. Spyer (eds), *The Handbook of Material Culture*, London, pp. 303–314.

Besteman, J.C. (1990) 'North-Holland AD 400–1200: turning tide or tide turned?', in: J.C. Besteman, J.M.H. Bos and H.A. Heidinga, (eds), *Medieval Archaeology in the Netherlands, Studies presented to H.H. Van Regteren-Altena*, Assen, pp. 91–120.

Besteman, J.C. and Guiran, A.J. (1986) 'De middeleeuwse bewoningsgeschiedenis van Noord-Holland boven het IJ en de ontginning van de veengebieden. Opgravingen in Assendelft in perspectief', in: M.C. Van Trierum and H.E. Henkes (eds.), *Rotterdam Papers V. A contribution to prehistoric, roman and medieval archaeology. Papers presented at the Symposium Landschap en bewoning rond de mondingen van Rijn, Maas en Schelde, Rotterdam* 5th–6th October 1984, Rotterdam, pp. 183–212.

Bisson, T.N. (1995) 'Introduction', in: T.N. Bisson (ed.), *Cultures of Power. Lordship, Status, and Process in Twelfth-Century Europe*, Philadelphia, pp. 1–8.

Blanchard, R. (1906) *La Flandre. Etude géographique de la plaine flamande en France, Belgique et Hollande*, Handzame.

Blok, D.P. (1984) 'Wie alt sind die ältesten niederländischen Deiche? Die Aussagen der frühesten schriftlichen Quellen', *Probleme der Küstenforschung im südlichen Nordseegebiet*, 15, pp. 1–7.

Boterberge, R. (1962) 'Historische geografie van het overstromingsgebied van de IJzer in de middeleeuwen', *Handelingen van de Maatschappij voor Geschiedenis en Oudheidkunde Gent,* 16, pp. 77–142.

Boterberge, R. (2000) *Geschiedenis van de landelijke parochies Klemskerke-Vlissegem en van de badplaats De Haan*, De Haan.

Brenner, R. (2001) 'The Low Countries in the transition to capitalism', in P. Hoppenbrouwers and J.L. Van Zanden (eds), *From peasants to farmers? The transformation of the rural economy and society in the Low Countries (Middle Ages-19th century) in light of the Brenner Debate*, Turnhout, pp. 275–338.

Buntinckx, W. (1968) 'De enquête van Oudenburg, Hervorming van de repartitie van de beden in het graafschap Vlaanderen (1408)', *Bulletin van de Koninklijke Commissie voor Geschiedenis*, 84, pp. 75–138.

Chouquer, G. (2000) *L'étude des Paysages. Essais sur leurs formes et leur histoire*, Paris.

Coles, J. and Hall, D. (1998) *Changing Landscapes: The Ancient Fenland*, Exeter.

Cook, H. and Williamson, T. (1999) 'Introduction: landscape, environment and history', in: H. Cook, and T. Williamson (eds), *Water Management in the English Landscape, Field, Marsh and Meadow*, Edinburgh, pp. 1–14.

Coornaert, M. (1974) *Knokke en het Zwin: de geschiedenis, de topografie en de toponimie van Knokke met een studie over de Zwindelta*, Tielt.

Darby, H.C. (1983) *The changing Fenland*, Cambridge.

Davies, W. and Fouracre, P. (1995) 'Introduction', in: W. Davies and P. Fouracre (eds), *Property and power in the early Middle Ages*, Cambridge, pp. 1–16.

Deckers, P. (2010) 'An illusory emporium? Small trading places around the southern North Sea', in: A. Willemsen and H. Kik (eds.), *Dorestad in an International Framework. New Research into Centres of Trade and Coinage in Carolingian Times*, Turnhout, pp. 159–168.

Deckers, P. (2011) *Leffinge Oude Werf. Prospectie met ingreep in de bodem, 30 augustus – 24 september 2010. Voorlopig rapport*, Brussels.

Declercq, G. (1997) 'Kamer van de hoofdredeninge (12de eeuw-1795)', in: W. Prevenier and B. Augustyn (eds), *De Gewestelijke en Lokale Overheidsinstellingen in Vlaanderen tot 1795*, Brussels, pp. 171–187.

De Coussemaker, I. (1882–1891) *Un cartulaire de l'abbaye de Notre-Dame de Bourbourg*, Lille.

Degryse R. (1987) *Nieuwpoort tot omstreeks 1302*, Nieuwpoort.

de Hemptinne, T. and Verhulst, A. (1988) *De oorkonden der Graven van Vlaanderen (juli 1128–september 1191). II-1. Regering van Diederik van de Elzas (juli 1128–januari 1168),* Brussels.

Dekker, C. (1971) *Zuid-Beveland, De historische geografie en de instellingen van een Zeeuws eiland in de Middeleeuwen*, Assen.

Dekker C. (1980) 'Resultaten van het historisch-geografisch onderzoek in Zeeland', in: A. Verhulst and M.K.E. Gottschalk (eds), *Transgressies en occupatiegeschiedenis in de kustgebieden van Nederland en België, Colloquium Gent 5–7 september 1978, Handelingen*, Ghent, pp. 93–106.

de Langen, G. (1992) *Middeleeuws Friesland. De economische ontwikkeling van het gewest Oostergo in de Vroege en de Volle Middeleeuwen*, Groningen.

De Langhe, J. (1967) 'Toponymica uit het kustland', *Biekorf*, 68, pp. 183–184.

De Meulemeester, J. (1979) 'De circulaire vesterking en de Warandemote te Veurne', *Conspectus MCMLXXVI*, pp. 152–156.

Dhondt, J. (1941) 'Het ontstaan van het vorstendom Vlaanderen', *Belgisch Tijdschrift voor Filologie en Geschiedenis*, 20, pp. 534–573.

Dhondt, J. (1942) 'Het ontstaan van het vorstendom Vlaanderen (einde)', *Belgisch Tijdschrift voor Filologie en Geschiedenis*, 21, pp. 53–94.

Dijkstra, J. and Zuidhoff, F.S. (eds) (2011) *Kansen op de Kwelder. Archeologisch onderzoek op negen vindplaatsen in het nieuwe tracé van de Rijksweg N57 en de nieuwe rondweg ter hoogte van Serooskerke (Walcheren)*, Amersfoort (ADC Monografie; 10).

Eddison, J. (1995) *Romney Marsh: The Debatable Ground*, Oxford.

Ervynck, A., Baeteman, C., Demiddele, H., Hollevoet, Y., Pieters, M., Schelvis, J., Tys, D. and Verhaeghe, F. (1999) 'Human occupation because of a regression or the cause of a transgression? A critical view on the interaction between geological events and human occupation history in the Belgian coastal plain during the first millennium AD', *Probleme der Küstenforschung im südlichen Nordseegebiet*, 26, pp. 97–121.

Fayen, A. (1906) *Liber Traditionum Sancti Petri Blandiniensis. Livre des donations faites à l'abbaye de Saint-Pierre de Gand, depuis ses origines jusqu'à XI*[e] siècle, avec des additions jusq'en 1273, Ghent.

Fenwick, H. (2001) 'Medieval salt-production and landscape development in the Lincolnshire Marsh', in: S. Ellis, H. Fenwick, M. Lillie and R. Van de Noort (eds), *Wetland Heritage on the Lincolnshire Marsh. An Archaeological Survey,* Kingston upon Hull, pp. 243–254.

Fockema-Andreae, S. (1960) 'L'eau et les hommes de la Flandre Maritime', *Tijdschrift voor Rechtsgeschiedenis*, 28, pp. 181–196.

Gelling M. (1998) 'Place-names and Landscape', in: S. Taylor (ed.), *The uses of placenames, Language, landscape, archaeology, environment, literature, history,* Edinburgh, pp. 75–100.

Gelling, M. and Cole, A. (2000) *The Landscape of Place-names*, Stamford.

Gottschalk, M.K.E. (1955) *Historische geografie van westelijk Zeeuws-Vlaanderen I: Tot de St-Elisabethsvloed van 1404*, Assen.

Gottschalk, M.K.E. (1977) *Stormvloeden en rivieroverstromingen in Nederland,* 3 vols., Assen.

Gottschalk, M.K.E. (1980) 'Subatlantische transgressiefasen en stormvloeden', in A. Verhulst and M.K.E. Gottschalk (eds), *Transgressies en occupatiegeschiedenis in de kustgebieden van Nederland en België, Colloquium Gent 5–7 september 1978, Handelingen*, Ghent, pp. 21–27.

Gysseling, M. (1950) *De toponymie van Oudenburg*, Brussels.

Gysseling, M. (1960) *Toponymisch woordenboek van België, Nederland, Luxemburg, Noord-Frankrijk en West-Duitsland*, Brussels.

Gysseling, M. (1969) 'Nederzettingsnamen in de Nederlanden, Noord-Frankrijk en Noord-West-Duitsland', in: M. Gysseling and A. Verhulst (eds), *Nederzettingsnamen en nederzettingsgeschiedenis in de Nederlanden, Noord-Frankrijk en Noord-West-Duitsland. Papers presented at the Commissie voor naamkunde en nederzettingsgeschiedenis der Koninklijke Nederlandse Akademie van Wetenschappen, 14 december 1968*, Amsterdam, pp. 5–35.

Gysseling, M. (1981) 'De voornaamste bestandsdelen van Zuidnederlandse plaatsnamen', *Naamkunde*, 13, pp. 75–78.

Gysseling, M. (1983) 'Inleiding tot de oude toponymie van West-Vlaanderen', *De Leiegouw*, 25, pp. 39–58.

Halbertsma, H. (1963) *Terpen tussen Vlie en Eems. Een geografisch-historische benadering*, Groningen.

Hall, R.A. (2011) 'Burhs and Boroughs: defended places, trade and tonw. Plans, defences and civic features', in: H. Hamerow, D. Hinton and S. Crawford (eds), *The Oxford Handbook of Anglo-Saxon Archaeology*, Oxford, pp. 600–621.

Hallewas, D.P. (1984) 'Mittelalterliche Seedeiche im Holländischen Küstengebiet', *Probleme der Küstenforschung im südlichen Nordseegebiet,* 15, pp. 9–27.

Hallewas, D.P. and van Regteren Altena, J.F. (1980) 'Bewoningsgeschiedenis en landschapsontwikkeling rond de Maasmond', in: A. Verhulst and M.K.E. Gottschalk (eds), *Transgressies en occupatiegeschiedenis in de kustgebieden van Nederland en België, Colloquium Gent 5–7 september 1978, Handelingen*, Ghent, pp. 155–207.

Heidinga, H.A. (1997) *Frisia in the First Millenium, An outline*, Utrecht.

Henderikx, P.A. (1987) *De Benden-Delta van Rijn en Maas. Landschap en bewoning van de Romeinse tijd tot ca. 1000*, Hilversum.

Henderikx P.A. (1995) 'De ringwalburgen in het mondingsgebied van de Schelde in historisch perspectief', in: R.M. Van Heeringen, P.A. Henderikx and A. Mars (eds), *Vroeg-Middeleeuwse ringwalburchten in Zeeland*, Goes, pp. 70–112.

Hermans, M. (2009) *Analyse van het aardewerk uit de archeologische opgraving van de site uitkerke Groenwaecke*, Unpublished Master's Thesis, Brussels Free University.

Hoppenbrouwers, P. and Zanden J.L. Van (2001), 'Restyling the transition from feudalism to capitalism. Some critical reflections on the Brenner thesis', in: P. Hoppenbrouwers and J. L. Van Zanden (eds), *From peasants to farmers? The transformation of the rural economy and society in the Low Countries (Middle Ages-19th century) in light of the Brenner Debate,* Turnhout, pp. 19–40.

Koch, A. (1951) *De rechterlijke organisatie van het graafschap Vlaanderen tot in de dertiende eeuw*, Antwerpen.

Koch, A. (1978) 'Het graafschap Vlaanderen van de negende eeuw tot 1070', in: *Algemene Geschiedenis der Nederlanden*, deel 1, pp. 354–383.

Lançon, M. (2005) 'Téteghem. Rue des Cuisinières', in: G. Fosse (ed.), *Bilan Scientifique 2004 Direction Régionale des Affaire Culturelles Nord-Pas-de-Calais, Service Régional d'Archéologie*, Lille, pp. 87–90.

Lebecq, S. (1983) *Marchands et navigateurs frisons du haut Moyen Âge*, Lille.

Loveluck, C. and Tys, D. (2006) 'Coastal societies and identity, AD 600–1200: rethinking the social distribution of material culture along the Channel and North Sea coasts', *International Journal for Maritime Archaeology*, 1, pp. 140–169.

Lyon, B. and Verhulst, A. (1967) *Medieval finance, A comparison of financial institutions in Northwestern Europe*, Ghent.

Meier, D. (1997) *Landschaftsentwicklung und Siedlungsgeschichte des Eiderstedter und Dithmarscher Küstengebietes als teilregionendes Nordseeküstenraumes, Untersuchungen der Arbeitsgruppe Küstenarchäologie am Forschungs- und Technologiezentrum Westküste der Christian-Albrechts-Universität Kiel*, Büsum.

Meijns, B. (2007) 'Les premières collégiales des comtes de Flandre, leurs reliques et les conséquences des invasions normandes (IXe-Xe siècles)', *Revue belge de philologie et d'histoire,* 85, pp. 539–575.

Miller, D. (1996) 'Artefacts and the Meaning of Things', in: T. Ingold (ed.), *Companion Encyclopedia of Anthropology*, London, pp. 396–419.

Monteyne, G. and Vandaele, R. (2008) *Wenduine. Van middeleeuws vissersdorp tot familiebadplaats*, Wenduine.

Moormann, F. (1951) 'De bodemgesteldheid van het Oudland van Veurne Ambacht', *Natuurwetenschappelijk Tijdschrift*, 33, pp. 1–124.

Pieters, M. (1995) 'Een 15de-eeuwse sector van het verdwenen vissersdorp te Raversyde', *Archeologie in Vlaanderen*, 4, pp. 219–236.

Platelle H. (1962) *Le temporel de l'abbaye de Saint Amand des origines à1340*, Lille.

Pype, P. (2002) 'Archeologisch onderzoek van een landelijke bewoningskern uit de Vroege en de Volle Middeleeuwen te Uitkerke (Blankenberge) (W.-Vl.)', *Archaeologia Medievalis*, 25, pp. 33–34.

Rippon, S. (2001) *The Transformation of Coastal Wetlands. Exploitation and Management of Marshland Landscapes in North West Europe during the Roman and the Medieval Periods*, Oxford.

Rippon, S. (2002) 'Infield and Outfield: The Early Stages of Marshland Colonisation and the Evolution of Medieval Field Systems', in: T. Lane and J. Coles (eds), *Through Wet and Dry. Essays in Honour of David Hall*, Exeter, pp. 54–70.

Schmid P. (1988) 'Die Mittelalterliche Neubesiedlung der Niedersächsischen Marsch', in: M. Bierma, O.H. Harsema, W. Van Zeist (eds), *Archeologie en Landschap, Bijdragen aan het gelijknamige symposium gehouden op 19 en 20 Oktober 1987, ter gelegenheid van het afscheid van H.T. Waterbolk*, Groningen, pp. 133–164.

Silvester, R. (1999) 'Medieval reclamation of marsh and fen', in H. Cook and T. Williamson (eds), *Water Management in the English Landscape, Field, Marsh and Meadow,* Edinburgh, pp. 122–140.

Soens, T. (2001) 'Het waterschap en de mythe van democratie in het Ancien Régime. Het voorbeeld van de Vlaamse Kustvlakte in de Late Middeleeuwen', *Jaarboek voor Ecologische Geschiedenis* 2001, pp. 39–56.

Soens, T. (2009) *De spade in de dijk? Waterbeheer en rurale samenleving in de Vlaamse kustvlakte (1280–1580),* Ghent.

Soens, T. and Tys, D. (2001) 'Het *ius notarii Flandrie*, een miskende getuige van de ontwikkelingen in de administratie van het graafschap Vlaanderen (12e–15e eeuw), *Handelingen van het genootschap voor geschiedenis*, 138, pp. 3–32.

Strubbe, E. (1942) *Egidius van Bredene (11–1270): Grafelijk ambtenaar en stichter van de abdij Spermalie*, Bruges.

Tavernier, R. (1938) 'De geologische ontwikkeling van de Vlaamsche kust', *Wetenschap in Vlaanderen*, 4–2, pp. 27–41.

Tilley, C. (2006) 'Introduction: Identity, Place, Landscape and Heritage', *Journal of Material Culture*, 11, pp. 7–32.

TeBrake, W.T. (2000) 'Hydraulic engineering in the Netherlands during the Middle Ages', in: P. Squatriti (ed.), *Working with Water in Medieval Europe. Technology and Resource-Use*, Leiden, pp. 101–127.

Thoen, E. (1993) 'The count, the countryside and the economic development of the towns in Flanders from the eleventh to the thirteenth century: Some provisional remarks and hypotheses', in: E. Aerts, B. Henau, P. Janssens and R. Van Uytven (eds), *Studia historica oeconomica: liber amicorum Herman Van der Wee,* Louvain, pp. 259–278.

Thoen, E. (2001) 'A "commercial survival economy" in evolution. The Flemish countryside and the transition to capitalism (Middle Ages – 19th century)', in P. Hoppenbrouwers and J. L. Van Zanden (eds), *From peasants to farmers? The transformation of the rural economy and society in the Low Countries (Middle Ages-19th century) in light of the Brenner Debate,* Turnhout, pp. 102–157.

Tys, D. (1997) 'Landscape and settlement: the development of a medieval village along the Flemish coast', in: G. De Boe and F. Verhaeghe (eds.), *Rural settlements in medieval Europe. Papers of the 'Medieval Europe Brugge 1997' Conference, Volume 6*, Zellik, pp. 157–167.

Tys, D. (1998) 'De heffing van het *superaratum* in *Vranckx Ambacht* (gemeente Middelkerke): de betekenis van enkele ruimtelijke gegevens op Testerep tussen de 9de en de 12de eeuw', *Handelingen van het genootschap voor geschiedenis*, 136, pp. 103–119.

Tys, D. (2001) 'De verwerping van het zgn. Duinkerke-transgressiemodel en nieuwe inzichten in de vroegste bedijking van de kustvlakte', in E. Huys and M. Vandermaesen (eds), *Polders en wateringen in maritiem Vlaanderen*, Brussels, pp. 17–53.

Tys, D. (2003) *Landschap als materiële cultuur. De interactie tussen macht en ruimte in een kustgebied en de wording van een laatmiddeleeuws tot vroegmodern landschap. Kamerlings Ambacht, 500–1200/1600*, Unpublished Dissertation, Brussels Free University.

Tys, D. (2004) 'Domeinvorming in de "wildernis" en de ontwikkeling van vorstelijke macht: het voorbeeld van het bezit van de graven van Vlaanderen in het IJzerestuarium tussen 900 en 1200', *Jaarboek voor Middeleeuwse Geschiedenis*, 7, pp. 34–87.

Tys, D. (2005) 'De inrichting van een getijdenlandschap. De problematiek van de vroegmiddeleeuwse nederzettingsstructuur en de aanwezigheid van terpen in de kustvlakte: het voorbeeld van Leffinge (gemeente Middelkerke, provincie West-Vlaanderen)', *Archeologie in Vlaanderen,* 8, pp. 257–279.

Tys, D. (2010) 'The Scheldt Estuary as a Framework for Early Medieval Settlement Development', in: A. Willemsen and H. Kik (eds), *Dorestad in an international framework. New research on centres of trade and coinage in Carolingian Times*, Turnhout, pp. 168–176.

Van der Leeuw, S.E. (1990) 'Rythmes temporels, espaces naturels et espaces vécus', in: J.L. Fiches and S.E. Van der Leeuw (eds), *Archéologie et espaces, X*[e] Rencontres Internationales d'Archéologie et d'Histoire, Antibes, Octobre 1989, Juan-les-Pins, pp. 299–346.

Vanhoutte, S. and Pieters, M. (2003) 'Archeologisch noodonderzoek op het toekomstig bedrijventerrein Plassendale III (Zandvoorde, stad Oostende, prov. West-Vlaanderen). Interimverslag 2000–2001', *Archeologie in Vlaanderen*, 7, pp. 95–110.

Vercauteren, F. (1938) *Actes des comtes de Flandre, 1071–1128*, Brussels.

Verhulst, A. (1959) 'Historische geografie van de Vlaamse kustvlakte tot omstreeks 1200', *Bijdragen voor de Geschiedenis der Nederlanden*, 14, pp. 1–37.

Verhulst, A. (1964) 'Die Binnenkolonisation und die Anfänge der Landgemeinde in Seeflandern', *Vorträge und Forschungen vom Institut für geschichtliche Landesforschung*, 7, pp. 447–460.

Verhulst, A. (1966) *Histoire du paysage rurale en Flandre de l'époque romaine au XVIII*[e] siècle, Brussels.

Verhulst, A. (1967) 'Initiative comtale et développement économique en Flandre au XIIe siècle: le rôle de Thierry et de Philippe d'Alsace (1128–1191)', *Miscellanea Mediaevalia in memeoriam J.F. Niermeyer*, pp. 227–240.

Verhulst, A. (1990) *Précis d'Histoire Rurale en Belgique*, Brussels.

Verhulst, A. (1995) *Landschap en landbouw in middeleeuws Vlaanderen*, Brussels.

Verhulst, A. (1998) 'Sheep-breeding and wool production in pre-thirteenth century Flanders and their contribution to the rise of Ypres, Ghent and Bruges as centres of the textile industry', in M. Dewilde, A. Ervynck and A. Wielemans (eds), *Ypres and the medieval cloth industry in flanders, archaeological and historical contributions*, Zellik, pp. 33–42.

Verhulst, A. (1999) *The rise of cities in North-West Europe*, Cambridge.

Verhulst, A. (2001) 'Polders en wateringen in Vlaanderen: Status Quaestionis van het historisch onderzoek', E. Huys and M. Vandermaesen (eds), *Polders en wateringen in maritiem Vlaanderen*, Brussels, pp. 11–16.

Verhulst, A. (2002) *The Carolingian economy*, Cambridge.

Verhulst, A. and Gysseling, M. (1962) *Le compte général de 1187, connu sous le nom de 'Gros brief' et les institutions financières du Comté de Flandre au XII*[e] siècle, Brussels.

Voet, L. (1942) 'De Graven van Vlaanderen en hun domein (861–1191)', *Wetenschappelijke Tijdingen*, 7, pp. 25–32.

Vos, P.C. and Heeringen, R.M. van (1997) 'Holocene geology and occupation history of the Province of Zeeland (SW Netherlands)', in: M.M. Fischer (ed.), *Holocene evolution of Zeeland (SW Netherlands),* Haarlem, pp. 5–110.

Warlop, E. (1968) 'De vorming van de grote schepenbank in het Brugse Vrije (11de-13de eeuw)', *Standen en landen*, 44, pp. 3–29.

11 Reclamations and submerged lands in the Ems River Estuary (900–1500)

Otto S. KNOTTNERUS, Zuidbroek, the Netherlands

Abstract

This article discusses dike-building and drainage in three adjoining regions south of the Ems River estuary (North-East Groningen) during the High and Late Middle Ages. It is based on scattered historical and archaeological evidence, as well as on ordnance and soil maps, aerial views, toponymical findings, and nineteenth-century field surveys. The first and second sections deal with the embankment of the Fivel tidal basin and on the systematic reclamation of the inland fens and bogs, particularly in the district of Duurswold. The final section then concentrates on peat-land reclamation and maritime intrusions in the Oldambt and Reiderland districts, which resulted in the formation of Dollard Bay. Relative success or failure of water management efforts is explained by a combination of ecological and social-political factors. It is also argued that the relative ability of supra-local authorities to bridge diverging interests may have been decisive in the continued existence of collective arrangements to restrain the risks of inundation and storm-surges.

I. Introduction

This article discusses dike-building and drainage measures in three adjoining regions south of the River Ems estuary (North-East Groningen), based on scattered historical and archaeological evidence, as well as on ordnance and soil maps, aerial views, toponymical findings, and nineteenth-century field surveys. As a rule, salt-marsh and peat-land reclamation went hand in hand, but as the former was quite profitable, the latter suffered from diminishing returns as soon as drainage measures could not keep up with surface subsidence caused by dehydration and surface erosion. Though peat-land communities could rely on collective arrangements that were far more elaborate than those of their coastal counterparts, in the long run they could not solve all of their problems by themselves. Consequently, coastal and inland districts had to cooperate in institutions that served the interests of both parties involved.

Our studies focus on the main actors of the reclamation process. Apparently, large-scale drainage and peat-land reclamation measures were applied at an early stage, at least since the tenth century. The embankment of the salt marshes presumably started during the eleventh century and the involvement of major landowners with widespread interests, who could force the other actors to comply, must have been decisive in this process. As such, the newly founded monasteries played an important part during the final stages of the reclamation process (thirteenth and fourteenth centuries), encouraging the establishment of supra-local institutions. Special attention has been given to contemporary myths supporting and legitimating collective arrangements.

Landscapes or seascapes? The history of the coastal environment in the North Sea area reconsidered, ed. by Erik THOEN, Guus J. BORGER, Adriaan M.J. DE KRAKER, Tim SOENS, Dries TYS, Lies VERVAET & Henk J.T. WEERTS, Turnhout, 2013 (*CORN Publication Series*, 13), pp. 241–266.

BREPOLS PUBLISHERS DOI: 10.1484/M.CORN.1.101555

The first part of this paper deals with the embankment of the Fivel tidal basin. Field evidence suggests a small-scale and localised approach, which became more offensive after the involvement of the large monastic orders around 1200.

The second part of the paper concentrates on the systematic reclamation of the inland fens and bogs, particularly in the district of Duurswold. There, a large-scale approach prevailed from the outset, thus indicating systematic organisation and the guiding hand of external experts. In order to facilitate settlement and the expansion of arable farming, local communities had to create alternative drainage systems, redirecting surplus water through a newly dug tidal shipping canal eastward into the River Ems estuary. In contrast, the lower course of the River Fivel silted up due to the declining water flow and was subsequently dammed. Only at a later stage did the monasteries become involved, instigating additional drainage measures and supervising regional water management boards.

The third part of this article shifts attention eastwards to the Oldambt and Reiderland districts. Unlike their neighbours from Duurswold, the inhabitants of those regions failed to neutralise the effects of peat-land subsidence and seasonal flooding. This led to growing pressure on drainage and embankment systems, apparently because the existing collective arrangements were fragile and subject to quarrels. The terrain, moreover, was quite vulnerable, as the Pleistocene valley systems underlying the bogs had been largely filled up with massive layers of peat. Once the sea-wall became neglected in the early fifteenth century the sea broke in, causing irreparable damage and leading to the destruction of extensive peat-land areas. As a consequence, a new tidal system came into existence, known as Dollard Bay. Additionally, as a result of the enlarged tidal volumes, the River Ems estuary widened and parts of the river banks and forelands had to be abandoned. The intertidal mudflats at the back of the tidal system gradually silted up and were subsequently embanked again.

II. The embankment of the Fivel tidal basin

The River Ems estuary took shape in the last millennium BC, when the tidal river mouth was narrowed down by salt-marshes and natural levees. Drowned reed-swamps and peat-bogs covered by huge sediments separated the main river bed from the submerged valley system of the Fivel tributary to the west. The salt-marsh ridges and levees in the River Ems estuary were settled by groups of pastoralists during the Early Iron Age. The settlers started to build their houses at ground level before raising mounds that sheltered them from storm surges.

The Fivel tidal basin began silting up behind expanding shore ridges during the first millennium AD (Vos and Van Kesteren, 2000; Beek and Vos, 2008). A sediment plain interlaced with tidal creeks started to build up from the west, offering fresh pastures to the populations of the adjacent village mounds. Salt-marsh ridges closed off the southern part of the tidal basin during the tenth and eleventh centuries, partly blocking the mouth of the Fivel River and cutting off the market town of Loppersum from the sea. Consequently, the drainage of inland fens and bogs was hampered. Slightly elevated cattle roads, parallel to the tidal creeks, connected the main villages with their outlying satellite settlements. Similar roads reached into the meadows, fens, and bogs of the interior (for an overview: Knottnerus, 2006; 2008a).

Scarce archaeological remains suggest that permanent settlement on village mounds in the western part of the tidal basin dates back to the seventh or eight centuries. Just like the older Iron Age village mounds they were raised up to a height of 3–5 m above ordnance level. In contrast, the outlying parish of Uithuizen was situated on top of a major shore ridge suited for arable farming. It is referred to in census documents from the tenth and eleventh centuries, compiled by the Rhineland abbey of Werden. The leeward salt-marsh ridge of Zeerijp was occupied during the twelfth century. Both villages were characterised by a chain of raised farmsteads known as *ripa* (shore, edge, linear settlement, cf. OE *sǽrima*, seashore, coast), which subsequently set the basis for a major dike (see also De Vries, 1946: 286–287; Remmers, 2004: 272). The abbey of Werden also acquired land in nearby villages whose toponyms cannot be identified with certainty. Other newly established salt-marsh villages such as Zandeweer and Eppenhuizen consisted of raised farmsteads around a churchyard mound, indicating that their territories were already protected by a minor sea-wall.

The outer salt marshes (around the village of 't Zandt), on the other hand, were settled during the twelfth century by dozens of scattered mounds measuring 1–2 m in height and locally known as *pollen* (knolls). Some of these small mounds may have been inhabited, while others served as a refuge for the cattle herds. Pastoral farming must have prevailed here prior to the embankment during the first decades of the thirteenth century. As late as the seventeenth century, the newly formed outer shore ridges (near Roodeschool) were occupied by rows of raised farmsteads, their infields barely protected by a minor summer dike.

Documentary evidence relating to the embankment of the Fivel tidal basin is scarce and largely dates to the fifteenth and sixteenth centuries. The famous thirteenth-century chronicle of the Norbertine Abbey of Wittewierum offers a number of scattered and rather ambiguous remarks (Jansen and Janse, 1991). To make matters worse, archaeological surveys cover only part of the area and most of the visual remains, ditches, and elevations were levelled as a consequence of extensive land consolidation programs in the 1980s and 1990s (Miedema, 1990; 2000). Ordnance and soil maps, aerial views, historical land surveys, toponymical findings, and observations by nineteenth-century antiquarians provide useful additions to our corpus of knowledge. Pioneer work has been done by the local farmer Pieter Mennes Bos and the engineer Johan Van Veen (celebrated as the architect of the Delta Plan), who published an important article on the history of the Fivel River estuary (Van Veen, 1930).

Generally, the construction of first large-scale embankments in the Wadden Sea area may be dated to *c.*1000 (Knol, 2010: 23–25; Ey, 2010). The system of coastal defences must been completed during the second half of the twelfth century with the construction of an uninterrupted sea-wall. Minor tributaries were shut down by sluice gates. The Wittewierum chronicle gives some evidence supporting the latter date (Ligtendag, 1995: 197). However, the question of who may have been responsible for such infrastructural measures is largely undecided. Conventional law historians such as Julius von Gierke tend to attribute them to corporate bodies or even tribal bonds of which the early existence has yet to be proven (Von Gierke, 1901–1917; see also Goldbeck, 1991). More convincing is the idea that overlords and state officials took the lead, forcing their subjects to comply (Tys, 2005). But since powerful external agents were largely absent in the area, a mixed model stressing the dynamics of collective action seems to be more appropriate (e.g. Olson, 1965; De Swaan, 1988).

Basically, the burden of drainage and dike building activities rested on individual families. As the scale of piecemeal intervention grew, so the number of people involved and the degree of collective involvement increased, thereby leading to growing group pressure to comply. So the system of coastal defences may be seen as a collective arrangement, exercising mounting pressure on individuals and leaving less and less room for withdrawal, except at high costs. Dike building, moreover, was paralleled by other collective arrangements, such as the establishment of parish churches, which were often collectively owned. Village chieftains and large landowners must have enforced compliance to collective efforts, as their stakes were bigger and their assets more widespread than those held by ordinary villagers. Local priests and church officials often became spokesmen for the community, invoking supernatural sanctions on those refusing to comply. The Wittewierum chronicle explains in detail how the nobles of the country compelled the unwilling inhabitants of six distant hamlets to repair the Ems River dikes after a major storm surge in 1219 (Jansen and Janse, 1991: 122–123; De Boer, 2009: 11–14).

To be sure, local authority was deeply embedded into supra-regional power structures, as the extent of foreign landownership (particularly by the abbey of Werden) shows. A major stakeholder was the bishop of Münster whose local representatives dominated the ecclesiastical courts. But even then, regional magnates who held ecclesiastical leases and dignities, royal estates, and coinage privileges, were largely dependent on the collaboration of their less powerful neighbours. As such, dike building efforts might have been inspired by medieval peace movements, intended to regulate feuds, promote communal authority, and advance land reclamation.

Apparently, the sea has been associated with the untamed forces of nature for long, as the name *Fivel* attests (sea-monster, giant – cf. OE *fifel-streám*, the sea) (Heeroma, 1955). Controlling the floods and draining the wetlands seemed to be part of a mission assigned to Christianity to defeat evil. As such, mythical stories played an important part in motivating people to engage in cooperative efforts. According to a medieval legend, Charlemagne sentenced twelve heathen judges to death, because they refused to put into writing their ancient customs. Lost at sea in a ship without oars or sail, they were saved by Christ, who set the men ashore and instructed them with the laws of a Christianised society. The landing place in the former salt marsh area near the village of Zeerijp had been demarcated since then by a shining beacon and a miraculous well (Knottnerus, 2007: 133–135). Another seventeenth-century legend tells the story of a local warlord, who murdered the village priest of Usquert, because the priest had celebrated the Easter mass without waiting for him. Obviously, the man was held up by the tide. As an ordeal the knight and his moated keep situated in the unembanked foreland of Uithuizermeeden were swallowed by a mud pool (Formsma, Luitjes-Dijkveld Stol, and Pathuis, 1987: 419–421).

Both legends reflect the customs of a society, locked behind its dikes, but looking back into an era in which the embankments were not yet evident. Older juridical texts, dating from the twelfth and thirteenth centuries, refer to a legendary era in which the local Frisians had been enslaved by Viking invaders. As Charlemagne was believed to have restored their freedom, it was also told that he had assigned the Frisians the task of building a sea wall to defend their country against the northern waves and the northern king, either with pitchforks, spades, and bearriers or with spears, swords, and shields (Schmidt, 2008: 203–204, 290).

The oldest embankments in the River Fivel estuary probably consisted of local dikes surrounding the village infields, as well as conducting embankments along creeks and canals in order to prevent seasonal inundation (see also De Boer, 2009: 10). The earliest sluices were hollowed trees, provided with a flap that functioned as a check valve. More complicated vaulted sluice gates have been known at least since the twelfth or thirteenth century. Some of them were large enough for passing vessels. Presumably, during the second half of the twelfth century the joint villages created a common sea-wall, thereby narrowing the Fivel tidal basin to a cove that was subsequently embanked. The sluice of Oldenzijl held a strategic position in the oldest sea-wall belonging to the western district of Hunsingo. Finds of tuff, possibly lost during transfer from ship to ship, suggest that it may date from the twelfth century, as tuff gave way to brick-building after that date. To the south, the dried up Fivel tributary was cut off from the sea by a dam near the village of Westeremden (from *a-muthon*, river mouth), where the main outlet had been located. The eastern banks of the Fivel Bay were situated on an elevated salt-marsh ridge. As subsequent embankments blocked the original drainage system, excess surface water had to be redirected towards inland canals indirectly leading to the River Ems.

Several roads and ditches crossing the district border between Hunsingo and Fivelgo seem to indicate that the offensive embankment of the salt marshes had already begun by the twelfth century. Newly founded monasteries such as the Benedictine Abbey of Feldwerd (*c.*1183) and the abbey of Wittewierum (*c.* 1214) played a major part in the process of salt-marsh reclamation. They were responsible for the establishment of extensive monastic granges in the newly embanked area. As a rule, monastic superiors stood at the head of the water management boards or *zijlvesten* (*zijl*, sluice, dike-lock, with suffix *veste*, jurisdiction), surveying the maintenance of sluices and waterways and using their authority to settle disputes. The abbot of Wittewierum became the chairman of the combined drainage board of the Drie Delfzijlen, his colleagues from nearby monasteries as well as the dean of Farmsum held similar positions in other drainage boards. The sluice of Oldenzijl was replaced in 1317 by another one closer to sea near the village of Zijldijk, probably under the authority of the abbot of Feldwerd. The name *Hooghersijl* (lofty sluice, 1428) suggests it may have been an partially open pound lock, suited for merchant shipping. New embankments caused the harbour to silt up swiftly, whereupon the lands concerned were allowed to discharge their water southward to the main sluices of the Drie Delfzijlen (1444) and Winsum (1464).

The newly embanked marshlands with their sabulous clay soils suited arable farming, contrary to the slightly heavier and less well-drained clays of the surrounding quarters, where cattle and horse breeding as well as dairy farming prevailed until the eighteenth century. In the nineteenth century arable and mixed farming got the upper hand. Since the 1970s local agriculture has been dominated by seed potatoes, as the constant sea breeze acting helps to contain diseases.

Salt-marsh occupation may be illustrated by the development of the village of Garsthuizen (Knottnerus, 2006: 34). Earliest settlement, probably dating from the eleventh century, consisted of scattered farmsteads on an outlying sandy levee in the Fivel Bay. The village infields may have been protected by a circular dike. Perhaps it could be identified with the village of *Reinbodashuson* near Westeremden (late tenth century). Apparently, the village chieftain and his relatives were responsible for the establishment of the first

Figure 11.1 Embankments and conjectural embankments around the Fivel Bay (*c.* 1000–1840)

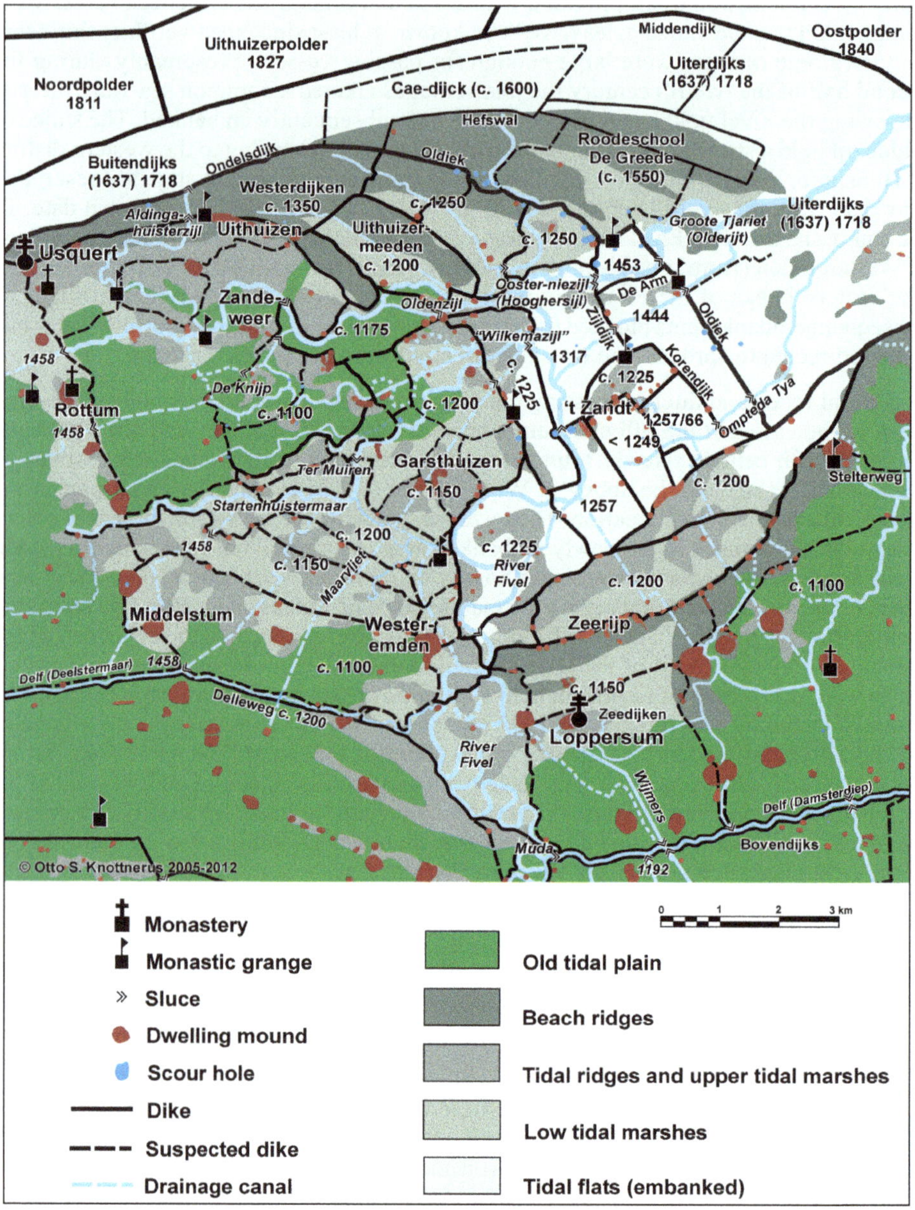

Figure 11.2 Interlocking ecclesiastical and domanial properties around the village of Garsthuizen (*c.* 1500)

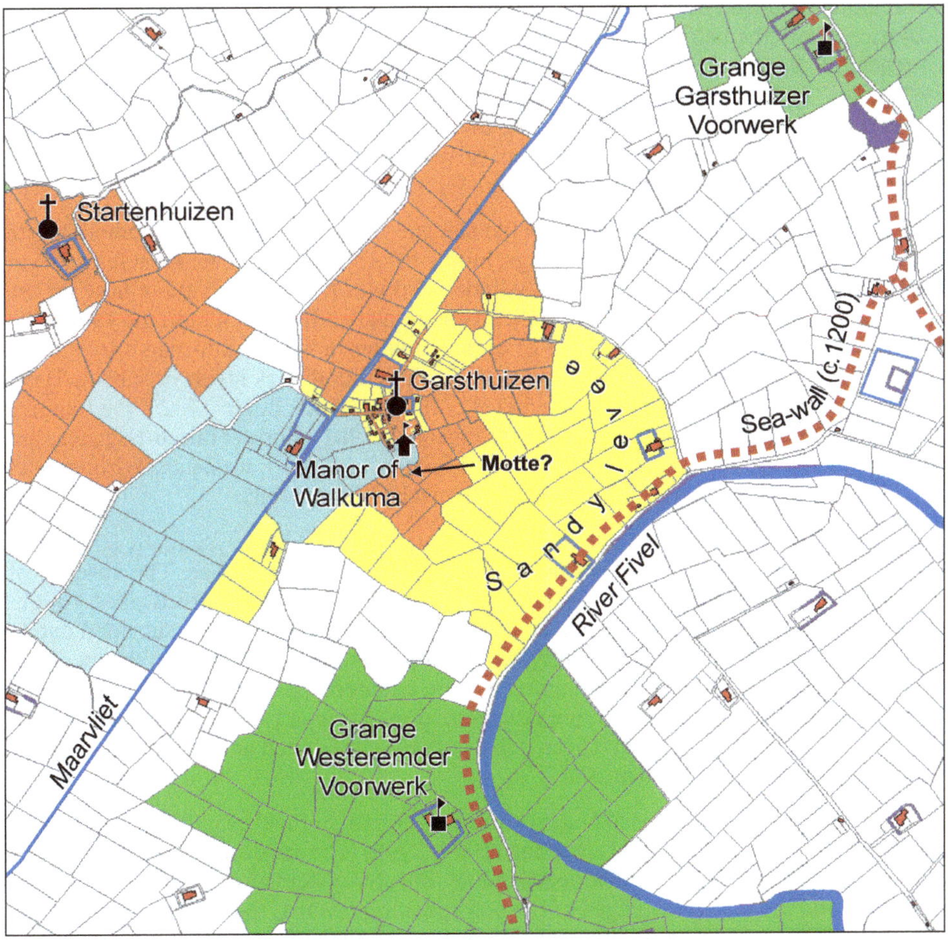

parish church, as the church domains had been split off from the manor of Walkuma. A circular meeting place near the manor suggests the presence of a motte (Formsma, Luitjes-Dijkveld-Stol, and Pathuis, 1987: 131–133). The original tuff church was replaced by an early thirteenth-century brick structure with a special lodge or westwork, designed for local officials. By then, other leading families had donated their outfields to the abbeys of Wittewierum and Feldwerd, which established two major granges on low mounds behind a recently constructed sea-dike. The village must have been renamed, as the oldest reference is to *Gershusen* (1283), indicating that the original name was recently exchanged for a new one, deduced from the Old Frisian **gershûs* (grange, domanial pastoral farm, grass house; Remmers, 2004: 261; but see De Vries, 1946: 27).

III. Peat-bog reclamation in Duurswold

Whereas the process of embankment began with small-scale and localised initiatives, the reclamation of raised bogs in the coastal hinterland seems to have involved large-scale operations from the beginning (Knottnerus, 2008a: 45–47; Molema *et al.*, 2010: 16–24; Groenendijk, Nip, and Van Schaïk, 2008: 118–120, 138–142; Schroor en Meijering, 2007: 69–72, 197–203). Clerics must have been involved in surveying the land, setting out the picket posts and assigning lots and tasks to the individual title-holders. Supra-local chiefs and regional church officials probably supervised the works in order to prevent feuds and to compel unwilling villagers to contribute.

It is well known that large parts of the Pleistocene landscape behind the southern North Sea coast has been covered by raised bogs, hiding its original surface under at least 2–5 m of oligotrophic *Sphagnum* peat. Largely undecided was the question about how far the bogs and fens originally extended into the marshland area. It was not until the 1970s that soil experts proved that high contents of organic matter in former salt-marsh soils often resulted from an eroded peat cover. Additionally, they found thick layers of subsided peat enclosed within the foundations of medieval buildings (Knol, 2010: 17–20; De Langen, 2011; Slofstra, 2008; Wassermann, 1985; Clingeborg, 1981). Former raised-bog areas are often characterised by the toponym *wold*, Old Frisian *wald* (carr, wasteland) (Philippa *et al.*, 2003–2009, vol. IV: 139, s.v. woud; Gildemacher, 2001: 163).

As a rule, it can be safely stated that every wetland area where systematic strip allotments were present has probably once been covered with raised bogs. The absence of recent clay deposits in extremely low lying areas – for instance in reclaimed lakes – might often be explained by the former existence of a bog or mire, which has subsided due to dehydration and erosion. Consequently, every subsequent reconstruction of the medieval landscape since the 1970s has presumed a more extended peat-land zone as compared with earlier reconstructions. Surface reconstructions are complicated, however, by the fact that subsoil peat layers have been subsequently dehydrated and compressed by the weight of more recent clay deposits.

The colonisation of the raised bogs surrounding the River Fivel estuary has been investigated by De Cock (1967) and Ligtendag (1994; 1995). They assumed that it occurred in a similar manner as in many other peat-land regions, by systematic reclamation from the riversides and the low-lying edges of the bogs pushing towards the core. For that reason they expected that the surviving linear villages on the Pleistocene ridges each once had a low-lying predecessor closer to the edges of the former bog. Archaeologists disagree, however, as they did not find sufficient evidence of earlier settlements (Molema and Clingeborg, 1996; Groenendijk, Nip, and Van Schaïk, 2008: 118–120).

Probably starting in the eight or ninth century the inhabitants of the coastal villages bordering the Fivel basin and the banks of the River Ems penetrated deeper into the back-barrier marshes and low-lying fens. The area largely consisted of clay-on-peat wetlands and isolated mires. Often the settlers erected satellite village mounds at the edges of the commons. The newly reclaimed meadows were made accessible by parallel cart tracks and drained by ditches, both leading farther inward. During the ongoing process of land reclamation and enclosure the individual settlers got a share in every subsequent block that was cleared, resulting in scalariform allotment schemes. Other settlements were founded along upland brooks and on isolated Pleistocene knolls and ridges partly hidden under the raised bog. One of them, *UUildiona* or *UUildonha* (Wilderhof, wildland stream, or wildland promontory) is referred to at the close of the tenth century, situated at the end of a minor ridge. The seniority of the parishes of Meedhuizen and Wagenborgen is backed by the choice of the patron saints (Lawrence and Peter), in the latter case also by pottery finds and a contemporary coin hoard. Demarcation lines suggest they were daughter settlements of the villages of Farmsum and Weiwerd (Künzel, Blok, and Verhoeff, 1989: 400; Van Lengen, 2003b: 101–102; Groenendijk, Nip, and Van Schaïk, 2006: 112, 115–118).

Apparently, the villagers jointly undertook extensive infrastructural measures in order to get rid of the excess water leaking out of the inland bogs. Remnants of a straight ditch suggest they dug a lengthy watercourse, leading from a mire at Hoeksmeer across several village territories towards Appingedam, where it probably discharged into a tidal creek called Heekt (both toponyms probably deriving from Old Frisian **haka*, meaning hook, indicating a swift bend in the watercourse). The latter must have been dug by humans as well because it cuts straight through the former salt marshes without having any connection to prior tidal systems (Knottnerus, 2008a: 41–42; Groenendijk, Nip, and Van Schaïk, 2008: 117–118; De Vries, 1946: 3, 54–55; Van Veen, 1951a; cf. De Langen, 2011: 73–74, 83, 90).

Peatland reclamation was largely facilitated by diverting the water of the River Fivel and its minor tributaries into a tidal canal leading towards the River Ems. As a result, downstream water runoff decreased and the original river course started to silt up. The canal banks harboured the eleventh-century market town of Garrelsweer and its twelfth-century successor Appingedam. The name *Delf* (also *Deel*, **Delle*, now Damsterdiep) indicates that at least parts of the canal had been dug (Gildemacher, 1993: 177–189). The energy of the tides must have been responsible for its regularly oscillating course, reflecting the situation before the main outlet was closed off at Delfzijl *c.*1300 (Ligtendag, 1995: 196–197).

The tidal canal was linked to a second one near the market-town of Winsum, thereby creating a safe shipping passage behind the elevated coastal marshes. Subsoil borings conducted by Van Veen during World War Two showed that both canals were interconnected (Van Veen, 1951b). This led to the idea that the canal may have been part of Carolingian defence measures against the Vikings (De Cock and Poelman, 1982; contra Formsma, 1976/77). According to Ligtendag, however, the Delf mainly served for the purpose of drainage and may have been dug in the last decades of the tenth century when the reclamation of the raised bogs was well under way. The dikes bordering the canal were subsequently constructed in order to reduce the impact of the tides and to direct the increased water runoff from the upland areas safely through the lowland zone.

The construction of the Delf could possibly be associated with the activities of Saint Walfridus, a local chief who took the lead in reclamation activities in the neighbouring quarter of Bedum. He and his son were killed during a Viking raid *c.*1000 and subsequently canonised. According to his *Vita Walfridus* he refused to participate in aristocratic family banquets, instead preferring to walk to the city of Groningen where he took the Holy Communion with his fellow colonists (Van Schaïk, 1985; Schmidt, 2008: 165–166, 201–202).

Figure 11.3 Peat-land reclamation in the Duurswold District (*c.* 900–1300). The oldest field alignments in the peat-land area are an extension of existing roads, drainage channels, field limits and demarcation lines (black arrows). When the settlers moved deeper into the bog, they tended to use distant churches and village mounds as landmarks (dark blue arrows). The basic allotment grid, however, seem to have been provided by experienced land-surveyors, who projected field alignments and demarcation lines up to 15 km away from the landmarks concerned (red arrows). The landmarks used provide important clues to the phasing of the reclamation process

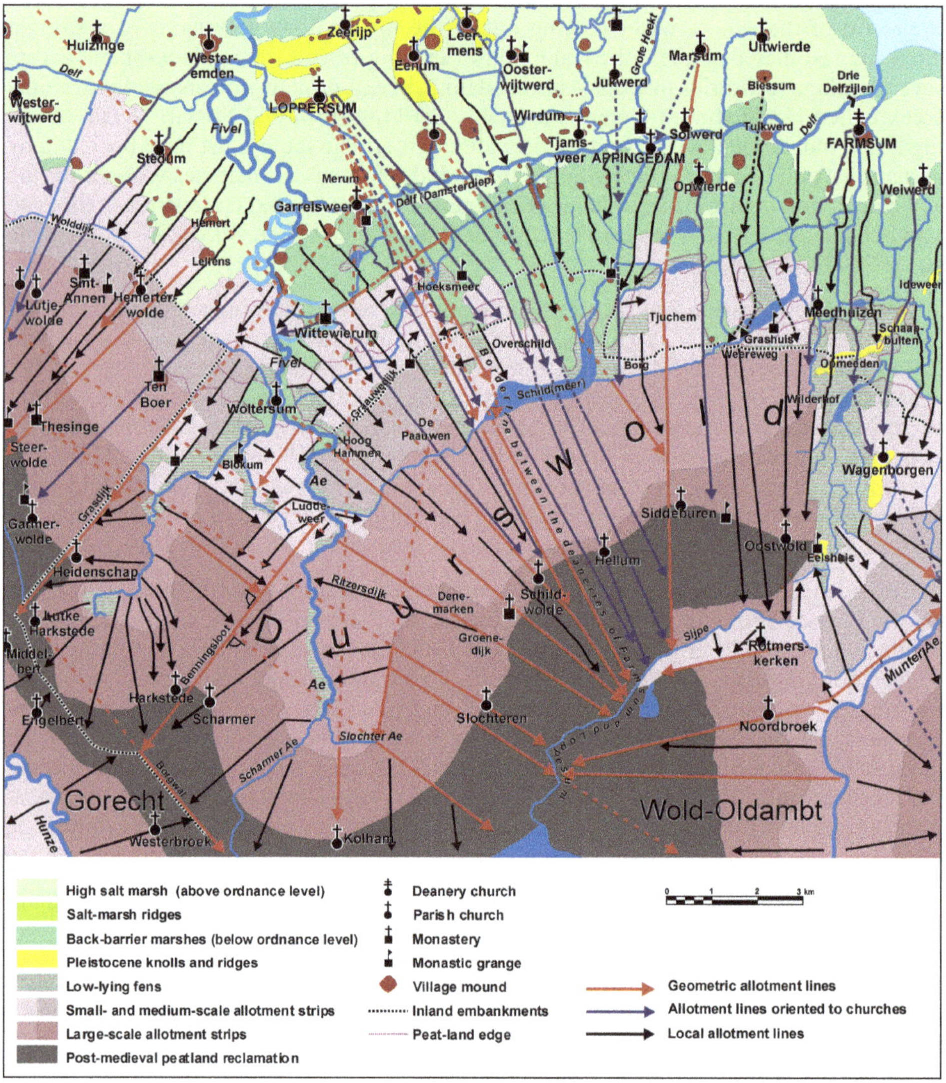

The systematic reclamation of the raised bogs may have started in the late ninth or early tenth century. The oldest settlement activities took place at the edges of the bogs, either setting out from the coastal marshes (*Tjuchem*, piece of (common) land), or from the Fivel riverbanks near the village of Wittewierum (Luddeweer, De Paauwen, Graauwedijk, Overschild), where the toponyms *Scalummawaldum* and *Fralemawaldum* (reclaimed wasteland of the Scalumma's and Fralema's) have been documented (Alma and Vries, 1990: 22–23).

Field arrangements may have been small-scale and partly overwritten by later developments. As a rule, each holding was endowed with a 30–60 m wide allotment strip called a *weer* (holding, share), or *heerd* (fireplace, holding), which extended as far as possible into the wasteland. Houses were erected at slightly elevated platforms or on former creek ridges that protected the inhabitants against seasonal flooding. Upland brooks and ditches, draining the central parts of the raised bogs, served as a rear boundary with neighbouring settlements reclaiming the bog from the rear. As soon the surface subsided, due to dehydration, erosion, and the effects of slash-and-burn agriculture, the farms were relocated to higher grounds, resulting in several linear village sites behind each other.

The individual holdings could not expand indefinitely into the wasteland, however, as they encountered an amphibious zone called *Schild* (from Old Frisian **scald*, wetland, shallow). Contrary to the general pattern, a new series of villages was created beyond the wetland zone on the flanks of the raised bog. The newly reclaimed area could be identified as the settlement of *Diurardasrip* (late tenth century, from *ripa*, linear settlement, maybe with the prefix *diār*, animal), from which the peatland district of Duurswold must have derived its name. As we saw before, wasteland reclamation was associated with a human mission to defeat evil and to tame the wild, as is witnessed by the choice of animal toponyms such as Wilderhof or the unidentified villages of *Wuldunun* (wildland dunes). *Uulfdalon* (valley of the wolves), and *Ottarfliaton* (otter brook, maybe Oterdum) (De Vries, 1946: 206–207, 283, 299; Künzel, Blok, and Verhoeff, 1989: 112, 315, 400, 406). A row of isolated farmsteads, reaching from Denemarken and Groenedijk near Slochteren to the older hamlets of Wilderhof, Opmeeden, Schaapbulten, and Ideweer could indicate the original location of the Duurswold villages.

Soon the villages were relocated to their present location on top of the major Pleistocene ridge, which must have been visible as a drainage divide underlying the raised bog. Parish churches were erected on the highest places, their foundation cutting through the peat-moor until they reached solid ground. The village of Schildwolde (reclaimed wasteland near the shallows) is first referred to in the late tenth century, Slochteren (probably meaning swamp) in the fourth or fifth decade of the eleventh century, together with the outlying meadows of *Extra Scaldmeda* (lowermost meadows near the shallows, maybe to be identified with Overschild). The last village founded here was Hellum, lying rather isolated on top of the ridge behind the wetland zone, the latter rapidly evolving into a full-grown lake known as Schildmeer. The choice of the patron saint suggests that its church was not founded before the canonisation of Walfridus, probably *c.*1100.

Land reclamations in Slochteren and further to the south (Kolham, Scharmer, Harkstede, and Klein-Harkstede) started from the upper reaches of the River Fivel. The latter were more piecemeal, but the choice of location was guided by the same principles that applied to the earlier settlements. By the middle of the thirteenth century, the total number of peat-land parishes in the Duurswold District (actually a subdistrict of Fivelgo) must have been eight or nine at least.

Settlement geometry shows that large-scale planning was predominant, particularly in the northern villages. Field alignments were aimed at churches and village mounds up to 15 km distant (see Van den Broek, 2007: 222–230, 394–398). Parish churches occupied a strategic location in the middle of the village, sometimes on the extended allotment strip of the mother church. The settlers organised themselves in village quarters or 'claws', each of them corresponding to an older village mound in the coastal area. Many quarters were named after the aristocratic clans that most likely took the lead in the reclamation process. The Wittewierum chronicle suggests that in the late thirteenth-century each local chieftain owned a moated brick keep or 'stone house'. Meanwhile, the village community as a whole often erected a defendable bell-tower, which served as a collective counterpart to the domanial strongholds.

All things considered, it is surprising that the population of a largely stateless society, which was deeply emerged into feuds and regional power-struggles, was still able to conduct such large-scale investments. As further information is lacking, we must assume that local magnates, acting on behalf of distant overlords and ecclesiastical institutions were a stabilising factor. They will have provided the expertise required and may have enforced peace during the working season. One of these leading families (or at least their descendants) could be identified as the (co-)founders of the abbey of Wittewierum, as it appears that the property of the abbey incorporated earlier royal domains near Garrelsweer and Leermens. The two distant village territories were combined into a single legal district. The unidentified parish of Henrikaskericon or *Æcclesia Henrici* (late tenth century) could refer to the village church of Wittewierum, which was later donated to the newly founded monastery. The foreign name Henrik (also documented in Leermens, *c.*1050) links its carrier to the Ottonian royal family. The abbey was endowed, moreover, with the village churches of Scharmer and Kolham. The benefactors considered themselves as vassals of the Count of Bentheim, whose ancestors administered the district in the eleventh century. Apparently, they had been able to cling to the royal domains and privileges obtained in the days when the reclamation started (Groenendijk, Nip, and Van Schaïk, 2008: 115, 127, 130; Knottnerus, 2008: 37, 41; Alma, 2001; Formsma, Dijkveld-Luitjes Stol, and Pathuis, 1987: 347–348, 503; Ehbrecht, 1974: 55, 125; De Vries, 1946: 305).

Another leading family resided in the coastal village of Farmsum, where their members owned the deanery. The (rare) name of their twelfth-century ancestor Sigrep could connect them to the supposed founder of the peat-land village of *Sigerdachurka* (Siddeburen). The borders of the deanery were diverted at an early date to include the newly founded parishes of Hellum and Schildwolde, indicating the influence of the dean and his lineage. The dean, moreover, owned several minor lakes or mires, indicating that his ancestors exerted the (royal) right to claim wastelands. The dean's estate was situated on a separate village mound that accommodated dozens of cottages, probably inhabited by craftsmen and merchants (Knottnerus, 2008: 48–49; De Vries, 1946: 316–318; Formsma, Dijkveld-Luitjes Stol, and Pathuis, 1987: 107).

The eastern parts of the reclamation area – around the later deserted village of Oostwold –could be linked to the possessions of the abbey of Werden, which for the most part were transferred to the Hospitallers of Saint John in 1284. Apparently, the

Commandery of Oosterwierum (near Heveskes) held several granges here; its properties included the hamlets of Eelshuis, Wilderhof, Opmeeden, and Grashuis (Siemens, 1962, but incomplete; Noordhuis, 1990: 75, 81, 115). As the coastal village of Heveskes was the seat of a powerful family whose offspring claimed to have founded the monastery, we may expect their ancestors to have played an essential part in the reclamation of the monastic domains in its hinterland. The monastery was strategically located on the main highway between Groningen and Emden (the so-called *Stadsweg*, municipal road); it accommodated a regional fair and controlled the drainage board of Oterdum. The village chief may also have possessed the domanial ferry crossing the River Ems. Archaeological research in Heveskes revealed an exclusive Carolinian gold brooch from the Rhineland area, restyled in the late eleventh or early twelfth century, indicating that its original owners had close ties with the Empire's highest nobility. Their domains were later usurped by the dean of Farmsum (Knottnerus, 2008a: 38, 51; Groenendijk, Nip, and Van Schaïk, 2008: 160–161; Formsma, Luitjes-Dijkveld Stol, and Pathuis, 1987: 503).

The process of reclamation was largely fostered by tapping Schildmeer (the central lake), first by means of a tidal outlet discharging into the Delf, then by digging a new canal (De Groeve) towards the newly founded town of Appingedam. Here it discharged in the Delf through a dike-lock that kept out the tides. The town benefited greatly from its location at a junction of roads and waterways, allowing it to accommodate regional fairs and diets (Knottnerus, 2008b). An extensive system of inland embankments (Graauwedijk, Ritzersdijk, Borg, Weereweg), apparently first mentioned in 1248, prevented excessive water from neighbouring quarters entered the lowlands around the basin (Knottnerus, 2008a: 55; contra Ligtendag, 1995: 188). The supervision of these infrastructural arrangements came into the hands of a drainage board called Woldzijlvest, headed by the abbot of the Norbertine Nunnery of Schildwolde (founded 1204). Two other drainage boards, the Slochter Zijlvest and Scharmer Zijlvest, controlled the upper reaches of the River Fivel and its tributary streams (Slochter Ae and Scharmer Ae).

Peat-bog cultivation and dehydration led to significant subsidence and further measures and extensive drainage schemes. These all cumulated into the establishment of a regional water management board by the early fourteenth century (Generale Zijlvest der Drie Delfzijlen). The board incorporated three smaller organisations: Scharmer Zijlvest, Slochter Zijlvest, and the newly established Dorpster Zijlvest, the latter combining about a dozen marshland villages directly discharging onto the Delf. The abbot of Wittewierum acted as a chairman, using his authority and administrative resources to settle disputes and maintain the precious balance between the divergent interests of all the stakeholders involved.

In a joint effort the participants took measures to shut off the entrance of the tidal canal. Each of them became responsible for the maintenance of one of three parallel sluices, which were first mentioned in 1303. One of the main sluices, probably that owned by Dorpster Zijlvest, served for shipping purposes. Other drainage boards, such as the Woldzijlvest and the coastal districts, were allowed to discharge into the Delf without additional payments, as they had to maintain their own infrastructure as well. In fact, the tidal canal had been transformed into a regulated storage basin, alternately receiving the runoff from its tributaries and discharging it during low tide (Ligtendag, 1995: 172–174, 186–191; Siemens, 1954).

As more and more areas became unable to discharge their waters directly into sea, the regional drainage board of the Drie Delfzijlen began to expand into the neighbouring constituencies. As early as 1370, two peat-land villages near the city of Groningen were allowed to discharge into a canal leading to the Delf. In 1424 the floodplains surrounding the city gained access. In order to facilitate runoff the city magistrate ordered a major shipping canal to be constructed. The latter (Damsterdiep) was accompanied by a new highway (the new Stadsweg), primarily intended for the cattle herds coming from East Frisia and replacing an older trail that may have been disrupted by seasonal flooding.[1] In winter, traffic had to take the alternative *Woldweg* (high road) following the Pleistocene ridges sticking out above the bogs. Around the middle of the fifteenth century the marshland villages bordering the embanked Fivel Bay were allowed access to the sluices of Delfzijl (Ligtendag, 1995: 128–133; Formsma, 1988: 23–35).

By the sixteenth century, the regional drainage board had become one the most powerful political organisations in the region. According to a 1531 covenant the Drie Delfzijlen and two other supra-local drainage boards were exempt from appeal to higher judges. When the monasteries were dissolved after 1594, the board came to be dominated by aristocratic landowners who were often more interested in status and income attached to their membership than in water management issues.

Yet, the drainage board's organisational and financial ability to bridge the gap between the coastal villages and their hinterland ensured that drainage and embankment systems continued to work, despite the fact that the upland villages were hardly able to meet their obligations. Until the thirteenth or fourteenth centuries the peat-lands were able to produce a substantial agricultural surplus, largely consisting of rye and other cereals. As they subsided, however, seasonal flooding increased and pastoral farming began to prevail. The older wetland settlements were gradually deserted, their subsiding outfields transferred to the inhabitants of neighbouring upland villages. The fact that the interlaced holdings came to belong to owners from different drainage boards led to much confusion and litigation[2].

In the 1570s some of the raised bogs at the back of the Pleistocene ridge near the village Siddeburen were still intact; by the eighteenth century, most of the Duurswold District had sunk far below ordnance level. Whereas occupation and arable fields concentrated on higher grounds, the central lowlands around Lake Schildmeer became largely marginalised. The main roads crossing the plain were blocked in winter. Large parts were accessible only by boat or – when the canals were frozen – by sledge or on skates. The remaining inhabitants resided on slightly elevated farmyards, reverting to dairy farming, fishing, fowling, and reed cutting.

Not before the 1770s did agricultural interest regain momentum. Then, local landholder associations erected dozens of drainage mills, thereby opening the way to arable farming. In the southern villages urban investors introduced peat dredging, and as a consequence were able to exploit peat reserves below the water surface. The construction of another

---

[1] See http://nl.wikipedia.org/wiki/Stadsweg with further archival references.

[2] Ligtendag's (1995) reconstruction of the drainage sub-districts has some serious flaws. See my contributions to www.wikipedia.org under the headings 'Generale Zijlvest der Drie Delfzijlen' and 'Woldzijlvest', both with further references.

drainage canal in the 1870s enabled a substantial lowering of the water levels within the storage basins. Several minor lakes were drained. Nevertheless, seasonal flooding caused many inconveniences until the construction of major pumping stations at Farmsum (1933) and Delfzijl (1974), making the natural runoff during low tide through the main sluices largely obsolete.

IV. Submerged peat-lands in the Dollard area

The reclamation of the Oldambt and Reiderland Districts south of the River Ems estuary occurred at practically the same manner as in the wastelands farther west (Groenendijk and Bärenfänger, 2008; Knottnerus, 2003; Wassermann, 1985; 1995; Stratingh and Venema, 1855). Basically, the entire area was part of an inaccessible raised bog, filling up the slopes and valleys of a plateau stretching from the River Ems up to the moraine hills of the Hondsrug and the isolated district of Westerwolde. Contrary to the Fivel basin, however, the Pleistocene valley systems to the east had been largely filled up with massive layers of peat, which made the area exceptionally vulnerable. Peat-bog cultivation led to severe flooding and growing pressure on drainage and embankment systems. This resulted, eventually, in the destruction of the large tracts of land and dozens of villages.

Initially, Frisian colonists from the River Ems banks created satellite village mounds in the back-barrier marshes. Some of these settlers moved further into the interior, as erratic ninth- or tenth-century pottery finds demonstrate. Several inland villages were mentioned in a contemporary census itinerary from the Abbey of Werden (*UUinghem* and *Garun*, first half of the ninth century). Their names may refer to the later submerged villages of Wynedaham (or else Wynham) and Bonewerda (near Bunde). Apparently these isolated settlements were located on riverside levees, which later offered an appropriate starting point for peat-land reclamation (Schleicher, 1997).

Other settlers occupied the Pleistocene knolls and ridges partly sticking out above the bogs. The heights of Bunde and Holtgaste must have been settled by colonists from the adjoining coastal marshes at an early date. This is supported by the choice of their patron saints (Martin and Ludger) and the discovery of a Carolingian coin hoard. The abbey of Werden, moreover, owned a manor in Holtgaste (Van Lengen, 2003b: 102, 105–107; Groenendijk, 1995: 212). Another ancient settlement was the later deserted village of Ulsda, situated on a moraine height bordering the River Aa. Its patron saint must have been Lawrence (not Lambert, as a sixteenth-century tradition mistakenly claimed) (Smith, 1901).

The coastal districts fell under the jurisdiction of the bishopric of Münster. Part of the settlers, however, came from the upland areas. As the latter's home district of Westerwolde had originally had been settled by Saxon colonists from the middle reaches of the River Ems, the newly occupied territories bordering the uplands (including Ulsda) were assigned to the bishop of Osnabrück. Several estates, such as the village of Winschoten were conferred to the Westphalian Abbey of Corvey. Apparently, most village sites were covered with trees. The place-names *Bunwyda* (1429, Bunde), *Olswyda* (1447, Ulsda), and *Winsewida* (1231, Winschoten), contain the suffix *-widu* (wood), contrasting with the more recent *-wold* toponyms (carr, wasteland) of the surrounding area (see Gildemacher, 2001: 164).

Figure 11.4 Peat-land reclamation in the Oldambt and Reiderland Districts (before the inburst of the Dollard Bay, *c.* 1413)

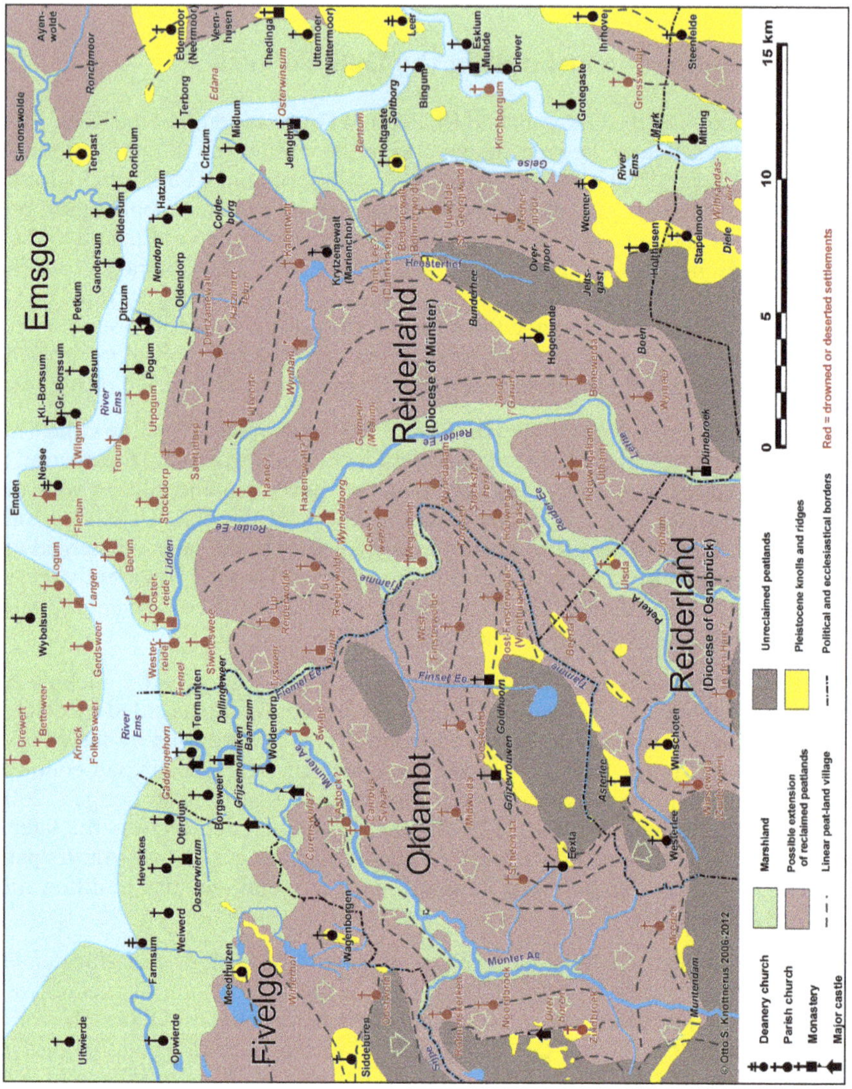

By the tenth century, Frisian colonists began to penetrate deeper into the hinterland, applying more systematic methods of land reclamation and integrating the older nuclear settlements into their own extensive field systems (see Schwarz, 2005). Settlement efforts started from the densely populated River Ems banks, resulting in a prolonged series of eight or nine daughter settlements, reaching from the mouth of the River Reider Aa to the village of Stapelmoor, 40 km upstream. As elsewhere, regional magnates holding foreign leases and dignities will have organised the earliest stages of the reclamation process. A tenth-century manor at Hatzum-Alte Boomborg has been excavated (Strahl, 2003: 357). Census documents from the abbey of Werden suggest that high-ranking noblemen from the interior had interests in the Lower Ems District. The duchess of Verdun, for instance, donated the village of Nesse (situated before Emden) *c.*1020. The domain was later transformed into a grange by the Norbertine abbey of Langen, which got hold of additional royal demesne north of the River Ems. The village of *Saleburc* (1189, Soltborg) near Holtgaste may have been the location of another estate (Jongbloed, 2006: 46; Friedländer, 1877: 20; Philippi *et al.*, 1969–1996: vol. I, no. 403).

Two major rivers and their tributary streams opened up the region, allowing the colonists to proceed landward from the edges of the bogs. Both river names were related to the market towns at their mouth, which may have served as main centres for the organisation of inland settlement. As soon as the reclamation process got underway the market towns and their hinterlands began to detach themselves from neighbouring districts. The villages of Reiderland withdraw from the Lower Ems District, establishing a separate district first mentioned in 1224. The villages farther west seem to have belonged to the Fivelgo District, from which they separated during the twelfth century. The new district's name Oldambt (*Aldaombechte*, the old subdistrict) is not documented before 1271 (Van Lengen, 2003a: 67–70; Halbertsma, 1955)[3].

The coastal town of *Hriadi* (Reide, late ninth century, from *hriād* = 'reed', *rēde* = 'berth, wharf', or **rīth* = 'brook, stream') became the starting point for the reclamation of the eastern peat-lands, giving its name to the newly founded district of Reiderland (Köller, 2006: 56–72). In fact, it consisted of two distinct village mounds, separated by a river outlet. The tenth-century domanial church of Oosterreide dominated the eastern river bank, facing the market town of Westerreide on the other side. From here, the River Reider Aa (nowadays Westerwoldsche Aa) and its tributaries (Tjamme, Rensel, and Pekel Aa) offered various opportunities to drain the bogs and cultivate their fringes. Census documents by the abbey of Werden from the early eleventh century mention the village of *Hredi in UUalda* (Reiderwolde) and the unidentified settlements of *UUalda* and *uppan UUalda* (maybe Sankt Georgiwold) (Wassermann, 1995: 99). By the thirteenth century, the central peat-land area of Reiderland counted at least seventeen parishes (as against twelve parishes on the adjoining Ems River banks). Most peat-land parishes, moreover, were much larger than their coastal counterparts.

---

[3] A fresco from the second half of the thirteenth century in Saint Pauls cathedral in Münster is supposed to have included the word *Smalagonia* (minor district) as an equivalent for the name Oldambt (Ehbrecht, 1974: 1–5). Photographs taken before the destruction of the painting in 1943 show that its nineteenth-century restorers read the faded inscription as *Emesgonia*. http://www.zi.fotothek.org/obj/obj19004092/Galerie.

The town of Termunten was the starting point for the reclamation along the river **Mente* or Munter Ae (Aa or Ee, stream), with its tributaries Sijpe and Finser Ae or Fiemel Ae. Just as its counterparts Farmsum and Reide, it was a dual village, consisting of a nucleated settlement around the church of Lutke Termunten and a distinct market settlement on the opposite eastern river banks (Groot Termunten). The latter could be identical with the tenth-century village of *Montanhae* (from the river name **munte*, **mente*, or **monta*, mouth, with the suffix *-hae*, meaning promontory' or 'berth, hide'). By analogy, the whole district came to be called Menterne (**mente* with suffix *-herne*, meaning quarter, neighbourhood or **-hurna*, 'headland in the marsh'). A local feud in 1201, perhaps caused by drainage problems, may have led to the secession of the peat-land district of Menterwolde (De Vries, 1946: 152–155; Miedema, 1970: 127; but see Künzel, Blok, and Verhoeff, 1989: 255, 487).

The colonisation efforts in the Oldambt District resulted in the foundation of nine parish churches at least (as against five churches on the Ems River banks). The most important of them was the late twelfth-century basilica of Saint John in Midwolda, where the regional diets were held. Demarcation lines suggest that it was a daughter settlement of Lutke Termunten. The undivided parish of Noordbroek and Zuidbroek with the presumed church of *Rotmerskerken (from **Rothmonnarkirika* or **Rothenseburan*, corrupted to *Rodendebord*, *c.*1475, *Rummelant* 1485, *Ruthmerlandt* 1609, *Rommelskerken* 1819) could have been a secondary centre (Molema, 1991–1992; Knottnerus, 2011: 8). Contrary to Duurswold, hardly any of these village churches were located on the Pleistocene ridges right from the onset. Some of them have been moved to higher grounds at an early date, however, because of their waterlogged location, as was the case in Noordbroek, Zuidbroek, Meeden, and Finsterwolde. Several other parishes, such as Rommelskerken, were dissolved (Molema, 2011: 12–15).

Just as in the neighbouring districts, the monasteries played an important part in the final phases of the reclamation process. Parts of the church holdings on the Aa riverbank in Midwolda were assigned to the Cistercian Abbey of *Campus Sylvae* or *Montrevalt* *c.* 1220. The monks took the lead in regulating the stream and supervising its sluices. The monastery moved to a new location near Termunten in 1299, probably because of frequent river flooding. By that time, the combined coastal and inland villages of the Oldambt District were accustomed to mutual cooperation. The abbot probably acted as chairman of the main drainage board Termunterzijlvest. The monastery founded a grange near the inland village of Muntendam, where the upper realms of the River Ae were dammed during spring time in order to prevent leakage water from the bogs flowing freely into the lowlands (Fockema Andreae, 1950: 8–10; Molema, 2012).

According a sixteenth-century tradition, the Reider Aa was closed off by seven parallel sluices. The peat-land settlement of *Westertili* (probably a clerical error for **Westercili*, with suffix *-siel*, sluice), listed in the last decades of the thirteenth century, may have been named after one of these sluices (Naarding, 1959; but see Künzel, Blok, and Verhoeff, 1989: 395; both with inadequate dating). Most likely, drainage arrangements were supervised by the Norbertine abbey of Palmar (founded about 1204). Diets were convened either on the dam between the sluices, in the monastery, or in the Dominican Nunnery of Oosterreide. The latter may have accommodated a regional fair as well: the remaining promontory of Reide harboured a Whit Monday fair well into the nineteenth century (De Blécourt, 1935: 325; Feith, 1846: 519; Stratingh and Venema, 1855: 30, 324).

The fact that peat-land villages outnumbered the older settlements had serious consequences. Apparently, the feudal elite lost its supremacy over the area at an early age. Foreign overlords suffered several bloody defeats; their representatives had to give in to local demands. The latter's powers were constrained, moreover, by a popular uprising in the 1270s, urging the bishop of Münster to take back ecclesiastical jurisdiction into his own hands. He took hold of the manor of Weener and (probably) the church of Nesse, whereas the remaining domains of the abbey of Werden (including the churches of Oosterreide, Holtgaste and presumably Marienchor) were transferred the Hospitallers of Saint John (Stüwer, 1973; Schmidt, 2008: 332–335, 381). One of the ruling families (Addinga) retreated to the inland district of Westerwolde, subsequently trying to impose a Westphalian style feudal regime on the country's freeholders. Other families shifted their interests to the opposite banks of the River Ems. The dean of Farmsum remained in function, but his jurisdiction in the Oldambt District was curtailed. The monasteries were barely able to intervene, as they also stood under the influence of the competing family clans (Renaud and Van Dijk, 1971: 52; Van Lengen, 1973; Boeles, 1866: 164).

As a consequence, the peat-land colonists and their leaders prevailed over the coastal nobility, compelling them to respect their newly acquired privileges. In some adjoining areas, notably in Brokmerland, popular movements enforced a temporary ban on fortified houses. In the Oldambt and Reiderland Districts, the reverse took place. Rivalry and display led to a building boom of brick keeps, of which 100 have been archaeologically documented by now. Initially, many villages may have been governed by a local headman who received a double plot. By the fifteenth century, most peat-land villages counted half a dozen fortified houses at least (Molema, 1995; see also Noomen, 2009).

The lack of a central authority induced a prolonged power struggle. Successful chiefs tried to eliminate their competitors and to infer on the rights of freeholders, seeking the assistance of distant allies and causing civil strife. Following one of these clashes in 1413 the main sluices of Reide were set on fire. Moreover, dike-repairs were postponed, allowing the sea to break in. A sixteenth-century tradition blames one of the district chiefs, who refused to act unless the waves should go as high as a lance over his property. Eventually, he had to rely on the charity of the abbey of Palmar. Other documents stated that people living near the dikes were unable to repair the damage; those living far off were unwilling to do so (Stratingh and Venema, 1855: 316–321). A contemporary chronicle leaves hardly any doubt as to the fact that the events of 1413 were decisive in making the area prone to floods (Rinzema, 1987: 71). Successful countermeasures had to wait until the 1430s and 1440s. By then, the cities of Groningen and Hamburg managed to pacify the region, making an end to mutual attacks and ongoing piracy.

According to the Renaissance myth, Dollard Bay was created after a storm surge on Saint Stephen's Eve 1277, during which thirty-three villages were supposed to have been lost (Knottnerus, in press; 2011: 3–7). This, however, was a scholarly fantasy, inspired by the biblical story of Lamech, whose seventy-seven descendants were said to have drowned in the deluge. The sacral number thirty-three reflected the lifetime of Christ. Additionally, Flemish chronicles introduced another flood 100 years later, which, according to local sources, would have affected the Reiderland District as well. Doubts about the accuracy of these statements led to a good deal of speculation about the actual date. Most researchers made use of proof by contradiction: because the Dollard Bay already existed in the fifteenth century, it must have originated at least half a century before. As a rule, Dutch scholars follow the work of Elisabeth Gottschalk, who clung to the Saint Lucia Flood of 1287 (Gottschalk, 1971–1977, vol. I: 49, 269–271, 485; vol. II, 40–41). Their German colleagues, on the other hand, tend go along with Carl Woebcken, who identified the Saint Marcellus Flood of 1362 as the starting point for the formation of the Dollard estuary (Woebcken, 1928; Homeier, 1976; Behre, 1999). In fact, there is no convincing evidence whatsoever to prove these early dates.

All things considered, the main cause of the disaster following the destruction of dikes and sluices must have been the unfavourable geological situation. Contrary to the Fivel basin, the Pleistocene valley systems to the east had been largely filled up with thick layers of peat, its base lying at 3.5–8 m below ordnance level. Peat-bog cultivation led to significant subsidence and growing pressure on the drainage and embankment systems. Substantial areas sank below sea level, making the region vulnerable to storm surges. Reports of frequent flooding accumulate from the end of the fourteenth century. Feuds became detrimental, as their effects could not be undone. As soon as the sea broke in, a tidal storage basin came into existence, acting as a natural engine, injecting oxygen and consuming the remaining peat-lands. The process of destruction continued until a new equilibrium was reached, in which the tidal energy generated by the funnel shaped river-mouth was absorbed by the expanding floodplain. As a result of the enlarged tidal volumes, the River Ems widened too. Some of the coastal embankments had to retreat landward, large tracts of tidal foreland and parts of the barrier islands were washed away, opening up the area for new storm surges (Oost, 1995: 213–214).

As late as 1429 the inhabitants deliberated about the repair of the River Ems embankments. Soon the tides reached as far as Winschoten (Knottnerus, 2011: 7–8). The central lowlands got drowned; its villages were abandoned or transferred to higher grounds, even though some of them (notably the village of Houwingaham, excavated in 1998–1999) held out until the early sixteenth century. A storm surge on Palm Sunday 1446 led to the evacuation of the abbey of Palmar. Yet, for several decades the raised bog on which the abbey was situated prevented the sea from breaching into the Oldambt District. A plan to erect a new dike from the mouth of the River Reider Aa to the Pleistocene heights of Finsterwolde was implemented in 1454. The Groningen magistrate seized the opportunity to take the matter in its own hands, but its desperate efforts proved to be in vain. Twelve years later the erosion of the peat-lands was apparently so well advanced that the sea could pass to the north, entering the (Termunter) River Ae and reaching the Pleistocene height of Wagenborgen. The village site of Scheemda, on the other hand, 10 km to the south, seems to have been largely unaffected. The church was relocated in the early sixteenth century, but archaeologists found no trace of a marine environment by then (Stratingh and Venema, 1855: 77–82, 327–31; De Blécourt, 1935: 322–323; Cleveringa, 1939: 144–146; Casparie and Molema, 1990).

Only when the heavily eroded River Ems dikes were destroyed by subsequent storm surges in the years 1507 to 1509 did the Dollard Bay reach its maximum extent. The River Ems district around Termunten remained unprotected from the tides until the dikes were reconstructed in the 1520s. Large-scale re-embankment of the salt marshes and intertidal mudflats started in the 1540s, but suffered a major setback due to wartime inundations in the 1580s. By the close of the sixteenth century about a third of the drowned lands were recaptured from the sea. The former peat-lands had been covered by 1–2 m of fertile clay. Ongoing sedimentation and subsoil subsidence ensured that each subsequent polder was raised higher than the previous one, causing, on the other hand, new drainage problems that had to be subsequently solved.

Severe flooding also occurred upstream, where the River Ems broke into the recently reclaimed fenlands, urging the villagers to retreat backward into the raised bog. Several dikes, most likely built under the authority of the Count of East-Frisia and reaching from the Ems River banks across the lowlands to the Pleistocene height of Bunderhee, prevented the river shifting its course westward. Nevertheless, following the Cosmas and Damian Flood in September 1509 the sea broke through the riverbanks farther north, creating a new tidal channel and subsequently destroying the remaining village mounds. By then, the total number of lost parishes must have been approximately twenty-four. Six or seven others had recently been relocated.

As a consequence of the break-through, the city of Emden came to be situated aloof from the mainstream; efforts to redirect the river's course by constructing piling failed. Only the peninsula of Nesse remained intact, until it was abandoned after a storm surge in February 1825. Thereafter, its remains were absorbed by the harbour site of Emden (Breuer, 1965; Knol, 2008).

Whereas the area's collapse was accelerated by the absence of central authority, its recovery was backed up by the rise of the modern state. The city-state of Groningen incorporated the districts of Oldambt and Westerwolde (including the adjoining bits and pieces of Reiderland), seizing large tracts of unembanked foreland and urging the inhabitants to cooperate in drainage and embankment matters. The Counts of East-Frisia did the same in their precinct (the remaining eastern parts of Reiderland). It appears that the 1277 flood myth played an important part in legitimising government intervention. In the end, the reclaimed Dollard marshes became a model area for large scale arable farming, whereas the unembanked flats and salt marshes came to be considered as an area of outstanding natural beauty.

V. Conclusion

Why is it that the districts of Reiderland and Oldambt were much less successful in managing surface subsidence and flooding then their neighbours of the Fivelgo District? Certainly, natural circumstances were quite unfavourable but the lack of an effective authority who could bridge the gap between coastal and inland interests certainly contributed to the desperate situation. The precarious balance between coastal aristocrats and peat-land colonists which remained intact in Fivelgo, was due to be disturbed farther east by the latter's sheer numbers. Even the monasteries could not bridge these differences, eventually leading to the erosion of collective arrangements and the destruction of the whole area.

Bibliography

Alma, R. (2001) 'Oostrum en het Eesterrecht', *Driemaandelijkse Bladen,* 53, pp. 123–128

Alma, R. and Vries, O. (1990) 'Summa agrorum in Slochtra: een gedeeltelijk Oudfriese tekst uit de Ommelanden', *Us Wurk: Tydskrift voar Frisistyk*, 39, pp. 1–48.

Behre, K.-E. (1999) 'Die Veränderungen der niedersächsischen Küstenlinien in den letzten 3000 Jahren und deren Ursachen', *Probleme der Küstenforschung im südlichen Nordseegebiet*, 26, pp. 9–34.

Blécourt, A.S. De (1935) *Oldambt en Ommelanden: rechtshistorische opstellen met bijlagen*, Assen.

Boeles, W.B.S. (1866) 'De hoofdbronnen eener geschiedenis van het kerkelijk regtswezen der Friezen onder het Munstersche bisdom', *Bijdragen tot de Geschiedenis en Oudheidkunde, inzonderheid van de Provincie Groningen*, 3, pp. 148–171.

Boer, E. De (2009) 'Verwikkelingen in de middeleeuwse dijkschepperij van Holwierde', *Historisch Jaarboek Groningen*, 16, pp. 6–23.

Breuer, H. (1965) 'Dollart und Ems. Die Folgen der Dollartbildung für das Gebiet der unteren Ems: ein Beitrag zur Geomorphologie und zur Hydrodynamik des Emsstromes', *Jahrbuch der Gesellschaft für bildende Kunst und vaterländische Altertümer zu Emden*, 45, pp. 11–90.

Broek, J. Van den (2007) *Groningen, een stad apart: over het verleden van een eigenzinnige stad (1000–1600)*, Assen.

Casparie, W.A. and Molema, J. (1990) 'Het middeleeuwse veenontginningslandschap bij Scheemda: de opgravingen op het 'Oud Kerkhof' van Scheemda in een veenkundig, hydrologisch en historisch perspectief', *Palaeohistoria*, 22, pp. 271–291.

Cleveringa Pzn., R.P. (1939) 'De Ommelander gastrechters', *Groningse Volksalmanak*, pp. 128–156.

Clingeborg, A.E. (1981) 'Het Groninger Woudgebied, een voormalig veenlandschap?', *Boor en Spade*, 20, pp. 184–205.

Cock, J.K. De (1967) 'Ontginningsgeschiedenis van de gemeente Slochteren', *Groningse Volksalmanak*, pp. 162–185.

Cock, J.K. De and Poelman, J.N.B. (1982) 'De Delf, een vroeg-middeleeuws kanaal in de provincie Groningen', *Driemaandelijkse bladen voor taal en volksleven in het oosten van Nederland*, 34, pp. 106–114.

Ehbrecht, W. (1974) *Landesherrschaft und Klosterwesen im ostfriesischen Fivelgo (970–1290)*, Münster.

Ey, J. (2010) 'Initiation of dike-construction in the German clay district', in: H. Marencic *et al.* (eds), *Proceedings of the 12th International Scientific Wadden Sea Symposium 'Science for Nature Conservation and Management'*, Wilhelmshaven, 30 March-3 April 2009, Wilhelmshaven, pp. 179–183 (Wadden Sea Ecosystem; 26).

Feith H.O. (ed.) (1846) 'Bericht van Reiderland', *Pro Excolendo Iure Patrio*, 6, pp. 518–521.

Fockema Andreae, S.J. (1950) *Studiën over waterschapsgeschiedenis, deel 6: Oostelijk Groningen*, Leiden.

Formsma, W.J. (1976/77) 'Delvinaria', *Groningse Volksalmanak*, pp. 28–36.

Formsma, W.J. (1988) *Geschiedenis tussen Eems en Lauwers*, Groningen.

Formsma, W.J., Luitjes-Dijkveld Stol, R.A. and Pathuis, A. (1987[2]) *De Ommelander borgen en steenhuizen*, Assen-Maastricht.

Friedländer, E. (ed.) (1877) 'Güterverzeichniss des Klosters Langen in Ostfriesland', *Jahrbuch der Gesellschaft für bildende Kunst und vaterländische Altertümer zu Emden*, 2, pp. 19–46.

Gierke, J., Von (1901–1917) *Die Geschichte des deutschen Deichrechts*, Breslau.

Gildemacher, K.F. (1993) *Waternamen in Friesland*, Leeuwarden.

Gildemacher, K.F. (2001), 'West Frisian place-names', in: H.H. Munske (ed.), *Handbuch des Friesischen: Handbook of Frisian Studies*, Tübingen, pp. 155–170.

Goldbeck, J. (1991) *Die Deichgenossenschaftene Eine Analyse ihrer historischen Entwicklung auf der Basis der Kollektivgüter- und Genossenschaftstheorie*, Münster.

Gottschalk, M.K.E. (1971–1977), *Stormvloeden en rivieroverstromingen in Nederland*, 3 vols, Assen.

Groenendijk, H.A. (1995) 'Niemandsland Ter Apel? Zuidelijk Westerwolde voor de kloostervestiging', in: P. De Rooy, J.W. Boersma, and C.J.A. Jörg (eds), *Ere-saluut: Opstellen voor mr. G. Overdiep*, Groningen, pp. 207–220.

Groenendijk, H.A. and Bärenfänger, R. (2008) *Gelaagd landschap: veenkolonisten en kleiboeren in het Dollardgebied*, Bedum (Archeologie in Groningen; 5).

Groenendijk, H.A., Nip, R.I.A., and Schaïk, R.W.M., Van (2008) 'Geschiedenis van Groningen', in: M.G.J. Duijvendak (ed.), *vol. I: Prehistorie en Middeleeuwen*, Zwolle.

Groenendijk, H.A. and Schwarz, W. (1991) 'Mittelalterliche Besiedlung der Moore im Einflußbereich des Dollarts: Ergebnisse und Perspektiven', *Archäologische Mitteilungen aus Nordwestdeutschland*, 14, pp. 39–68.

Halbertsma, H. (1955) 'Smalagonia: bijdrage tot de geschiedenis van het Oldambt', *De Vrije Fries*, 42, pp. 118–131.

Heeroma, K. (1955) 'Populieren aan de Fivel', *Driemaandelijkse bladen voor taal en volksleven in het oosten van Nederland*, 7, pp. 135–143.

Homeier, H. (1976) 'Einbruch und weitere Entwicklung des Dollart bis um 1600', *Forschungsstelle für Insel- und Küstenschutz der Niedersächsischen Wasserwirtschaftsverwaltung. Jahresbericht*, 28, pp. 39–81 and Appendix B 2, nrs. 1–14.

Jansen, H.P.H. and Janse, A. (1991) *Kroniek van het klooster Bloemhof te Wittewierum*, Hilversum.

Jongbloed, H.H. (2006) Wanburtich: Heinrichs II. Beteiligung an der Wahl von Kamba (1024)', *Deutsches Archiv für die Erforschung des Mittelalters*, 62, pp. 1–63.

Knol, E. (2008) 'Nesserland, de meest oostelijke punt van Reiderland', *Historisch jaarboek Groningen*, pp. 30–35.

Knol, E. (2010) 'Het verleden van kwelders, wierden en dijken: werk voor natuurwetenschappers', in: G.J. Borger, P. Breuker, and H. De Jong (eds), *Van Groningen tot Zeeland: geschiedenis van het cultuurhistorisch onderzoek naar het kustlandschap*, Hilversum, pp. 11–27, 65–68, 145–149.

Knottnerus, O.S. (2006) *Fivelboezem: de erfenis van een verdwenen rivier*, Bedum (Archeologie in Groningen; 2).

Knottnerus, O.S. (2007) 'Eine gefahrvolle Existenz: zur inhärenten Ambivalenz der frühneuzeitlichen Küstengesellschaft', in: N. Fischer, S. Müller-Wusterwitz and B. Schmidt-Lauber (eds), *Inszenierungen der Küste*, Berlin, pp. 107–149.

Knottnerus, O.S. (2008a) *Natte voeten, vette klei: Oostelijk Fivelingo en het water*, Bedum (Archeologie in Groningen; 3).

Knottnerus, O.S. (2008b) *Tussen twee grijpgrage vogels: boeren en burgers in de Renaissance: 16e Fransemalezing*, Appingedam.

Knottnerus, O.S. (2011) 'Verdronken dorpen', *Groninger kerken*, 28, pp. 3–8.

Knottnerus, O.S. (in press) 'Dollardgeschiedenis(sen): mythe en realiteit'.

Köller, A.R. (2006) *Reiderland oder Rheiderland?*, Weener.

Künzel, R.E., Blok, D.P. and Verhoeff, J.M. (1989), *Lexicon van Nederlandse toponiemen tot 1200*, Amsterdam.

Langen G.J., De (2011) 'De ontginning van de (klei-op-)veengebieden in Fryslân gedurende de late ijzertijd, Romeinse tijd en middeleeuwen (van ca. 200 v. Chr. tot ca. 1200 n. Chr.)', in: M.J.L.Th. Niekus (ed.), *Gevormd en omgevormd landschap van Prehistorie tot Middeleeuwen: DPV 95 jaar, 1913–2008*, Assen, pp. 70–97.

Lengen, H. Van (1973) *Geschichte des Emsigerlandes*, Aurich.

Lengen, H. Van (2003a) 'Tota Frisia: sieben Seelande und mehr. Die territoriale Gliederung des freien Frieslands im Mittelalter: ein Überblick mit einer Karte', in: H. Van Lengen (ed.), *Die Friesische Freiheit des Mittelalters: Leben und Legende*, Aurich, pp. 56–89.

Lengen, H. Van (2003b) 'Karl der Große, Jungfrau Maria und andere Heilsbringer als Garanten und Patrone Friesischer Freiheit. Zu den Siegeln der Landesgemeinden Frieslands im Mittelalter', in: H. Van Lengen (ed.), *Die Friesische Freiheit des Mittelalters: Leben und Legende*, Aurich, pp. 90–133.

Ligtendag, W.A. (1994) 'De ontginningsgeschiedenis van de gemeente Slochteren opnieuw bezien', *Gronings historisch jaarboek*, 1, pp. 9–26.

Ligtendag, W.A. (1995) *De Wolden en het water: de landschaps- en waterstaatsontwikkeling in het lage land ten oosten van de stad Groningen vanaf de volle middeleeuwen tot ca. 1870*, Groningen.

Miedema, H.T.J. (1970) 'Norther Hee (*Oudfriesche oorkonden I, 6*), Hee op Skylge en Hêthe oan de kust van Kent', in: T.B. Hoekema *et al.* (eds), *Flecht op é koai: stúdzjes oanbean oan Prof. Dr. W.J. Buma ta syn sechstichste jierdei*, Groningen, pp. 121–129.

Miedema, M. (1990) 'Oost-Fivelingo 250 v.C.-1850 n.C.', *Palaeohistoria*, 32, pp. 111–245.

Miedema, M. (2000) 'West-Fivelingo 600 v.C.-1900 n.C.: archeologische kartering en beschrijving van 2500 jaar bewoning in Midden-Groningen', *Palaeohistoria*, 41/42, pp. 237–445.

Molema, J. (1991–1992) 'Van de Mieden, Egeste en Broke: de middeleeuwse nederzettingsgeschiedenis van het zuidwestelijk Wold-Oldambt in kort bestek', *Palaeohistoria*, 33–34, pp. 311–320.

Molema, J. (1995) 'Een steenhuis te Midwolda (Oldambt, Gr.)', *Paleo-Aktueel*, 6, pp. 129–132.

Molema, J. (2011) 'Verdwenen kerken van veenontginningsnederzettingen', *Groninger kerken*, 28, pp. 9–15.

Molema, J. (2012) 'De eerste Muntendammers', *Veenkoloniale Volksalmanak*, 24, pp. 11–22.

Molema, J. and Clingeborg, A.E. (1996) *Een aanvullende archeologische inventarisatie in het landinrichtingsgebied Midden-Groningen*, Groningen.

Naarding, J. (1959) 'Groninger toponiemen: I. in Westercili, II. Nunspete', *Driemaandelijkse bladen voor taal en volksleven in het oosten van Nederland*, 11, pp. 118–121.

Nieuwhof, A. (2010) 'Living in a dynamic landscape: prehistoric and proto-historic occupation of the northern-Netherlands coastal area', in: H. Marencic *et al.* (eds), *Proceedings of the 12*th International Scientific Wadden Sea Symposium 'Science for Nature Conservation and Management', Wilhelmshaven, 30 March-3 April 2009 (Wadden Sea Ecosystem, Vol. 26), Wilhelmshaven, pp. 173–178.

Noomen, P.N. (2009) *De stinzen in middeleeuws Friesland en hun bewoners*, Hilversum.

Noordhuis, G.F. (1990) *De Johannieters in Stad en Lande: geschiedenis van de Johannieters in de provincie Groningen (13de-17de eeuw)*, Warffum.

Olson, M. (1965) *The logic of collective action: Public goods and the theory of groups*, Cambridge.

Oost, A. (1995) *Dynamics and sedimentary development of the Dutch Wadden Sea, with emphasis on the Frisian inlet: A study of the barrier islands, ebb-tidal deltas, inlets, and drainage basins*, Utrecht.

Philippa, M. *et al.* (2003–2009) *Etymologisch woordenboek van het Nederlands*, Amsterdam.

Philippi, F. *et al.* (eds) (1969–1996) *Osnabrücker Urkundenbuch*, 7 vols, Osnabrück.

Remmers, A. (2004) *Von Aaltukerei bis Zwischenmooren: die Siedlungsnamen zwischen Dollart und Jade*, Leer.

Renaud, J.G.N. and Dijk, E. Van ([2]1971) *Het huis te Wedde*, s.p.

Rinzema, A. (1987) 'Een burger vertelt: de kroniek van Johan van Lemego', in: B. Ebels-Hoving, C. G. Santing and C.P.H.M. Tilmans (eds.), *Genoechlicke ende lustige historiën: laatmiddeleeuwse geschiedschrijving in Nederland*, Hilversum, pp. 61–78.

Schaïk, R.W.M. Van (1985) *Sint Walfridus van Bedum: een duizend jaar oude Groninger overlevering in woord en beeld,* Groningen.

Schmidt, H. (2008) *Ostfriesland und Oldenburg: gesammelte Beiträge zur norddeutschen Landesgeschichte*, Aurich.

Schroor, M. and Meijering, J. (2007) *Golden Raand: landschappen van Groningen*, Groningen.

Schwarz, W. (2005) 'Morsaten, Moorsiedler im frühmittelalterlichen Norder- und Brokmerland', in: H. Schmidt, W. Schwarz, and M. Tielke (eds), *Tota Frisia in Teilansichten. Hajo van Lengen zum 65. Geburtstag*, Aurich, pp. 13–40.

Siemens, B.W. (1954) 'Het generale zijlvest der Drie Delfzijlen', *Groningse Volksalmanak*, pp. 118–152.

Siemens, B.W. (1962) *Historische atlas van de provincie Groningen*, 1962.

Slofstra, J. (2008) 'De kolonisatie van de Friese veengebieden ca. 900–1200', in: K. Kuisman (ed.), *Diggelgoud: 25 jaar Argeologysk Wurkferbân: archeologisch onderzoek in Fryslân*, Leeuwarden, pp. 206–241.

Smith, A.J. (1901) 'Het eiland Ulsda', *Groningse Volksalmanak*, pp. 196–224.

Strahl, E. (2003) 'Reiderland', *Reallexikon der Germanischen Altertumskunde*, 24, pp. 348–361.

Stratingh, G.A. and Venema, G.A. (1855) *De Dollard of geschied-, aardrijks- en natuurkundige beschrijving van dezen boezem der Eems*, Groningen.

Stüwer, W. (1973) 'Zur Werdener Besitzgeschichte in Friesland', *Westfalen: Hefte für Geschichte, Kunst und Volkskunde*, 51, pp. 57–66.

Swaan, A. De (1988) *In care of the state: Health care, education and welfare in Europe and the USA in the Modern Era (Europe and the international order)*, New York and Cambridge.

Tschan, F.C. (ed.) (2002) *Adam of Bremen: History of the archbishops of Hamburg-Bremen*, New York.

Tys, D. (2005) 'Landscape, settlement and dike building in coastal Flanders in relation to the political strategy of the counts of Flanders, 900–1200', in: M. Fansa (ed.), *Kulturlandschaft Marsch: Natur, Geschichte, Gegenwart*, Oldenburg, pp. 106–126.

Veen, J. Van (1930) 'De Fivel en hare verzanding; bewerkt uit de nagelaten aanteekeningen van P.M. Bos', *Tijdschrift van het Koninklijk Nederlandsch Aardrijkskundig Genootschap*, 57, pp. 673–692; pp. 773–801.

Veen, J. Van (1951a) 'Tidal gullies in the youngest peat layer of Groningen', *Proceedings of the 3rd* International Congress of Sedimentology, Groningen-Wageningen, The Netherlands, Wageningen, pp. 257–266.

Veen, J. Van (1951b) 'De Delf', *Tijdschrift van het Koninklijk Nederlandsch Aardrijkskundig Genootschap*, 68, pp. 368–375.

Vos, P.C. and Kesteren, W.P. Van (2000) 'The long-term evolution of intertidal mudflats in the northern Netherlands during the Holocene: natural and anthropogenic processes', *Continental Shelf Research*, 20, pp. 1687–1710.

Vries, W. De (1946) *Groninger plaatsnamen*, Groningen and Batavia.

Wassermann, E. (1985) *Aufstrecksiedlungen in Ostfriesland: ein Beitrag zur Erforschung der mittelalterliche Moorkolonisation*, Aurich.

Wassermann, E. (1995) 'Siedlungsgeschichte der Moore', in: K.-E. Behre and H. Van Lengen (eds), *Ostfriesland: Geschichte und Gestalt einer Kulturlandschaft*, Aurich, pp. 93–111

Woebcken, C. (1928) *Die Entstehung des Dollart*, Aurich.

Websites

Beek, J.L. and Vos, P.C. (2008) *Regio Noord-Groningen: gemeenten De Marne, Winsum, Bedum, Ten Boer, Loppersum, Eemsmond, Appingedam en Delfzijl: archeologische verwachtingskaart en beleidsadvieskaart*, Weesp (Raap-Rapport; 1732) http://www.noordgroningen.org/archeologie.html.

Knottnerus, O.S. (2003) 'Historische geografie van het Dollardgebied', http://www.xs4all.nl/~ottoknot/dollard/.

Molema, J. *et al.* (2010) *Nota archeologiebeleidgemeente Slochteren*, Groningen http://decentrale.regelgeving.overheid.nl/cvdr/images/Slochteren/i67669.pdf

12 Monasteries and water management in the Frisian coastal plain. The reconstruction of landed property as a trigger for new research on the chronology of embankment and drainage

Johannes A. Mol, Leiden University and Fryske Akademy

Abstract

This paper examines the role that monasteries played in the transformation of the medieval landscape. It tries, for the Frisian coastal area, to reassess the general claim of post-war German and French historians that the new religious orders of the twelfth century were not the great land reclaiming pioneers they were thought to be. For this purpose a method has been developed to enable a reconstruction of monastic property in a retrospective way, making use of GIS, so as to analyze the acquisition of land in detail against the background of relief and soil value. In doing so, special attention was paid to the foundation, formation, and parceling of granges as uncommonly large farms that were exploited by the monks themselves. Other themes discussed are the contribution of monasteries to dike building, and the draining of large hinterland areas through the construction of sluices and canals. The paper establishes that the Frisian monks were very active in this field between 1150 and 1350. This was not the first or even the main phase of dike construction but rather the age of damming the remaining streams, the construction of central water outlets, and of the reorganisation and intensification of the use of agricultural space. The data suggests that the different religious orders contributed significantly to the transformation of the landscape in this later phase, much more so than modern historians have previously assumed.

I. Introduction

One of the medieval success stories we are familiar with is the speed with which the monastic reform movement spread throughout Europe in the eleventh and twelfth centuries (Lawrence, 2001: 146–206). In less than 200 years orders like the Cistercians and the Premonstratensians founded monasteries in every corner of Christendom, built impressive churches, and acquired vast estates. Although their demesnes never became as large as those of the great Benedictine abbeys of the Carolingian era, taken together they vastly increased the landholding wealth of the Church. A well-known explanation for the extraordinary growth of these communities, given by the orders themselves and by their first historians, is that they settled far away from the sinful world in waste lands, that were then transformed into fertile soils by the hard labor of their monks and lay brothers

Landscapes or seascapes? The history of the coastal environment in the North Sea area reconsidered, ed. by Erik Thoen, Guus J. Borger, Adriaan M.J. de Kraker, Tim Soens, Dries Tys, Lies Vervaet & Henk J.T. Weerts, Turnhout, 2013 (*CORN Publication Series*, 13), pp. 267–285.

BREPOLS PUBLISHERS DOI: 10.1484/M.CORN.1.101556

(Rösener, 1982a: 118–120). In this way, the great European reclamation movement of the same period would have received a new and vibrant impetus.

This rosy picture of ecclesiastic toil amidst wildernesses was tarnished in the decades after the Second World War by several German and French economic historians (Wiswe, 1953; Higounet, 1965; Duby, 1962: 146–153). With their research on the agricultural expansion of the High Middle Ages, they showed, or at least tried to show, that most of the new religious houses were founded in places that had long been occupied, and that they did not play a leading role in opening up new territories, reclaiming forests and the like. It has to be stressed, though, that their studies – even if they only had a regional scope – were carried out on a general basis. They had to rely on a rough location of monastic properties, rather than on precise reconstructions of granges and other parcels of land. This is understandable, since the techniques for mapping medieval monastic property on the basis of modern or early modern administrations still had to be developed by then. Besides, for many regions and countries, no combinable detailed maps were available on historical topography, soil conditions, relief etc.

Since tools and research methods in both historical geography and agrarian history have improved enormously since the nineties, particularly thanks to GIS-based research (Gregory and Ell, 2007), it seems rewarding to examine the role of the monasteries anew. It is, however, not so much the pioneer question that I would like to address in this chapter. Instead, the study of the monasteries and their property is here discussed to contribute to our knowledge on the transformation of the landscape; in this case, particularly the chronology and the process of embankment and drainage in the Frisian coastal area. This region covers the two northern Dutch provinces of Friesland and Groningen which contained a number of religious houses in the period between 1100 and 1600. The landed properties of these houses can be reconstructed in good detail. The parcel-based historical GIS that has been developed for both provinces by the Fryske Akademy in Leeuwarden since *c.*1997 (Mol and Ekamper, 2009), makes it possible to link this reconstruction of the monastic demesnes to all kinds of data sets and mapped interpretations of Frisian landscape history.

II. GIS, Frisia, the reclamation of the fens and the embankment process

Before presenting a few case studies, some introduction is required about GIS and the geophysics of Frisia. Let us begin with the historical GIS of Friesland and Groningen, which can be consulted via the internet portal www.hisgis.nl. Its base layer consists of the digitised Napoleonic cadaster of 1812/1832. This is structured in such a way so that information on the size, the owners of, the value etc. of the various parcels is attached to the polygons representing them on the computer map and stored in a database. So the system is parcel based, with a central focus on the beginning of the nineteenth century. While it can be used for all kinds of research concerning that period, it also serves as a starting point for investigations into earlier historic periods. For the province of Friesland in particular, the retrospection possibilities are favorable because the Frisian land registers from the eighteenth and seventeenth centuries can be linked to the cadaster of 1832. Taking the oldest of these (1640) as a stop-over, one can make a further jump back in time to some sixteenth-century property land surveys. These, in turn, provide a view on the landownership of the area towards the end of the Middle Ages.

Three categories are of special interest to us here: the possessions of the monasteries, the property of the parish churches, and the lands of the Frisian nobility. The monastic possessions can be delineated via the sales surveys that were made up by the Estates of Friesland for all the religious houses within their territory when they were seized at the dissolution of 1580 (Mol, 1991: 325–328). The same can be done with the glebes and other lands of the parish churches by linking the eighteenth-century registers in which they are described in great detail, with the cadastral surveys. The landed property of the nobility can, with some difficulty, be traced along inheritance lines (Noomen, 2009: 15–16). As for the province of Groningen, the reconstruction perspectives for the latter two categories are less promising. But just as in Friesland, it has been possible to reconstruct the former monastic lands and store them into GIS on the basis of registers and maps of the Estates of Groningen, who sequestrated these lands after the suppression of the monasteries in 1594 (Mol, 2010: 18–19). All of this means that we now have a clear view of the location and spread of all monastic property, including the granges, in the whole region between Vlie and Eems.

The large Frisian coastal plain consists of fertile clay zones along the sea, intersected by gullies and tidal streams, with a hinterland of high but poor sandy soils, partly covered by peat (Figure 12.1). Taking a closer look at these zones, it is obvious that they are composed of two elements: a series of comparatively small sandy clay ridges near the seaside, and basins of heavy clay, deficient of lime, as a top layer on a thick blanket of peat, which therefore tended to sink through soil compaction (De Langen, 1992: 17–67). Originally, these basins were drained by small rivers breaking through the ridges. In both areas people had settled on artificially raised dwelling mounds, the so-called *terpen* or *wierden* (*Wurten* in German), to protect themselves against floods. The older dwelling mounds on the ridges have been dated to pre-Roman times; the younger ones were built in the basins in the early Middle Ages. Archaeological finds suggest that they were all densely populated *c.* 900, long before the appearance of the first monasteries (Knol, 1993: 239–242).

South of these clay basins, peat came to the surface, originally forming thick and high moor blankets that crept up against the slopes of the Pleistocene sand plains of Southeast Friesland and Drenthe. Like everywhere in the Low Countries, these fens and moors were gradually reclaimed for agricultural use by draining them with long and straight ditches. The chronology of the earliest reclamations is still not clear. As for the Frisian regions, archaeologists now think that the large scale reclamation process had started by the ninth and tenth centuries, at least in the region south of Sneek and in the so-called Wold area to the northeast of Groningen city (De Langen, 2007; Ligtendag, 1994: 33–35, 66–76). One of the less pleasant effects of the drainage of the fens was that their surface gradually began to sink, which in turn brought the danger of prolonged inundation by either salt or fresh water. This, in its turn, called for the construction of embankments or levies alongside the watercourses.

At the same time, the inhabitants of the clay zones, who had been accustomed to the regular flooding of their lands, began to protect these with low dikes (Borger, 1985). At first, in the same way as it had been done in Roman times, quays of barely one meter

Figure 12.1 The landscape of Friesland west of the Lauwers *c.* 1200 (source: *De Bosatlas van Fryslân*, 2009: 68)

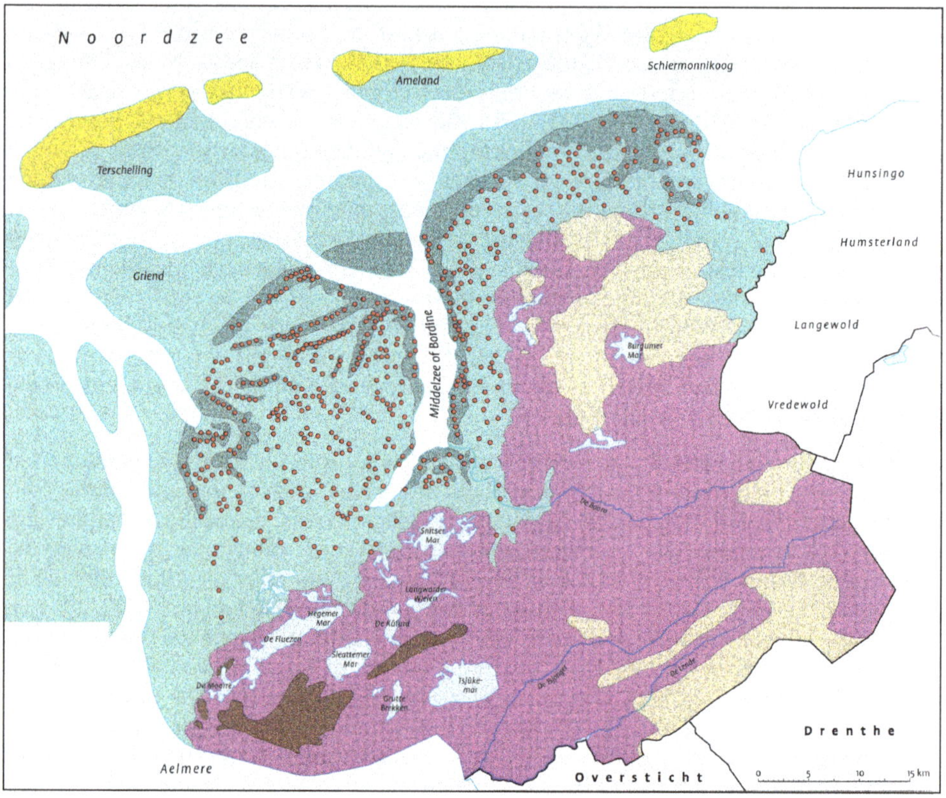

high were raised as partial dikes, closing the smallest tidal creeks and forcing the salt water sideways. Secondly, these dikes were subsequently interconnected, after which partial embankments could be enclosed. As people were not then capable of damming up the major creeks, the quays often ran along the streams. For a long time they were only meant to give protection against summer flood and simple outlets in the form of hollowed trees ensured that superfluous water was discharged. In a third phase, the dikes were then heightened and raised to one and a half or sometime to two meters to protect both men and the land behind the dikes against winter flood, though they still could not prevent the fields becoming inundated by storm surges occasionally. The dikes had gentle slopes, so that the waves could do limited damage. Only in this later phase were the farms moved from the slopes of the dwelling mounds to the flat lands nearer to the pasture complexes that belonged to them. This also meant that more efficient valve locks had to be constructed in the dikes, in combination with feeding channels, to get rid of rain and other seeping water.

The interaction between the reclamation, sinking, and embankment of the fen lands on the one hand, and the damming and draining of the clay zones on the other, has not yet been studied in detail, possibly because the chronology of the different phases is hard to detect. Historical data is sparse, and archaeological and geophysical research of the dikes themselves (Meijer, 1992) has not yet been carried out in Friesland and Groningen. In the Northern Netherlands, only the Westfriese Omringdijk near Enkhuizen and some embankments in the Meuse estuary, west of Rotterdam, have been studied in this way (Hallewas, 1984). All that can currently be said is that the process of defensive embankment in both the *terpen* areas and the earliest reclaimed fens alongside the rivers started sometime during the eleventh century.

III. The Frisian monasteries

Obviously, the foundation dates of the Frisian monasteries indicate that they cannot have initiated this process (Mol, 2003b: 221–226). The new monks arrived in Frisia only in the middle of the twelfth century, half a century later than in Flanders and Zealand, and in a period when the agricultural space in the long-inhabited *terpen* area was being rearranged and repartitioned. The main question therefore has to be in what way the monks have contributed to these processes. The first four monasteries that came into being in the province of Friesland were the Benedictine abbey of Staveren (1132), the abbey of Augustinian Canons of Ludingakerke near Harlingen (1158), the Cistercian abbey of Klaarkamp, not far from Dokkum (1163), and the Premonstratensian abbey of Mariëngaarde (1163). The earliest monasteries in the province of Groningen appeared a few years later. The Premonstratensian Oldeklooster in the Marne, for instance, was founded some time before 1170, the Benedictine abbey of Feldwerd near Appingedam *c.*1180 and the Cistercian abbey of Aduard, west of Groningen city, in 1192. These dates suggest that the foundation movement went from west to east across the region (Figure 12.2).

The main orders, which appeared more or less simultaneously in Frisia, were all trendy at that time. Not only the Cistercians and Premonstratensians but also the Benedictines from the Affligem-Oostbroek filiation and the Augustinian Canons from Kloosterrade presented themselves as sober and austere, although the Cistercians were considered to be the most severe because they strived for the utmost simplicity of diet, architecture, and liturgy. The distribution pattern of the first twenty monasteries suggests that every order tried to plant a central abbey in one of the Frisian districts. The Benedictines were the first to settle in Suthergo, the Canons regular built the first monastery in Westergo, whilst the Cistercians founded the first one in Oostergo. Thus it looks as if each order feared the competition of the others and tried to avoid becoming close neighbors. Later on, when it had become clear that there was enough space for several religious houses within the same district, nearly all orders founded houses in every region. Hence, the number of abbeys and priories rapidly increased in Frisia in the first half of the thirteenth century.

After 1220 the above mentioned orders also got new competitors in two military and hospital orders (the Hospitallers and the Teutonic Order), and two mendicant orders – which of course, were less interested in acquiring landed property. Around 1300 the province of Friesland counted more than thirty-five abbeys, priories and commanderies, whilst the smaller province of Groningen had no less than twenty religious houses within its territory. In the fifteenth century a new wave of monastic foundations would reach both regions, but only a few of these succeeded in becoming rich landowners. Therefore, these new convents had no chance to put their own imprints on the landscape.

Figure 12.2 Monasteries in Friesland and Groningen 1100–1350 (source: *Der Friesische Freiheit des Mittelalters – Leben und Legende,* 2003: 223)

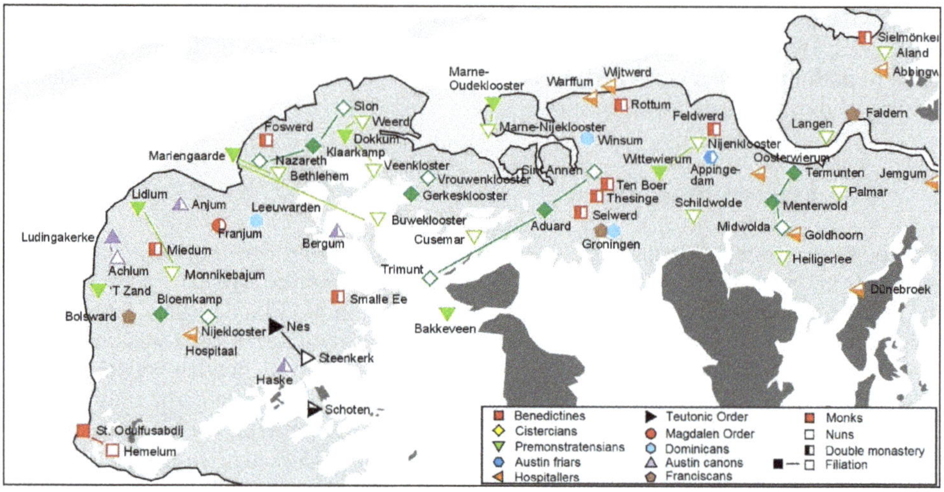

IV. Granges and the transformation of the Frisian landscape

As all these institutions were prayer houses, whose main function it was to make the route to heaven as easy as possible for their members and supporters, it can be questioned what turned them into possible transformers of the landscape. In general, up to the middle of the fourteenth century they all showed themselves to be active landowners (Roehl, 1972; Despy, 1974/1975; Lekai, 1977: 83–113; Brittain Bouchard, 1991). They acquired land in three ways. First, they attracted major foundation grants and subsequent donations from both lay and ecclesiastical benefactors. Second, they created funds for new investments by commuting the entry fees of their novices. Third, they earned considerable sums of money by selling agrarian surpluses on the market. The demesnes they thus built up were not exploited by domestic servants, as in the older Benedictine abbeys. The new monks preferred to cultivate the land themselves– according to the Benedictine rule that the true monk should live from the work of his own hands. Since the choir monks could not save much time between the prayer services, they admitted *conversi* (lay brothers) in astonishingly large numbers, to take up both the management of the farming and the manual labor that was required. Thereafter, the farming was organised into large units of exploitation (150–400 acres), the so-called granges or *curiae*. In theory, these had to function within an autarkical system, providing the abbey with all the products it needed. But in practice any surplus from these large exploitation units was sold. For this end, many abbeys founded town granges as depots where grain and other products were stored to bring them to the market when prices were favorable.

This way of organising monastic economy is known as the grange system (Rösener, 1982b) and it is commonly found throughout Europe, particularly in relation to Cistercian and Premonstratensian foundations. In Frisia, both orders were represented and nearly each of their convents founded granges and had *conversi* to run them. But the Benedictines, Augustinian Canons, and the military orders also admitted lay brothers and set up granges (Mol, 2003a). Although their granges were generally smaller than those of the Cistercian and Premonstratensian orders, and although they had less granges per convent, it looks as though the Benedictines copied the Cistercian system and adapted it to their own needs.

The Frisian granges can often be recognised by their original names because they got 'fossilised' in the names of the later farms. In the province of Friesland most of them were called *monnikhuizen* (monk houses) after the lay brothers or working monks that were seen to be active at them. In the province of Groningen the majority were known as *voorwerken* (Germ. *Vorwerke*). To identify them, the place-names with these suffixes and their locations have to be compared with both the cartographic property reconstructions and all of the data on granges that can be found in the monastic archives. Using these methods, no less than eighty-five granges can now be localised in Friesland and sixty in Groningen. Because there is a general lack of documentation for the early period of the Frisian monasteries, these figures should be considered minimal.

It is well-known that all over Europe the grange system broke down in the course of the fourteenth century because of demographic factors and a diminished religious attraction of the lay brother status (Donnelly, 1954; Lekai, 1977: 323–324, 343–344; Toepfer 1984, 60–62). As a result, the *monnikhuizen* in Frisia were gradually split up. In most cases they were divided into three or four separate farms that could be leased out to wealthy tenants (Mol, 1992: 90–91). In sixteenth- and seventeenth-century lists of farms, a lot of these units can be recognized more in detail as elements that together once comprised former granges. It is now comparatively easy to bring these elements and parcels together, because those granges that had once been geographical entities in their own right stood apart from the rest of the monastic demesne lands. This is the case with two granges in the parish of Holwerd in the northeast part of Friesland: the *Zwarte* (Black) *Monnikhuis* and the *Witte* (white) *Monnikhuis*, that had once respectively belonged to the Cistercian abbey of Klaarkamp and the Premonstratensian abbey of Dokkum. Both *monnikhuizen*, the former later comprising five and the latter seven tenant farms, are assumed to have been constructed in the last decades of the twelfth century. Figure 12.3 shows the grange lots against the background of estimated tax revenue value in 1832.

The images encourage us to question whether or not the grange complexes were created as new structures – not on *terpen* but on the flat land that had been embanked with winter dikes – and to what extent their parceling had been the work of the monks. The Cistercian *Zwarte Monnikhuis* is located just behind the dike on the sandy clay ridge on which the Holwerd terp had been raised. But the rectangular form of its lands and its position on a site where the high yielding ridge gets smaller, indicate that at least a part of its parcels must have been newly partitioned at its start. The same goes for the Premonstratensian *Witte Monnikhuis* east of Holwerd, with the observation that this grange was erected not on the ridge itself but some distance to the south. In addition, the toponyms *Zwarte* and *Witte Monnikhuis* suggest that they were new establishments, otherwise they probably would have taken over the previous names of the older farms. The names of *terp* settlements can easily be recognised by their *-werd*, *-ingi* (or *-ens*) or *-um* suffixes (Gildemacher, 2008: 297–299). The fact that more than half of the *c.*145 monastic granges identified in

Figure 12.3 Granges near Holwerd (source: J.J. Feikens, Fryske Akademy)

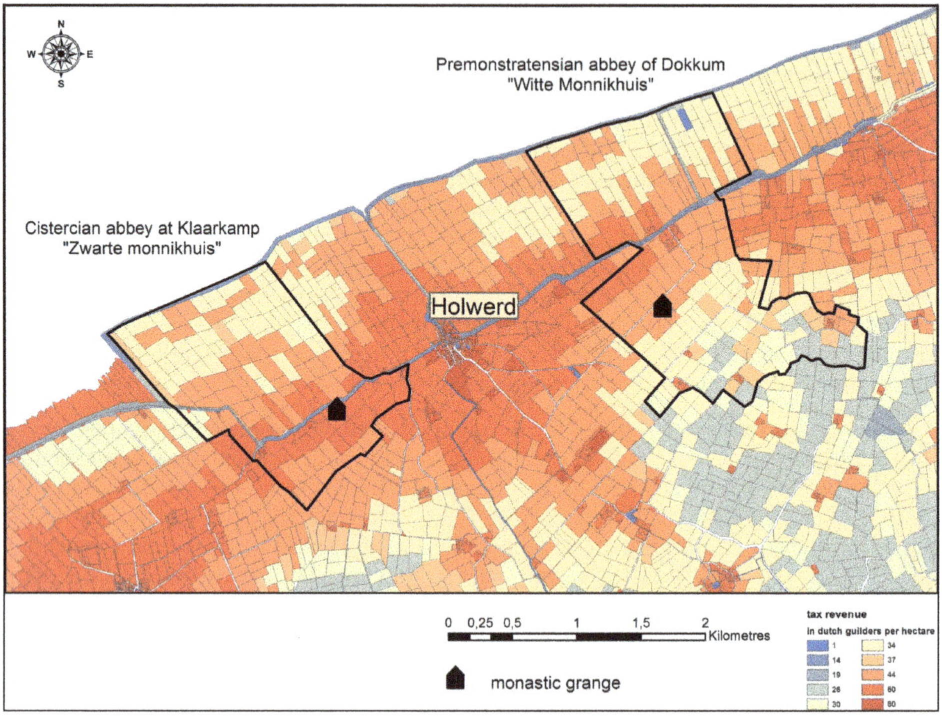

Friesland and Groningen do not contain such suffixes, but instead a *-house*, *-voorwerk*, *-hof*, or *-heem* toponym, gives a clear indication that they were newer creations. Perhaps a more obvious sign of their relative lack of antiquity is the fact that many of them are not located upon the older dwelling mounds.

However, we have to be careful with our deductions in this matter. Some farmsteads that did not lie on mounds but nevertheless had elevated positions on a sandy clay ridge are known to have already been in use in the eleventh century; if a grange can be identified and located there, one can be sure that it was not a completely new structure. In fact, the best way to determine the age of the granges is to have them investigated archaeologically. Unfortunately, hardly any Frisian granges have received archaeological attention that has gone beyond the dating of scattered pottery finds. Thus far, only one of them has been excavated (Kortekaas 2006; Huys in't Veld 2011). This site was the so-called *Hoogeheem* at Roderwolde, west of Groningen, which was owned by the Cistercian nuns of Mariënkamp (*Maria in Campis*) near Coevorden in the south eastern part of the province of Drenthe (Figure 12.4). They probably received it from their father abbot of Aduard, when they moved their abbey to the nearby village of Assen in northern Drenthe in the middle of the thirteenth century. *Hoogeheem* lies on a site where the clay basin touches a fen reclamation zone (Zomer, 2009: 54–55), and it was discovered that it was

Figure 12.4 The location of the Hoogeheem Grange, west of Groningen (source: J.J. Feikens, Fryske Akademy)

constructed upon a thin layer of clay that had probably been deposited on a fen settlement as the result of one or several storm surges in the twelfth century. Because no occupation traces were found in the top layer, it seems that the grange was newly built from scratch. It is striking that the *Hoogeheem* site neatly matches the large core complex of Aduard lying on the more fertile clay lands to the north. There, investigations have shown that to construct their grange the monks first made a platform of clay, measuring 50 x 40 m (Figure 12.5).They subsequently extended this first site by digging a moat around both the platform and an adjacent plot, so they eventually created a separate space measuring 200 x 100 meters in size. From the datable finds it is inferred that the complex was built between 1250 and 1300 (Huys in't Veld, 2011). As such, it certainly was one of the latest Cistercian granges in the north and it's structure demonstrates how systematically and

Figure 12.5 The Hoogeheem excavation: traces from the thirteenth and fourteenth centuries (source: Huys in't Veld, ill. 197)

thoroughly the *conversi* built their larger-scale exploitation units, even during a later period. Hopefully an opportunity will soon present itself to investigate several other granges, both archaeologically and geographically. For instance, those of the already named abbey of Aduard, which – with its demesne of nearly 6000 hectares *c.*1594 – can be qualified as the richest monastery of Frisia (Mol, 2010). Six out of its eight identifiable granges in the immediate surroundings of this abbey were not located on mounds but on newly created moated sites (Figure 12.6). In fact they seem to have had the same design as the Cistercian and Premonstratensian *monnikhuizen* near Holwerd.

V. Acquiring fertile drainable lands

To discover the rationale behind the acquisition policy of the monasteries and the organisation of their exploitation system, it is best to focus on Aduard's mother house, the Cistercian abbey of *Clarus Campus* or Klaarkamp, which was founded in central Oostergo *c.* 1163. The outline of the abbey enclosure shows that the complex of convent buildings had been erected on an already existing dwelling mound. Soon after its foundation this complex was extended to the north with a rectangular plot called the *Hofland* to provide space for the industrial and agricultural operations of the abbey. There are indications that its core property was formerly part of a former count's demesne at Rinsumageest

Figure 12.6 The landed property and granges of the Cistercian abbey of Aduard, against the background of the elevation map (source: J.J. Feikens and J.A. Mol, Fryske Akademy)

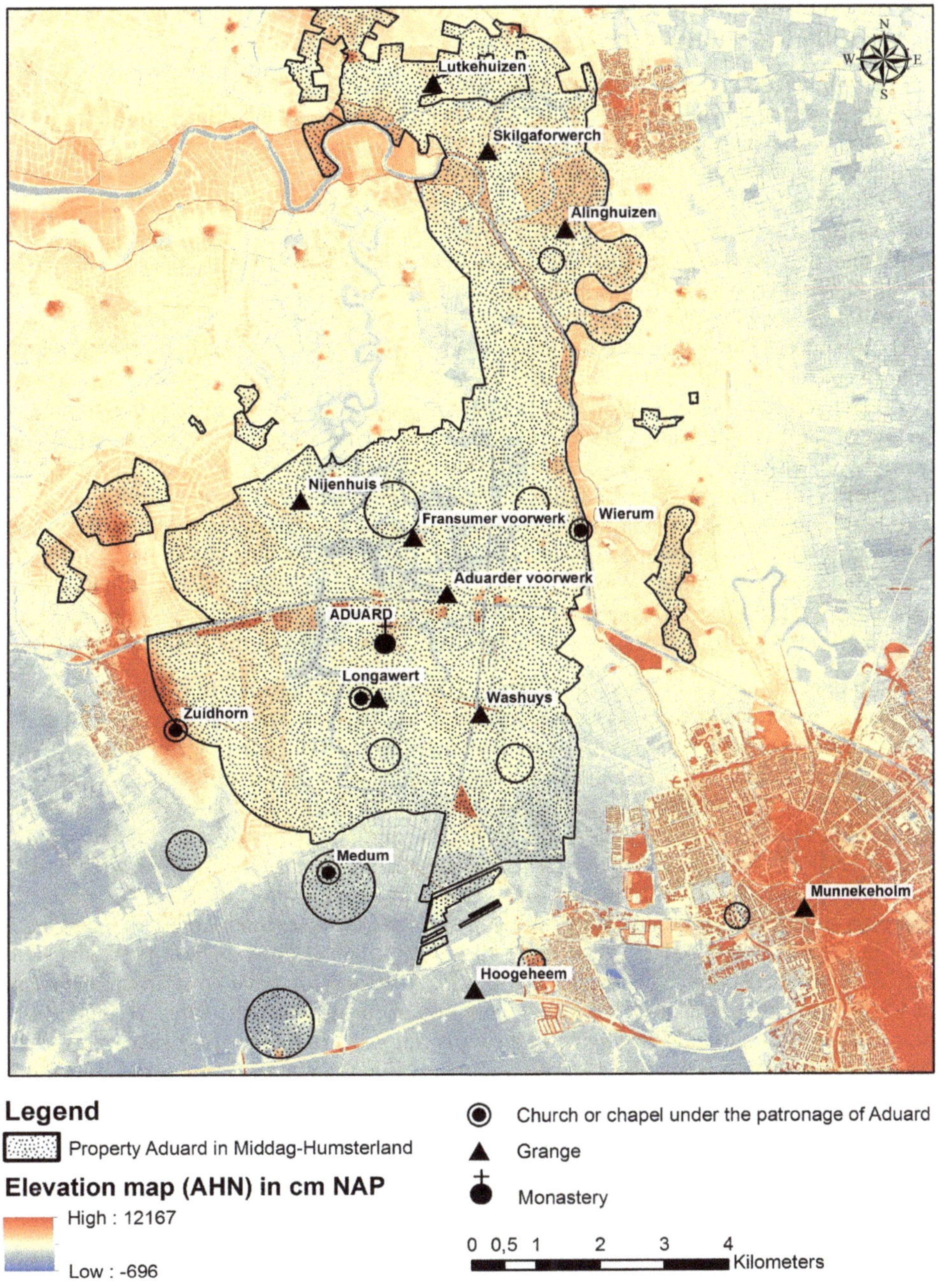

(Noomen, to appear). Its foundation could therefore have been encouraged by descendants or former *ministeriales* of the count of Braunschweig who had exercised the comital rights in Oostergo before 1088. There is no doubt, however, that the convent owed its fast expansion to the massive support of the local aristocracy (Mol, 1996). The reconstruction of Klaarkamps core property against the background of the estimated soil value (1832) demonstrates that the abbey land is concentrated on a highly valued inverted ridge.

The monks soon obtained lots of land alongside the River Paesens, which originally discharged into the Waddensea to the north, draining both the clay basin and the higher sandy soils and peat bogs in the south. On both of its banks a string of Cistercian lands and granges can be identified (Figure 12.7). This string of properties can even be extended to the Isle of Schiermonnikoog (*Grey Monks Eye or Isle*) where, as the name suggests, the monks also possessed a grange. Most interesting, however, is that to the west of Dokkum there is a small canal that connects the River Paesens with the Dokkumer Diep, an older water course that had been dug to reach the trading place of Dokkum (Mol, 1992). In constructing this new canal, it was possible to reverse the direction of the silted up River Paesens north of Dokkum and steer its water into the Dokkumer Diep. It provided also a shorter drainage route for the excess water from the basin west of Klaarkamp. Unfortunately, there are no documents relating to this early piece of water engineering, which I would like to date at the end of the twelfth century. However, since Klaarkamps largest grange complex was located very close to the spot where the river and this canal were linked to each other, and where a drainage sluice was built, it seems likely that both the canal and the sluice were been constructed by the monks. We know in any case that in the fourteenth century Klaarkamp was responsible for the maintenance of the Leppa sluice, through which the bulk of the water of Southern Oostergo was drained. In acquiring wet but fertile pieces of land along a silting up river, the monks of Klaarkamp must have previously decided that they could get rid of the excess water by creating a new outlet to the southeast.

Another early piece of Cistercian water engineering can be found in Westergo near the parish of Easterwierum, half way between the well-known towns of Leeuwarden and Sneek. Here, the abbey of *Floridus Campus* or Bloemkamp (1192), Klaarkamps first daughter house, owned a grange called *Vogelzang*. Bloemkamp had acquired the core of its property in or near the silted up estuary of the Middelzee basin (Figure 12.8). The GIS image of the soil revenues in 1832 demonstrates how highly the embankment ridges and the newly accreted clay lands were valued there. On a map this former estuary has the shape of a bottle, with its neck between Oosterwierum on the Westergo side, and Rauwerd at the opposite Oostergo side. The relief map clearly shows the course of the former stream that had once brought the tide far inland. It is precisely at this spot that a dam was built between 1200 and 1230 to prevent the seawater from going in. Although we have no written proof that the Cistercians of Bloemkamp have been involved in constructing the dam, the location of their grange with its lands alongside it, strongly suggests that they were responsible. By closing up the estuary, they had a lot to gain in securing their recently acquired lands both near the abbey and its nunnery of Nijeklooster, founded in 1234 (Mol, 1992).

The same development can be observed in the Fivel estuary in northeast Groningen (Figure 12.9). Here, both the Benedictine abbey of Feldwerd (*c.*1180) and the Premonstratensian abbey of Wittewierum (1204) founded a series of enormous granges in the first half of the thirteenth century (Knottnerus, 2005: 45–48). It is likely that the abbey of Wittewierum took the lead in this process. Her eldest grange in the Fivel-estuary, the

Figure 12.7 The landed property and granges of the Cistercian abbey of Klaarkamp and its dependent nunnery of Sion (source: J.A. Mol)

Westeremder voorwerk (*c.* 1219), was located upstream at the east side of the village of Westeremden, albeit on a high marsh flat that was probably still open to the tides when it was founded. Shortly thereafter, Feldwerd founded the *Garsthuizer voorwerk* on a bank just to the north. At approximately the same time (*c.* 1225) the *Zandster voorwerk* of Wittewierum came into existence a bit farther to the northeast, on a sandy flat between two branches of the Fivel gully that, with the outgoing tide, sought its twisting way to the sea. As for the northerly granges of the Benedictines of Feldwerd: Kolhol on the Zijldijk, and De Hoorn, east of Oosternieland, we do not know when they came into being. Looking more generally at the development of *Feldwerd*, I assume that this occurred sometime between 1225 and 1250. The location of all five granges has to be linked to successive phases of embanking along with the construction of sluices and water outlets. But the exact sequence of this process will not be clear until it is researched. Were the granges really constructed on high silted-up marshes before they were embanked or was it the

Figure 12.8 The landed property and granges of the Cistercian abbey of Bloemkamp and its nunnery of Nijeklooster, against the background of 1. the estimated soil revenue of 1832 and 2. the altitude map (J.J. Feikens and J.A. Mol, Fryske Akademy)

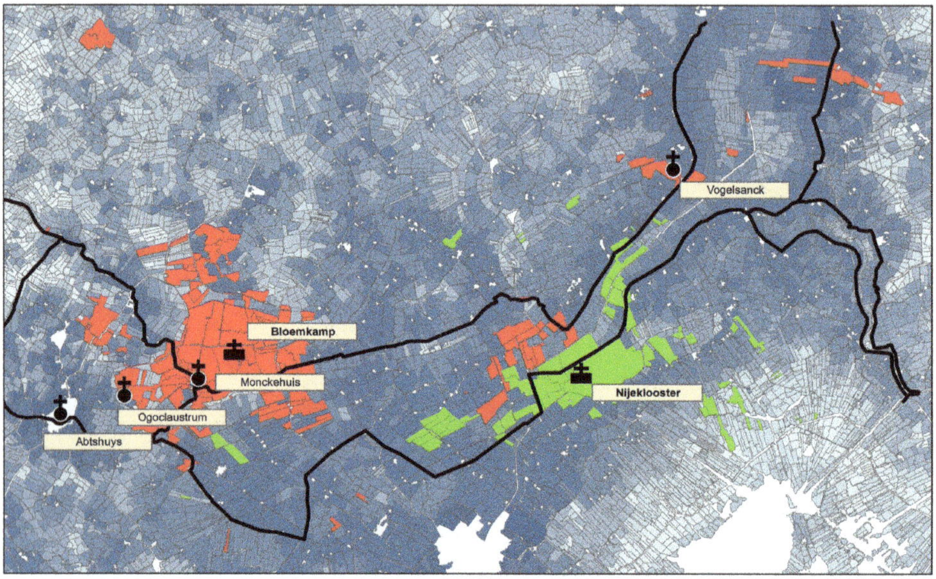

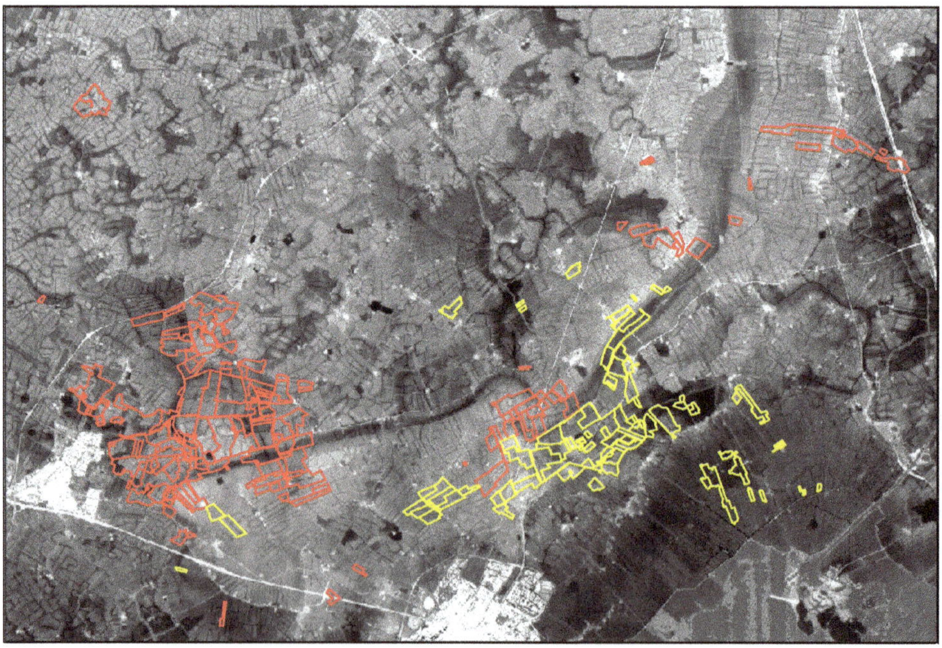

Figure 12.9 Reconstruction of the granges of the Benedictine abbey of Feldwerd and the Premonstratensian abbey of Wittewierum in the Fivel Estuary (Northeast-Groningen) (source: J.J. Feikens, Fryske Akademy)

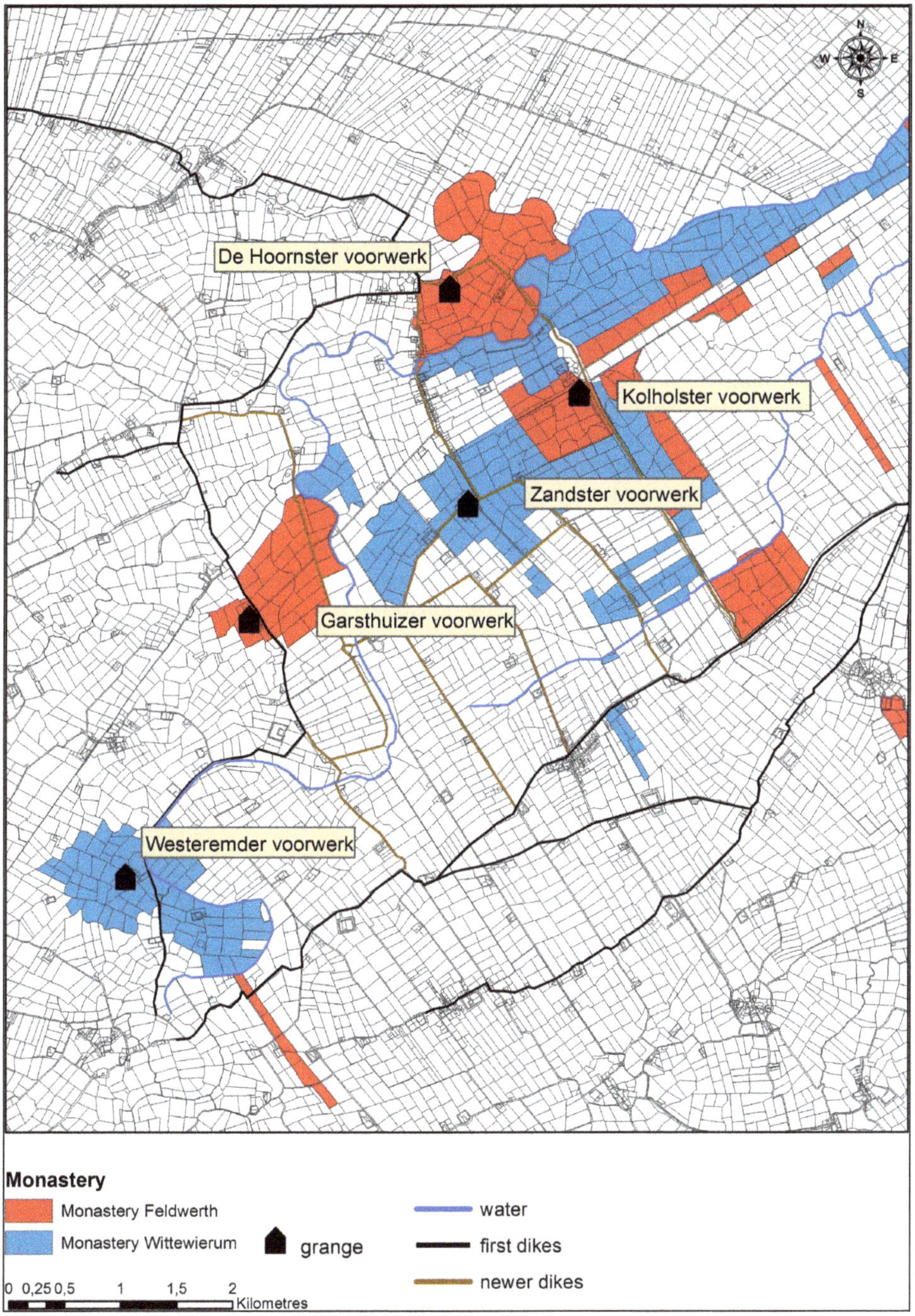

other way round? In any case it is clear that once a grange like the *Zandster voorwerk* was functioning, within a complex of enclosed lands, its lay brothers did their best, working in common with the neighbouring landowners, to acquire new properties by constructing adjoining polders. Such was the case in 1266 when, according to the chronicle of Wittewierum, a high and firm sea dike (the Korendijk) was built between the village 't Zand on the one side and the embankment around the *Zandster voorwerk* on the other (Jansen and Janse, 1996: 411). Seven years later this dike had to be provided with a sluice in the eastern branch of the Fivel near Ompteda borg.

These cases suggest that the Cistercians and Premonstratensians were particularly keen on acquiring wet and thus comparatively cheap lands whose productivity could be considerably raised by damming gullies, embanking water outlets, and constructing drainage works like sluices and canals, both in the already embanked clay basins and on the newly silted up tidal marshes. In general the monasteries were not interested in possessions in the higher but poorer sand and peat districts in the hinterland, except for some specific purposes like raising sheep, growing rye, and extracting peat for fuel.

There are many more cases that could equally illustrate the efforts of the monasteries in the field of water management in the twelfth and thirteenth centuries. And yes, the achievements of the Cistercians attract most attention. After all, on a European scale they enjoyed the greatest fame as water engineers and in Frisia too, they were the most successful of all the new orders in extending their properties. Their ranking, at least according to the surface of their property in the provincial surveys that were made after the suppression of the monasteries, is revealing. It has to be stressed, however, that the other orders also achieved a lot. The property and grange reconstruction by the Augustinian convent of Bergum, for example, clearly demonstrates the concern of the canons and their lay brothers to regulate the draining of the excess water from the sand and peat district around their monastery via a sluice in the mouth of the Alddjip (Mol, 2004). At this very spot, where the river broke through the sandy clay ridge south of Leeuwarden to discharge its water in the Middelzee, they had built their main grange, called *Barrahuis*. When the mouth of the Alddjip silted up, they changed the drainage to the north through a new, or at least deepened and broadened, water course in the direction of the great Leppa sluice near the Cistercian abbey of Klaarkamp.

VI. Conclusion

Reconstructing monastic property in a retrospective way, making use of GIS, enables us to analyze the acquisition of land in (parcel) detail against the background of relief and soil value. This, in turn, makes it possible to reassess the role that monasteries played in the transformation of the landscape. In this, the foundation, formation, and parceling of granges as uncommonly large farms that were exploited by the monks themselves, grab our attention. Other themes were the contribution of the monasteries to dike building, and the draining of large hinterland areas through the construction of sluices and canals. As for chronology, it has to be noticed that the main activity of the Frisian monks in this latter field generally took place within a century after the founding of their convents, that is – depending on the starting point – in the last decades of the twelfth century until *c.*1350. This was not the period of the first, or even the main phase of dike construction.

Rather, it was the age of damming the remaining streams, the construction of central water outlets, and of the reorganisation and intensification of the use of agricultural space. Our historical and geographical data suggests that the different monastic orders have contributed significantly to the transformation of the Frisian landscape in this later phase, much more so than modern historians, including myself, assumed until recently. It has to be stressed, however, that our data does not yet tell the whole story. More detailed research by historians and archaeologists is required, in which the granges are compared with the other, non-monastic farms that were founded outside the dwelling mounds in the same period.

Bibliography

Borger, G.J. (1985) 'De ouderdom van onze dijken. Een nieuwe discussie over een oud vraagstuk', *Historisch Geografisch Tijdschrift*, 3, pp. 76–80.

Despy, G. (1974/1975) 'Les richesses de la terre. Cîteaux et Prémontré devant l'économie de profit aux XIIe et XIIIe siècles', *Problèmes d'histoire du christianisme*, 5, pp. 58–80.

Duby, G. (1962) *L'Economie rurale et la vie des campagnes dans l'occident medieval*, vol. 2, Paris.

Gregory, I.N. and Ell, P.S. (2007) *Historical GIS: Techniques, methodologies and scholarship,* Cambridge.

Ey, J. (2007) 'Early dike construction in the coastal area of Lower Saxony' in: J.J.J.M. Beenakker *et al.* (ed.), *Landschap in ruimte en tijd*, Amsterdam, pp. 92–99.

Gildemacher, K.F. (2008) *Terpen en terpnamen. Een naamkundig en historisch-geografisch onderzoek naar vijf naamtypen in Fryslân*, Leeuwarden.

Hallewas, D.P. (1984) 'Mittelalterliche Seedeiche im Holländischen Küstengebiet', *Probleme der Küstenforschung im südlichen Nordseegebiet*, 15, pp. 9–27.

Higounet, Ch. (1965) *La grange de Vaulerent: structure et exploitation d'un terroir cistercien de la plaine de France, XIIe* - XVe siècle, Paris.

Huys in't Veld, J.Y. (2011) *Het Hoogeheem, een Drents kloostervoorwerk in Groningen.Een opgraving van een podium uit de late middeleeuwen bij Hoogkerk*, Groningen (Stadse Fratsen; 28).

Jansen, H.P.H., and A. Janse (eds) (1991), *Kroniek van het klooster Bloemhof te Wittewierum*, Hilversum.

Knol, E. (1993) *De Noordnederlandse kustlanden in de Vroege Middeleeuwen*, Groningen.

Knottnerus, O. (2005) *Fivelboezem, de erfenis van een verdwenen rivier*, Bedum.

Kortekaas, G., e.a. (2006) 'Matsloot 12, opgraving (Archisnr. 13777)', *Hervonden Stad*, 11, pp. 20–21.

Langen, G.J. de (1992) *Middeleeuws Friesland. De economische ontwikkeling van het gewest Oostergo in de vroege en volle Middeleeuwen*, Groningen.

Langen, G.J. de (2007) 'Archeologische kroniek: een Karolingische veenontginning onder afslag Akkerwinde', *De Vrije Fries*, 87, pp. 261–263.

Lawrence, Ch.H. (2001) *Medieval Monasticism,* Harlow.

Lekai, L.J. (1977) *The Cistercians. Ideals and Reality*, Ohio.

Ligtendag, W.A. (1994) *De Wolden en het water. De landschaps- en waterstaatsontwikkeling in het lage land ten oosten van de stad Groningen vanaf de volle middeleeuwen tot ca. 1870*, Amsterdam.

Meier, D. (1992) 'Archäologie in den Nordseemarschen. Untersuchungen an Warften und Deichen in Norderditmarschen und Eiderstedt' in: M. Müller-Wille and D. Hoffmann (eds), *Der Vergangenheit auf der Spur. Archäologische Siedlungsforschung in Schleswig-Holstein*, Neumünster, pp. 63–82.

Mol, J.A. (1991) *De Friese huizen van de Duitse Orde. Nes, Steenkerk en Schoten en hun plaats in het middeleeuwse Friese kloosterlandschap,* Leeuwarden.

Mol, J.A. (1992) 'Besitzerwerbungen der friesischen Zisterzienserklöster Klaarkamp, Bloemkamp und Gerkesklooster', in: K. Elm (ed.), *Erwerbspolitik und Wirtschaftsweise mittelalterlicher Orden und Klöster*, Berlin, pp. 67–96 (Berliner Historische Studien; 17, Ordensstudien; 7).

Mol, J.A. (1996) 'De stichting fan it Cistersiënzer kleaster Klaarkamp by Rinsumageast', *It Beaken*, 58, pp. 1–13.

Mol, J.A. (2003a) 'Die Wirtschaftsführung der Ritterorden in den friesischen Ländern' in: R. Czaja en J. Sarnowsky (ed.), *Die Ritterorden in der eutopäischen Wirtschaft des Mittelalters*, Ordines Militares – Colloquia Torunensia Historica, Torun, pp. 107–128.

Mol, J.A. (2003b) 'Friesische Freiheit in Kirchspiel und Kloster' in: H. van Lengen (ed.), *Die Friesische Freiheit des Mittelalters – Leben und Legende*, Aurich, pp. 194–245.

Mol, J.A. (2004) 'Het uithofsysteem van het klooster Bergum in Friesland', *Historisch Geografisch Tijdschrift*, 22, pp. 73–85.

Mol, J.A. (2010a) 'Bezitsverwerving en goederenbeheer van de abdij Aduard', in: J. van Moolenbroek and J.A. Mol (eds), *De abtenkroniek van Aduard. Studies, editie en vertaling*, Hilversum-Leeuwarden, pp. 173–202.

Mol, J.A. (2010b) *HISGIS Groningen. Opzet, verrijkingsmogelijkheden en prioriteiten*, Leeuwarden.

Mol, J.A., and Ekamper,P. (2009) 'De kadastrale kapstok: HISGIS Fryslân als model voor een Nederlands basissysteem' in: O. Boonstra en A. Schuurman (eds), *Tijd en ruimte: nieuwe toepassingen van GIS in de alfawetenschappen*, Utrecht, pp. 198–209.

Mol, J.A., and Delvigne, J. (2010) 'Het klooster, het land en het water' in: J. van Moolenbroek and J.A. Mol (eds), *De abtenkroniek van Aduard. Studies, editie en vertaling*, Hilversum-Leeuwarden, pp. 153–172.

Noomen, P.N. (2009) *De stinzen in middeleeuws Friesland en hun bewoners*, Hilversum.

Noomen, P.N. (to appear 2013), *Koningsland en heiligenland. Geografische aspecten van middeleeuws Friesland*, Dissertation, Rijksuniversiteit Groningen.

Roehl, R. (1972) 'Plan and Reality in Medieval Monastic Economy: The Cistercians', *Studies in Medieval and Renaissance History*, 9, pp. 83–113.

Rösener, W. (1982a) 'Zur Wirtschaftstätigkeit der Zisterzienser im Hochmittelalter', *Zeitschrift für Agrargeschichte und Agrarsoziologie*, 30, pp. 117–148.

Rösener, W. (1982b) 'Grangienwirtschaft und Grundbesitzorganisation südwestdeutscher Zisterzienserklöster vom 12. bis 14. Jahrhundert', in: K. Elm, P. Joerissen and H. Roth (eds), *Die Zisterzienser. Ordensleben zwischen Ideal und Wirklichkeit, Ergänzungsband*, Cologne, pp. 137–163.

Southern, R.W. (1970) *Western Society and the Church in the Middle Ages*, Harmondsworth.

Toepfer, M. (1983) *Die Konversen der Zisterzienser. Untersuchungen über ihren Beitrag zur mittelalterlichen Blütes des Ordens*, Berlin (Berliner Historische Studien; 10).

Wiswe, H. (1953) 'Grangien niedersächsischer Zisterzienserklöster. Entstehung und Bewirtschaftung spätmittelalterlicher landwirtschaftlicher Grossbetriebe', *Braunschweigisches Jahrbuch*, 34, pp. 5–134.

Zomer, J. (2009) *De landschapsgeschiedenis van Roderwolde. Een interdisciplinair onderzoek naar de natuurlijke landschapsgenese van een woldgebied in de Kop van Drenthe en de kolonisatie en ontginning in de late Middeleeuwen en de Nieuwe Tijd*, unpublished Masters Thesis, Rijksuniversiteit Groningen.

13 The origins of the Western Scheldt. Environmental transformation, storm surges and human agency in the Flemish coastal plain (1250–1600)

Tim SOENS, University of Antwerp

Abstract

Between the twelfth and the sixteenth centuries AD a small peat river called the 'Honte' turned into the 'Western Scheldt' – the main river mouth of the River Scheldt and one of the largest commercial waterways of North-West Europe. This spectacular environmental transformation underpinned the rise of Antwerp as main maritime gateway of Europe in the sixteenth century, but at the same time caused long-lasting environmental and social disequilibria in adjacent territories. This contribution reconsiders both the chronology and the causes of this transformation, framing the origins of the Western Scheldt in the general evolution of coastal and estuarine wetlands in the North Sea Area. Whereas during the early and high medieval period the estuary was characterised by a gradual drowning of the peat, a gradual increase of embanked territory and overall rather stable environmental conditions, in the fourteenth and fifteenth centuries AD abrupt changes occurred forced by repeated flooding, but eventually caused by increasingly aggressive embankment practices and the pronounced impact of the late medieval crisis in the coastal wetlands.

I. Introduction

The medieval history of the Scheldt Estuary offers a fascinating example of environmental transformation triggered by a combination of natural and anthropogenic forces. The general geological processes steering the evolution of the Scheldt Estuary during the Holocene period are well known. As explained elsewhere in this volume, regional changes in the land-water balance since the beginning of our era are mainly controlled by the compaction and oxidation of soft Holocene clays and peat on the one hand; and the relative sediment budget on the other (Weerts, this volume; Vos *et al.*, 2011). From *c.* 3850 BC to *c.* 500 BC the whole of the present-day Western Scheldt area was gradually filled with peat, profiting from the slowing down of sea-level rise. In the meanwhile, sand from the sea was deposed in front of the beach and formed huge coastal barriers that protected the expanding peatlands from inundation. Around 500 BC, the coastal barrier in front of the Zeeland coast increasingly suffered from a lack of sediment, initiating a process of coastal erosion. The sea started to invade and drown the peatlands, a process which was hugely accelerated by human agency: particularly in the Roman period, the drainage and exploitation of peatlands for agricultural or industrial activities led to considerable compaction and oxidation of the peatlayers, which were subsequently invaded

Landscapes or seascapes? The history of the coastal environment in the North Sea area reconsidered, ed. by Erik THOEN, Guus J. BORGER, Adriaan M.J. DE KRAKER, Tim SOENS, Dries TYS, Lies VERVAET & Henk J.T. WEERTS, Turnhout, 2013 (*CORN Publication Series*, 13), pp. 287–312.

BREPOLS PUBLISHERS DOI: 10.1484/M.CORN.1.101557

by the sea and covered by marine deposits (Vos and Van Heeringen, 1997; Kiden, 2006; Lascaris and de Kraker, this volume). By the fourth century AD, most of the peatlands near the mouth of the River Scheldt had turned into coastal marshlands, criss-crossed with tidal channels. Uncovered peat only remained in the highest, more inland parts of the estuary (Jongepier, Soens *et al.*, 2011)

The drowning of the peatlands in the Scheldt area, however, is not the end of the story. In the medieval period another important transformation took place, the chronology and causes of which remain highly unclear: the estuary changed from a large floodplain, transected by many minor tidal channels, to a single channel, and the mouth of the river also shifted southwards. Around 1000 AD, the River Scheldt still discharged the majority of its water through a branch of the estuary north of the Zeeland isles of Walcheren and Beveland. This northern branch is nowadays called the Eastern Scheldt, but in medieval times was simply known as 'the Scheldt', separating 'Zeeland west of the Scheldt' and Zeeland east of the Scheldt'. Maritime traffic from both the North Sea and the German world followed this road towards Flanders and Brabant, converging near the town of Yerseke where custom duties on this traffic were levied by the count of Holland (Unger, 1939). The Western Scheldt, situated between the Zeeland islands of Walcheren and Zuid-Beveland to the north and the Flemish mainland to the south, through which the River Scheldt discharges today, did not then exist. Medieval sources only mention a secondary tidal inlet, 'the Sincfal', near what would become the mouth of the Western Scheldt on the one hand, and a small peat river, 'the Honte', discharging into the River Scheldt to the north of Antwerp on the other. The connection between Honte and Sincfal and their transformation into the Western Scheldt remain one of the great mysteries in the environmental history of medieval Flanders, both in terms of chronology and in underlying causes. In historiography, the 'Honte connection' is alternatively dated to between the eight and the sixteenth centuries AD, and blamed on sea-level rises, increases in storm activity, peat subsidence, embankments, and on growth or decline in the coastal economy (for a recent overview see Lases and de Kraker 2009 and 2010; Leenders, 2010). What is certain, is that the birth of the Western Scheldt completely reshuffled the economic and political geography of the Scheldt region to such an extent that it gave rise to long lasting border-disputes between the three neighbouring principalities of Holland/Zeeland, Brabant and Flanders. These disputes are known as the 'Scheldt Question' (Bindoff, 1945).

This contribution does not pretend to solve all the debates surrounding the origins of the Western Scheldt. Instead, it uses the medieval history of the Western Scheldt to tackle two of the central issues in this volume: the interaction between natural and anthropogenic forces in shaping the maritime and estuarine landscape; and the respective role of land-side and sea-side processes in steering this development. As we will argue, the medieval breakthrough of the Western Scheldt was above all an example of large-scale estuarine transformation, *unintentionally* produced by human intervention in the coastal wetlands. In this process, two main stages can be discerned: the progressive embankment that started in the eleventh and twelfth century slowly changed the size and the characteristics of the tidal floodplain, which however remained shallow and criss-crossed with mainly small tidal channels. Only in a second stage, in the later Middle Ages, a more radical transformation occurred, forged by an increased number of flood catastrophes, the impact of which was accelerated by the economic crisis of the later Middle Ages and more 'aggressive' embankment practices, which created highly unstable environmental

Figure 13.1 Map of the present-day Scheldt Estuary with the main places mentioned in the text
(1: Bruges; 2: Wulpen; 3: Groede; 4: Braakman; 5: Ghent; 6: Antwerp; 7: Saeftinghe; 8: Rilland; 9: Yerseke)

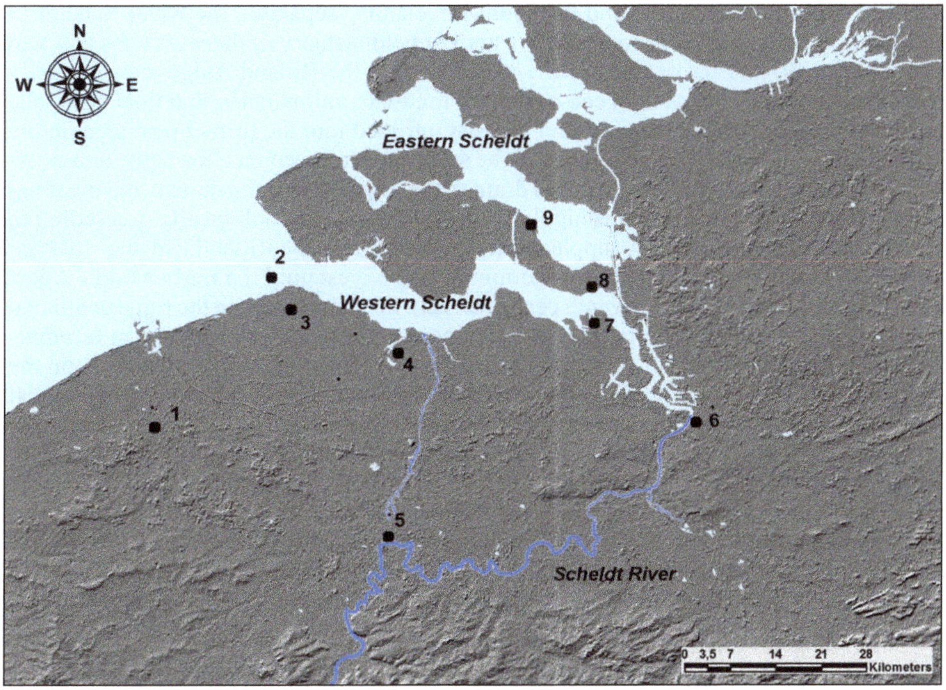

conditions. As such the creation of the Western Scheldt is just one example of the kind of rapid environmental change that occurred all over the North Sea Area in the later Middle Ages: a result of deteriorating environmental and economic conditions. On the one hand this was the ultimate consequence of the embankment process itself, a process that had started centuries before. On the other hand, by the fourteenth century coastal communities all over the North Sea Area had become economically and politically vulnerable, which is exemplified by the numerous devastating revolts in these areas (Van Bavel, 2010; Thoen and Soens, forthcoming) which significantly hampered an adequate reaction to new environmental challenges.

II. The 'Honte-connection' and the medieval transformation of the Scheldt Estuary: a *blurred* chronology

Thanks to excellent studies by *inter alia* Vos and Van Heeringen (1997) and Kiden (2006) we are well informed on the long-term evolution of the River Scheldt and its estuary until the Early Middle Ages. For the period after *c.* 1550 AD, this is complemented by detailed historical reconstructions (de Kraker, 2002; de Kraker and Aalbersberg, 2005). However, on the crucial period between *c.* 800 AD and *c.* 1550 AD, when the present-day Scheldt Estuary originated, many uncertainties still persist. According to most authors, a Pleistocene clay ridge covered by peat bogs – the so called Rilland ridge, stretching from Saeftinghe in Flanders to Rilland on South-Beveland – separated the River Scheldt in the east from what would become the Western Scheldt estuary in the west (Vos and Van Heeringen, 1997). After *c.* 800 AD the raised bog on the Rilland Ridge was gradually drowned and replaced by a tidal eco-system of mudflats, saltmarshes, and tidal channels. Tidal channels both from the west and the east intruded into the former peat area on the Rilland Ridge. When they eventually met, the so-called Honte-connection came into existence, and the water of the Scheldt could drain to the west. The establishment and meaning of this Honte-connection remains highly difficult to frame chronologically. According to Vos and Van Heeringen (1997, supplemented by Vos, 2000) the Rilland Ridge obstructed the water from the River Scheldt from draining to the west until the Early Middle Ages/ Merovingian period (sixth to seventh centuries AD). Peat formation on the ridge continued at least until the eighth century AD, which apparently provides us with a firm *terminus post quem* for the creation of the 'Honte Connection'. However, in Ellewoutsdijk, on the south coast of Borsele, only 20 to 30 km west of the Rilland ridge, recent archaeological investigations revealed marine influence from the Roman period onwards (de Clercq and van Dierendonck, 2008: 8–9). Sea water might have reached Ellewoutsdijk either from the old Scheldt mouth through a channel in between the islands of Borsele and Walcheren (the Zwake), or perhaps from some kind of Roman precursor to the Western Scheldt estuary. A strong marine influence in Ellewoutsdijk during the Roman period might suggest an earlier date for the drowning of the Rilland Ridge peat bog (or alternatively, marine influence must have diminished again temporarily after the end of the Roman period).

As *terminus ante quem* for the Honte-connection, the twelfth century AD is usually mentioned. By that time, historical sources have become more abundant, and we see Flemish abbeys active in reclaiming *werplant* – newly deposited land – all along the Flemish coast of the Honte, from Ossenisse in the west to Saeftinghe in the east, which seem to confirm that by that time, the Rilland ridge had disappeared, and an integrated tidal system had come into place[1]. However, well beyond the twelfth century AD, there was no single tidal channel that could unequivocally be considered as the boundary between Flanders in the south and the Zeeland island of Walcheren in the North. In 1167, for example, a treaty between the counts of Holland and Flanders placed the boundary between the smaller islands of Koezand (Zeeland) and Wulpen (Flanders), separated from each other by the *Heidensee* (Verhulst, 1995: 31–32). Because Koezand was situated close to the Flemish mainland, and would later on be connected to Wulpen (see below), this *Heidensee* cannot then have been very large. Even *c.* 1380, the counts of Holland (Albert of Bavaria)

[1] Vos and Van Heeringen, 1997: 76, the latter argument based on Gottschalk 1971: 120 and Brand 1983. Most explicit is the grant of *'werplant sive sant'* (newly deposited land or sand) between Ossenisse and Hontenisse, by the Flemish count first to St. Peter's abbey in Ghent and its dependency Cosforth in 1183; subsequently transferred to the Cistercian abbey Our Lady of the Dunes in 1200. In the latter charter, the land is granted '*sicut extenditur usque ad extremum recessum maris quod dicitur grundebbe*' – as far as it emerges during extreme low tide.

Figure 13.2 The Honte-connection on the so-called 'Scheldt map' (sixteenth-century copy of a map of 1494, Antwerp, City Archives).
The Honte or Western Scheldt is located between the castle of Saeftinghe and the island of Rilland/Bath below left, and the Eastern Scheldt between South-Beveland and Tholen above left. On this late fifteenth-century Scheldt-map, both the Honte and the Eastern Scheldt are depicted as broad estuarine branches frequented by a multitude of ships. The navigability of the Honte has probably been exaggerated as the map originated from a lawsuit concerning custom rights. In the State Archives of Brussels, an older version of the Scheldt-map has been preserved (*c.* 1468) albeit rather damaged

and Flanders (Louis of Male), still quarrelled on the ownership of a *scor* (salt marsh) between Biervliet and Zeeland (ed. Meijers, 1940: 86). Further east, the *Honte* was not a boundary, which is reflected in the fact that the Flemish *Vier Ambachten* south of the Honte were part of the northern bishopric of Utrecht. It was only in the later fourteenth century AD[2] that sea-faring merchants began to explore the possibilities of the Honte as maritime link between the North Sea and Antwerp, introducing a number of long lasting juridical discussions between the neighbouring principalities of Flanders, Brabant, and Holland/Zeeland on the question of who owned this 'newly created' waterway (Bindoff, 1945; Doeleman, 1974; Wyffels, 2002). And even then the Honte connection remained highly problematic for larger – Hanseatic – ships well into the sixteenth century, not least because of several sand flats and bars which persisted between Flanders and Zeeland (Lases and de Kraker, 2010: 72). Between 1530 and 1532, the partial drowning of South Beveland created a huge intertidal area between the Honte and the (Eastern) Scheldt, criss-crossed by several mostly shallow tidal channels. This left the sea-faring community in great uncertainty on exactly which route to Antwerp was to be preferred. In the next decades the Eastern Scheldt gradually silted up and became less accessible to large ships. At last, the Western Scheldt unequivocally became the major branch of the Scheldt Estuary. But even as late as 1572 the well-known dike expert Andries Vierlingh still considered the Eastern Scheldt as the 'usual' waterway for ships descending from Holland to Antwerp (Dekker and Baetens, 2010: 200).

From this overview, it becomes clear that the precise chronology of the estuarine evolution remains fairly uncertain. Kiden (2006: 288) even doubts that such a chronology will ever be possible, as most relevant sediment layers have either been eroded or dredged. The main uncertainties have recently been inventoried by Lases and de Kraker (2009 and 2010) and Leenders (2010). These are related to the increase of the tidal range (how reliable are pre-1800 figures like those cited by Guicciardini in the sixteenth century?); the gradual salinisation of the estuary (did salt peat exist in Saeftinghe in the fifteenth century AD and, if yes, what was its origin?); the path of the flood-water during storm surges (using the Eastern or the Honte-route?) and the impact of embankment (limiting tidal volumes but increasing high water levels?) and sea-level rise (limited in the last two millennia, at least compared to the compaction and oxidation of the peat), and last but not least, the end of the peat growth. This last point is important, because ^{14}C dating of the top of the peat often provides the only reliable way of dating an increase in marine or fluvial influence at a certain spot. In some places peat growth seems to have ended in the Roman period, but in other locations – near the Rilland-ridge (Hinkele/Waarde) and on Belgian territory (Deurganckdoel/Doel and Oosterweel) – this was much later, in the sixth to eighth centuries AD (Gelorini *et al.*, 2006: 246; Leenders, 2010: 68). Furthermore, the end of peat growth was often followed by several centuries of very low or 'non'-deposition until the Later Middle Ages, when renewed deposition occurred (in the thirteenth/fourteenth centuries in Doel). This intriguing gap between the end of the peat growth in the Early Middle Ages and the late medieval sediments, is mainly explained by the very limited tidal influence in the Belgian part of the estuary until the eleventh and twelfth centuries AD (Kiden, 1991: 294 and 2006: 88). On the other hand, it is also tempting to see the influence of human activities, either draining the peat and hence causing oxidation of the top-layer or making it suited to extraction, or protecting the peat regions from inundation (and the deposition of sediment) by building dikes.

---

[2] One of the first indications is a prohibition in 1387–88 by the bailiff of Zeeland for herring boats to take the Honte and not the Scheldt, (probably) when heading to Antwerp from the north (ed. Meijers, 1940: 192).

Perhaps the conclusion of this short historiographical overview is, that research has probably focused a bit too much on the creation of the 'initial' Honte-connection and the drowning of the Rilland Ridge. It seems that for more than five centuries – between the seventh/eight and the thirteenth/fourteenth century AD, the Scheldt Estuary presented itself as a relatively 'stable' environment, with a lot of smaller tidal channels invading a large and shallow flood plain, confined by a decreasing area of peat marshes, which were gradually drowned or reclaimed by men. For the reasons mentioned above, the precise configuration of these channels will never be retraced, but it is plausible that indeed some of them connected the 'eastern' Scheldt Estuary to the estuary that separated Flanders from Walcheren and South-Beveland in the west. In the later Middle Ages however, these relatively 'stable' conditions were dramatically altered, which can be noticed by looking in more detail at the late medieval patterns of storm flooding.

III. Patterns of storm flooding in the late medieval Western Scheldt area

For the late medieval Western Scheldt area, a tremendous amount of documentary evidence on embankments, floods, and re-embankments can be found in the regional landscape studies by Gottschalk (1955–58; 1977–79; 1984), Dekker (1971), and more recently de Kraker (1997).

Nevertheless it remains difficult to put them together, as much of the historical landscape in question has since disappeared, either through flooding and renewed sedimentation or, more recently, through the expansion of harbour activities. Furthermore, none of the above mentioned regional studies covers the whole of the estuary, making it very difficult to provide cartographical reconstructions of the Western Scheldt at different stages of its late medieval development. Apart from the above mentioned palaeogeographic maps by Vos and Van Heeringen, K.J. Brand (1983) attempted a fairly hypothetical reconstruction covering the whole period from *c.*1000 AD to present. More reliable reconstructions have recently been made available for the period after 1550 AD (de Kraker and Aalbersberg, 2005). For the period before 1550 AD, any reconstruction of the medieval landscape will remain tentative. On the micro-topographical level, the possibilities for reconstructing the medieval landscape underneath the younger marine deposits have progressed enormously during the last decade, mainly thanks to the advances in geophysical survey methods (digital elevation models, resistivity imaging, electromagnetic induction etc.), and close interdisciplinary collaboration (Vanslembrouck, Lehouck and Thoen, 2005). However, such detailed reconstructions of late medieval landscapes in the Western Scheldt Estuary are still very scarce, and the results of individual area-studies are often difficult to extrapolate to the whole of the estuary.

Nevertheless, even a superficial plotting of the available documentary evidence on flood damage during the major medieval storm surges in GIS, can uncover a lot about the development of the Western Scheldt. When plotting the impact of catastrophic fourteenth-century storms, like those of 1334 and 1375, it becomes clear that the Western Scheldt as such did not yet present a unified pattern of flood risk. A catastrophic storm like the one of 28 November 1334 caused huge damage along the North Sea Coast (including the island of Wulpen and the South Coast of Walcheren), as well as in the *Waasland* in the east of

Figure 13.3 Documentary evidence of flood damage after the storms of 28 November 1334 (first map) and 8–10 October 1375 (second map) (based on Gottschalk 1955–58; 1977–79 and 1984 and Dekker 1971)

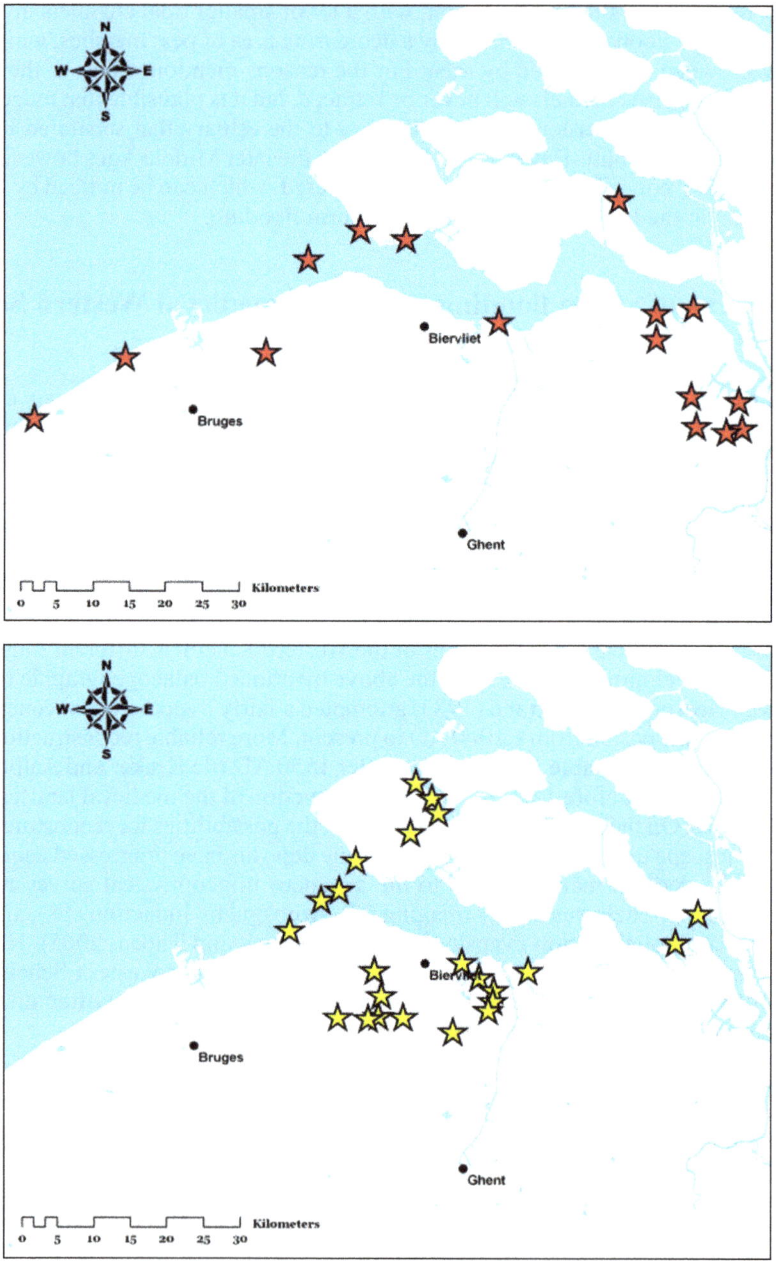

Figure 13.4 Documentary evidence of flood damage after the storms of 10 April 1446 (first map) and 26–27 September 1509 (second map) (based on Gottschalk 1955–58; 1977–79 and 1984; Dekker 1971 and de Kraker 1997)

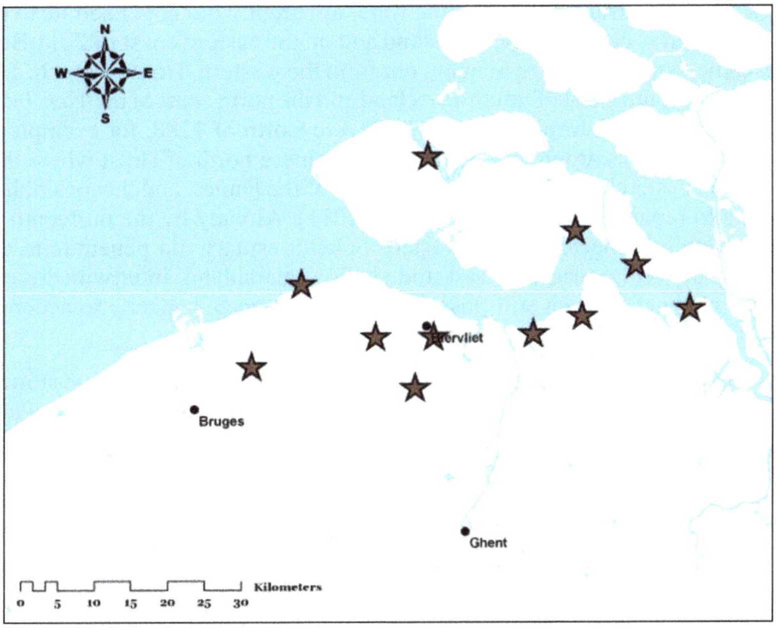

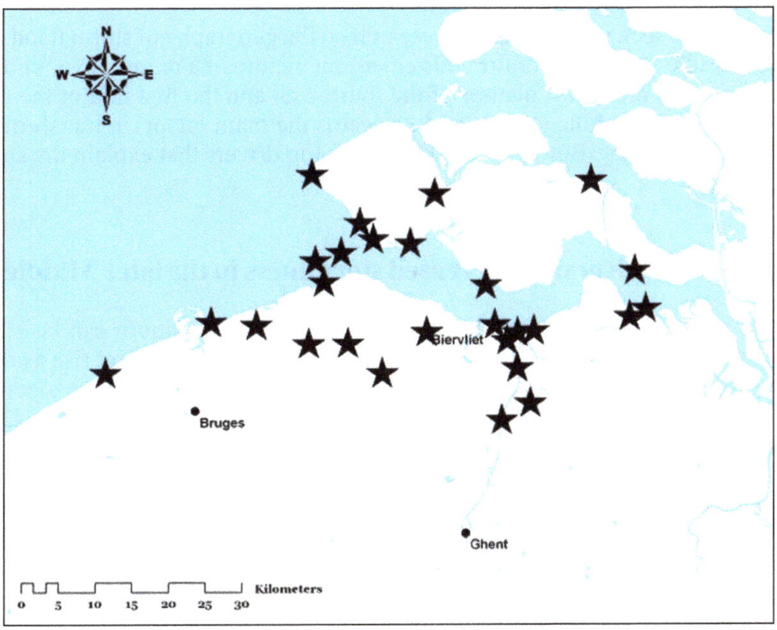

Flanders, but little or no damage is mentioned in the central part of what would become the Western Scheldt Area. In 1375 and 1404 storms again hit the coastal islands, the east coast of Walcheren and the region of Biervliet in the North of Flanders, where a huge tidal inlet (the Braakman) was created, leading to the destruction of tens of villages. In all three cases, the destructive force of the water apparently did not penetrate to the east. The damage that was caused in the Waasland and on the eastern coast of Zuid-Beveland, obviously came from the Eastern Scheldt, not from the western 'Honte' branch. This does not mean that the south coast of South-Beveland and the north coast of the *Vier Ambachten* were completely free of storm damage. The severe storm of 1288, for example, caused huge flooding problems around the village of Ossenisse north of Hulst where the count of Flanders, the Cistercian abbey called Our Lady of the Dunes, and the local inhabitants needed years to repair the dike bursts (Soens, 2011). Already by the thirteenth century AD tidal channels connected to the Western Scheldt estuary did penetrate to the east, but they probably still ended in a vast and shallow marshland, intertwined with many smaller tidal channels, which still possessed enough storage capacity to accommodate most floods without causing huge damage.

In the course of the fifteenth century this would change. One of the first storm surges to show a different pattern of flood damage was the storm that occurred on Palm Sunday, 10 April 1446. For the first time we see a pattern of damage that was seemingly randomly distributed along both shores of the Western Scheldt estuary, in other words, where the flood water seemed to have followed more or less the present-day course of the Western Scheldt Estuary. This was repeated in most sixteenth century floods (1509, 1511, 1530, 1532, 1552 etc.), which concurrently wreaked havoc in the western part of the Scheldt Estuary (Walcheren and the western part of Zeeland Flanders) and in the east (South-Beveland and the Vier Ambachten), suggesting a stronger connection between both parts of the estuary.

It remains to be seen whether further research on the geography of storm flood damage in the later Middle Ages, can confirm this emerging picture of a progressive 'unification' of the flood pattern in the last quarter of the fourteenth and the first half of the fifteenth centuries. In the meanwhile, storm flooding seems the main 'actor' in transforming the Western Scheldt Estuary. But what are the underlying drivers that explain the agency of storm flooding?

IV. Exogenous drivers: an increased storminess in the later Middle Ages?

First of all, we have to question whether the changing flood pattern can be related to changes in the physical environment, with climate change and sea level rise as the main driving forces. In the historiographical debate on the 'Scheldt Question', exogenous environmental processes were already identified as prime movers in the post-WW II period, when geologists tried to identify periods of increasing and decreasing sea-influence related to periods of rising sea-level – so-called transgressions and regressions – during the first and second Millennium AD. In this view, the late medieval inundations, and hence the origins of the Western Scheldt, could be related to a 'late medieval transgression period'. However, at least since the 1980s, a considerable variation in mean sea level in the North Sea Area after 1000 AD is no longer accepted (already in Verhulst and Gottschalk, 1980;

Vos and Van Heeringen, 1997 and many contributions in this volume). Nevertheless, the rejection of the 'transgression' theory does not imply that natural or exogenous environmental processes might be completely excluded as driving forces behind the formation of the Western Scheldt Estuary. Thanks to the recent 'global warming' debate, the idea of climate as an independent historical agent is again gaining importance (e.g. Brown, 2001; Campbell, 2010a). The late medieval transformation of the Western Scheldt Estuary more or less chronologically corresponds with the transition of the Medieval Warm Period into the Little Ice Age. This not only implies a drop in mean temperatures (particularly from 1250 onwards and again in the fifteenth century), but also an increased climatic variability or unsustainability – measured in terms of variation in the mean temperature, in crop yields and in tree growth – in the periods 1275–1350 and 1400–1475 (Campbell, 2010b). With regard to coastal and estuarine change, this could have provoked an increased storm activity, a late medieval 'Age of Storms' as it is sometimes labelled in historiography (e.g. Bailey, 1991).

In the southern part of the North Sea area, historical references to inundations following storm surges were particularly numerous from the second half of the fourteenth century to the end of the sixteenth century (Gottschalk, 1971–7; Buisman, 1995–2006; Zeiler and Vos, 2008). Remarkably, in England, an 'Age of Great Storms' is usually situated somewhat earlier than in the Low Countries: notably between 1250 and 1320, although with some prolongations into the late fourteenth and early fifteenth centuries (e.g. Gross and Butcher, 1995; Galloway and Potts, 2007). Apart from a certain bias in the historical source material which underestimates the number of storm surges before 1300, there is yet another problem with this increased storm activity: medieval sources usually do not inform us on the actual occurrence of a storm, but only on the damage that it provoked. By contrast to the storm itself, the impact of the storm is largely determined by human conditions: technological shortcomings, bad maintenance of infrastructure, economic, political, or military disturbances that turned a storm surge into a disastrous inundation (de Kraker, 2006). From the late fourteenth century onwards there is increasing historical evidence for a diminished willingness to invest in dike maintenance (Soens, 2011a; Galloway, 2009), which might have led to an inflation of the number of 'catastrophic storm surges', particularly in those regions where economic and environmental conditions had already deteriorated (see below). Furthermore, even if we admit an increased storm activity in some decades – the evidence all over the North Sea area is quite convincing – this does not explain the important regional variations within different parts of the North Sea area, and even within the Scheldt Estuary itself, as we will discuss below.

V. Environmental and economic problems as triggers of flooding

When looking in more detail at the areas that were permanently flooded through the expansion of the Western Scheldt Estuary in the later Middle Ages, we see that at the moment of their inundation many of them had become particularly vulnerable to flooding due to environmental or economic problems. First, the largest late medieval extensions of the Western Scheldt Estuary were realised in areas where important peat-reclamations took place in the medieval period, the Braakman area and the *moeren* (moorlands) of

Hulst, Beveren, and Saeftinghe on the Flemish left bank of the estuary, and the so-called *'Wilde Moeren'* on South-Beveland on the right bank. Local exploitation of the peat for fuel had already started before 1000AD. With the boom of the Flemish and Brabantine urban industries in the eleventh and twelfth centuries and the parallel disappearance of forests in this area, peat had been exploited very intensively (Soens and Thoen, 2009; Jongepier *et al.*, 2011). For the peatlands near Beveren, Augustyn calculated an average annual production of *c.*5000 'lasts' in the fourteenth century – one last being 10,000 blocks and corresponding to a volume of 22.75 m^3 and a dry weight of 5100 kg, which gives an average total production of 25,500 tonnes of peat (dry weight) (Augustyn, 1992, II, 492 ff). As the peat-industry was labour intensive and perfectly fitted for seasonal employment, the peat areas attracted a dense population of smallholders who combined farming activities with peat-digging. By the middle of the fourteenth century peat exploitation had virtually ceased in both the Braakman region and the Wilde Moeren of South-Beveland. According to Dekker, the *Wilde Moeren* had turned into a vast wasteland that was useless except for fishing and hunting (Dekker, 1996: 62). In the Braakman region too, peat reclamation was at a maximum in the thirteenth century before declining in the fourteenth century (Soens and Thoen, 2009). Only in the north-eastern part of Flanders (Moerbeke-Wachtebeke, Hulst, Saeftinghe and Beveren) did peat exploitation continue well into the sixteenth century, but the peak production of the fourteenth century would never be reached again (Augustyn, *op.cit.;* Jongepier *et al.*, 2011). In the beginning of the fifteenth century the counts of Flanders, who owned a major part of the peatlands in the Beveren area, were still willing to invest considerable amounts of money in the maintenance of the water control system which was necessary to drain and protect the peatlands. As their income from the remaining peatlands diminished, so did their willingness to invest in the maintenance of drainage and flood protection. The latter became increasingly challenging as most peat lands had become very vulnerable to flooding after centuries of peat extraction and subsidence of the peat layers. Along the Western Scheldt, the peat had developed continuously on top of the Pleistocene surface (situated between 0 and -8 m TAW) from the sixth or fourth millennium BP until the early Middle Ages. In this period sea level had risen slowly but steadily, so the former peat districts had to catch up with at least 1–2 m of sea level rise, and a huge accommodation space for renewed sediment deposition was created. This process does not explain the flooding itself (improved sea-walls and drainage systems could still guarantee flood protection), but once the dike was broken vast spaces of peatlands were rapidly inundated and proved very hard to recover, at least until sedimentation had filled the accommodation space that been created due to the oxidation, compaction and extraction of the peat in the previous centuries.

Apart from the peatlands, the floods of 1375 and 1404 also dismantled the coastal barrier islands of Wulpen, Koezand and Waterdunen, which once stretched from the Flemish coast all the way to Zeeland, as indicated by the so-called Dampierre map (a sixteenth-century forgery pretending to depict Flanders *c.* 1300) and also believed by most contemporaries in the later Middle Ages (Gottschalk, 1958: 130). The original extent of these islands is highly enigmatic, but it seems that the largest part of them – including three out of four parishes on Wulpen – had been swept away by the sea in 1375 and 1404. The rest of these islands would be progressively abandoned during the course of the fifteenth century and this process is reflected in the fief registers of the count of Flanders which

Figure 13.5 Jan de Hervy's view of the Scheldt Estuary (*c.* 1501) (source: Groeningemuseum inv. nr. 1382)

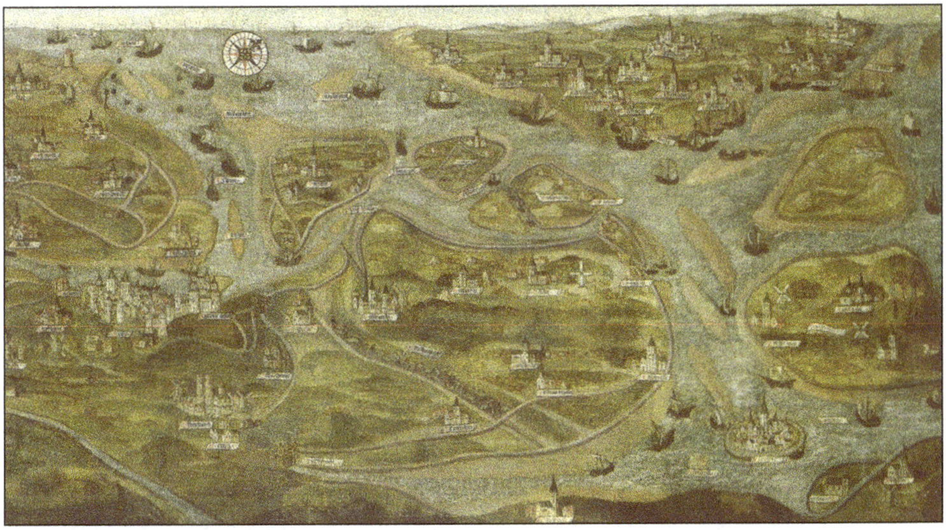

mention forty fiefs on Wulpen *c.* 1300, eight fiefs in 1435 and 1468, and only five in 1501 (Gottschalk 1958: 34 and 138). The remaining church of Wulpen – St Lambert – was probably abandoned in the aftermath of the 1516 storm surge (idem: 140), although contemporary cartography indicates that some embankments on Wulpen persisted even at the very end of the sixteenth century. Even after the flooding, parts of the islands subsisted as sand banks or mudflats, leased out by their owners, which is very well documented for the Comtesse Hospital in Lille, owner of several hundreds of hectares of (inundated) land on the former island of Koezand until 1544 (Gottschalk 1958: 137–139; 166–168; 182). As mentioned above, the underlying causes of the disappearance of coastal barrier islands like Wulpen are related to a lack of sediment supply from the sea, which already was a 'problem' before the beginning of our era (see above). The deterioration of coastal barriers accelerated in the fourteenth and fifteenth centuries can be related, at least partly, to human agency: over-exploitation of the natural dune barriers on the islands through the introduction of rabbits and an insufficient financial base on the islands either to fix the dune sands or to construct alternative barriers (Augustyn, 1995, and more nuanced, Van Dam, 2010)

The late medieval expansion of the estuary however, was not solely realised at the expense of vulnerable ecosystems like peat or dune lands. Huge land losses also occurred on silted-up marshlands, most of which had been occupied continuously since the High Medieval Period. On the rural economy in this part of the estuary, we are particularly well informed for Flanders on the left bank on the estuary. The drainage and embankment of the Flemish Polder Area took place at an early date. The largest part of the polder area was reclaimed before the twelfth century, not – as has long-time been thought – by Cistercian abbeys, but mainly by free peasant communities in collaboration with the counts of Flanders (see also Tys, this volume). The strong position of the count of Flanders in this area also blocked the rise of independent feudal lords, and guaranteed in a certain sense the high degree of personal freedom enjoyed by the population of the Polder area. Recent studies by Soens (2009); Thoen and Soens (2009) and Tys (2003) made clear that from the twelfth to the fourteenth centuries peasant smallholdings were predominant in the Flemish Polders. Co-existing with a strict minority of big or very big farms, both landownership and landholding were extremely dispersed, with a majority of holdings smaller than 5 hectares, and in some parts of the coastal plain even smaller than 2 hectares. These small farms practised a mixed agriculture, combining arable farming and cattle breeding with important para-agrarian and non-agrarian seasonal activities: apart from the peat-exploitation mentioned above, many coastal peasants devoted part of their labour and land to the cultivation of tinctorial plants (like madder, which produces a red pigment). Salt making industries flourished in the fourteenth and fifteenth centuries, and there is also is evidence for a rural textile (woollen) industry, at least until 1400. Finally, fishing villages or hamlets were scattered all over the dunes. For one of these villages – Raversyde – archaeological evidence made clear that many inhabitants combined small-scale farming with seasonal fishing (Tys, 2006).

This coastal peasant economy proved extremely vulnerable to the economic and demographic crisis of the later Middle Ages, in striking contrast with the rest of Flanders (Soens, 2006; Soens and Thoen, 2008). Between the fourteenth and the sixteenth centuries, many of the peasant smallholders living in the area disappeared. Their land was acquired by larger farmers, and increasingly by absentee landowners, many of them living in towns. By 1550, 60 to 70% of the polder land in the wide surroundings of Bruges was owned by urban dwellers and urban institutions (Soens, 2009) The decline of peasant smallholding was a general tendency all over coastal Flanders, but it was perhaps most pronounced along the Western Scheldt Estuary where the environmental stress was the highest. To give but one example, in the Old Polder of Groede near the mouth of the Western Scheldt, the number of landowners declined from 219 in 1415, to 135 in 1552, and then to fifty-four in 1617 (see Figure 13.6).

The concentration of landownership was a gradual process but it accelerated in the second half of the fifteenth century and in the first half of the sixteenth century. In the latter period, the majority of farms became larger mid-sized units. Social polarisation created a class of full-time resident farm labourers who for most of their lives worked on a large farm. Protoindustrial or non-agricultural income strategies such as fishing, peat digging, and perhaps also salt making, which were still widespread in the fourteenth century, subsequently lost importance. Moreover, as farms grew larger and became more commercially oriented, non-agrarian subsidiaries interested them less. More than before, they turned to breeding cattle, a tendency reflected in growing consumption of dairy products and meat – especially beef – in this area (Soens and Thoen, 2010).

This restructuring of the coastal economy in turn had a profound impact on the investments in the water control system, which can be reconstructed from the late thirteenth century onwards in the neighbourhood of Bruges and from the second half of the

Figure 13.6 Landownership in the Groede Polder near the mouth of the Western Scheldt Estuary, 1415–1617: concentration of land (in hectare) per group of landowners (source: Soens, 2009: 78; Lamon, 2008)

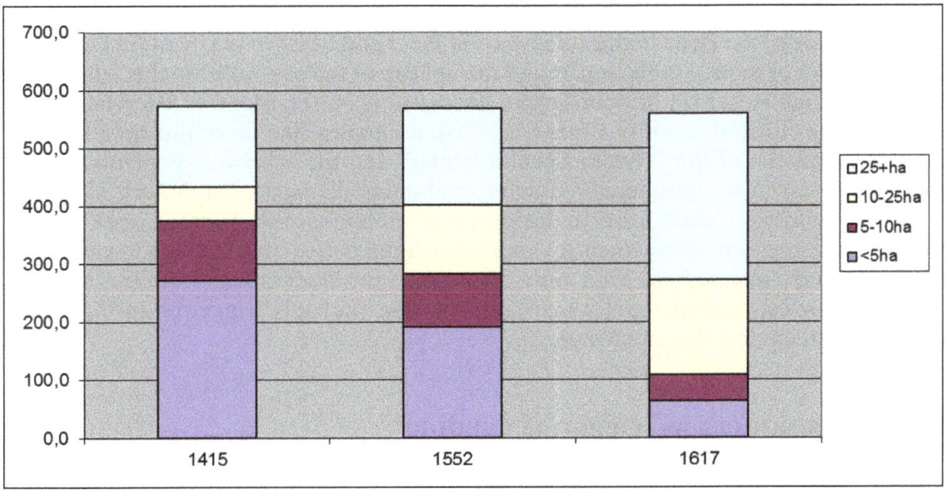

fourteenth century along the Western Scheldt Estuary (Soens, 2011a). For different test-regions both along the Flemish North Sea Coast and the Western Scheldt Estuary (the region of Oostburg), we could demonstrate a period of high investment from *c.*1375 to *c.*1425, followed by a significant drop in investment levels, both in nominal and in real terms. Only from 1530 onwards did investments rise again, with new peak levels in the 1560s and the 1570s. In the Oude Yevene district near Oostburg, for instance, the nominal investment level of the 1380s –70.8 Flemish *groten* per hectare on average – was only reached again in the 1560s. By that time, however, coin debasements and price inflation (the so-called sixteenth-century 'price revolution') had substantially diminished the real value of these investments. In the 1350s the taxes levied for the maintenance of the water control system in the Blankenbergse district west of Bruges typically yielded the equivalent of more than 40,000 days of unskilled labour, whereas after 1450 they seldom exceeded 10,000 days (increasing to 20,000 days after 1540) (Soens, 2009: 123–124). Apparently the enthusiasm to invest in the maintenance of the sea-walls and the drainage system of the coastal plain dropped in the course of the fifteenth century, which significantly amplified the impact of storm surges. Defaults in dike maintenance are associated with most catastrophic dike breaches throughout history, and there is reason to believe that the upkeep of dikes and sea-walls was increasingly problematic from the fifteenth century onwards, parallel with the transition of the coastal economy from a peasant smallholding economy to a commercialised society dominated by absentee landowners. The latter faced a significant reduction of their income from leased out properties in the coastal plain, as typical lease prices per hectare declined from more than 400 litres of wheat per hectare in the last quarter of the fourteenth century to about 150 litres in the first half of the sixteenth century, which did not favour high investment levels in the water control system (Soens 2009: 138). In such a bad investment climate, landowners, including major ecclesiastical landowners like abbeys often preferred to abandon their properties most threatened by flooding. We can see this, for instance, around the city of Biervliet on the left bank of the Western Scheldt Estuary. Biervliet became entirely surrounded by water after the flooding of the Braakman area in 1375. The two oldest abbeys of Flanders, the Benedictine abbeys of St Peter and St Bavo in Ghent first invested huge amounts of money in the repair and the upkeep of the sea-walls on this island, but eventually tried to get rid of their landed possessions on the island of Biervliet in the course of the fifteenth century (Soens, 2011b). Examples like these illustrate that the gradual expansion of the Western Scheldt Estuary and the economic transition of the coastal region can be considered mutually reinforcing elements. On the one hand, the increased flood risk clearly fostered the economic problems of the peasant-smallholders and their progressive replacement by capitalist farmers and absentee landowners. On the other hand – and perhaps even more important – the latter groups were less and less inclined to invest in proper water management, thus probably triggering further flood problems and environmental change.

VI. Embankment as trigger of flooding

In some areas, there clearly was a problem of underinvestment in flood protection by the later Middle Ages, which triggered flooding, and estuarine transformation. However, this was not due to a general shortage of capital or an unwillingness to invest. In the very same periods when investments in the old lands were dropping and flood problems increased, massive amounts of money were invested in water management projects, although not in the improvement of the existing drainage systems or the repair of endangered sea-walls. In the fifteenth and sixteenth centuries new *polders* (embankments) were started almost every year, often in areas that had been flooded in the past, but also on newly silted-up saltmarshes in the estuary (Gottschalk, 1955–58; 1984; de Kraker, 1997). As a result, total capital input in dike maintenance in the estuary remained high. For some elite groups, investing in this kind of embankment project apparently was more attractive than spending money on existing dikes elsewhere, both for economic and extra-economic reasons. On the one hand, re-embankment permitted them to create a new landscape perfectly adapted to the needs of a large scale commercial agriculture. In the re-embanked *polders*, infrastructures, settlement patterns, and field systems had been planned in a rational and rectilinear way, so much different from the scattered and chaotic organisation of the medieval landscape (see Vanslembrouck, Thoen and Lehouck, 2005). Furthermore, the re-embanked lands were reputed to be extremely fertile, as they were silted up with marine clay. In reality, this fertility was highly variable from one *polder* to another, but the prospect of high yields was undoubtedly attractive for investors. On the other hand, the new embankments also generated social prestige for investors with lots of capital but a lack of ancestors. There are many fifteenth- and sixteenth-century examples of investors, part of them belonging to the urban bourgeoisie of Bruges, Ghent, or Antwerp, others officials of the Burgundian-Habsburg state, who grasped the opportunity of embankments to create estates and lordships, to build churches and villages, and to obtain privileges and titles, as necessary complements to their steep rise in society. On the Flemish west bank of the estuary this was, for instance, the case in the Braakman area, where small cities like

Table 13.1 Embankments and land losses on the island of South-Beveland (1250–1570, in *'gemeten'* of *c.* 0,4 hectare) (Based on: Dekker, 1971: 312–313)

Period	1250-99		1300-49		1350-99		1400-49		1450-99		1500-70	
	gained	lost	gained	lost	gained	lost	gained	lost	gained	lost	gained	lost
Zwake	485		3146		2281	140	861		1230	416	1555	835
West-Coast	1204		339		1000	950	2105		294	294	950	
South-Coast			513		1539		1625		56	619	1298	385
South-East	600		1526		1027		2292		1830			7275
South-West	1657		674		456	330	277	26		117		634
Other	1100		25		1008		886	200	98	615	20	1176
	5046	0	6223	0	7311	1420	8046	226	3508	2061	3823	10305

Watervliet and *Philippine* originated in this kind of embankment and land-transformation project (Soens, 2009: 226–233). Particularly from the late fifteenth century onwards, these private embankment projects were strongly encouraged by the central government of the Habsburg Low Countries who granted tax exemptions – often including exemption from tithes – to the investors. This stood in sharp contrast with the absence of any structural policy let alone financial aid with regard to the flooding problems in the 'old lands' (de Kraker, 1997; Soens, 2007).

On the Zeeland-Isles on the right bank of the Western Scheldt Estuary the same dynamics could be found. Compared to the Flemish west bank, the Zeeland Isles of Walcheren and South-Beveland were much less urbanised. The overlordship of these islands had long time been contested between the count of Flanders and the count of Holland. Eventually the latter had prevailed, but the real power on the islands long time remained with the *ambachtsheren* (local nobility). From the thirteenth to the fifteenth centuries, these *ambachtsheren* engaged in a real '*polder* contest', trying to embank as much land as possible as quickly as possible: on the Zeeland island of South-Beveland alone, 165 new *polders* were created in the later Middle Ages. In the first half of the fifteenth century, a time of massive land losses in other parts of the estuary, more than 8000 hectares of land were reclaimed (Table 13.1).

In regions where the estuary expanded in the later Middle Ages, new embankments also continued to be constructed. In the *Braakman*-area on the Flemish west bank, for instance, a major expansion of intertidal land took place between 1375 and 1404. A large part of the inundated land consisted of former peat land, which through drainage and peat digging had become extremely vulnerable for flooding (Gottschalk, 1955). After the flooding, re-embankment activities initially remained limited and only a few minor *polders* were reclaimed until 1488. In the latter year, a major dike burst near *Nieuwersluys* in the eastern part of the Braakman-area created a new tidal area of *c.* 11,000 to 13,000 hectares (de Kraker, 1997: 25–27). In part related to the civil war that was raging in Flanders at that moment, the dike burst could be not repaired and most of the inundated area remained flooded. As the tidal dynamics now increased in the eastern part of the Braakman area they apparently diminished in the western part where, between the 1490s and 1530, several major embankment projects succeeded and the area of estuarine land significantly shrank.

Figure 13.7 Embankments in the Western part of the Braakman area (Flanders, 1400–1570) and maximal extent of the 1488 flooding (based on Soens, 2009 and de Kraker, 1997)

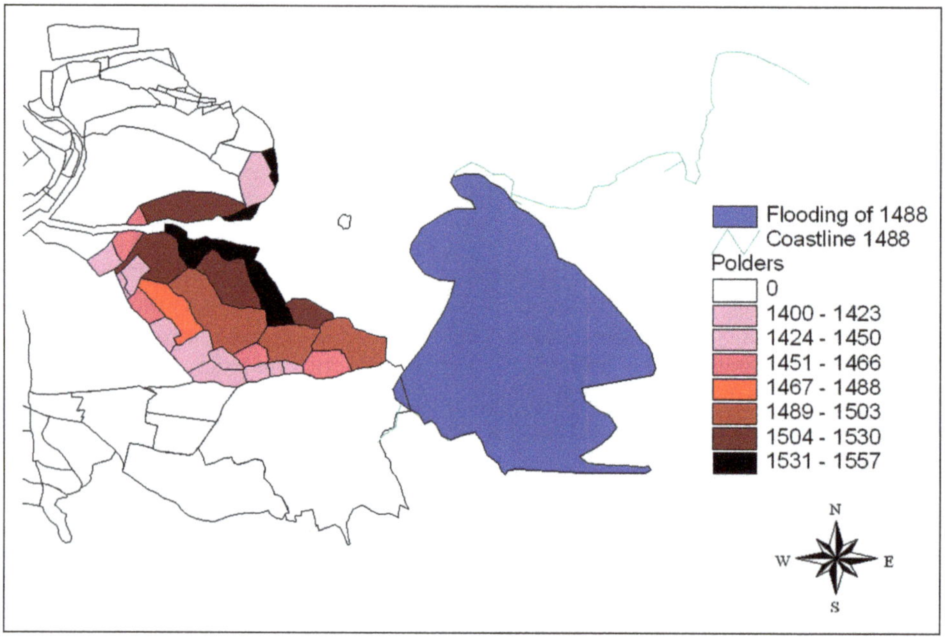

Both in South-Beveland and on the Flemish west bank, the opportunity to realise a huge return-on-investment in a relatively short period of time was the major driving force in this process. Of course the financial risk was high as well: not all projects were equally successful. Some did not succeed technically; others proved to be unfit for arable farming; and yet another group of *polders* was flooded again after a short period of time. Apparently this did not stop the late medieval '*polder* capitalists' of Flanders and Zeeland.

Unintentionally, the massive embankment projects of the late medieval period had a major impact on the estuarine environment of the Western Scheldt. As saltmarshes were progressively embanked, an important 'buffer' in case of high water and storm surges disappeared. With every new *polder* the sea-walls came closer to the main tidal channels and the natural forelands disappeared. Not surprisingly, tidal channels changed their course and became increasingly dangerous. Equally important most remaining secondary tidal channels were dammed in the course of the late medieval period. In regions like South-Beveland this process clearly accelerated in the later Middle Ages: every bit of saltmarsh that seemed high enough to embank, was indeed embanked. This 'aggressive' embankment fostered by the financial and political interests of the Zeeland nobility and the Flemish bourgeoisie, fundamentally disturbed the environmental equilibrium of the estuary.

The pre-1300 Scheldt and Zwin estuaries had consisted of innumerable smaller and larger tidal channels, many of them connected to small fresh water rivers draining the inland peat bogs and the uplands. Between the tenth and the twelfth century, embankment mostly consisted of the building of dikes parallel to these channels, the channels themselves however, often remained open for quite some time, not at least because they enabled maritime traffic. From the thirteenth century onwards, however, increasingly more channels were dammed, often with a lock being built in the dam to ensure drainage. In the Zwin estuary, for instance, the Zwin itself was dammed in 1180 near Damme (Verhulst, 1995: 56–57); the so-called Old Zwin or Reigarsvliet west of the Zwin *c.* 1160 (Coornaert, 1974: 25–26); and the *killa* between the islands of Zuidzande and Kadzand after 1290 (Meyer, 1992: 144–145). Kadzand was probably attached to Wulpen in the first half of the fourteenth century (Meyer, 1992: 387–388); the Zwarte Gat between Kadzand and the mainland was dammed in 1424 (but re-opened) and again in 1515, and the Havele between Breskinszand and the mainland *c.*1510 etc.

Building dams was both technically difficult and capital intensive, which was the reason why local lords on the Zeeland Isles called upon the Cistercian Abbey of Ter Doest to close the *Hinkele* lying close to the junction of the Honte and the (Eastern) Scheldt in between the islands of Rilland and South-Beveland. After several failures and a fierce juridical struggle between different stakeholders, the Hinkele was finally dammed in 1263 (Dekker, 1971; Obreen, 1932). In the fourteenth century the closing of the Hinkele was followed by other projects, including in 1330 the circular embankment of a silted up marshland which would be called the 'island of Bath'. On this island a settlement was founded and turned into an independent parish in the early fifteenth century. The remaining tidal channels surrounding the island were dammed during the next decades, so that the island of Bath was joint to both Rilland and South-Beveland (see Dekker, 1971: 122–123; 284–290). In the same region other major channels were dammed like the Zwake between Baarland and South-Beveland in 1445, and the Dierick between Borsele-Baarland and Stuivezand in 1525. A regular pattern can be noticed in the closing of all these tidal channels: at first dikes were erected on both sides of the channel, which resulted in the creation of new *polders* in a rectangular or oblong shape. Only in a second phase was the channel itself dammed and often turned into a waterway used for draining the freshwater from the area. The disappearance of all these minor tidal channels, must have had important consequences, especially in times of high water and storm surges: the water masses entering the estuary were increasingly forced into one single channel, resulting in higher and higher water levels with a positive impact on navigation but equally with an increased risk of flooding. When the next major storm surge struck the coasts of the Low Countries, the water was swept up against the dikes and disastrous flooding followed. One of the last channels to be dammed before the storm surge of 1530 was the '*Dierick*' which had separated the small island of Stuivezand from Baarland. The Dierick was dammed in 1525 by the lords of Stuivezand who were hoping to *decrease* the flood-risk and at the same time gaining an additional *polder* of *c.* 200 hectares (Dekker, 1971: 266–270). Five years later, exactly the opposite happened: the complete island of Stuivezand which had formed one of the last barriers in the middle of the estuary was flooded.

Figure 13.8 Middle part of the Honte on the 'Scheldt map' (sixteenth-century copy of a 1494 map, Antwerp, City Archives).
It shows several secondary tidal channels which had been dammed (the Zwake, above in the middle) or would be over the next years (the Dierick, separating the island of Stuivezand (on the left) from Baarland)

Around 1500, the oldest figurative depictions of the Scheldt Estuary – the anonymous Scheldt-map (Figure 13.2 and 13.8) or the painting of Jean de Hervy (Figure 13.5) – still provide us with an impression of the medieval estuary, which was a highly fragmented floodplain, with a lot of islands and sand banks, and dispersed by smaller and larger tidal channels. At that precise moment this 'medieval' estuarine landscape was about to vanish: coastal and estuarine islands either disappeared or were connected to the mainland, and secondary tidal channels were dammed or became so narrow that they silted up. At the end of the sixteenth century, cartographers like Pieter Pourbus, Gerard Mercator, or Christian Groten would show us another estuary: a broad single branched Western Scheldt, with the contours of its present-day shape already clearly visible.

VII. Conclusion

The genesis of the Western Scheldt remains a spectacular example of radical estuarine change resulting from the complex interaction between human interventions and endogenous estuarine dynamics. In her 1955–58 synthesis of the landscape history of the Flemish Westbank of the Western Scheldt Estuary, Elisabeth Gottschalk strongly objected to the suggestion of geographer R. Blanchard in 1906 that the late medieval changes in the Western Scheldt Estuary might have been provoked by speculating investors who had embanked and drained salt marshes that were still 'immature', not yet silted up high enough to be turned into farming land, and often situated too close to the main tidal channels. According to Gottschalk, extra-ordinary storm surges and a rise in sea-level were to be blamed for the expansion of the estuary (Gottschalk, 1955–58, II: 175 versus Blanchard, 1906: 177 and 200). Our analysis of the late medieval estuarine transformation shows that neither of them was fully correct, although they both provided important elements in the discussion. Blanchard was right in pointing out that embankment increased flood risk, and hence could rather paradoxically trigger the expansion of the estuary. Embankment diminished storage capacity of excess water and hence increased water levels, especially storm levels. Storms, as advanced by Gottschalk, certainly were agents of change that provoked the transformation of the estuary. But they merely hit areas that were already vulnerable, either because they were environmentally degraded after centuries of intensive human exploitation or because their population lacked the economic capacity to ensure a sufficient degree of flood protection. As explained elsewhere in this volume, most scholars no longer accept the occurrence of variations in sea-level rise during the historic period. Sea-level rise however still plays a role in the transformation of the Western Scheldt Estuary, but only in a very indirect way: the accumulation of peat from the sixth or fourth millennium BP onwards, followed by embankments in the historic period, impeded sedimentation from keeping pace with the slow but steady sea level rise of that period. Following subsidence of the peat or its removal, a huge accommodation space for renewed sediment deposition was created.

The initial drowning of the peat in what would become the Western Scheldt region, might be in part a natural phenomenon, forged by diminishing supplies of sediment from the North Sea in front of the Flemish coast, which led to the degradation of the coastal dune barriers and the intrusion of the sea in the peatlands. But even at that early stage, natural dynamics interfered with human interventions, as the Roman Age for the first time saw an increasing drainage and exploitation of the peatlands. In the early Middle Ages, the estuary took the shape of a shallow floodplain, with a multitude of smaller tidal channels finding their way in between drowning peat lands, and increasingly embanked sand- or claylands. In this contribution we argued that these conditions prevailed until the very end of the Middle Ages. Based on an analysis of flood damage in the Scheldt Estuary, we saw that the flood water was not forced into a single tidal branch well into the fifteenth century. Only in the latter half of that century – which is later than often assumed – the flood water started to follow the present-day trajectory of the Western Scheldt.

This final transformation of the Scheldt Estuary into one single, massive and funnel-shaped tidal outlet, was also the result of what could be labelled a long-term environmental path dependency: by 1400 AD the area had become extremely flood prone after three or four centuries of intensive occupation and exploitation of peat and salt marshes. However, environmental vulnerability does not necessarily lead to flooding. Peat area's like the Wilde Moeren on South-Beveland had been exhausted by 1300 AD, but were only flooded two centuries later. In order to explain the accelerated expansion of the Western Scheldt in the fifteenth and sixteenth century, two further developments have to be taken into account. On the one hand the fifteenth century was characterised by underinvestment in flood protection, as the peasant economy of the coastal area suffered severely from the economic crisis of the later Middle Ages. On the other hand, embankment practices were increasingly 'aggressive', aiming at the rapid expansion of the embanked area at the expense of the intertidal area, which hugely affected the estuarine hydrography, not at least because of the progressive closing of many secondary tidal channels, These shifting land-use strategies – declining flood protection on old lands paralleled by a reduction of the tidal floodplain – in the end provoked large scale inundations, and the final restructuring of the estuary.

As such, the medieval history of the Western Scheldt cannot be explained by pointing at exogenous drivers like storm surges, but rather by the analysis of human activities and the unintended estuarine dynamics they unleashed. In actual plans for 'ecosystem' restoration, geomorphologists have developed complex models to simulate the estuarine change that will occur when formerly reclaimed lands are returned to the estuary (e.g. Vandenbruwaene, Meire and Temmerman, 2012). However, over a longer period of time the precise dynamics of the estuary – the growth rate of tidal marshes; the erosion of tidal channels, and the patterns they follow – are very difficult to predict. Not surprisingly the massive land losses and aggressive embankment projects in the late medieval period had an even more unpredictable – and for medieval observers without doubt highly mysterious – impact on the estuary. In future, more detailed topographical reconstructions of the geographic and chronological extent of medieval and post-medieval floodings and re-embankments could provide us with highly valuable insights in how river estuaries respond to large scale human-induced modifications to its size, shape, and functions over a longer period of time.

Bibliography

Augustyn, B. (1995) 'De evolutie van het duinecosysteem in Vlaanderen in de Middeleeuwen: antropogene factoren versus zeespiegelrijzingstheorie', *Historisch Geografisch Tijdschrift*, 13, pp. 9–19.

Bailey, M. (1991) 'Per impetum maris: natural disaster and economic decline in eastern England, 1275–1350', in: B.M.S. Campbell (ed.), *Before the Black Death*, Manchester, pp. 184–208.

Bavel, B.J.P. van (2010), 'Rural Revolts and Structural Change in the Low Countries, thirteenth–early fourteenth centuries', in: R. Goddard, J. Langdon and M. Müller (eds), *Survival and discord in medieval society. Essays in honour of Christopher Dyer,* Turnhout, pp. 249–268.

Bindoff, S.T. (1945) *The Scheldt question to 1839*, London.

Blanchard, R. (1906) *La Flandre. Étude géographique de la plaine flamande en France, Belgique et Hollande*, Lille.

Brown, N. (2001) *History and Climate Change: a Eurocentric Perspective*, London-New York.

Brand, K.J.J. (1983) 'Over het ontstaan en de ontwikkeling van de Hont of Westerschelde', *Zeeuws Tijdschrift*, 33, pp. 99–110.

Buisman, J. (1995–2006) *Duizend jaar weer, wind en water in de Lage Landen*, Franeker.

Campbell, B.M.S. (2010a) 'Nature as Historical Protagonist: environment and society in pre-industrial England (the 2008 Tawney Memorial Lecture)', *Economic History Review*, 63, pp. 281–314.

Campbell, B.M.S. (2010b) 'Physical shocks, biological hazards and human impacts. The crisis of the fourteenth century revisited', in: S. Cavaciocchi (ed.), *Le interazioni fra economia e ambiente biologico nell'Europa preindustriale. Secc. XIII-XVIII, Proceedings of the XLI settimana di studi of the Datini Institute,* Prato, pp. 14–32.

Clercq, W. De and Dierendonck, R. van (2008) 'Extrema Galliarum. Zeeland en Noord-west-Vlaanderen in het Imperium Romanum', *Zeeuws Tijdschrift* 58, pp. 6–34.

Coornaert, M. (1974) *Knokke en het Zwin, de geschiedenis, de topografie en toponymie van Knokke met een studie over de Zwindelta*, Tielt.

Dam, P. van (2010) 'Rabbits Swimming Across Borders. Micro-Environmental Infra-structures and Macro-Environmental Change in Early Modern Holland', in B. Scott (ed.), *Economies and Ecologies in Medieval and Early Modern Europe*, Leiden, pp. 63–92.

Dekker, C. (1971) *Zuid-Beveland. De historische geografie en de instellingen van een Zeeuws eiland in de Middeleeuwen*, Assen.

Dekker, C. (1988) 'Tussen twee vloeden. De strijd tegen het water in Zeeland bewesten Schelde tussen 1530 en 1532', *Bijdragen en Mededelingen betreffende de geschiedenis der Nederlanden*, 103, pp. 607–621.

Dekker, C. (1996) 'De moernering op de Zeeuwse eilanden', *Tijdschrift voor Water-staatsgeschiedenis,* 5, pp. 60–66.

Dekker, C. and Baetens, R. (2010) *Geld in het water: Antwerps en Mechels kapitaal in Zuid-Beveland na de stormvloeden in de 16e eeuw*, Hilversum, pp. 219–223.

Doeleman, F. (1975) 'Zeggenschap op de Honte', *Legal History Review*, 43, pp. 23–43.

Donkersloot-De Vrij, M. (1981) *Topografische kaarten van Nederland vóór 1750. Hand-getekende en gedrukte kaarten, aanwezig in de Nederlandse Rijksarchieven*, Groningen.

Engelen, A.F.V van, Buisman, J. and IJnsen, F. (2001) 'A Millennium of Weather, Winds and Water in the Low Countries', in: Ph.D. Jones *et al.* (eds), *History and Climate: Memories of the Future?,* New York, pp. 101–124.

Galloway, J.A. and Potts, J.S. (2007) 'Marine flooding in the Thames Estuary and tidal river, *c.* 1250–1450: impact and response', *Area,* 39, pp. 370–379.

Galloway, J.A. (2009) 'Storm flooding, coastal defence and land use around the Thames estuary and tidal river c.1250–1450', *Journal of Medieval History*, 35, pp. 171–88.

Goldschmitz-Wielinga, L.C.J. e.a. (2004) 'Verdronken kerkdorpen in Zeeland: een over-zicht', J.J.B. Kuipers (ed.), *Sluimerend in Slik. Verdronken dorpen en verdronken land in zuidwest Nederland*, Middelburg, pp. 48–49.

Gottschalk, M.K.E. (1955–1958; ²1983) *Historische geografie van Westelijk Zeeuws Vlaanderen*, Assen.

Gottschalk, M.K.E. (1984) *De Vier Ambachten en het Land van Saeftinge in de Middeleeuwen. Een historisch-geografisch onderzoek betreffende Oost-Zeeuws-Vlaanderen*, Assen.

Gottschalk, M.K.E. (1971–1977): *Stormvloeden en rivieroverstromingen in Nederland*, Assen.

Gross, A. and Butcher, A. (1995) 'Adaptation and investment in the Age of the Great Storms: agricultural policy on the manors of the Principal Lords of the Romney Marshes and the Marshland fringe, *c.* 1250–1320', in: J. Eddison (ed.), *Romney Marsh: the debatable ground*, Oxford, pp. 107–117.

Jongepier, I., Soens, T., Thoen, E., Eetvelde, V. Van, Crombé, Ph. and Bats, M. (2011) 'Brown gold: a reappraisal of medieval peat marshes in Northern Flanders (Belgium)', *Water History*, 3, pp. 73–93.

Kiden, P. (1991) 'The Lateglacial and Holocene Evolution of the Middle and Lower River Scheldt, Belgium', in: L. Starkel, K.J. Gregory and J.B. Thornes (eds), *Fluvial processes in the temperate zone during the last 15,000 years: temperate paleohydrology*, Chichester, pp. 283–299.

Kiden, P. (2006) 'De evolutie van de Beneden-Schelde in België en Zuidwest-Nederland na de laatste ijstijd', *Belgeo*, 3, pp. 279–294.

Kraker, A.M.J. de (1997): *Landschap uit balans. De Vier Ambachten en het Land van Saeftinghe tussen 1488 en 1609*, Utrecht.

Kraker, A.M.J. de (ed.) (2002), *De Westerschelde, een water zonder weerga: ontstaansgeschiedenis en kaartbeeld, havens, handel en scheepvaart, verkeer, verdronken dorpen, oorlog en verdedigingswerken, natuur en milieu en andere aspecten van de Westerschelde*, Kloosterzande.

Kraker, A.M.J. de and Aalbersberg, G. (2005) 'Digitized reconstruction of the Westerschelde basin, 1550–2000', in: Landschafsverbandes Rheinland (ed.), *Beiträge zur Landesentwicklung 58. Kulturlandschaft digital – Forschung und Anwendung. 15. Fachtagung des Umweltamtes (Landschaftsverband Rheinland). 11. Jahrestagung der Arbeitsgruppe Angewandte Historische Geographie (Arbeitskreis für historische Kulturlandschaftsforschung ARKUM), 2. bis 5 März 2005 in Aachen. Tagungsdokumentation*, Cologne, pp. 136–144.

Kraker, A.M.J. de (2006) 'Flood events in the southwestern Netherlands and coastal Belgium, 1400–1953', *Hydrological Sciences*, 51, pp. 913–929.

Lamon, R. (2008) *Landschapsreconstructie en eigendomsverhoudingen in de Oude Groedsche Polder van de vijftiende tot de zeventiende eeuw*, unpublished Master thesis, Ghent University.

Lases, W.B.P.M. and Kraker, A. de (2009) 'De Westerschelde, natuurlijk? Verdieping van en ontpoldering langs de Westerschelde in historisch perspectief geplaatst', *Tijdschrift voor Waterstaatsgeschiedenis*, 18, pp. 25–39.

Lases, W.B.P.M. and Kraker, A. de (2010) 'De Verdieping van een artikel over de Westerschelde. Discussiedossier. Respons van Lases en de Kraker', *Tijdschrift voor Waterstaatsgeschiedenis*, 19, pp. 70–74.

Leenders, K.A.H.W. (2010) 'De Verdieping van een artikel over de Westerschelde. Discussiedossier. Reactie van K.A.H.W. Leenders', *Tijdschrift voor Waterstaatsgeschiedenis*, 19, pp. 67–70.

Meijers, E.M. (1940) 'Des graven stroom', *Mededelingen der Koninklijke Nederlandse akademie van wetenschappen. Afdeling letterkunde,* 3, pp. 103–205.

Meire, P. and Damme, S. Van (eds) (2005) *Ecological structures and functions in the Scheldt estuary: from past to future*, Berlin.

Mertens, J. (1981) 'De meekrapteelt in de omgeving van Brugge vooral in de 14de en 15de eeuw', *Het Brugs Ommeland*, 21, pp. 259–282.

Meyer, H.D. (1992) 'Nieuwe Getuigenissen over Stormvloedschade in Zeeuws-Vlaanderen, in de jaren 1375 tot 1412', *Handelingen der Maatschappij voor Geschiedenis en Oudheidkunde te Gent*, pp. 5–46.

Obreen, H. (1932) 'De bedijking der Hincline op Zuid-Beveland (1263–1269). Bijdrage tot de kennis van het 13e-eeuwsche dijkrecht', *Tijdschrift voor Rechtsgeschiedenis. Revue d'Histoire du Droit,* 11, pp. 56–77.

Rippon, S. (2000) *The transformation of coastal wetlands: exploitation and management of marshland landscapes in North West Europe during the Roman and medieval periods,* Oxford.

Roberts, A.-M. (1991) 'The landscape as legal document: Jan de Hervy's "View of the Zwin"', *Burlington Magazine,* 133, 1055, pp. 82–86.

Ryckaert, M. (1995) 'The Zwin offers a new way out to the sea', *Bruges and Zeebrugge. The city and the sea*, London, pp. 24–30.

Soens, T. (2005) 'Explaining deficiencies of water management in the late medieval Flemish coastal plain, 13th-16th centuries', *Jaarboek voor Ecologische Geschiedenis*, pp. 35–61.

Soens, T. (2009) *De spade in de dijk? Waterbeheer en rurale samenleving in de Vlaamse kustvlakte (1280–1580)*, Ghent.

Soens, T. (2011a) 'Floods and money. Funding drainage and flood control in coastal Flanders (13th-16th centuries)', *Continuity and Change*, 26, pp. 333–365.

Soens, T. (2011b) 'Threatened by the sea, condemned by man? Flood risk and environmental inequalities along the North Sea Coast (1200–1800 AD)', in: G. Massard-Guilbaud and R. Rodger (eds), *Environmental and social inequalities in the city. Historical perspectives*, Cambridge, pp. 91–111.

Soens, T. and Thoen, E. (2009) 'Mais où sont les tourbières d'antan? Géographie, chronologie et stratégies économiques du tourbage en Flandre Maritime (XIIe-XVIe siècles)', *Aestuaria. Cultures et développement durable*, 14, pp. 45–60.

Soens, T. and Thoen, E. (2010) 'Vegetarians or Carnivores? Standards of living and diet in late Medieval Flanders', in: S. Cavaciocchi (ed.) *Le interazioni fra economia e ambiente biologico nell'Europa preindustriale. Secc. XIII-XVIII – Economic and Biological Interactions in the Pre-industrial Europe from the 13th to the 18th Centuries*, Prato, pp. 495–527.

Thoen, E. (2004) '"Social Agrosystems" as an economic concept to explain regional differences. An essay taking the former county of Flanders as an example (Middle Ages-19th Century)', in: B.J.P. van Bavel and P. Hoppenbrouwers (eds), *Landholding and land transfer in the North Sea area (late Middle Ages-19th century)*, Turnhout, pp. 47–66.

Thoen, E. and Soens, T. (2009) 'The origins of leasehold in the former county of Flanders', in: B.J.P. van Bavel and Ph. Schofield (eds), *The development of leasehold in northwestern Europe, c. 1200–1600*, Turnhout, pp. 31–56

Thoen, E. and Soens, T. (forthcoming) 'The family or the farm. A Sophie's choice? The late medieval crisis in Flanders', unpublished manuscript.

Tys, D. (2003) *Een middeleeuws landschap als materiële cultuur: de interactie tussen macht en ruimte in het kustgebied en de wording van een laatmiddeleeuws tot vroegmodern landschap. Kamerlingsambacht, 500–1200/1600*, Unpublished Dissertation, Free University Brussels (VUB).

Tys, D. (2005) 'Landscape, settlement and dike building in coastal Flanders in relation to the political strategy of the counts of Flanders, 900–1200', in: M. Fansa (ed.), *Kulturlandschaft Marsch – Natur, Geschichte,* Oldenburg, pp. 106–126.

Tys, D. (2006) 'Walraversijde: een lastig parket? Dynamiek en identiteit van een laatmiddeleeuwse kustnederzetting in een proto-kapitalistische omgeving', in: M. Pieters *et al.* (eds), *Fishery, trade and piracy: fishermen and fishermen's settlements in and around the North Sea area in the Middle Ages and later. 1: Papers from the Colloquium at Oostende-Raversijde, 21–23 November 2003*, Brussels, pp. 19–40.

Unger, W.S. (1939) *De tol van Iersekeroord: documenten en rekeningen, 1321–1572,* Den Haag.

Vandenbruwaene, W., Meire, P. and Temmerman, S. (2012) 'Formation and evolution of a tidal channel network within a constructed tidal marsh', *Geomorphology*, 151/152, pp. 114–125.

Vanslembrouck, N., Lehouck, A. and Thoen, E. (2005) 'Past landscapes and present-day techniques. Reconstructing submerged medieval landscapes in the western part of Sealand Flanders', *Landscape History*, 27, pp. 5–18.

Verhulst, A. and Gottschalk, M.K.E. (eds) (1980) *Transgressies en occupatiegeschiedenis in de kustgebieden van Nederland en België. Colloquium 5–7 september 1978. Handelingen,* Ghent.

Verhulst, A. (1995) *Landbouw en landschap in middeleeuws Vlaanderen,* Brussel.

Vos, P. and Heeringen M. van (1997) 'Holocene geology and occupation history of the province of Zeeland', in: M.M. Fischer (ed.), *Holocene evolution of Zeeland (SW Netherlands),* Haarlem, pp. 3–109.

Vos, P. *et al.* (2011), *Atlas van Nederland in het Holoceen. Landschap en bewoning vanaf de Laatste IJstijd tot nu,* Amsterdam.

Wyffels, A. (2002) 'Émergence et engloutissement judiciaires du "Brabant maritime". Les prétentions territoriales sur l'Escaut occidental durant la seconde moitié du XVe siècle', *Entre Royaume et empire: frontières, rivalités, modèles. Rencontres de Porrentruy (27 au 30 septembre 2001). Publication du centre Européen d'Études Bourguignonnes (XIVe-XVIe s.),* 42, pp. 55–74.

Zeiler, F.D. and Vos, P. (2008) 'Overstromingsgeschiedenis van Zuidwest-Nederland. Interactie tussen natuurlijke en antropogene processen', *Grondboor en Hamer*, pp.86–95.

14 The evolution of the landscape and the social and political organisation of water management: the Po Valley in the Middle Ages (fifth to fourteenth centuries)*

Michele CAMPOPIANO, University of York

Abstract

This paper contributes to the history of the social and political framework of water management organisation in the Po Valley. It discusses the dominant social and political cadres who influenced the evolution of the landscape by setting goals relating to land clearance and water management. The affirmation of landownership of large areas from the eighth/ninth centuries, the subsequent development of rural lordship and, from the end of the eleventh century, the political affirmation of the Italian city-states, created social and political forces that were increasingly able to subordinate the evolution of these tasks to their own particular interests, thereby accelerating the process of land reclamation and intensifying land use. This resulted in a highly anthropogenic landscape that was particularly vulnerable to environmental pressures, including flooding.

I. Introduction

The history of water management in Medieval Italy is a topic that has scarcely attracted the attention of scholars. Yet water management had a crucial function in the history of Italian agriculture. The role of irrigation in the diffusion of new crops in Sicily after the Arab conquest, for example, is fairly well known to scholars, but probably because it better fits the cliché of linking water management in the Mediterranean basin to the necessity of bringing water to a dry environment (Watson, 1974). It is perhaps less well known that water management played a key role in the history of the agriculture of the Po Valley in Northern Italy, by far the richest area of the country during this period.

The central role of water in the history of Northern Italy has not gone unobserved by Italian historians: Carlo Cattaneo, Milanese historian and politician (1801–1869), compared the Lombard landscape to the Dutch countryside, and attributed to the Lombards the same merit of having extracted their land from the water (Cattaneo, 1844). But, in contrast to the Netherlands, and to the Low Countries in general, Italy does not have the same tradition of studies on water management as can be found in Belgian and Dutch historiography, particularly as concerns the history of water management in the

* I would like to thank Professors Chris Wickham and François Menant for their suggestions.

Landscapes or seascapes? The history of the coastal environment in the North Sea area reconsidered, ed. by Erik THOEN, Guus J. BORGER, Adriaan M.J. DE KRAKER, Tim SOENS, Dries TYS, Lies VERVAET & Henk J.T. WEERTS, Turnhout, 2013 (*CORN Publication Series*, 13), p. 313–332.

BREPOLS PUBLISHERS DOI: 10.1484/M.CORN.1.101558

Middle Ages[1]. Literature specifically dealing with water management and its relationship with the evolution of the landscape in the medieval Po Valley is restricted to a few titles. Among them, we should particularly note the studies of Vito Fumagalli (Fumagalli, 1970; 1976; 1985a; 1985b; 1990; 1991; 1992), an excellent discussion of land clearance and irrigation in Lombardy in the High Middle Ages in Menant's *Campagnes lombardes au moyen âge* (Menant, 1993: 217–266), and the more recent works of Paolo Squatriti (Squatriti, 1992; 1998; Magnusson-Squatriti 2000). Squatriti's works are, from several points of view, ground-breaking, but they mainly focus on the history of mentality, rather than on the social and economic structure of the society in which land clearances took place. For example, Squatriti describes the hatred of swampland expressed in the works of Roman agronomists, such as Varro, Columella, and Palladius, and explains that their attitude reflects the interests of wealthy Roman agriculturalists interested in, 'a marketable surplus capable of maintaining the landowner (probably resident in a city) in the style to which he was accustomed' (Squatriti, 1992: 6). But it is evident that such an assumption raises several questions concerning the property rights of the landowning elite on land and water and their capability of extracting surplus from this land. Mentality cannot be separated from the social and economic structure in which it arises. Accordingly, there is a need to discuss the role of the human factor in the evolution of the landscape by analysing the particular social structure in which humans operate. It is necessary to analyse the relevant aspects of the social structure that could influence water management policies, such as property rights on land and water, control of the workforce, and political organisation. These aspects of the social structure were also the battleground where conflicts between different social groups were fought, influencing the evolution of water policies (see, for example: Soens, 2009).

This paper intends to contribute to writing the history of the social and political framework of water management organisation in the Po Valley by discussing how these social and political cadres influenced the evolution of the landscape by setting the goals and founding the organisations that undertook water management tasks. When speaking of water management tasks, I do not exclusively mean the routine maintenance of canals and dykes, but also the land clearances that led to a modification of the relationship between water and the landscape, such as deforestation. Social and political structures also influenced the scale of the land clearance process, since this depended on the goals that had been set by the politically and socially dominant groups. I intend, therefore, to reject any explanation of the rise and the extension of land clearances based purely on demographic models. I do not want to deny that in a given historical situation demographic pressure can encourage the extension of arable land in order to feed a larger population, but the social structure in that situation influences the need to feed a growing population, since it affects, for example, the level of surplus extraction, the kind of crops grown on the soil, and access to other resources like hunting and fishing in woods, marshes, and rivers (for criticism of the 'demographic model', see: Brenner, 1976: 32–42).

The time span analysed in this work, from the fifth to the fourteenth centuries, saw a dramatic change in the relationship between human society and the landscape. At the

---

[1] Ciriacono's work on the Venetian countryside (Ciriacono, 1994), which also benefits from Dutch and Belgian studies discussing the role of water management in the development of agriculture in Veneto, focuses on the Early Modern Period (late fifteenth-eighteenth centuries).

Figure 14.1 Locations mentioned in the text

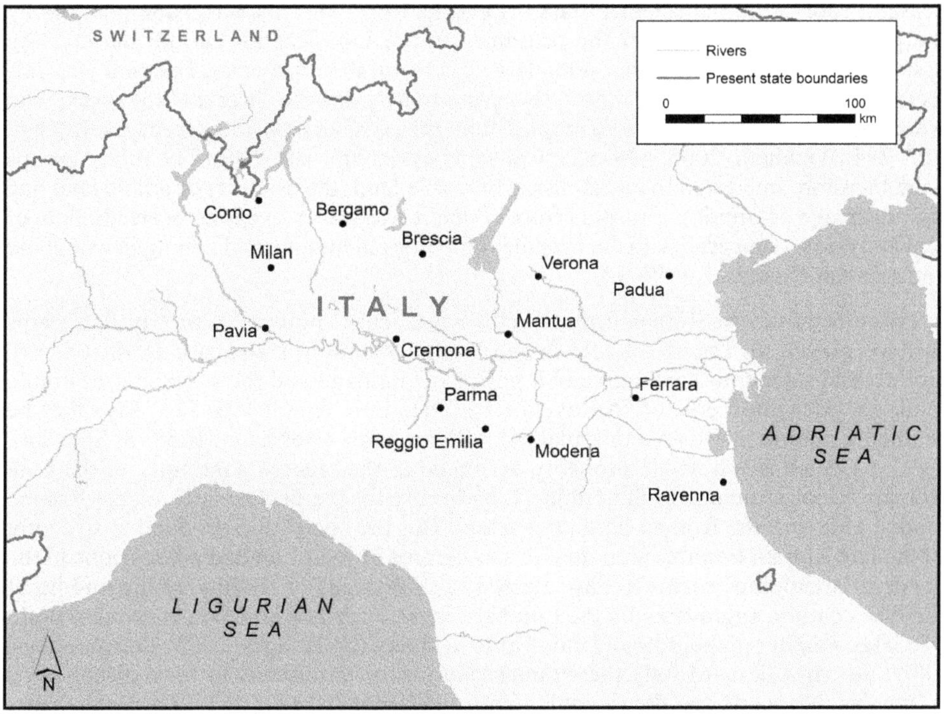

beginning of the period, in the late Roman era, there was substantial intervention from central authority in order to keep the hydrology of the Po Valley under control, and an attempt by landlords to minimise the presence of swampland. Later, in the early medieval period, there was a move towards the expansion of woodland and swampland and a less-regulated hydrology, while, from the end of the eighth century, the augmentation of land clearance led to attempts to more drastically intervene in the hydrology. This intervention was intensified by the rise of the Italian city-states from the twelfth century onwards (which caused further works of colonisation and land clearance, and dyke building to retain the riverbeds). The thirteenth and fourteenth centuries saw an increase in the number and violence of river floods, due to the more heavily human-regulated hydrology. At that time the Po Valley started to assume some of the features that still characterise its landscape, with highly productive agriculture sustained by irrigation and drainage works and protected from river flooding by a system of dykes. This system of dykes and canals did not (and does not) prevent the insurgence of major floods, as the recent floods have reminded us.

II. The Late Roman period and Early Middle Ages

The Late Roman Empire (fourth and fifth centuries) was able to invest in major hydraulic works to control the hydrology of the Po Valley and protect its arable land (Squatriti,

1998: 68–69; Marchetti, 2002: 367). This was because the State was interested in the extraction of surplus through land tax. The State was also able to use the labour of the peasantry for water management tasks (Leveau, 1993: 3). This was made possible by the political subordination of the peasantry in the Late Roman Period: the taxation system tied the *coloni* to work on the large estates of the aristocrats. Taxation was collected through landlords, creating a social union between the State and the lords, who were both interested in extracting surplus from the peasantry (Whittaker-Garnsey, 1998: 287–294; Wickham, 2005: 524–525; Constambeys, 2009: 108–109). The ruling groups were therefore interested in an extension of arable land: the products of arable land had the advantage of offering a surplus from which it was easier to extract a steady flow of rent and taxes, compared with the irregular outcomes of hunting and fishing in woodland and marshes (Squatriti, 1992: 6).

After the fall of the Roman Empire in the west, central political authority also seems to have played an important role in the Gothic kingdom (Squatriti, 1998: 68–69). Gothic kings seem to have pursued a policy of drainage and the extension of arable land: an inscription erected in Ravenna by Theodoric (ruled 493–526) says that he drained a swamp in the area to replace it with orchards (Bormann, 1888: 8; Squatriti, 1992: 9). Such a policy can probably be linked to the relative continuity of the Late Roman social structure in the Gothic kingdom, with the permanence of the Roman landed elite and the Roman land tax system. But the social and productive structure of the late Roman Empire, with its villas as centres of major landed possessions in the hands of aristocrats, seems to have started to show signs of decline by the middle of the fifth century, aggravated by the Lombard invasion in 568 and the consequent flight of a large number of the Roman landed elite (Francovich-Hodges, 2003; Constambeys, 2009: 96). In Lombard Italy the Roman taxation system seems to have disappeared in the second quarter of the seventh century (Gasparri, 1990: 262–268; Pohl, 2001; Wickham, 2005: 117).

The weakening of the structure of social control and surplus in the Late Roman Empire hindered the possibilities for radical intervention in the regulation of hydrology. An important example of this is the *fossa Augusta,* a large canal whose digging had been ordered by Emperor Augustus to transport water from the main branch of the Po into the lagoon of Ravenna in the Po Delta, in order to flush the lagoon with fresh water from the Po. This canal required maintenance, provided by the Roman State, but it ceased to be navigable in the fifth or sixth century (Fabbri, 1975: 16; Squatriti, 1992: 4–8). Between the beginning of the seventh and end of the eighth centuries radical intervention in canal digging and embankment building seems to have disappeared, leaving more room for marshland and woodland (Magnusson-Squatriti, 2000). The Po Valley experienced a process of reforestation and the diffusion of marshes: Roman centuriation tracks are often found under palustrine deposits (Veggiani, 1983; Ravazzi, 1989; Castaldini, 1989; Marchetti, 2002: 367). Marshland and woodland represented vital sources of food for the rural population: forests could feed pigs and marshes were used for hunting and fishing (Montanari, 1979: 34–35, 48–49 and 268–270). Human communities adopted an approach of, to use the concept elaborated by Stephen Rippon regarding coastal marshlands, 'exploitation' of marshlands and woodlands, 'simply harvesting the natural produce' of these areas (Rippon, 2001: 141; see also Rippon, 2000) whilst allowing these changes in the landscape to take place.

Demography also played a role in determining this new evolution of the landscape (Beloch, 1937; Bellettini, 1973; Del Panta *et al.,* 1996; Lo Cascio-Malanima, 2005). The population declined from the middle of the sixth century, with the Greek-Gothic war (535–553), a series of famines between 538–542, and an epidemic plague in 541 (Pinto, 1996: 27). But the influence of demography on the evolution of the Early Medieval landscape has probably been overestimated. The extent to which population decline or population growth influences the relationship with the landscape is determined by the structure of the social relationship. For example, the level of surplus extraction and access to pastures influences the relationship with natural resources, determining the need to reclaim more arable soil to increase the cultivation of grain or the possibility of consuming more expensive calories (from husbandry, for example). Furthermore, the size of the demographic decline in Early Medieval Italy is questionable: recent estimates by Elio Lo Cascio and Paolo Malanima, for example, point to a growth from nine million around 300 AD to eleven million people around 500 AD, followed by a decline in the population to eight million people in the seventh to ninth centuries. This differs from previous estimates that suggested a continuous decline from eight and a half million in 200 AD to about four million in 700 (Lo Cascio-Malanima, 2005: 12–16).

Therefore, the evolution of the landscape in the seventh and eighth centuries should be linked to the changes in the political and social structure. Although Lombard law reaffirmed the authority of the State over the major rivers, any significant attempts to introduce land clearance policies or dyke building seem to have come to an end. The crisis of the system of surplus extraction from the politically subordinated *coloni* (tenant farmers), centred on land tax and large landownership, led to the affirmation of different relationships between the rural population and a landscape where woodland and marshland were becoming an increasingly important source of food.

III. The increase of large landownership and the diffusion of land clearance (end of the eighth century onwards)

From the end of the eighth century, written and archaeological sources show us increasing evidence of land clearance. For example, attempts to clear forests are mentioned in a charter of 772 for the area of Carpi, in Emilia-Romagna. The same charter also mentions a 'canal of the men of the village of Bedollo', probably a drainage canal (Brühl, 1973: 241–243). In the area around Verona, the development of settlements in marshland and woodland, and the probable intensification of the economic use of this area, during the ninth century is attested by archaeological evidence (Saggioro, 2005: 91–94). Archaeology also discloses an interesting settlement near Verona, the village of Bovolone which, from the ninth or tenth century, occupied an area of six to seven hectares: this settlement was made possible by the digging of several canals that separated it from the surrounding swamps (Saggioro *et al.,* 2004; Brogiolo-Chavarria, 2005: 122).

The efforts made towards land reclamation seem to be related to the strengthening and extension of large landownership patterns and its reorganisation in the form of the manorial system (in Italian: *sistema curtense, azienda curtense*). Lords (in our sources especially ecclesiastical institutions) encouraged and controlled the process of land reclamation. Wealthy and powerful landholders pushed human communities in the Po Valley towards, to use another of Stephen Rippon's categorisations, a full-scale transformation of the landscape, turning both marshes and woodlands into arable areas (Rippon, 2001). The *Silva de Ostilia,* a large area (about eighty square kilometres) covered by swamps, woods and meadows between Mantua, Padua and Verona, belonging to the *fiscus* (Public treasure) in the eighth century, was the object of a dispute between the monastery of Nonantola and the Count of Verona between 820 and 833 (Manaresi, 1955: 95–98; Manaresi, 1957: 566–568; Rossini, 1979: 37–42). Once the monastery of Nonantola obtained the *ius possidendi* (right of possession) of the *Silva de Ostilia,* the monastery pursued a policy of land clearance by renting out farms of about 100 m by 1500 m to colonists with contracts of twenty-nine years, in order that they could clear the land and build and maintain drainage canals. These contracts date to between 837 and 867 (Castagnetti, 1977: 49; Rossini, 1979: 42–43, 79–86; Baruzzi-Montanari, 1988: 101). It is interesting to see how a great expanse of marsh and woodland, at least in theory previously under the control of the crown, was disputed between two large landowners (the monastery and the count) in order to proceed with the colonisation of the area.

The efforts made regarding land clearance on large estates can be also seen in the inventories of Italian *curtes* (manors; sing. *curtis*) surviving from the end of the eighth and beginning of the ninth centuries (Pasquali, 2002a: 19–20). Some of these inventories show a reduction of the *incolto* (untilled areas) to make room for grain crops and vineyards. The earliest surviving inventory comes from the manor of Migliarina near Modena and dates from *c.*800. This can be compared with an inventory inserted in the polyptych of the monastery of Santa Giulia in Brescia, the owner of the manor, from 879–906. In the second inventory, the manor's forest seems to have become much smaller, only supporting 2000 pigs instead of 4000 (Andreolli-Montanari, 1983: 147–160; Fumagalli, 1990: 101). The use of the woodland was allowed, according to the second inventory, in return for a payment of five *modia*[2] of wheat, five *anforae*[3] of wine and thirty days of *corvées* (labour duties on the lordly domain-the payment was probably owed by the tenant farmers of the *curtis*). The payment of rent in wheat and wine compelled the rural community to expand the cultivation of vineyards and grain if they wanted to preserve their rights to the commons; the *corvées,* of course, were used to work the demesne. This seems to have been reorganised to increase cereal and vine cultivation. The number of oxen needed to plough the demesne increased from two (first inventory) to eight (second inventory) (Castagnetti *et al.,* 1979: 85–86 and 203–204; Pasquali, 2002b: 87). The extension of arable land also seems to be apparent in an analysis of the two polyptychs of the monastery of San Colombano of Bobbio, from 862 and 883 respectively. The polyptych of 883 reveals that part of the woodland owned by the monastery had been cleared in order to create thirty-two new farms. The polyptych of 883 mentions that even in Calice, in the mountains near Parma, there were twenty-three tenants instead of the nineteen documented in 862 (Castagnetti *et al.,* 1979: 127–144, 147–165, 133 and 154).

[2] A dry measure whose name derives from the Latin *modius,* this was equivalent to *c.* 8.7 liters (Cardarelli, 2003).

[3] From the Roman liquid measure *amphora,* this corresponded to *c.* 26 litres (Cardarelli, 2003).

The process of colonisation appears to be related to the affirmation of large estates and their reorganisation under the manorial system. The aristocracy strengthened its influence on untilled areas in order to increase the cultivation of grain and vines. The demesnes of the *curtes* often seem to have been formed from previous swamps or woods, and *corvées* and colonisation contracts were used to increase the tilled surface area (Pasquali, 2002a: 57–58). Arable land could supply a fairly steady flow of surplus, easier to measure and extract, in comparison with the more irregular output (in terms of hunting and fishing) from marshes. The *curtes* were probably able to collect a large surplus: it has been estimated, on the basis of the polyptych of the monastery of Santa Giulia of Brescia from 879–906, that this religious institution was able to collect a surplus of at least 20% of grain and wine (subtracting self-consumption and seed) (Pasquali, 1981). The aristocracy probably aimed to commercialise this surplus: Cinzio Violante, in his famous work on Milan and its surroundings, observed the diffusion of markets, usually based in towns but also on manors, and the grants by kings and emperor,of market and fiscal rights (Violante, 1953: 3–40). The diffusion of vineyards, a product of commercial value, in the Po Valley from at least the eighth century is a clear sign of this evolution (Pini, 1989; Pini, 1990; Cortonesi, 2002: 217–219). Many of the *corvées* required by tenants on the demesne were related to the maintenance of vineyards (Pasquali, 1990). The monastery of Nonantola encouraged the colonists of the *Silva de Ostilia* to plant vineyards by reducing the rent on this product to 25% of the output (Violante, 1953: 79–80).

In the past, demographic pressure has often been given as an explanation for the process of colonisation and the clearance of marsh and woodland. Popularised by Georges Duby's *Guerriers et paysans. VII*e*-XII*e *siècle. Premier essor de l'économie européenne,* the idea of a demographically driven–aristocratically directed colonisation process has found its way into the historiography of Italian agricultural expansion and water management. It is found, for example, in the works of both Fumagalli and Squatriti (Duby, 1975). It is clear that in the High Middle Ages Italy experienced a period of demographic growth: according to the recent estimates of Lo Cascio and Malanima, from nine million in 900 AD to ten and a half million in 1100 (Lo Cascio-Malanima, 2005: 12–16). But demography is not an independent variable. For example, the diffusion of arable land in place of marshes and woodlands, which, as we have seen, was largely induced by the interests of lords, could have encouraged demographic growth, since it provided a more reliable food supply (Squatriti, 1992: 6; Mitterauer, 2003: 34–41). It is also simplistic to assume that demographic growth would have lead automatically to land clearance: as I affirmed in the introduction, this would have depended on access to resources, which was determined by social structure.

The social relationships between lords and peasants would therefore play a key role, not only determining the level of surplus extraction through taxes and rent, but also the control of the *incolto,* marshes and woods that were the main source of income for Italian peasants in the Early Middle Ages. Rights to the *incolto* could be the object of disputes, not just among lords, as in the case of the dispute between the monastery of Nonantola and the count of Verona, but also between lords and rural communities. For example, in 824 a dispute arose between the monastery of Nonantola and some inhabitants of Fiesso, who were presumably acting as representatives for a group of *consortes* (associates), concerning fishing and pastoral rights in the forest near Fiesso. The inhabitants produced a document that decreed that the Lombard king Liutprand had granted pastoral rights to

'Renato, Adriano, Leoni et Mauro, omnibus consortibus vestris habitantibus pleve sancti Laurenti' ('Renato, Adriano, Leone and Mauro, and to all their associates living in the plebs of San Lorenzo') (Manaresi, 1955: 109–113). We have seen how the monastery of Santa Giulia of Brescia asked for rent in exchange for the use of the woods in Migliarina. This rent probably had to be paid by the local peasant community who would have paid collectively to use its resources.

In some areas, rural communities could play an important role in defining access to the *incolto* (Bognetti, 1978: 8–32; Castagnetti, 1983: 10–12; Menant, 2004: 250). It is possible that this regulative function of rural communities grew stronger through the interplay with the strengthening power of the aristocratic lords: common subjection increased the need for collective action among the peasantry (Wickham, 2001: 138–140). As we shall see, the interplay between these communities and the lords played a key role in the further evolution of the process of land clearance.

IV. Lordships, rural communities and water management (tenth-twelfth centuries)

The tenth century saw the expansion of the land clearance process, both for swampland and woodland. The studies by François Menant even show the appearance of a number of irrigation canals in Lombardy between the Adda and Serio Rivers in the tenth century: the generalisation of the irrigation process became one of the characteristic elements of Lombard agriculture in the Late Middle Ages. The Seriola d'Aulene, an irrigation canal in the Bergamo area, was drawn from the Brembo River shortly before 909 (Menant, 1993: 183–193). Irrigation and land reclamation activities were both related to a general effort to extend arable areas, an effort which was supported by having recourse to irrigation for drier soils and drainage for marshland, which, in both cases, resulted in a dramatic transformation of the landscape (Menant, 1993: 171–172).

Written documents allow us to see this dramatic transformation of the landscape, increasingly characterised by irrigation and drainage canals. Even the stereotyped formulae that appear in charters, such as those concerning the cession of rights on land *cum usibus aquarum et aquarum ductibus* (with the use of water and water conduits) or similar, which are very frequent from the tenth century onwards, can be considered a reference to the frequent presence of canals in the Po Valley (Menant, 1993: 76; Squatriti, 1998: 184). The use of toponyms such as *Ronco* or *Runco* was also quite common by the tenth century (Saggioro, 2005: 90; Saggioro, 2006: 79 and 81). These toponyms, etymologically deriving from the Latin word *runcare* (also spelt *roncare*), refer to areas which originated both in deforestation and the uprooting of bushes, and from the reclamation of swampland (Baruzzi-Montanari, 1988: 101). It can be instructive to look at different descriptions of the same place, in this case *Villole,* probably Valverde near San Benedetto Po, in two different periods to see how the landscape was transformed. The charter by which Marquis Tedaldo of Canossa founded and granted some possessions to the monastery of San Benedetto in Polirone near Mantua (1007), mentions four *mansos* (cultivated areas allotted to tenant farmers), four *runcarii* (i.e. areas recently cleared) and three *mulinarii* (millers) in a place called *Villole.* In a charter of 1105, in which Marquise Matilde of Canossa confirmed several grants by her predecessors, *Villole* is called a *curtis,* or manor: it represents a unit of production. The borders of this productive unit are drawn up by human intervention: 'Villola habet a mane fines aggerem antiquum; a sera silvam que dicitur Castagnola, et silva Solamen, et Carpaneta, et sicut percurrit Rivolus usque in Piscarolam; a meridie fossa Piscarola usque in Perotolam, a septentrione fluvium Padi et in parte Sancti Benedecti'('Villole has as borders, from the east the ancient dyke, from the west the forest that is called Castagnola and the forests of Solamen and Carpeneta, and the course of the Rivolus to the Piscarola; from the south from the Piscarola canal till the Peretola canal, from the north the Po River and the land of Saint Benedict (the monastery)' Rinaldi-Villani-Golinelli, 1993: 99 and 216).

The aristocracy was able to impose works of ground amelioration and digging of canals on the colonists, as well as encouraging colonisation by granting rent exemptions, especially for the cultivation of crops intended to ensure a tradeable surplus, such as vineyards. In the lands of the abbey of Pomposa we also find, at least from the second half of the tenth century, exemption from the payment of rent in the case of the establishment of new vineyards or where new land was put under tillage: a charter dated 976, for example, grants an exemption of three years for newly tilled land, which is extended to nine years if the land is planted with vineyards (Bacchi, 1985: 137–138). The diffusion of this kind of obligation to put land under cultivation became very common in the twelfth century, and the contracts even included fines and the loss of usufruct rights if the peasant did not put the land under tillage (Panero, 1984: 84–94; Fumagalli, 1985a: 32–33). For example, we read in a charter dated 1173 that the bishop of Modena granted land to some families *ad laborandum, extirpandum, et meliorandum* (to work it, clear it and bring in ground amelioration'), but the bishop did this on the following condition: anyone among the renters who has not cleared (*extirpaverit*) his portion after two years from the next Saint Michael's day, will lose his portion which will be returned to the sacristy with full rights (*pleno iure*), and he will pay twenty imperial *solidi* for each *iugerum* to the sacristan) (Vicini, 1936: 70).

Many of these works of ground amelioration required co-operation between the colonists to dig canals and build dykes. An example can be seen in the land concessions made to colonists in the area of Ferrara in the eleventh and twelfth centuries, particularly near the Goro River. There, landlords systematically co-ordinated the grants of single fields in order to settle colonist communities in homogeneous areas (Bacchi, 1985: 144–149). Even so, the small fields that had been granted to individual farmers appear to have been structured into larger farming units and the farmers thus had to co-ordinate their activities to be able to clear the land, largely swamp, which appears to have been the main aim of these land grants in the first instance. Another interesting example comes from the area of Mantua, where the monastery of San Benedetto in Polirone pushed the rural communities to clear land near the Po and repair and maintain the dykes (Castagnetti, 1983: 7). In a charter dated 24 April 1197, Albertus, the abbot of San Benedetto in Polirone, granted the inhabitants of Villabona (and their heirs) some land near the Po. The charter stated that *Aggeres quisque faciat pro defensione tenute eius* (every farmer had to build dykes to protect his field) (Torelli, 1914: 365–366). This expression seems to suggest individual land management; however, we should not forget that this land had been collectively granted to the community.

Rural communities also proceeded with apparently autonomous initiatives of land clearance. The repopulation of the area of Brescello, near Reggio Emilia, is an important example of this (Settia, 1984: 139–140 and 255–256; Saggioro, 2005: 94). The presence of an interesting narrative source, the *Cronica sancti Genesii,* probably written at the beginning of the eleventh century, allows us to take a closer look at the process of land reclamation. This source tells us that during the tenth century the *ruricolae* (peasants) were trying to occupy areas which had been abandoned during the Lombard invasion. According to the chronicle, they had to reclaim the land from brambles and bushes (Cappelletti, 1859: 451). It is interesting to see that they were able to build a fortification *instar castelli* (in the manner of a castle) near the Po, although this structure proved to be weak (at least according to the chronicle) and unable to stand up to the assaults of their enemies (Cappelletti, 1859: 451). The colonisation process was probably possible because the peasants had some form of self-organisation, illustrated by the fact that they built a castle together. However, Marquis Adalberto-Atto was able to assert his authority over the colonists, imposing on the peasants to build a new castle (Fumagalli, 1971: 23; Provero, 2001).

The ability to impose such labour interventions may have played a role in the organisation of water management tasks, since it meant that the landlords could include work necessary to water management, such as the maintenance of canals. Lords probably took advantage of ancient communitarian and collective duties, which included the management of community structures like pits and roads, to impose their will on these forms of collective organisation (Provero, 2004: 375–377; Bognetti, 1978: 143–150). This intervention should be linked to the affirmation, from the end of the tenth century, of territorial lordship, where the local lords filled the empty places left by the crisis of central authority, establishing political dominion, with rights of ban, over particular territories. According to Wickham, 'The result of these trends was that by 1100, local lords, i.e. substantial landowners, often called *milites* in our texts, had much more authority over their poorer neighbours, *rustici,* than they had in 800 or 900' (Wickham, 2001: 137). But we should not consider this to be a passive acceptance of lordly authority by a submissive rural population. The rural communities often tried to achieve recognition of their rights to pastures, water, and woods (Fumagalli, 1976: 62–63; Menant, 2004: 239–240). In a charter dating from 920, twenty-five free men, *paterfamilias* living in the castle of Nogara near Verona, received confirmation that they could use the pastures on the lands belonging to the monastery of San Silvestro of Nonantola and collect wood in the forest belonging to the monastery (Andreolli-Montanari, 1983: 194–196).

We also have evidence of disputes between rural communities and lords concerning the possession of cleared land. This is illustrated, for example, by the dispute that arose between the community of Pegognaga and the monastery of San Benedetto in Polirone, which, from the eleventh century, was trying to claim as its own some of the land that had recently been reclaimed (Fumagalli, 1976: 34). In 1178 a witness reported on to what extent and by whom the land had been reclaimed: *parum iuxta Padum runcatum erat per homines Pigognage ante quam per S. Benedictum runcaretur* (a little had been cleared by the men of Pegognaga before it was cleared by Saint Benedict) (Torelli, 1914: 261–262). The monastery invested in the land's reclamation, but the land clearance had already been started by the men of Pegognaga.

Rural communities seemed to be trying to react against the strengthening of the political rights of lay and clerical lords over men and property rights by asking for recognition of their rights to marshes and forests and/or by proceeding to work independently on land clearance, thereby aspiring to the acquisition of stronger property rights on the newly cleared land. The key factor in this development was the opportunity for rural communities to act collectively. The general movement towards organisation and political reaction in rural communities would lead, from the middle of the eleventh century, to the granting of freedom charters. These charters were concerned with the limitation or suppression of seigniorial rights over free men, and they represented an important step towards the affirmation of the rights of autonomy and the self-rule of rural communities (Menant, 2004: 240–244; Panero, 2008). Rural communities seem to have been able to counterbalance the lords' strengthening control of property rights on land and water and of political rights over men that had been taking place between the ninth and the eleventh centuries. However, thus far we have only considered two players, the lords and the rural communities, but by the end of the eleventh and beginning of the twelfth century there was a definite increase in the importance of a third player: the cities.

V. Cities, water and land

Italy in the High Middle Ages was characterised by increasing urbanisation. Early estimates for the tenth century propose an urbanisation rate of 8.4%, but the urban population at least doubled between 1000 and 1300, while in the rest of Europe there was almost no growth in the rate of urbanisation (Malanima, 2005: 99–102). Perhaps more telling than these figures are the political and social developments that were taking place in the Italian cities. From the eleventh century, communal institutions were slowly taking shape, emerging from the pre-communal forms of organisation, giving the Italian cities a new political structure and strengthening their autonomy. The ruling elite of the early communes was quite heterogeneous. However, although the development of trade had created new sources of wealth and prestige, the communal elite of the eleventh and twelfth centuries still had strong connections to the countryside, often holding seigniorial rights outside the city (Tabacco, 1979; Jones, 1997: 103–151). The communes, as a collective body, increased their authority over the countryside, extending their jurisdiction, levying contributions in kind and in nature and acquiring rights to the *incolto* and water. An important step in the extension of the rights of the city to the countryside was represented by the Peace of Constance in 1183, in which Emperor Fredrick Barbarossa *de facto* recognised the legal entity of the Lombard communes and recognised their rights to pastures, woods, bridges, water and mills, thereby acknowledging their jurisdiction outside the city walls (Rapetti; 2003; Raccagni, 2008). However, the power of the cities had already started to extend into the countryside from the beginning of the twelfth century. For example, Milan and Como had already subordinated some rural communities under their jurisdictions to taxation and contributions in grain. Agents of Milan collected grain contributions from the communities in the basin of the River Serpio from 1111, if not before; between 1119 and 1125 Como attempted to repress a rebellion that had arisen among the rural communities, probably in objection to the fiscal impositions made by the commune (Manaresi, 1919: 2–8; Grillo, 2003). The cities could also use the men of the rural communities to provide labour for various works: for example, in the second half of the eleventh century Pavia asked people from the local communities to work on the locks of the Ticino River (Grillo, 2003: 45–46).

Italian communes in the Po Valley used their political authority in the countryside to push land clearance policies and to encourage (or impose) the plantation of crops that could feed the urban population (grain) or furnish a tradable surplus (vineyards) (Piccinni, 2002: 136–148). An interesting early example of the colonisation policies driven by the cities in the countryside is the commune of Verona. The city had food-supply problems in the last quarter of the twelfth century, evidenced by the famine of 1178 (Castagnetti, 1974). As a result of this, the *comune* tried to increase the tilled surface area, having recourse to irrigation or drainage according to the different soil conditions. In an arid area, the *Campanea maior,* the *comune* founded the free borough of Villafranca in 1184. A community of peasant families was settled on this land and an irrigation canal was dug. Each colonist was given a *mansus,* a cultivable unit measuring thirty-three Veronese 'fields' with one for a house and thirty-two *pro laborare* (to work). Six hundred and fifty-six additional fields were given to Villafranca Veronese as woodland or pasturage: in all 2,088 fields (Castagnetti, 1974: 365–366; 1977: 62–63; Sereni, 83–84). A different strategy was used to reclaim the area of the *Palus comunis Verone* (the swamp of the commune of Verona) where the settlement of Palù is now located. Here it was necessary to reclaim the land and undertake extensive canalisation. The *comune* had recourse to a consortium of private individuals, all Veronese citizens. Among them we find people belonging to the so-called *aristocrazia consolare* (the consular aristocracy), the upper class of the early *comune,* and even people who held fiefs and castles, those from juridical professions, such as two judges and ten notaries, as well as craftsmen. These people received extensive land grants: they were citizens who were able to invest in the reclamation process and sublet the land to farmers. As Castagnetti has shown, the participation of a consortium of private citizens in land reclamation can be seen as part of the concentration of landholding in the hands of the urban upper class in the area of Verona, where wealthy urban dwellers repeatedly purchased recently reclaimed swamplands, taking advantage of the high productivity of the soil. The *comune* would have initially received rent from the grantee in cash, and the grantee would also have paid a tithe to the *comune* for the maintenance of the canals. In 1216 the entire swamp was sold to the consortium. It is worth noting that the *comune* did not ask for any rent in kind. The food supply would have been provided by the members of the consortium themselves, who were interested in selling the products from the reclaimed land on the city market (Castagnetti, 1977: 62–84; 1974: 399–416).

The efforts of colonisation were also sustained, especially in the thirteenth and fourteenth centuries, by compelling the rural communities to clear woods and bushes in order to plant grain crops or vineyards. This is attested in the city statutes (Fumagalli, 1985b: 104–105). Additions from 1268 to the Statutes of Reggio Emilia (1265) order the cutting down of all the poplars, elms and oaks situated near the cultivated fields of the community of Rivalta (Campanini, 1997: 31–58). Interestingly, we also find obligations to till vineyards in the area controlled by the *comune*: the Statutes of Parma of 1255 force the people of Berceto to plant vineyards, and another article in the same statutes compels whoever owns land in the mountains to do the same (Ronchini, 1860:342–343). The cities also invested in the building of canals to expand irrigation (Balestracci, 1992: 442; Menant, 1993: 197–200; Boucheron, 2001). Cremona and Milan started to dig great irrigation and navigation canals in the second half of the twelfth century. The Naviglio Grande of Milan, which is now 49.9 km long, was started in 1177.

Rural communities also played an important role in the construction and maintenance of drainage and irrigation canals. In the second half of the eleventh century, Cremona established free boroughs near the points where the canals reverse the water in the main water collectors (rivers or other canals) to help with the management of these points (Menant, 1993: 174–175). The Statutes of Parma (1266–1304) give us another interesting example. They state that the *potestas* (supreme communal magistrates) had to repair and maintain the canal that ran across the lands of Sant'Ilario, Taneto and Prato Ottesola. The communities that benefited from this canal were required to maintain and supervise it (Ronchini, 1860: 382). The Statutes of Ferrara of 1287 establish that the officials of the cities were supposed to discuss matters of water management organisation with the inhabitants of the local communities in order to achieve greater efficiency. The rural communities were also supposed to provide large amounts of manpower to work on the river's embankments (Zupko-Laures, 1996: 69).

The rural communities played a key role in the development of land clearance and water management, but their actions appear to have been increasingly subordinated to the aims of the city-states. The history of the role of rural communities in water management is therefore part of the larger history of the subordination of the countryside to the political and economic aims of the cities. Although between the mid-twelfth and mid-thirteenth centuries the communes may have encouraged the emancipation of rural communities from seigniorial rights and the establishment of the rights and autonomous administration of communities in freedom charters (in contrast to the affirmation of rural lordships in the countryside as autonomous centres of power) (Chiappa Mauri, 2003: 241–247; Menant, 2004: 240–7 and 253), the rural communities became largely subordinated to the policies of the city-states. The communes used the rural communities to frame the government of their *contado* (countryside area surrounding a city-state), to fulfil their administrative tasks in the countryside, such as fiscal or judicial administration (Grillo, 2003; 67–81: Menant, 2004: 259–260), and also, as we have seen, to carry out water management and land clearance policies. In particular, land colonisation accelerated rapidly from the twelfth century, in an attempt to increase the tilled area for the cultivation of grain crops, vineyards and other crops that could be traded by the urban elites.

VI. Conclusions: The transformation of the landscape

This article has shown how human communities, radically changing the landscape of the Po Valley during the High and Late Middle Ages, moved from a strategy of exploitation to a strategy of transformation of wasteland areas. To take this comparison further, in the examples analysed in this paper there is no evidence of any clear adoption of a strategy of modification, defined by Rippon as the utilisation of marshes by making them: 'more suited them to agriculture through improving the drainage and the provision of rudimentary measures to prevent summer floods, while retaining their essentially intertidal character' (Rippon, 2001:141). In the absence of the modification approach it is possible to see an essential difference between hinterland and coastal areas: in these areas it was possible to adopt a policy of protecting areas of cultivation during the summer growing season against low high tides, without making any attempt to control higher winter floods. Such a strategy retained the essentially intertidal character of the area.

The dramatic transformation of the landscape of the Po Valley in the high and Late Middle Ages deeply affected the dynamics of natural risks in the area, undermining the sustainability of the interaction between man and nature. The disappearance of swampland (it would be more accurate to say the reduction of swampland) deprived the rivers of a precious relief valve: riverine marshes absorbed floods, functioning, as Magnusson and Squatriti wrote, 'as a sustainable flood control system'. Since trees contribute to soil cohesion, deforestation favoured the insurgence of landslides and floods. The progression of land clearance and the attempts to protect the recently tilled land brought with it the implementation of embankment systems along the river (Fumagalli, 1985b: 110–113; Hoffman, 1996: 638–639; Magnusson-Squatriti, 2000; Marchetti, 2002: 368–369). The water policies of the city-states were increasingly concerned with the construction and maintenance of these embankments: the Statutes of Ferrara from 1287 are very particular about the correct maintenance of the embankments, which were placed under the control of a 'judge of the embankments' (Zupko-Laures, 1996: 68–70). Similar officials also existed in other cities, for example the *arçinerius* of Modena, attested in 1306, was in charge of the maintenance of embankments and the regulation of waterways (Fumagalli, 1985b: 113). However, embankments constrained rivers in somewhat unnatural flows and contributed to the aggradation of river beds, an effect that would have been worsened by the increase in the number of canals discharging their water into rivers, leading to an increase in the magnitude of floods (Camuffo-Enzi, 1996: 439; Marchetti, 2002: 367–369) and flooding frequency (Pavese *et al.,* 1995: 159–165; Camuffo-Enzi, 1996: 444–44). Comparable consequences of increased discharge and extensive embankments can be observed also in some areas of Northern Europe during the Late Middle Ages, like the effects of increased sediment carriage on the Tay river estuary in Scotland analysed by Richard Oram, or the influence of embankment works on the genesis of the Wetstern Scheldt investigated by Tim Soens, both illustrated in this volume.

The Franciscan chronicler, Salimbene de Adam, describes the flood of the Crostolo, a small river near Reggio Emilia, on 23 June 1276: 'a very big rainfall came, so that the Crostolo grew to such a point that all the land between Rivalta (in the municipality of Reggio Emilia) and Bagnolo in Piano (8 km north of Reggio Emilia) was under water, and many people were killed or suffocated and died' (Salimbene de Adam, 1966: 723). In 1293, a third of Mantua was flooded (Fumagalli, 1985b: 119).

These radical changes in the landscape, which affected the environmental equilibrium of the valley, were related to changes in the social and economic structure. The affirmation of large landownership from the eighth/ninth centuries and its consolidation in the manorial system accelerated the process of land colonisation. The political affirmation of the Italian city-states subordinated the evolution of the landscape to urban economic and social interests, leading to a further expansion of colonisation. The result may have been the creation of a more 'humanised' landscape, but it was a landscape that was more vulnerable to environmental events like floods. The response to this evolution was the further 'humanisation' of the landscape through the establishment of dykes and embankments. These countermeasures could offer protection to the increasingly colonised landscape, but also accelerated a vicious circle of floods of increasing magnitude and frequency.

Bibliography

Andreolli, A. and Montanari, M. (1983) *L'azienda curtense in Italia. Proprietà della terra e lavoro contadino nei secoli VIII-XI,* Bologna.

Bacchi, T. (1985) 'Conquista del territorio e modificazione dei modelli impeditive. Le aziende fondiarie nel Ferraresei (secoli XI-XII)', in: B. Andreolli, V. Fumagalli and M. Montanari (eds), *Le campagne italiane prima e dopo il Mille. Una società in trasformazione,* Bologna, pp. 135–161

Balestracci, D. (1992) 'La politica delle acque urbane nell'Italia comunale', *Mélanges de l'Ecole Française de Rome. Moyen Âge,* 104, pp. 431–479

Baruzzi, M. and Montanari, M. (1988) '"Silva runcare". Storie di cose, di parole, di immagini', in: B. Andreolli and M. Montanari, *Il bosco nel Medioevo,* Bologna, pp. 101–109.

Bellettini, A. (1973) 'La popolazione italiana dall'inizio dell'età volgare ai nostri giorni', in: R. Romano and C. Vivanti (eds), *Storia d'Italia,* vol. 5/I, Turin, pp. 487–532.

Beloch, K.J. (1937) *Bevölkerungsgeschichte Italiens,* Berlin-Leipzig.

Bognetti, G. (1978) 'Sulle origini dei comuni rurali con speciali osservazioni pei territori milanese e comasco', in: F. Sinatti d'Amico and C. Violante (eds), *Studi sulle origini del comune rurale,* Milan, pp. 3–262.

Bolpagni, R., Bartoli M. and Viaroli P. (2003) 'Caratterizzazione di acque, sedimenti e idrofite nella Riserva Naturale Paludi di Ostiglia', *Studi Trentini di Scienze Naturali,* 80, pp. 169–174.

Bormann, E. (1888) *Corpus inscriptionum Latinarum,* vol. 11/1, Berlin.

Boucheron, P. (2001) 'Water and power in Milan, *c.*1200–1500', *Urban History,* 28, pp. 180–193.

Brenner, R. (1976) 'Agrarian Class Structure and Economic Development in Pre-Industrial Europe', *Past & Present,* 70, pp. 30–75.

Brogiolo, G.P. and Chavarria, A. (2005) *Aristocrazie e campagne nell'Occidente da Constantine a Carlo Magno,* Florence.

Brühl, C. (1973) *Codice diplomatico longobardo,* vol. 3/1, Rome.

Campanini, A. (1997) I rubricari degli statuti comunali di Reggio Emilia (secc. XIII-XVI), Reggio Emilia.

Camuffo, D. and Enzi, S. (1996) 'The analysis of two bi-millennial series: Tiber and Po River floods', in: Ph.D. Jones, R.S. Bradley and J. Jouzel, *Climatic Variations and Forcing Mechanisms of the Last 200 Years,* Berlin Heidelberg-New York, pp. 433–450.

Cappelletti, G. (1859) *Le chiese d'Italia dalla loro origine ai nostri giorni,* vol. 15, Venice.

Cardarelli, F (2003) *Encyclopaedia of Scientific Units, Weights, and Measures: Their SI Equivalences and Origins,* New York.

Castagnetti, A. (1974) 'Primi aspetti di politica annonaria nell'Italia comunale. La bonifica della "palus comunis Verone" (1194–1199)', *Studi medievali,* 3, 13, pp. 363–481.

Castagnetti, A. (1977) 'La Pianura Veronese nel Medioevo', in: G. Borelli (ed.), *Una città e il suo fiume,* vol. 1, Verona, pp. 33–138.

Castagnetti, A. *et al.* (1979), *Inventari altomedievali di terre, coloni e redditi,* Rome.

Roma Castagnetti, A. (1983) *Le comunità rurali dalla soggezione signorile alla giurisdizione del comune cittadino,* Verona.

Castaldini, D. (1989) 'Evoluzione della rete idrografica centropadana in epoca protostorica e storica', *Insediemanti e viabilitá nell'alto ferrarese dall'Etá Romana al Medioevo,* Ferrara, pp. 115–134.

Castellarin, A., Di Baldassarre, G. and Brath A. (2010) 'Floodplain management strategies for flood attenuation in the river Po', *River Research and Application,* 27, 1037–1047.

Cattaneo, C. (1844) *Notizie naturali e civili su la Lombardia,* vol. 1, Milan.

Chiappa Mauri, L. (2003) '*Statuti rurali* e autonomie locali in Lombardia (XIII-XIV secolo)', in: L. Chiappa Mauri (ed.), *Contado e cittá in dialogo. Comuni urbani e comunitá rurali nella Lombardia medievale,* Milan, pp. 227–268.

Ciriacono, S. (1994) *Acque e agricoltura. Venezia, l'Olanda e la bonifica europea in età moderna,* Milan.

Cortonesi, A. (2002) 'Agricoltura e tecniche nell'Italia medievale. I cereali, la vite, l'olivo', in: A. Cortonesi (ed), *Uomini e campagne nell'Italia medievale,* Rome-Bari, pp. 193–270.

Costambeys, M. (2009) 'Settlement, Taxation and the Condition of the Peasantry in Post-Roman Central Italy', *Journal of Agrarian Change,* 9, pp. 92–119.

Del Panta, L., Livi Bacci, M., Pinto, G., Sonnino E. (1996) *La popolazione italiana dal Medioevo a oggi,* Rome-Bari.

Duby, G. (1975) *Le origini dell'economia europea. Guerrieri e contadini nel Medioevo,* Rome-Bari (first French edition Paris, 1973).

Fabbri, P. (1975) *Il Padenna; L'Uomo e le acque nel ravennate dall'antichitá al medioevo,* Ravenna.

Fantoni, G. (2008) 'Water management in Miland and Lombardy in medieval times: an outline', *Journal of Water and Land Developmnet,* 12, pp. 15–25.

Francovich, R., Hodges, R. (2003) *Villa to Village. The Transformation of the Roman Countryside in Italy,* c.400–1000, London.

Fumagalli, V. (1970) 'Colonizzazione e insediamenti agricoli nell'Occidente altomedioevale: la Valle Padana', *Quaderni storici,* 14, pp. 319–338.

Fumagalli, V. (1971) *Le origini di una grande dinastia feudale. Adalberto-Atto di Canossa,* Tübingen.

Fumagalli, V. (1976) *Terra e società nell'Italia padana. I secoli IX e X,* Turin.

Fumagalli, V. (1985a) 'L'evoluzione dell'economia agraria e dei patti colonici dall'alto al basso Medioevo. Osservazioni su alcune zone dell'Italia settentrionale', in: B. Andreolli, V. Fumagalli and M. Montanari (eds), *Le campagne italiane prima e dopo il Mille. Una società in trasformazione,* Bologna, pp. 13–42

Fumagalli, V. (1985b) 'Il paesaggio si trasforma: colonizzazione e bonifica durante il Medioevo. L'esempio emiliano', in: B. Andreolli, V. Fumagalli and M. Montanari (eds), *Le campagne italiane prima e dopo il Mille. Una società in trasformazione,* Bologna, pp. 95–131.

Fumagalli, V. (1990) 'Il paesaggio delle campagne nei primi secoli del Medioevo', *L'ambiente vegetale nell'alto Medioevo,* vol. 1, Spoleto, pp. 19–53 (Settimane di studio del Centro italiano di studi sull'alto Medioevo; 37).

Fumagalli, V. (1991) 'Conquiste di nuovi spazi agrari', *Il secolo di ferro: mito e realtà del secolo X,* vol. 2, Spoleto, pp. 615–635 (Settimane di studio del Centro italiano di studi sull'alto Medioevo; 38).

Fumagalli, V. (1992) *L'uomo e l'ambiente nel Medioevo,* Rome-Bari.

Gasparri, S. (1990) *Il regno longobardo in Italia. Strutture e funzionamento di uno stato Altomedievale,* in: S. Gasparri S. and P. Cammarosano (eds), *Langobardia,* Udine, pp. 237–305.

Grillo, P. (2003) 'Comuni urbani e poteri locali nel governo del territorio in Lombardia (XII-inizi XIV secolo)', in: L. Chiappa Mauri (ed), *Contado e cittá in dialogo. Comuni urbani e comunitá rurali nella Lombardia medievale,* Milan, pp. 41–82.

Hoffman, R. (1996) 'Economic Development and Aquatic Ecosystems in Medieval Europe', *American Historical Review,* 101, pp. 631–669.

Jones, Ph. (1997) *The Italian City-State. From Commune to Signoria,* Oxford.

Laurence, R. (1999) *The roads of Roman Italy: mobility and cultural change,* London.

Leveau, P. (1993) 'Mentalité économique et grands travaux hydrauliques: Le drainage du lac Fucin. Aux origins d'un modèle', *Annales ESC,* 48, pp. 3–16.

Lo Cascio, E. and Malanima, P. (2005) 'Cycles and stability. Italian population before the demographic transition (225 BC-AD 1900)', *Rivista di Storia Economica,* 21, pp. 5–40.

Magnusson R. and Squatriti, P. (2000) 'The Technologies of Water in Medieval Italy', in P. Squatriti (ed.), *Working with Water in Medieval Europe: Technology and Resource-Use,* Leiden-Boston-Cologne, pp. 217–266.

Malanima, P. (2005) '*Urbanisation and the Italian Economy during the Last Millennium', European Review of Economic History,* 9, pp. 97–122.

Manaresi, C. (1919) *Gli atti del comune di Milano fino all'anno MCXVI,* Milan.

Manaresi, C. (1955) *I placiti del 'Regnum Italiae*', vol. 1, Rome.

Manaresi, C. (1957) *I placiti del 'Regnum Italiae*', vol. 2/1, Rome.

Marchetti, M. (2002) 'Environmental changes in the central Po Plain (northern Italy) due to fluvial modifications and anthropogenic activities', *Geomorphology,* 44, pp. 361–373.

Menant, F. (1993) *Campagnes lombardes au Moyen Âge,* Rome.

Menant, F. (2004) 'Les Chartes de franchises de l'Italie communale', in: M. Bourin and P. Martínez Sopena (eds), *Pour une anthropologie du prélèvement seigneurial dans les campagnes médiévales (XI*[e]*-XIV*[e]*).Réalités et représentations paysannes,* Paris, pp. 239–267.

Mikhailova, M.V. (2002) 'The hydrological regime and the peculiarities of formation of the Po river delta', *Water Resources,* 29, pp. 370–380.

Mitterauer, M. (2004) *Warum Europa? Mittelalterliche Grundlagen eines Sonderwegs,* München.

Montanari, M. (1979) *L'alimentazione contadina nell'alto Medioevo,* Naples.

Panero, F. (1984) *Terre in concessione e mobilitá contadina. Le campagne fra Po, Sesia, e Dora Baltea (secoli XII e XIII),* Bologna.

Panero, F. (2008) 'Signori e servi: una conflittualitá permanente', in M. Bourin, G. Cherubini and G. Pinto, *Rivolte urbane e rivolte contadine nell'Europa del Trecento: un confronto,* Florence, pp. 305–321.

Pasquali, G. (1981) 'I problemi dell'approvvigionamento alimentare nell'ambito del sistema curtense', *Archeologia medievale,* 8, pp. 93–116.

Pasquali, G. (1990) 'Viti e vino in Piemonte (secoli VIII-XII)', *Vigne e vini nel Piemonte medievale,* Cuneo, pp. 17–33.

Pasquali, G. (2002a) 'L'azienda curtense e l'economia dei secoli VI-XI', in: A. Cortonesi, *Uomini e campagne nell'Italia medievale,* Rome-Bari, pp. 5–71.

Pasquali, G. (2002b) *Le condizione degli uomini,* in: A. Cortonesi, *Uomini e campagne nell'Italia medievale,* Rome-Bari, pp. 76–122.

Pavese, M.P. *et al.* (1995) 'Three Historical Data series on Floods and Anomalous Climatic Events in Italy', in: R.S. Bradley and Ph.D. Jones (eds), *Climate since 1500 AD,* London, pp. 155–170.

Piccinni, G. (2002) 'Le campagne e le cittá (secoli XII-XV)', in: A. Cortonesi, *Uomini e campagne nell'Italia medievale,* Rome-Bari, pp. 125–189.

Pini, A. I. (1989) *Vite e vino nel Medioevo,* Bologna.

Pini, A.I. (1990) 'Vite e olio nell´alto Medieovo', *L'ambiente vegetale nell'alto Medioevo,* vol. 1, Spoleto, pp. 329–370.

Pinto, G. (1996) 'Dalla tarda antichità alla metà del XVI secolo', in: L. Del Panta, M. Livi Bacci, G. Pinto and E. Sonnino, *La popolazione italiana dal Medioevo a oggi,* Rome-Bari, pp. 15–71.

Pohl, W. (2001) '"Per hospites divisi". Wirtschaftliche Grundlagen der langobardischen Ansiedlung in Italien', *Römische Historische Mitteilungen,* 43, pp. 179–226.

Provero, L. (2001) 'Il sistema di potere carolingio e la sua rielaborazione nei comitati di Parma e Piacenza (secoli IX-XI)', in: R. Greci (ed.), *Studi sull'Emilia occidentale nel Medioevo: società e istituzioni,* Bologna, pp. 43–64.

Provero, F. (2004) 'Le corvées nelle campagne dell'Italia settentrionale', in: M. Bourin and P. Martínez Sopena (eds), *Pour une anthropologie du prélèvement seigneurial dans les campagnes médiévales (XIe-XIVe). Réalités et représentations paysannes,* Paris, pp. 365–380.

Raccagni, G. (2008) 'Il diritto pubblico, la pace di Costanza e i "libri iurium" dei comuni lombardi', in: D. Quaglioni and G. Dilcher, *Gli inizi del diritto pubblico, 2, da Federico I a Federico II,* Bologna-Berlin, pp. 309–340.

Rapetti, A.M. (2003) 'L'organizzazione distrettuale in Lombardia tra impero e cittá (IX-XII secolo)', in: L. Chiappa Mauri, *Contado e cittá in dialogo. Comuni urbani e comunitá rurali nella Lombardia medievale,* Milan, pp. 15–40.

Ravazzi, C. (1989) 'Analisi polliniche del riempimento del fossato. Dati paeoecologici sulle variazioni ambientali', in: M. Bernabó Brea and M. Cremaschi (eds), *La terramara di Poviglio,* Reggio Emilia, pp. 31–36.

Rinaldi R., Villani C. and Golinelli P. (1993) *Codice diplomatico polironiano (961–1125),* Bologna.

Rippon, S., (2001) 'Reclamation and regional economies of medieval marshland in Britain', in: B. Raftery and J. Hickey (eds), *Recent development in wetland research,* Dublin, pp. 139–158.

Rippon, S. (2000) *The transformation of coastal wetlands: exploitation and management of marshland landscapes in North West Europe during the Roman and medieval periods,* London.

Ronchini, A. (1860) *Monumenta Historica ad provincias Parmensem et Placentinam pertinentia,* vol. 2/5, Parma.

Rossini, E. (1979) *Contributi alla storia dell'agricoltura veronese,* Verona.

Saggioro, F. *et al.* (2004) 'Insediamento ed evoluzione di un castello della Pianura Padana (Bovolone VR) (1995–2002), Località Crosare e Via Pascoli', *Archeologia Medievale,* 32, pp. 169–186.

Saggioro, F. (2005) 'Insediamenti, proprietà ed economie nei territori di pianura tra Adda e Adige', in: G. P. Brogiolo, A. Chavarria Arnau and M. Valenti (eds), *Dopo la Fine delle Ville,* Mantua, pp. 80–107.

Saggioro, F. (2006) 'Ricognizioni, paesaggi ed esperienze di ricerca nei territori di pianura tra Veneto e Lombardia', in: N. Mancassola and F. Saggioro, *Medioevo, paesaggi e metodi,* Padua, pp. 65–86.

Salimbene de Adam (1966) *Cronica,* in: G. Scalia (ed.), vol. 2, Bari.

Sereni, E. (1997) *History of the Italian agricultural landscape,* Princeton (first Italian edition Bari, 1961).

Settia, A.A. (1984) *Castelli e villaggi nell'Italia padana,* Naples.

Scardia G., Muttoni G. and Sciunnach D. (2006) 'Subsurface magnetostratigraphy of Pleistocene sediments from the Po Plain (Italy): constraints on rates of sedimentation and rock uplift', *Geological Society of America bulletin,* 118, pp. 1299–1312.

Soens, T. (2009) *De spade in de dijk? Waterbeheer en rurale samenleving in de Vlaamse kustvlakte (1280–1580),* Ghent.

Squatriti, P. (1992) 'Marshes and Mentalities in Early Medieval Ravenna', *Viator,* 23, pp. 1–16.

Squatriti, P. (1998) *Water and Society in Early Medieval Italy, AD 400–1000,* New York.

Tabacco, G. (1979) *Egemonie sociali e strutture del potere nel Medioevo italiano,* Turin.

Torelli, P. (1914) *Regesto mantovano. Le carte degli archivi Gonzaga e di Stato in Mantova e dei monasteri Mantovani soppressi. (Archivio di Stato in Milano),* vol. 1, Rome.

Veggiani, A. (1983) 'Degrado ambientale e dissesti idrogeologici indotti dal deteroriramento climatico nell'Alto medioevo in Italia. I casi riminesi', *Studi Romagnoli,* 34, pp. 123–146.

Vicini, E.P. (1936) *Regesto della Chiesa Cattedrale di Modena,* vol. 2, Rome.

Violante, C. (1953) *La societá milanese nell'etá precomunale,* Bari.

Watson, A.M. (1974) 'The Arab Agricultural Revolution and Its Diffusion, 700–1100', *The Journal of Economic History,* 34, pp. 8–35.

Whittaker, C.R. and Garnsey, P. (1998) 'Rural Life in the Later Roman Empire', in: A. Cameron and P. Garnsey (eds) *The Late Empire, A.D. 337–425. Cambridge Ancient History,* vol. 13, Cambridge, pp. 277–311.

Wickham, Ch. (2001) 'Rural economy and society', in: C. La Rocca, *Italy in the early Middle Ages,* Oxford, pp. 118–43.

Wickham, Ch. (2005) *Framing the Early Middle Ages: Europe and the Mediterranean, 400–800,* Oxford.

Zupko, R.E. and Laures, R.A. (1996) *Straws in the Wind. Medieval Urban Environmental Law. The Case of Northern Italy,* Boulder.

15 Human impact on the coastal wetlands of Britain in the medieval period*

Stephen Rippon, University of Exeter

Abstract

This paper considers the impact of human communities upon the coastal wetlands of southern Britain during the medieval period. The way in which people changed from simply exploiting the rich natural resources of these environments, though modifying them by limited embankment/drainage, and then finally transforming them through permanently excluding the sea, is considered. The social context of reclamation is explored, with the evidence suggesting that the early stages were carried out piecemeal: in southern Britain there were no equivalents to the counts of Flanders and Holland, and the bishops of Utrecht, who assumed the regalian rights of unoccupied areas and subsequently transformed them through large-scale engineering projects. Monasteries became major land-owners in British coastal wetlands, with much of their land being reclaimed, but the reason why early Christian communities were attracted to these areas in the first place had more to do with their remoteness than their agricultural potential. It is also suggested that even on the estates of the great ecclesiastical institutions, it was often local communities who transformed the landscape, rather than the land-owners themselves. Some areas of coastal wetland in southern Britain were, however, left unreclaimed in the medieval period, and this reflects the very high value that was placed upon their rich natural resources. Key conclusions, therefore, are that it is social agency that predominantly determines the character of the cultural landscape in wetlands, and that the expansion of settlement into wetlands was part of a wider intensification in how the landscape of southern Britain was used during the 'long eighth century'.

I. Introduction

During the medieval period, most of the coastal wetlands of southern Britain (Figure 15.1) were transformed from intertidal saltmarshes into freshwater, reclaimed landscapes. This initial act of reclamation, and the subsequent maintenance of the drainage and flood defence systems, was a remarkable human achievement which mirrors that seen in the adjacent parts of mainland Europe. This work to transform coastal wetlands reflects the ingenuity and adaptability of past human communities in the face of difficult environ-

* I would like to thank the many archaeologists, historians and geologists with whom I have had fruitful discussions over the years, and in particular Robert van de Noort and Dries Tys who kindly commented on an earlier draft of this paper. In my work on the South Essex marshes I was assisted by Adam Wainwright. I would also like to thank the anonymous readers of this paper for their very helpful comments.

Landscapes or seascapes? The history of the coastal environment in the North Sea area reconsidered, ed. by Erik Thoen, Guus J. Borger, Adriaan M.J. de Kraker, Tim Soens, Dries Tys, Lies Vervaet & Henk J.T. Weerts, Turnhout, 2013 (*CORN Publication Series*, 13), pp. 333–351.

BREPOLS PUBLISHERS DOI: 10.1484/M.CORN.1.101559

Figure 15.1 Location of the wetlands referred to in the text

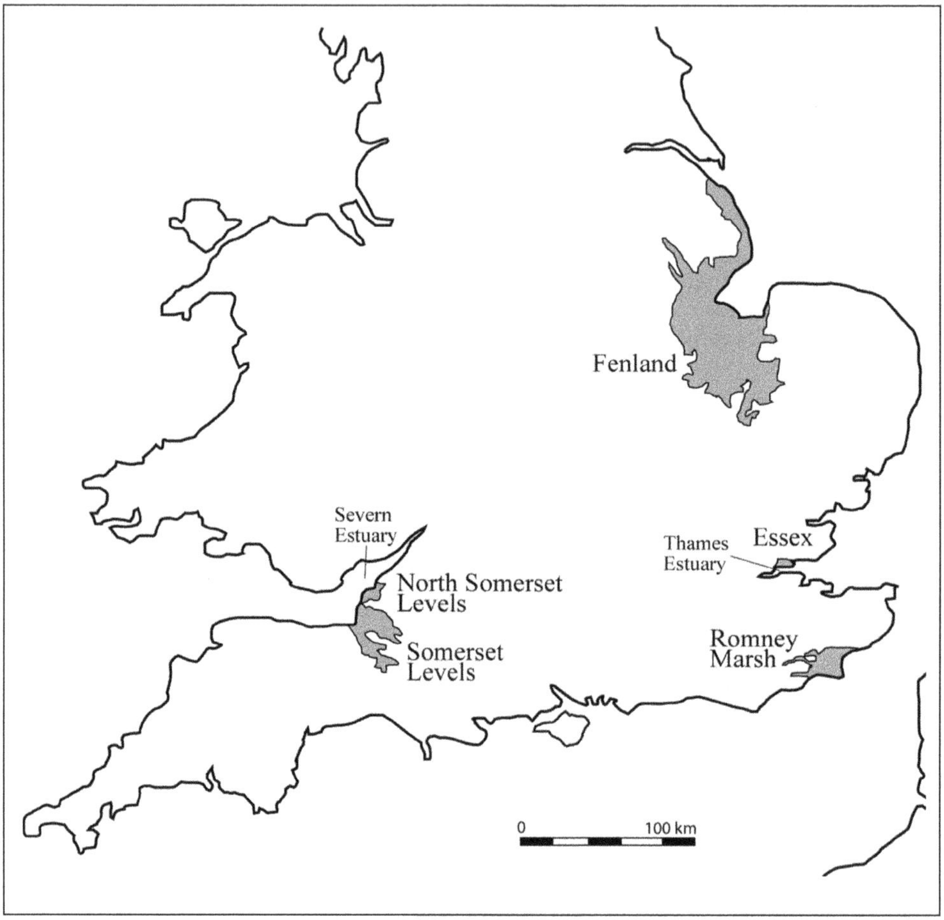

mental conditions, notably the need to exclude the sea from large areas of the landscape, and deal with the potential of freshwater flooding. Traditionally, such an intensification in land-use has been seen as a 'push into the margins', as rising populations meant that new agricultural land had to be found, and in many areas this was indeed the case as reclaimed wetlands became highly productive agricultural landscapes. Elsewhere, however, extensive areas of saltmarsh were left unreclaimed, and this suggests that instead of all human communities responding to pressures such as rising population in the same way, the approaches they adopted towards managing wetlands reflects the importance of social agency, with careful decision-making at a local level determining whether an area was reclaimed or not. This in turn reflects past perceptions of wetlands, their potential value, and the perceived challenges and risks of different strategies towards their utilisation, which is an example of how we have to understand the social context within which any landscape was created and then managed. Across medieval studies, there has been much debate about who was responsible for major landscape changes, with some scholars seeing powerful individuals or institutions as being the prime movers, while others have argued that it was peasant communities who undertook the greatest amount of change. This paper will therefore examine the increasing intensity with which human communities used coastal wetlands within its social context, and an example of the previously neglected unreclaimed marshes will remind us of the diversity of ways in which human communities faced up to the challenges of life in wetlands.

II. The exploitation, modification, and transformation of coastal wetlands

The creation of intensively managed wetland landscapes as they exist today was a long process, that saw human communities change from simply exploiting their rich natural resources, through modifying the environment in order to increase its productivity, to a complete transformation (i.e. reclamation) of the wetlands (Figure 15.2: Rippon 2000). This progression from exploitation, through modification, to transformation reflects the experience of communities in southern Britain up until the thirteenth century when the predominant theme was expansion and the increasing significance of human agency in the relationship between communities and their natural environment. This is in contrast to the late medieval period when there was flooding and erosion in some areas which led to a limited retreat both from the freshwater backfens and coastal areas, although in southern Britain this was on a very limited scale compared to the Low Countries.

The experience of exploitation, modification, and transformation is exemplified in the Fenland region of eastern Britain which has seen particularly extensive archaeological survey, excavation, and palaeoenvironmental analysis (Figure 15.3: Hall and Coles 1994; Crowson *et al.*, 2005). As is the case with many of the coastal wetlands in southern Britain, including those around the Thames Estuary in South East England and around the Severn Estuary in the South West (Rippon 1996; 2000; 2006), Fenland consisted of extensive areas of intertidal (i.e. below Mean High Water Spring Tide) saltmarshes (vegetated) and mudflats (unvegetated) that were open to the sea (i.e. there was no coastal barrier of sand or shingle), with lower-lying freshwater reed swamps, sedge fens, and raised sphagnum bogs in the inland 'backfens' that lay beyond the limit of tidal flooding (this lack of natural coastal barriers in many British wetlands being in sharp contrast to the wetlands of mainland North West Europe). Although there had been extensive settlement in the Roman period, in common with other British coastal wetlands most of Fenland was abandoned during the early medieval period and previously occupied land was sealed beneath a layer of silt and peat suggestive of a rise in natural sea level, creating a *tabula rasa* upon which medieval communities started afresh.

The wetlands of Fenland, and elsewhere in southern Britain, were recolonised later in the early medieval period (that in eastern Britain is sometimes referred to as the 'Early Saxon' period). Settlements were initially typically located at the interface of the alluvial saltmarshes and the freshwater fens, giving them access to a wide range of natural resources, and a series of excavations has revealed sites of an ephemeral nature that may

Figure 15.2 A model for the different strategies that human communities could adopt towards the utilisation of coastal wetlands.
This is based upon an earlier version published by the author in *The Transformation of Coastal Wetlands* (Rippon, 2000, Figure 19) but has been modified to reflect the better understanding that we now have of the 'modification' stage following a programme of survey and excavation on the North Somerset Levels (Figure 5; Rippon, 2006)

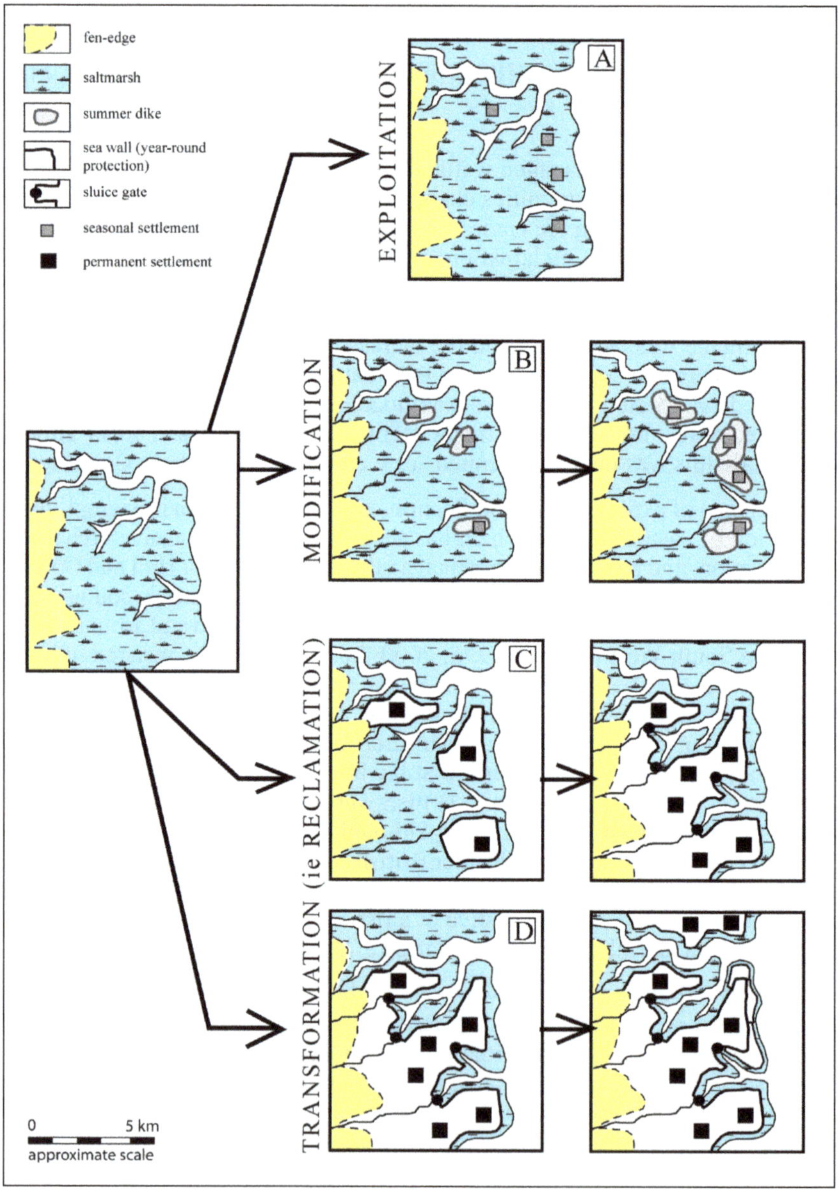

Figure 15.3 The development of the landscape in the Norfolk Marshland area of Fenland (source: Silvester, 1988 and Rippon, 2000, fig. 59)

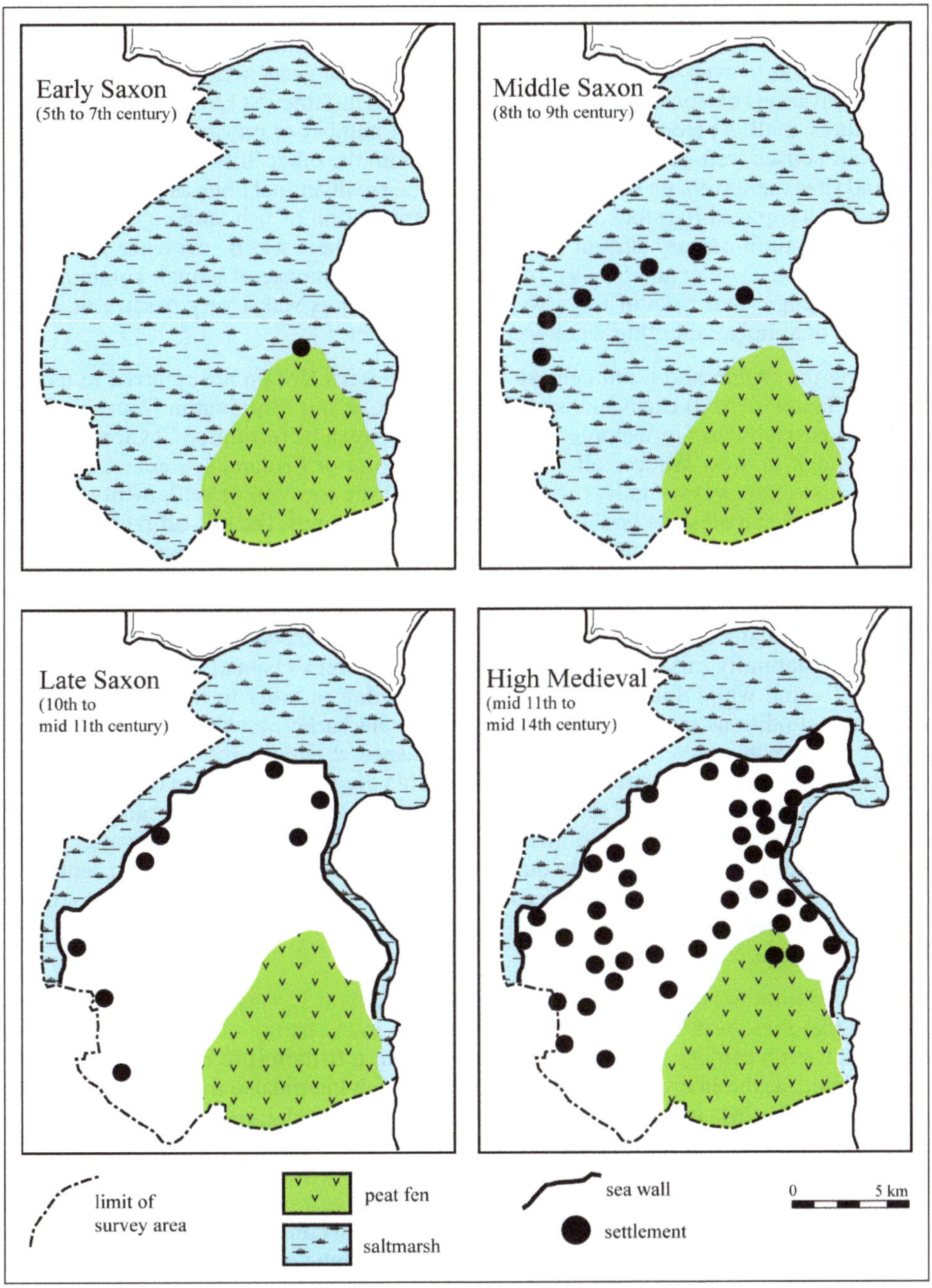

only have been occupied seasonally. At Gosberton Third Drove, for example, excavation of a pottery scatter that had been recorded through fieldwalking revealed two sub-rectangular ditched enclosures measuring *c.* 5 m by 7 m. The ditch of one of these enclosures was recut on several occasions, but a sparsity of domestic refuse from both enclosures suggests they may not have been occupied permanently. The intercutting and recutting of similar small gullies at Leaves Lake Drove in Pinchbeck also suggests periodic use of the site, while excavations of a pottery scatter at Hirdling Field in Tilney St Lawrence revealed no cut features at all. At Hoe Hills in Dowsby, a small number of pits and several possible post-built houses were recorded, but in common with the other fifth- to seventh-century Fenland sites there were no substantial ditches. The small animal bone assemblages may reflect locally grazed livestock, as well as the exploitation of fish, although in contrast to eighth- to ninth-century sites in the area (see below) there was no evidence for arable cultivation (Crowson *et al.*, 19–69). In this period the response of human communities to the challenges and risks of these wetland environments was a very passive one that consisted of simply exploiting what was easily on offer.

During the eighth and ninth centuries (that in eastern Britain are referred to as the 'Middle Saxon' period) excavations at Chopdike Drove and Mornington House in Gosberton, Rose Hall Farm in Walpole St Andrew, Hay Green in Terrington St Clement, and Ingleborough in West Walton, all show that the nature of Fenland settlement changed in three important ways: firstly, there was a shift in the location of settlement onto the higher coastal marshes; secondly, the complexity of features, the kill-off pattern of the livestock, and the large amounts of domestic refuse all suggests these were now long-lived and permanently occupied settlements; and thirdly, there was a dramatic change in the impact that human communities had on their environment through the digging of ditched drainage systems. These eighth- to ninth-century settlements were located on areas of naturally raised ground – silted up former creek channels, known locally as 'roddens' – and while ditches appear to have been dug in order to improve the drainage, water conditions remained brackish. Salt-tolerant barley was the main crop grown locally, along with some beans, peas, and flax/linseed, while the livestock consisted of mainly sheep, with some cattle and a few pigs. This digging of drainage ditches, the far more substantial nature of the settlements, and clear evidence for arable cultivation, all clearly distinguish this eighth- to ninth-century phase of landscape modification from the simple exploitation of natural resources that was seen in the fifth to seventh centuries (Crowson *et al.*, 2005: 69–190). In this period the response of human communities to the challenges and risks of these environments was therefore a more active one that consisted of changing their environment.

Palaeoenvironmental samples from the latest contexts at Terrington and West Walpole indicate that there was a further significant change in the nature of Fenland settlement around the tenth century (at the start of what in eastern Britain is referred to as the Late Saxon period) as the environment changed from brackish/intertidal to freshwater (Crowson *et al.*, 2005: 233, 261) indicating that reclamation had occurred. Excavations at Banklands in Clenchwarton have established that the sea wall there was indeed constructed around the tenth century, and palaeoenvironmental analysis shows that the bank was constructed on the surface of a saltmarsh that was dissected by a creek system and dotted with salt pans (Crowson *et al.*, 2005: 199). The response of human communities to the challenges and risks of these coastal environments was now a dramatic one that involved removing the landscape from the intertidal zone and thereafter carefully managing the watertable to prevent flooding.

Most other major coastal wetlands in southern Britain also saw this progression from exploitation, through modification, to transformation in the eighth to tenth centuries (e.g. Romney Marsh: Rippon, 2002; Barber and Priestley-Bell, 2008; Somerset: Rippon, 2006), and expansions of settlement such as this have traditionally been explained in terms of a change in the natural environment, notably a period of observed marine regression, and/ or an increase in population that forced people to live in more 'marginal' environments. Indeed, there has been a tendency to see such expansions of settlement as evidence for sea level change and/or an increase in population. While it is possible that both factors may have played a part, the context of this major intensification in how wetlands were used is more likely to be the far wider changes seen in the landscape of southern Britain during what historians now refer to as the 'long eighth century' (the late seventh to early ninth centuries: Hanson and Wickham, 2000; Rippon 2010). The distribution of eighth- to ninth-century settlement in Fenland, for example, suggests some degree of planning with scattered fifth- to seventh-century settlements being replaced by a far smaller number of eighth- to ninth-century sites distributed evenly along the higher coastal marshes. It is tempting to see some coordination behind this regularity, and if we look more widely at what was happening in the dryland landscapes of eastern and southern Britain at this time, this is precisely what we see. From around the eighth century, settlement patterns characterised by large numbers of isolated farmsteads and small hamlets were replaced by fewer, larger, villages, while there was an intensification of agriculture including an expansion in arable cultivation (Rippon, 2008). The context of this appears to have been the emergence of large estates owned by the crown, the church, and the aristocracy on which there was considerable investment in infrastructure such as water mills to provide power, causeways to improve communications, and the construction of large fish weirs to exploit new sources of food (Rippon, 2010). The intensification of wetland management therefore appears to have been part of a far wider and essentially dryland phenomenon of a social elite that was increasing its wealth through controlling the land and its resources, and could well have happened without there being any a period of observed marine regression.

Before the construction of sea walls along the coast, these early medieval settlements in Fenland were located on the naturally raised ground afforded by silted up former tidal channels (roddens). Elsewhere on the coastal marshes of southern Britain, however, such naturally raised areas did not exist due to the fine grained nature of the marsh and sub-tidal sediments, and in these areas a different solution had to be found to protect settlements from flooding. Around the Severn Estuary for example this response was to enclose small oval-shaped areas of marsh by low embankments ('ring dikes'), and at Puxton in Somerset, excavation and palaeoenvironmental analysis shows that the bank was built on the surface of a high saltmarsh (Figure 15.4: Rippon, 2006: 193–203). These small-scale 'ring dikes' may have been sufficient to have afforded year-round protection from flooding, or may simply have acted as 'summer dikes' that protected the enclosed areas from unseasonally high summer tides, but made no attempt to protect the enclosed area from winter flooding. Similar oval-shaped enclosures are found all around the Severn Estuary wetlands, suggesting it was a common human response to the challenge of cultivating crops in a marshland landscape, and another example of landscape modification in the early medieval period.

Figure 15.4 Aerial view of the former 'summer dike' at Puxton, on the North Somerset Levels

Until the thirteenth century, all the evidence from southern Britain points to a unilinear progression from exploitation, through modification, to transformation, although subsequently this trend was reversed in some areas. In the late medieval period there were increasing problems with flooding in many of the coastal wetlands as an increase in storminess marked the onset of a climatic deterioration. With the exception of Romney Marsh, however, where the breaching of a natural shingle barrier led to extensive land loss, the affects of this climatic deterioration were limited. In Fenland (Rippon, 2001) and the Somerset Levels (Rippon, 1997: 210–212) there appears to have been a retreat from some of the lowest lying backfens, and in various places coastal sea walls were set back in the face of erosion (e.g. Gwent Levels: Rippon, 1996: 97–99; Norfolk Marshland: Rippon, 2001: fig. 2). This loss of land was not, however, on anything like the same scale as seen in the Low Countries, in part because there had not been such extensive peat cutting and the resultant land subsidence, but also because sea defence in many areas relied wholly on man-made structures (sea walls) rather than natural coastal barriers (sand dunes etc) that were prone to breaching. What we see in southern Britain is that where human communities exerted the greatest control over their environment – for example where their wetlands were protected by sea walls rather than natural coastal barriers – they proved to be most resilient to environmental change resulting in only very limited reversal of the progression through exploitation, modification to transformation. The extensive survival of medieval drainage systems right across the wetlands of southern Britain reflects the resilience of these communities in the face of environmental change, and the sustainability of the engineering solutions that they created. In contrast, where human communities sheltered behind natural barriers – as in Romney Marsh – they were extremely vulnerable to environmental change, and it was here that the greatest setbacks occurred.

III. The social context of wetland colonisation

Across the coastal wetlands of southern Britain, the early medieval phase of landscape modification, associated with the rodden settlements in Fenland and 'ring dikes' around the Severn Estuary, was followed by their transformation through reclamation: the construction of a sea wall that protected the marsh from tidal inundation all year round, and its subsequent drainage through digging a network of ditches. In the model of exploitation, modification, and transformation as first proposed (Rippon, 2000: fig. 19), it was assumed that when 'summer dikes' were replaced by embankments that protected an area all year round, the sea walls ran continuously along the coast as they do today (Figure 15.2:D), with minor creeks blocked by dams through which water could flow via a sluice. The construction of such a sea wall would have been a major undertaking suggesting a significant change in the organisation of society that lay behind these wetland landscapes, with piecemeal local community effort being replaced by the vision and resources of a powerful landowner. In the wetlands of mainland North West Europe this does appear to have been the case, with the Counts of Flanders and Holland, and the Bishops of Utrecht, assuming the regalian rights to unoccupied areas and then coordinating their colonisation and drainage. In early medieval Britain, in contrast, there are no equivalents to the comital authorities and regalian rights were retained by the crown. Although there were powerful landowners – both secular and ecclesiastical – their estates tended to be scattered across wide areas, and by the tenth century the wetlands of southern Britain were divided between a multiplicity of land-owners: the Somerset Levels, for example, were divided between the abbeys of Athelney, Muchelney, and Glastonbury Abbey, the bishops of Winchester and Bath & Wells, and a large number of minor secular lords (Rippon, 2004a: figs. 27.13–14). It comes as no surprise, therefore, that a closer examination of the early phases of reclamation on the North Somerset Levels, suggests that there was an important stage between the 'ring dikes' and the existence of a continuous sea wall all the way along the coast and the major tidal rivers. Initially, individual marshes appear to have been embanked with sea walls running along side, but not across, the tidal creeks and it was only later that these creeks were blocked by dams in order to create a single continuous sea wall (Figures 15.2.B and 15.5).

IV. The perception of wetlands

Whilst the role of monasteries in draining marshland is well known, the reasons why they came to be located in wetland landscapes in the first place has received less attention, but it appears that rather than Christian communities being attracted to wetlands by their agricultural potential, it was actually the way that they were perceived as empty and challenging places that was important. Felix's eighth-century 'Life of St Guthlac', for example, suggests that his dwelling was situated within 'a most dismal fen of immense size ... now consisting of marshes, now of bogs, sometimes of black waters overhung by

Figure 15.5 The sequence of exploitation, modification and transformation of a part of the North Somerset Levels (after Rippon, 2006).
When the summer dikes (A) were first replaced by more substantial embankments that offered year-round protection the sea walls encircled discrete areas of marsh and did not cross major tidal creeks (B). A sea wall only ran all the way along the coast when these various individual reclamations were joined up (D)

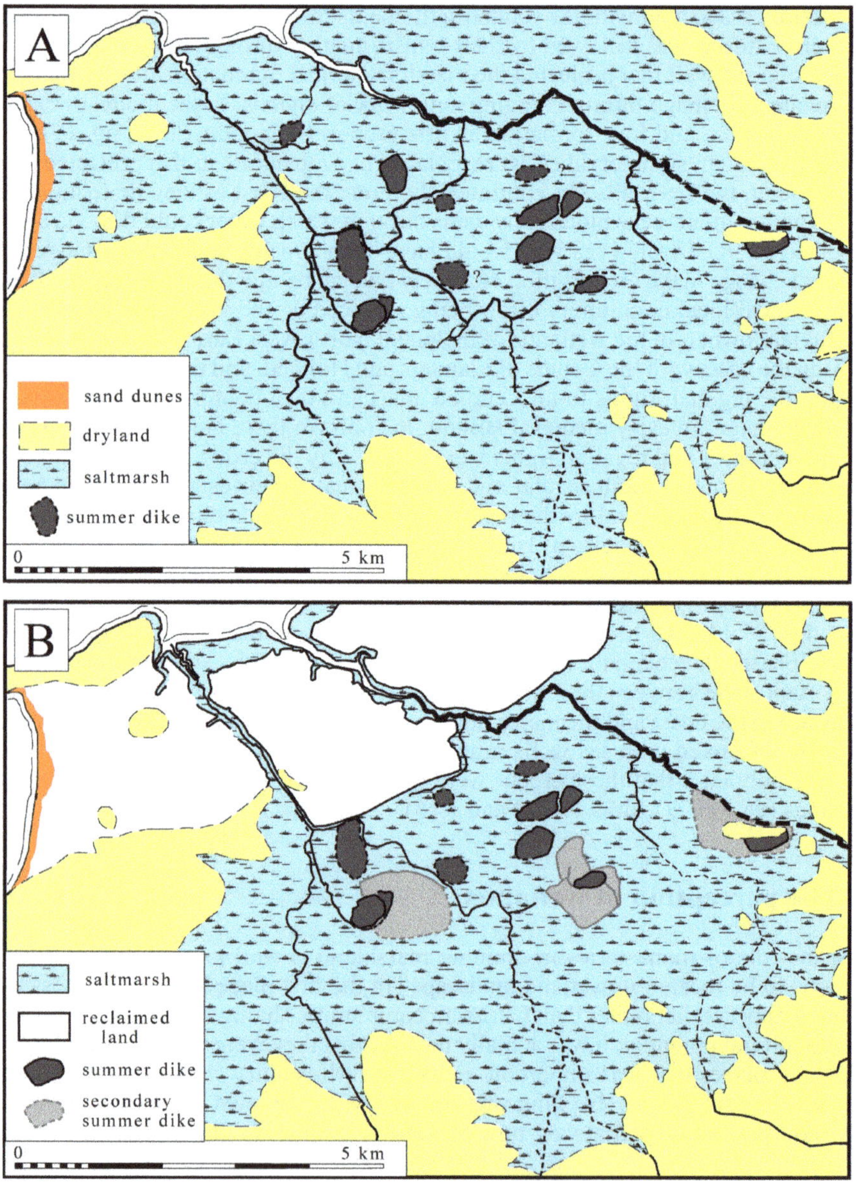

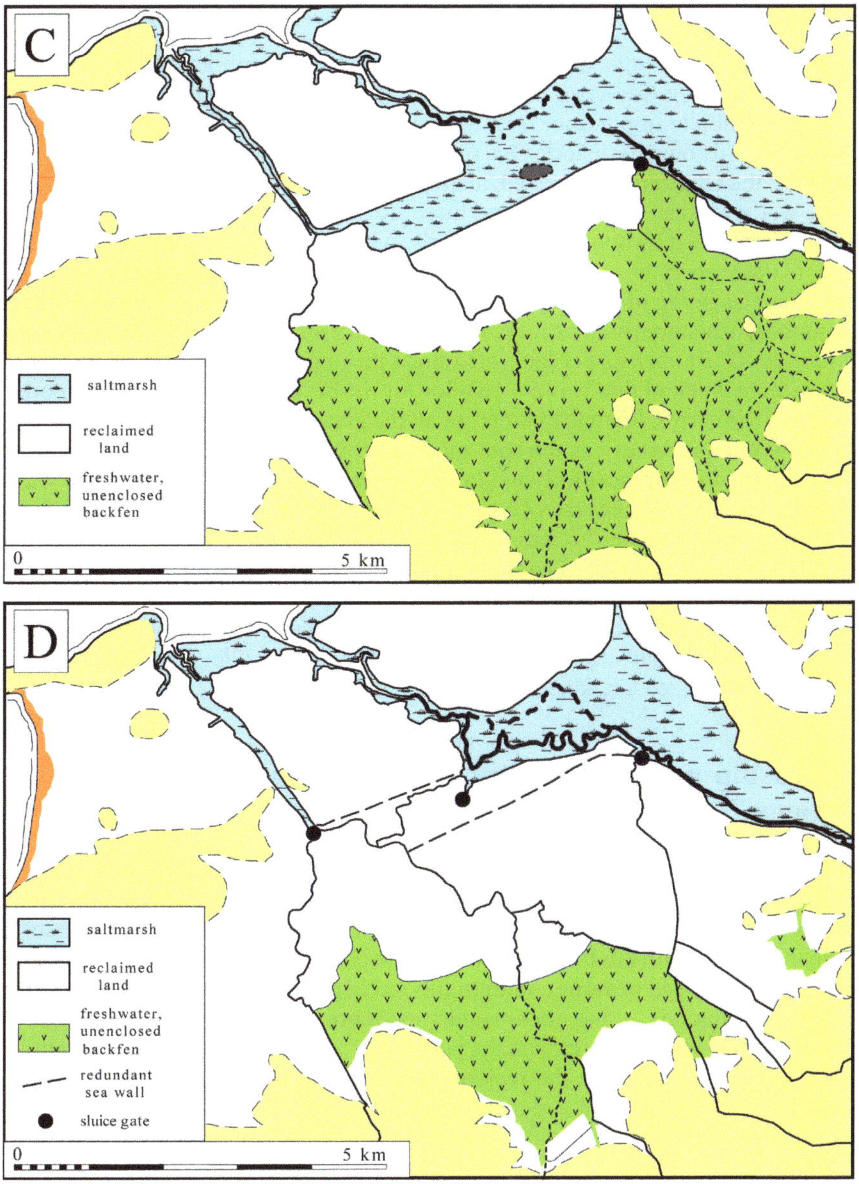
C
saltmarsh
reclaimed land
freshwater, unenclosed backfen
0
5 km
D
saltmarsh
reclaimed land
freshwater, unenclosed backfen
redundant sea wall
sluice gate
0
5 km

fog, sometimes studded with wooded islands, and traversed by the windings of tortuous streams' (Colgrave, 1956: 86–89; Hines, 2004: 62–70). Felix also describes how this 'dismal fen' was haunted by demons and phantoms, and other early medieval accounts of wetlands make it clear that they were also initially viewed as hostile environments, such as in the story of Beowulf in which the demon Grendel is said to have lived in a fen (Gardiner, 2008: 299–300). For early Christian communities these wetlands appear to have been hostile environments equivalent to the deserts sought out by the early founding fathers of the Church, and this may explain why so many early monastic foundations were on islands in wetland landscapes (Stocker and Everson, 2003; Rippon, 2004b).

One example of a major early Christian foundation within a wetland landscape is Glastonbury Abbey in Somerset, and its early association with an unoccupied wetland increased in symbolic importance over time which is reflected in the way that it created and embellished its own its history. One manifestation of this was the creation of a special jurisdiction – the Glastonbury Twelve Hides' – that embraced the peninsula of Glastonbury itself, the wetlands that surrounded it on three sides, and a series of small but prominent islands that protruded through these wetlands. Although some of the fertile drylands to the east of Glastonbury were included, it was clearly the islands and wetlands to the west that were seen as being of primary importance to the monks (Rippon, 2004b). Accounts of Glastonbury's history, such as John of Glastonbury's mid-fourteenth-century *Cronica sive Antiquites Glastoniensis Ecclesie*, also stress the importance of Glastonbury as having lain in a remote wetland that was a metaphor for the deserts which were so important in the history of the early Christian church: 'of all the places he [Patrick] might have chosen, he settled upon Glastonbury as the spot most apt for his triumphs over the devil, where he might be able to earn most fully the joys of a heavenly reward; for the place was then suited to heavenly vigils because of its remoteness from mankind being almost inaccessible because of the marshes' (Carley, 2001: 101). In this early medieval period Clarke (2006: 68) has suggested that 'The topographical reality of Glastonbury's situation as a pleasant island amid the floods and marshes of the Levels provides a consistent focus for mythologisation of the abbey's spiritual identity and destiny' with 'the practicalities of land reclamation and cultivation … transformed into literary allegories of spiritual work and protection' (and see Clarke, 2008).

V. Lordship and community: who was responsible for landscape change?

There has been much debate in Britain over who was responsible for the physical restructuring of medieval landscapes with Campbell (1981), for example, arguing that replanning could only have occurred where there was strong lordship, with villagers being either unable or unwilling to bring about dramatic change, in contrast to Dyer (1985) and Harvey (1989) who see peasant communities as being more important than lordship in shaping the countryside, citing documentary evidence for the remoteness of landowners from their often scattered estates, and the power and organisation that a community could exert locally. Recent work on the reclamation of wetlands is informing this debate in showing that both landowners and their communities were important in different ways, and in different places. It is particularly clear that different landowners adopted different strategies towards managing their estates. If we take the example of Glastonbury Abbey, in Somerset, we see a pattern of landscape management that is common to much of central Britain with settlement nucleated into often carefully planned villages, and the cultivated land around them arranged in two or three large open fields (sub-divided into strips and furlongs). A good example is Meare that lay on a small bedrock island at the centre of the Somerset Levels, and where a planned block of tenements lay next to the church/manor complex and to the north of an extensive open field (Rippon, 2004; 2008). This pattern of nucleated villages and open fields is seen right across the estates of Glastonbury Abbey which clearly preferred this approach towards managing its resources, and the repetitive nature of some village plans makes it clear that Glastonbury was closely involved in shaping the landscape on its estates. This was not, however, the case on the nearby lands belonging another great ecclesiastical land-owner in Somerset: the Bishops of Bath and Wells. They held particularly extensive estates on the wetlands of the North Somerset Levels, but interestingly markedly different settlement patterns and field systems were created through the reclamation process, with nucleated villages and open fields being laid out in close proximity to highly dispersed settlement patterns and predominantly enclosed field systems. It appears, therefore, that rather than the Bishops of Bath and Wells dictating the ways that settlement patterns and field systems should be created in the post reclamation landscape, it was left to local communities to make these decisions.

In dryland areas, differences in medieval landscape character such as this could be accounted for in a number of ways such as variations in the natural environment (geology, soils, and topography) or the cultural landscape that preceded the medieval period (what Roberts and Wrathmell (2000; 2002) have termed the 'antecedent landscape). In the North Somerset Levels, however, the natural landscape was uniform – the whole area was former saltmarsh – and as this had been flooded in the late/post-Roman period the 'antecedent landscape' was of uniform character too (an entirely natural saltmarsh with no surviving Romano-British settlement patterns or field systems to influence the form taken by the medieval landscape). Crucially, the area was also all owned by the same institution (the Bishop of Bath and Wells), and so the fact that some areas saw the creation of villages and open fields following their reclamation, while others had dispersed settlement patterns and enclosed field systems, suggests that local communities were left to carry out reclamation in the way that they preferred, rather than being told how to do so by the landowner (Rippon, 2006; 2008). This was in sharp contrast to the estates of Glastonbury Abbey, which appears to have had a very strong role in shaping how landscapes were managed, and shows how local and regional variation in the character of the British landscape in part results from its highly fragmented pattern of land-holding which is in marked contrast to the great comital estates seen in parts of mainland North West Europe.

This more piecemeal approach towards reclamation in Britain is also reflected in how the drainage and flood defences were maintained. During the medieval period the lord of the manor had responsibility for surveying the condition of the major embankments and watercourses on his land, but the responsibility for carrying out any necessary maintenance was part of the customary tenure and services owed to the lord of the manor by his tenants (Baildon, 1914: 78; Hembry, 1967: 20). In 1497, for example, John Payne held 100 acres of land in Banwell off Woodspring Priory in Somerset by service of finding two men for a day to scour a weir beside Banwell called Kinges River (*CIPM* Hen VII Vol I, No.1150). The major rivers and embankments were divided into 20 foot stretches known as 'ropes', each of which was the responsibility of a specific tenement. In 1654, for example, John Ewens of nearby Congresbury surrendered a customary tenement and twenty-eight acres in Wick, the services including scouring two ropes 'in the common river called the main Yeo' (BRO BMC/4/376). While there was some royal oversight through the establishment of periodic surveys carried out by the 'Commissioners of Sewers', this was a landscape in which responsibility for drainage and flood defence was highly fragmented with numerous landowners each having their own stretches of sea wall and drainage channels, and responsibility for these various flood defences being further sub-divided between each of the manorial tenants. There was a heavy reliance upon communities working together.

VI. Unreclaimed wetlands

In the examples discussed so far, the dominant theme in how medieval communities met the challenges of living in coastal wetlands was by first exploiting them, then modifying them, and then finally transforming them through extensive reclamation that permitted large-scale arable cultivation. In some regions, however, this pattern was not followed and large areas of coastal wetland were left unreclaimed as human communities responded to the challenges of living in wetlands in a different way. One example is the marshes of southern Essex, on the northern banks of the Thames estuary, east of London (Figure 15.1). Our first insight into the character of this landscape comes in the Domesday Book of 1086, which records many inland manors as having 'pasture for *x* sheep'. Later sources show that these manors lay in parishes that had detached parcels on the Thameside marshes and these would appear to reflect the rights that these communities had in what was once common land (Rippon, 2000: 201–207). Parish boundaries in southern Britain were generally established between the tenth and twelfth centuries AD (Blair, 2005: 368–425), and it was probably the need to identify parochial jurisdictions, coupled with a desire on the part of major land-owners to define property rights, that led to the break up of these commons and their division between the various communities who had held rights there.

Although there is some documentary evidence for reclamation and arable cultivation on the marshes of southern Essex in the twelfth and thirteenth centuries, particularly close to London, many areas around the outer Thames Estuary appear to have been left as intertidal saltmarshes and used for grazing sheep. These were not, however, areas of 'waste' land but highly prized grazing for which there was an active land market, with St Paul's Cathedral in London acquiring particular extensive estates such that 'One cannot fail to be impressed by the evidence of steady expansion from the early coastal holdings along the low-lying marshy shores of the adjacent river estuaries... It appears probable that throughout the [seventh to eleventh centuries] the economy of the St. Paul's estates was dominated by the sheep-farming industry of the Essex marshlands, for which there was a ready market in London, easily accessible by river and land' (Hart, 1957: 40). Westminster Abbey (also in London) also built up extensive holdings in the largest area of marshland, Canvey Island (*CPR 1354–58*, p.5; *CPR Rich. II 1391–6*, 133; Kirk, 1899–1910: 50; Kirk, 1929–49: 102; Harvey, 1977: 418–20), as did the Cathedral Priory of Christ Church in Canterbury that held a 'sheepwalk' on Canvey that amounted to 240 acres of 'salt marsh', providing pasture for 120 sheep (Nichols, 1926–8: 26, 39; Nichols,

1930: 28; Nichols, 1932a: 121 n1, 123, 139). A number of other monastic houses also held land on the southern Essex marshes, and alongside the land-holdings of secular estates, this led to a highly fragmented pattern of land-ownership.

The emphasis on sheep grazing and other natural resource exploitation is also illustrated in a court case in 1201 in which Thomas de Camville, lord of the manor of Fobbing was in dispute with Robert de Sutton over the marsh of 'Richeresnes' (Russellhead) in Canvey Island, alleging that during the reign of Henry II his grandfather had taken the profits from 'cheeses and wool and rushes' (Baildon, 1890: 36; Grieve, 1959: 5). William Camden's description from the late sixteenth century also paints a revealing picture of a pastoral landscape devoted to sheep in that Canvey Island was 'so low that often times it is quite overflowne, save for the hillocks cast up, upon which the sheepe have a place of refuge ... it keepeth about foure thousand sheepe, whose flesh is of a most sweet and delicate taste, which I have seene young lads taking women's function, with stooles fastened to their buttocks to milk, yea and to make cheeses of ewe's milk in those dairy sheedes [sheds] of theirs that they call there 'wickes' (quoted in Cracknell, 1959: 13).

This pattern of wetland management on Canvey Island and the adjacent Thameside marshes is quite unlike that seen in areas such as Fenland and the Somerset Levels, and would appear to reflect the landowners' perceptions of three things: the costs, risks, and benefits of reclamation. Reclamation involves a considerable initial capital investment in both labour and material, and then a recurrent cost in maintaining the flood defences. There is also a constant risk of flooding which could destroy that investment, although the potential return was great as agricultural productivity on a reclaimed marsh was high. In the case of these southern Essex marshes, however, the perception was clearly that the potential benefits of reclamation were not worth the costs and risks as the location of the marshes so close to London – by far the largest city in medieval and early modern Britain – meant that there was a large local market for fresh dairy produce. It would appear that until the seventeenth century the perception of local landowners was that the profits from simply grazing sheep on unreclaimed marshes and selling butter and cheese in the nearby market of London, would be greater than the profits from arable and the high costs and risks that the required reclamation would involve. Local people found a local solution to the problem of how to most effectively utilise this wetland environment.

VII. Conclusions

The idea that the coastal wetlands of southern Britain were first exploited, then modified, and finally transformed during the medieval period (up to the thirteenth century) was first published in Rippon 2000, and since then a series of archaeological and historical studies have both confirmed and refined that model. In Fenland (Crowson *et al.*, 2005) and on the North Somerset Levels (Rippon, 2006), extensive excavation and palaeoenvironmental analysis have added detail to the picture with the different character of the settlements and their economies in each of the stages becoming particularly clear. What we are also now seeing, however, is that the history of our coastal wetlands in the medieval period was not just one of unilinear progress towards reclamation, and that in different areas human communities responded to the challenges and risks of living in wetland areas in different ways. Traditionally, the human relationship with wetlands has been seen in a

Figure 15.6 The sequence of reclamation on the South Essex marshes, on the northern bank of the Thames Estuary.
Unlike many other British wetlands, the medieval reclamation here was extremely limited as the area's proximity to the urban market at nearby London meant that grazing livestock and the associated dairy production was more profitable than arable cultivation

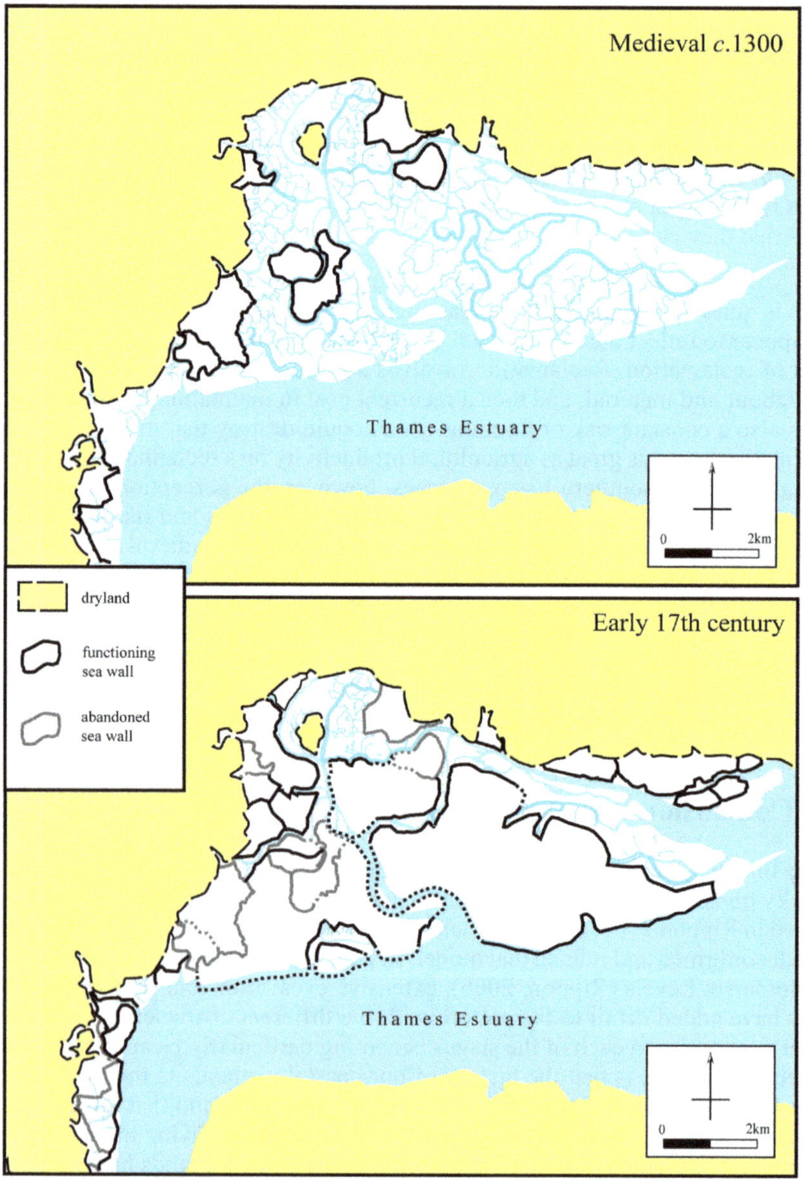

functional way with a strong emphasis on resource maximisation, but the expansion of settlement into these areas was not just a matter of progressive economic intensification but also reflects the ways that these environments were initially perceived and valued by early Christian communities as locations that were remote from the rest of society; only later was their reclamation seen as a means of increasing agricultural productivity. In many ways the development of wetland landscapes reflects wider patterns within medieval society, most notably in providing an example of the intensification of landscape management around the eighth century, and again in the tenth to thirteenth centuries, as major landowners tried to maximise the potential of their estates. The different histories of our coastal wetlands also reflect the relatively fragmented nature of medieval society and land-holding in Britain, a fragmentation that is particularly marked when compared to the near continent. As examples of landscapes with relatively uniform topography and soils, and a lack of 'antecedent' cultural landscapes to influence the form taken by medieval settlement patterns and field systems, these reclaimed wetlands also reveal something of the social context within which landscapes evolved. Some landowners were clearly highly proactive and interventionalist, leading to a common approach towards managing land, while on the estates of other landowners it appears to have been local communities who were left to decide whether they should live in nucleated villages or more dispersed settlement patterns. A key conclusion from this review is, therefore, that the marked local and regional variation in landscape character seen across medieval Britain resulted in part from the relatively fragmented patterns of landownership, and the different strategies towards landscape exploitation and management that these landowners and their tenant communities adopted. That so many of these medieval landscapes survive today reflects the resilience of the human communities living there, and the sustainability of the landscapes they created.

Abbreviations used for primary sources

BRO: Bristol Records Office

CRP: *Calendars of Patent Rolls.* London: Her Majesty's Stationary Office.

CIPM: *Calendars of Inquisitions Post Mortem.* London: Her Majesty's Stationary Office.

Bibliography

Baildon, W.P. (ed) (1890) *Select Civil Pleas 1200–1203,* vol. 1.

Baildon, W.P. (1914) *Calendar of the Manuscripts of the Dean and Chapter of Wells*, vol. 2, London.

Barber, L. and Priestley-Bell, G. (2008) *Medieval Adaptation, Settlement and Economy of a Coastal Wetland: The Evidence from Around Lydd, Romney Marsh, Kent*, Oxford.

Blair, J. (2005) *The Church in Anglo-Saxon Society*, Oxford.

Campbell, B.M.S. (1981) 'Commonfield origins – the regional dimension', in: T. Rowley (ed), *The Origins of Open Field Agriculture*, London, pp. 112–129.

Carley, J.P. (2001) *The Chronicle of Glastonbury Abbey. An Edition, Translation and Study of John of Glastonbury's 'Cronica sive Antiquites Glastoniensis Ecclesie'*, Woodbridge.

Clarke, C. (2006) *Literary Landscapes and the Idea of England, 700–1400*, Woodbridge.

Clarke, C. (2008) 'The allegory of landscape: land reclamation and defence at Glastonbury Abbey' in: M. Carr, K.P. Clarke and M. Nievergelt (eds), *On Allegory: Some Medieval Aspects and Approaches from Chaucer to Shakespeare*, Newcastle, pp. 87–103.

Colgrave, B. (1956) *Felix's Life of Saint Guthlac*, Cambridge.

Cracknell, B.E. (1959) *Canvey Island: the History of a Marshland Community*, Leicester (Department of English Local History, University of Leicester, Occasional papers; 12).

Crowson, A., Lane, T., Penn, K. and Trimble, D. (2005) *Anglo-Saxon settlement on the Siltland of Eastern England*, Sleaford (Heritage Lincolnshire, Lincolnshire Archaeology and Heritage reports series; 7).

Dyer, C. (1985) 'Power and Conflict in the Medieval Village', in: D. Hooke (ed.), *Medieval Villages*, Oxford, pp. 27–32.

Gardiner, M. (2008) 'The wider context', in: L. Barber and G. Priestley-Bell (eds), *Medieval Adaptation, Settlement and Economy of a Coastal Wetland: the Evidence from around Lydd, Romney Marsh, Kent*, Oxford, pp. 297–304.

Grieve, H. (1959) *The Great Tide: The Story of the 1953 Flood Disaster in Essex*, Chelmsford.

Hall, D. and Coles, J. (1994) *Fenland Survey: An Essay in Landscape and Persistence,* London.

Hanson, L. and Wickham, C. (eds) (2000) *The Long Eighth Century: Production, Distribution and Demand*, Leiden.

Hart, C. (1957) *The Early Charters of Essex: The Norman Period*, Leicester.

Harvey, B.F. (1977) *Westminster Abbey and its Estates in the Middle Ages*, Oxford.

Harvey, P.D.A. (1989) 'Initiative and authority in settlement change', in: M. Aston, D. Austin, D. and C. Dyer (eds), *The Rural Settlements of Medieval England*, Oxford, pp. 31–43.

Hembry, P.M. (1967) *The Bishops of Bath and Wells 1540–1640: Social and Economic Problems*, London.

Hines, J. (2004) *Voices in the past: English Literature and Archaeology*, Woodbridge.

Kirk, R.E.G. (1899–1910) *Feet of Fines for Essex I (1182–1272)*, Colchester.

Kirk, R.E.G. (1913–1928) *Feet of Fines for Essex II (1272–1326)*, Colchester.

Kirk, R.E.G. (1929–1949) *Feet of Fines for Essex III (1327–1422)*, Colchester.

Nichols, J.F. (1924) 'New Light on History of Milton Hamlet', *Transactions of the Southend-on-Sea and District Antiquarian and Historical Society*, I, pp. 172–181.

Nichols, S.J.F. (1926–1928) 'Milton Hall: The Extent of 1309 and an inventory of 1278', *Transactions of the Southend-on-Sea Antiquarian and Historical Society*, II, pp. 7–33 and 39–40.

Nichols, S.J.F. (1930) 'Farming operations in the fourteenth century', *Transactions of the Southend-on-Sea Antiquarian and Historical Society*, II, pp. 25–28.

Nichols, J.F. (1932a) 'Milton Hall: The Compotus of 1299', *Transactions of the Southend-on-Sea Antiquarian and Historical Society*, II, pp. 113–167.

Nichols, J.F. (1932b) 'Southchurch Hall', *Journal of the British Archaeological Association (New Series)*, 37, pp. 101–117.

Rippon, S. (1997) *The Severn Estuary: Landscape Evolution and Wetland Reclamation*, London.

Rippon, S. (2000) *The Transformation of Coastal Wetlands*, London.

Rippon, S. (2001) 'Adaptation to a changing environment: the response of marshland communities to the late medieval "crisis"', *Journal of Wetland Archaeology*, 1, pp. 15–40.

Rippon, S. (2004a) *Historic Landscape Analysis: deciphering the palimpsest*, York.

Rippon, S. (2004b) 'Making the most of a bad situation? Glastonbury Abbey and the exploitation of wetland resources in the Somerset Levels', *Medieval Archaeology*, 48, pp. 91–130.

Rippon, S. (2010) 'Landscape change in the "Long Eighth Century"', in: N. Higham and M.J. Ryan (eds) *The Landscape Archaeology of Anglo-Saxon England*, Woodbridge, pp. 39–64

Rippon, S. (2006) *Landscape, Community and colonisation: the North Somerset Levels During the 1st to 2nd Millennia AD*, York (Council for British Archaeology Research Report; 152).

Roberts, B. and Wrathmell, S. (2000) *An Atlas of Rural Settlement in England*, London.

Roberts, B. and Wrathmell, S. (2002) *Region and Place*, London.

Silvester, R. J. (1988) *The Fenland Project Number 3: Marshland and the Nar Valley, Norfolk* (East Anglian Archaeology; 45).

16 Estuarine environments and resource exploitation in eastern Scotland *c.* 1125 to *c.* 1400. A comparative study of the Forth and Tay Estuaries

Richard ORAM, University of Stirling

Abstract

The two major eastern Scottish river estuaries – the Forth and the Tay – have been the scene of dynamic human-natural environmental interaction across six millennia but with the most rapid and extreme changes before the late eighteenth century occurring in the period c. 1125 to c. 1400. Differences in estuarine hydrology – especially salinity and silt levels – contributed significantly to the development of divergent exploitation strategies around the two inlets. While the non-mercantile regional economy around the Tay estuary remained primarily agriculture and fisheries based, that of the Forth was marked by development of a major salt manufactory. This divergence had different effects on the extensive estuarine plains of both rivers – the Carse of Gowrie and the Carse of Stirling – with the peatlands of the former being drained largely for agriculture and the latter originally exploited chiefly for fuel. This essay considers factors influencing the different processes of exploitation in the two estuaries, the socio-economic consequences which these operations stimulated, and the short and long term physical impacts which resulted – principally exposure to flooding through increased run-off and heightened risks of marine transgression in the upper reaches of the estuaries.

I. Introduction: the physical setting

Scotland's North Sea coastline is broken by two major inlets, the Firth of Tay and the Firth of Forth. Barely 50 km apart along the length of their tidal zone and sharing certain fundamental topographical characteristics, the two firths are significantly different in form and scale. The similarities and contrasts between the two, however, provides an excellent case for a comparative study which can potentially illuminate several of the research questions posed in this current project from the perspective of the north-western limits of the coastlands which border the North Sea. This northern perspective, from a region which has experienced a positive post-glacial isostatic adjustment rather than the negative adjustments experienced in the regions at the southern end of the North Sea, provides an alternative vision of the consequences of sea-level changes and associated processes of coastal erosion, deposition and inundation. As estuaries, they also provide a case study for exploration of different terrestrial and hydrological processes than are experienced in maritime zones proper, while still sharing the characteristics of the wider

Landscapes or seascapes? The history of the coastal environment in the North Sea area reconsidered, ed. by Erik THOEN, Guus J. BORGER, Adriaan M.J. DE KRAKER, Tim SOENS, Dries TYS, Lies VERVAET & Henk J.T. WEERTS, Turnhout, 2013 (*CORN Publication Series*, 13), pp. 353–377.

BREPOLS PUBLISHERS DOI: 10.1484/M.CORN.1.101560

Figure 16.1 The Tay and Forth basins, general location and associated hydrology (source: W. Jamieson)

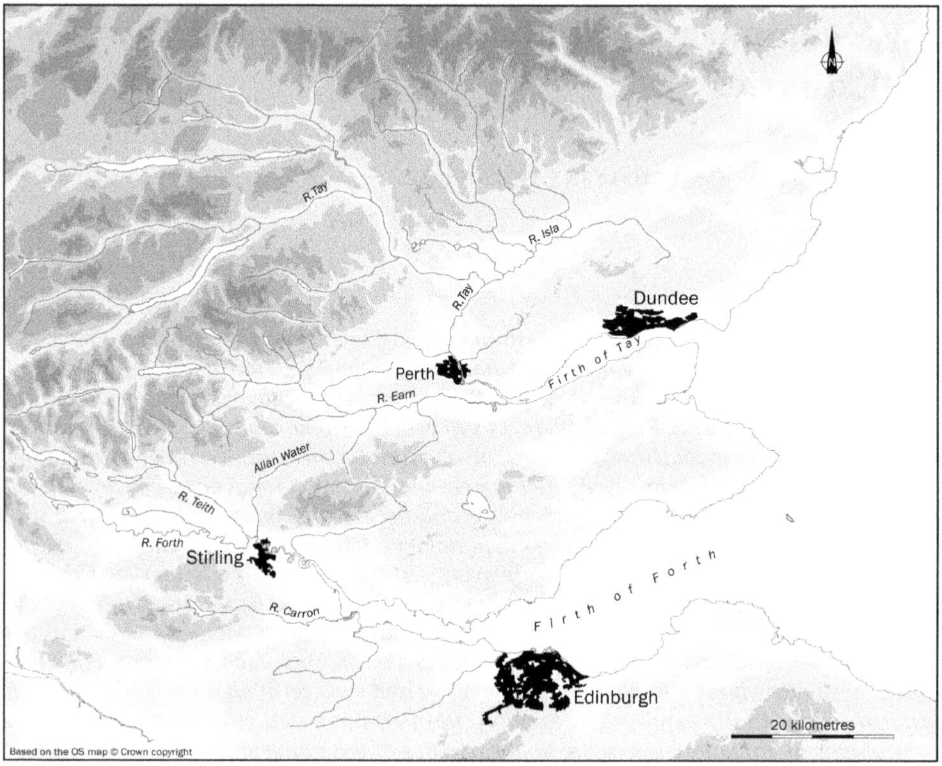

coastal hinterlands which they penetrate. The more northerly latitudes and direct exposure of the region to Atlantic weather regimes, moreover, offers the potential for further comparison and contrast with climate-related impacts in territories around the southern part of the North Sea. More northerly latitudes additionally offer contrasts with the range and types of natural resources available for exploitation, while geological and hydrological differences between the two estuaries offer a sharper infra-regional comparison. From a human perspective, differences in terms of available resource base are matched by differences in rates, manner and intensity of development in the Middle Ages, a situation which arises from the place of one estuary at the heart of the Early Medieval kingdom of *Alba* and that of the other as the axis upon which *Alba*'s High Medieval successor state, the kingdom of Scotland, was focussed. This shift in the political and economic centre of gravity from one to the other permits the two estuaries to be used as test cases to illustrate the variable experience of commercialisation, population movement and its consequences for land-use and natural resource management and exploitation across a long time-span.

Figure 16.2 The Tay estuary region (source: W. Jamieson)

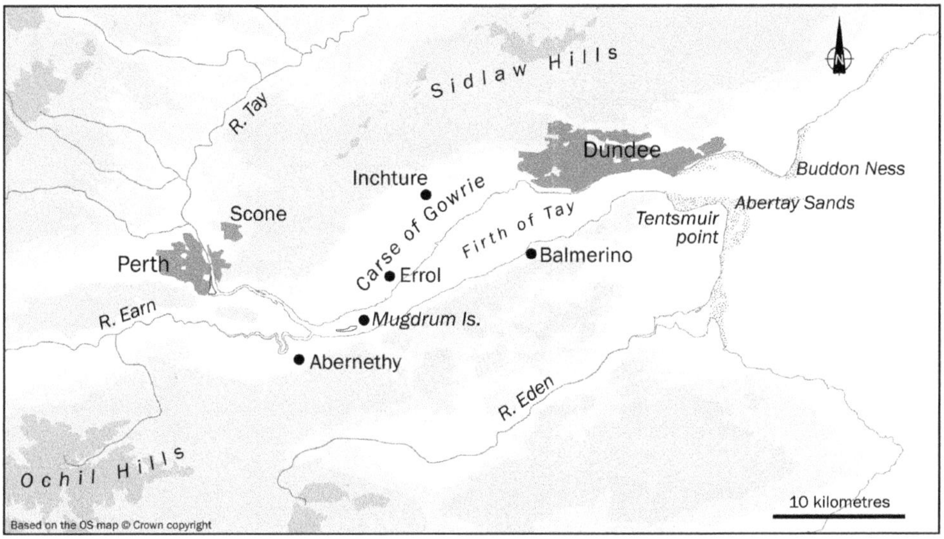

Although the Tay has a catchment of only 4970 km^2, some sixty per cent smaller than the Thames basin, its average discharge of 175 m^3/s[1] at the head of its tidal reach renders it the largest river in Great Britain by volume of discharge (http://www.tdsfb.org/river-tay.htm [accessed 10 March 2010]). A further 30 m^3/s is added to its estuary by the River Earn, which joins the Tay 6 km below Perth. This combined discharge is three times greater than that of the Thames (65.8 m^3/s) and more than twice that of Scotland's next largest river catchment, the Tweed (79 m^3/s). At 193 km in length from remotest headwaters to rivermouth, it is also the longest river in Scotland; around 30 km of that length is the tidal, estuarine zone. This zone, extending from Perth to the North Sea between Buddon Ness in Angus and Tentsmuir Point in north-east Fife, has a maximum width of around 5 km between Invergowrie on the north bank and Balmerino on the south before narrowing to an average of 2 km for its final 20 km. The volume of fresh water discharging through this estuary significantly lowers the average salinity level of the tidal zone.

At 1712 km^2 the Forth catchment is equivalent to only forty per cent of the area of the Tay basin, its non-tidal length extending barely 45 km west of Stirling, with a tidal stretch between Stirling and Kincardine of seventeen kilometres. East of Kincardine, the river broadens into an estuary which widens to 31 km in its outer zone, 90 km to the east. The estuarine zone has two distinct sections, an inner firth of 30 km between Kincardine and the Queensferry Narrows, and an outer firth of 60 km from Queensferry to the Isle of May; this outer zone is more marine than estuarine in character, with higher general salinity levels than the inner zone and both sectors having a higher average level

[1] m^3/s = a cumsec or cubic metre of discharge per second.

Figure 16.3 The Forth Estuary region (source: W. Jamieson)

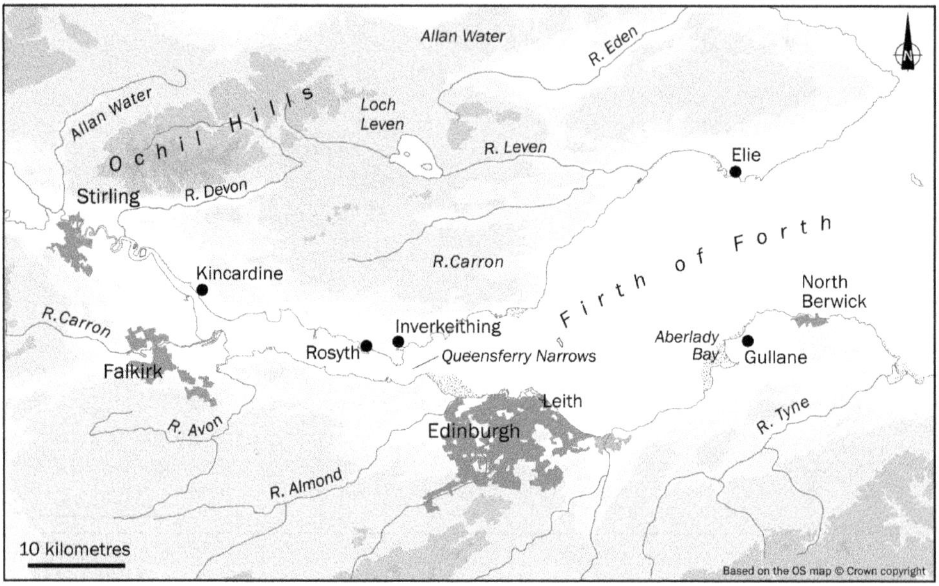

than the Firth of Tay. Indeed, the wide firth of the Forth was regarded as a sea in its own right – the Scottish Sea (Skene (ed.), 1872: volume ii, 32) – with the land to its north being *Scotia ultra marina* to the thirteenth-century St Alban's monk Matthew Paris, and as late as the 1500s it was described as 'the sea of Forth' (Anonymous (ed.), 1840: Appendix, document 37).

As sediment delivery mechanisms carrying materials washed down from the hilly hinterlands of their basins, the Tay is substantially more productive than the Forth. Consequently, the Tay estuary experiences a far higher level of silting and sand-bank formation than the Forth and dredging is necessary to keep a navigable deep-water channel open to connect the still functioning port of Perth with the North Sea. A second consequence of the Tay's high volume of sediment carriage is new land formation at the rivermouth, where deposition east of Tentsmuir Point has created the fastest-growing area of land in Scotland. The river-bar and sand-bank system of the Abertay Sands which extends east from Tentsmuir is a rapidly shifting complex and navigation hazard for vessels entering the estuary. There is no equivalent to these characteristics in the Forth estuary, which is edged instead by more stable tidal sand and mudflats except at the outflows of the larger streams entering the firth and in Aberlady Bay and near Gullane in Lothian, or Elie in Fife, where deposition is creating new areas of salt-marsh or where there were formerly mobile dune systems.

There are broad but superficial similarities between the topography of the land adjoining the tidal stretches of the Tay and the Forth (west of Queensferry) in that both estuaries are flanked for much of their length by expanses of carseland (Scots, *kerse*, low-lying, alluvial land), those on the north banks being backed by steep south-facing escarpments which rise to over 300 m OD in less than half of 1 km. These carses, the result of a geological uplift through post-glacial isostatic adjustment, were formerly areas of mixed wetland (in the Forth mainly raised peat-mosses over alluvial and estuarine clays) and scrub woodland with higher 'inches' (Gaelic, *innis*, an island) or 'tullachs' (Gaelic, *tulach*, a mound or knowe) of gravel. The Carse of Gowrie, which extends for 30 km along the north bank of the Firth of Tay, is now one of the prime agricultural areas of eastern Scotland. Carseland extends east and west of Stirling on both sides of the Forth, most now bearing mixed agricultural land and extensive urban settlement. The deep channel of the Tay lies closer to the south bank for most of the length of the estuary and its scouring effect has prevented the development of carse and flats. Instead, low cliffs and steep shorelines here back on to broad ledges containing pockets of alluvial soils before the ground rises steeply into a north-eastern extension of the Ochil Hills. On either side of the rivermouth, hillier ground yields to expanses of sandy links interspersed with areas of salt-marsh and dune system extending east of Dundee as far as Buddon Ness on the north and Tentsmuir Point on the south. The long outer firth zone of the Forth is more complex and generally maritime in character, but it has significant zones of links systems extending east from Edinburgh past Aberlady and Gullane bays as far as North Berwick, broken by expanses of salt-marsh around those bays.

Modern land-route-dominated perspectives present both estuaries as barriers, and simple consideration of crossing-points seems to support that view; until the nineteenth century the lowest bridging (and fording) points of the Tay and the Forth were Perth and Stirling respectively, at the head of the tidal and navigable reaches of the rivers. Bridges were possibly established in the vicinity of both places during the two main phases of Roman military occupation in the later first century AD and mid-second century AD, but there is no clear evidence for the presence of built river-crossings at either location again until *c.* 1200. But we must consider alternative transport mechanisms and below these bridging-points the estuaries were already carrying extensive ferry traffic from at least the late eleventh century when St Margaret established a pilgrim crossing at Queensferry on the Forth (Anderson (ed.), 1922: volume ii, 77). Again, however, the Tay and inner firth of the Forth with their shorter, more protected crossings presented less of an obstacle with this regard, but the outer zone of the Forth was more problematical. It is recorded in the mid-nineteenth century that before the introduction of steam-powered ferries, passengers could be trapped onboard for days by contrary winds which prevented their vessels from entering the small ports and havens in the outer Forth.

Neither estuary provides much weather protection other than in the lee of the southern escarpment of the Tay or one of the major islands of the Forth. The only significant natural havens on either river are at Inverkeithing and Rosyth on the Forth and Dundee on the Tay (Perry, 2005: 3, 21–22). At the former, the deeply set bay was a landing-place associated with Dunfermline which evolved into the royal burgh of Inverkeithing, while at Rosyth St Margaret's Bay was apparently a landing-place by *c.* 1070 (Skene (ed.), 1872: volume ii, 201). Apart from the partial shelter offered by promontories at, for example, Aberdour, Blackness and Elie, most other shelter for vessels before new harbour construction in the sixteenth century was in river-mouth havens, including at Scotland's most important late medieval port, Leith, or simply by drawing vessels up on cleared areas of a shelving beach (Graham, 1968–9). On the Tay estuary, there are no large tributaries to provide shelter.

Figure 16.4 The Carse of Gowrie and the Tay estuary looking south from Kinnaird, showing the broad zone of arable land on the north bank of the river and the steep escarpment rising on the south side of the estuary (Photograph: R. Oram)

II. Pre-twelfth century political context

Topographical and hydrological features combined with human political circumstances had already created divergent development patterns before the beginning of the period under consideration here. By the early 900s, the lower Tay valley between Dunkeld and Perth was the political and cultural heartland of *Alba*, the core of the later medieval kingdom of Scotland. Dunkeld and Abernethy were major monastic centres and a third, Monifieth, lay close to the river-mouth, Scone was the royal inauguration site and assembly place, and Forteviot in Strathearn (12 km south-west of Perth) the location of a major royal palace. Scottish kings possessed important estates in the district around Scone and at the eastern end of the Carse of Gowrie near to modern Dundee, some of which they gave as the founding endowment of Scone abbey in 1115. South of the estuary, the monastery at Abernethy had land at the east end of the carseland flanking the Earn and extending eastwards from there along the Tay, but also in the Carse of Gowrie north of the river (http://arts.st-andrews.ac.uk/corpusofscottishchurches/site.php?id=125608 [accessed 10 March 2010]). This status of the region as a nexus of early political and ecclesiastical power probably acted as a stimulus to the more intensive exploitation of its resources, both directly to produce the agricultural surpluses which supported the lavish feasting and hospitality of the elite households clustered there and indirectly to maintain the craftsmen who serviced the needs of a noble society founded on a culture of gift-giving and display.

In sharp contrast to the central place of the inner reaches of the Tay estuary in the early Scottish kingdom, until late in the tenth century, the Forth was a political boundary between Gaelic *Alba* to the north and Anglian Lothian to its south. Since the late 600s, Lothian had been the northernmost territory of the kingdom of Northumbria and the plain along the south side of the Forth estuary had been settled by English-speaking communities with decreasing intensity from east to west. The carseland at the confluence of the River Carron with the Forth near modern Falkirk, referred to as 'the Plain of Manau' (Anderson (ed.), 1922: volume i, 213), was a disputed borderline in the eighth century between rival kingdoms. This marchland character retarded the economic development of the district before the second half of the tenth century, when the kings of *Alba* extended their control over Lothian. Even then, however, the country between the two major Early Historic power centres south of the river, Edinburgh and Stirling, remained thinly settled and under-developed economically into the 1100s. The consolidation of Scottish political control over the populous and fertile Lothian plain, however, caused a southward shift in the locus of royal power as kings sought to exploit the wealth of their new territory while maintaining their power-base in *Alba*'s traditional heartland north of the Forth. This shift in political weighting saw Dunfermline emerge as a royal centre and perhaps stimulated the economic development of the Forth's inner estuarine zone to service the new focus of the royal household and its dependents.

III. Resource exploitation and trade

How might these power centres have affected the wider patterns of settlement and resource exploitation along the estuaries? Traditional historiography has long argued that urban communities and international trade were innovations in Scotland of the reign of David I (1124–53), but an increasing body of archaeological and historical evidence indicates the existence of both before the first formal awards of privileged status as royal burghs (towns with regional monopolies over certain trade-goods and in all international trade) (Hall, Hall and Cook, 2005). Early eleventh-century shellyware pottery of a style known from London and Bergen has been recovered in secure archaeological contexts in Perth, pointing to involvement in North Sea trade. A charter of King Alexander I (1107–24) to the canons of Scone granted them the *càin* (tribute due to the king) from one ship coming to Perth each year and extended his protection to the foreign merchants in them (Anonymous (ed.), 1843: documents 1, 3). Alexander's contemporary grant to Scone of property in Stirling, Edinburgh and Inverkeithing suggests that surplus production on its estates was sufficient to stimulate an external trade through these places before 1125 (Anonymous (ed.), 1843: document 1). Incidental comments in the *vita* of St Godric of Finchale, a former merchant, and the *vita* of St Margaret, imply that traders came to Scotland from England and Flanders in the later 1000s with luxury items to exchange for raw and semi-processed commodities, such as fish, hides and wool, and more high-value items like pelts and fresh-water pearls (Anderson (ed.), 1922: volume ii, 68; Anonymous (ed.), 1847a: 28–30). Perth and Stirling, which emerge clearly in later medieval records as the channels through which the produce of hinterlands which extended far to the north and west were controlled, and with their proximity to important royal residences and manorial complexes, most likely developed early as entrepôts to which traders were drawn.

Evidence for trading settlements functioning before the formalisation of their status as self-governing communities in the mid-twelfth century does not change the basic fact that these were rather insignificant and isolated anomalies in a landscape which was agricultural in character with an economy which functioned at primary level. While Scotland shared in the general rise in north European population levels between the eleventh and late thirteenth centuries, and experienced significant levels of immigration, the urban component in that population growth was minor. Rising rural population levels in eastern Scotland saw the expansion of settlement through assarts into areas that were more marginal in terms of arable productivity. In both estuarine zones, population growth and developments in agricultural technology and practice led to extension of arable and pastoral land-use beyond traditional limits.

IV. Woodland and woodland clearance

The low-lying carselands of the Tay and Forth estuaries and the woodland along the higher ground flanking the latter presented many opportunities for clearance and drainage for agriculture. Place-names at the eastern end of the Carse of Stirling suggests that the process was underway between the ninth and early twelfth centuries and possibly earlier still. Menstrie, for example, is a P-Celtic[2] place-name meaning 'farm or settlement on a plain or open field' (Taylor, 2004: 15). P-Celtic languages ceased to be used in name-forming in this region before 850 AD, which implies that agriculture was pushing into the carseland in the Early Middle Ages. Around Clackmannan a cluster of Gaelic place-names contain the element *gart*, meaning an enclosure or enclosed field, typically for arable cultivation (Taylor, 2004: 17). This name form occurs in a broad band across the isthmus between the rivers Forth and Clyde, but is clustered particularly where significant clearance for agriculture was underway before the twelfth century, especially on the fringes of the raised bogs in the Carse west of Stirling. The twelve surviving *gart*-names around Clackmannan represent intakes from woodland covering the strip between the Forth and the steep escarpment of the Ochil Hills. There is no early documentary record for when this work began – *gart*-settlements simply appear as existing communities in charters of the post-1150 period (for example Fraser (ed.), 1872: document 127) – or how it was directed, but the small scale of the settlements involved suggests that these were essentially assarts undertaken by peasant cultivators.

Charter evidence indicates that the rising ground at the eastern end of the carse on either side of the Forth was still heavily wooded around 1125, mainly with oak. Already by that date, however, the landscape was one of open country within which there were wooded areas rather than general woodland with named cleared areas. A series of discrete

---

[2] P-Celtic is the so-called Brythonic linguistic group represented nowadays by Welsh, Kernow (Cornish) and Breton. Q-Celtic or Goidelic languages are represented by modern Irish and Scots Gaelic and Manx.

Figure 16.5 The Carse of Stirling and the Ochils escarpment, looking east from Stirling Castle (Photograph: R.Oram)

woods are named on royal demesne in the region, from which gifts of timber were made towards building-work at the new monasteries being erected across central Scotland, and of dead-wood and windfall for fuel. Charters from 1125–65 record that the monks of Dunfermline had a general grant of timber for buildings and fires from the royal woods in this region (Barrow (ed.), 1999: document 33); that the Wood of Clackmannan was the source of timber for work at St Andrews Cathedral-priory, May Island Priory and Holyrood Abbey (Barrow (ed.), 1999: documents 126, 147, 154; Barrow (ed.), 1971: document 7); Cambuskenneth Abbey received the land of Cowie with its woods on the south side of the Forth and easements in the Tor Wood east of Stirling (Barrow (ed.), 1999: documents 159, 213); and Newbattle Abbey and the nuns of Manuel near Linlithgow were given access to the king's woods on the Carse of Callendar for all wood, building-timber, fencing and fuel needs (Barrow (ed.), 1999: document 144; Barrow (ed.), 1960: documents 109, 270). That St Andrews was still bringing building-timber in the 1160s from woods 90 km distant need not imply that there were no areas of woodland in Fife closer to the priory, but simply that access to none had been granted (Barrow (ed.), 1960: document 174). Nevertheless, the crown possessed extensive properties in eastern Fife and the failure to grant rights in woodland there is suggestive of the absence of significant tree-cover by *c.* 1125. The Lothian coast of this outer Forth zone was equally lacking in substantial woodland by the twelfth century. Here it survived in named pockets, usually controlled by lay lords and mainly located well south of the intensively cultivated coastal plain. Excavation of the twelfth- to fifteenth-century village at Eldbotle close to the East Lothian coast revealed limited evidence for timber use in construction of the byres and houses and both hearth-debris and charter references indicate that peat was the main fuel here by the 1170s (Hindmarch, 2006).

On the Tay estuary, evidence for significant woodland is lacking even in the early twelfth century. This situation probably reflects the intensive economic development of this region before the eleventh century. Scone Abbey had extensive property interests in the Tay valley above Perth and in the estuarine zone below the town, but the only specific woodland mentioned in its early charters lay upstream to the north and east of the abbey (Barrow (ed.), 1999: document 225; Barrow (ed.), 1960: document 244). Coupar Angus Abbey, one of the greatest reclaimers and developers in the Carse of Gowrie, likewise received timber rights and woodland easements in royal forests along the southern fringe of the Highlands, but none in the carse (Barrow (ed.), 1960: document 226). The general absence of significant woodland along the north side of the Tay estuary in the twelfth century is confirmed in 1175 by the failure of King William (1165–1214) to grant Arbroath Abbey any nearby woodland and the closest building-timber source which it eventually received lay nearly 55 km away (Barrow (ed.), 1971: document 513; Anonymous (ed.), 1848: documents 65, 66). The land along the southern side of the Firth of Tay also lacked significant stands of woodland by the later twelfth century; when Earl David, brother of King William, founded Lindores Abbey in 1195, no local timber resources were allotted to it (Dowden (ed.), 1903: document II). Grants of building-timber and firewood made to it 50 km to the west in Strathallan were specifically for operations on their estates there, and sources of firewood and rods and hoops of hazel, alder and birch were granted 70 km away (Dowden (ed.), 1903: documents XXIV, XXV, LXXIII, CXI, CXII). A dearth of local timber south of the estuary is given circumstantial confirmation in 1228 when King William's widow, Ermengarde, founded Balmerino Abbey; like Lindores, it received no grants of timber resources and remained chronically under-supplied throughout its existence (Oram, 2008: 74–77).

Around both estuaries it is clear that before 1300 substantial areas of woodland containing trees suitable for constructional use were very scarce as unregulated cutting without any policy of replanting had led to a sharp contraction in tree-cover. Excavated evidence from Perth shows a gradual increase in the use of scrub and hedgerow species in domestic buildings from the twelfth to fourteenth centuries, a trend which suggests that local sources of structural timber were extremely limited (Crone and Watson, 2003: 66–67). Timber shortages may be reflected in ordinances designed to remove financial disincentives from merchants and dealers who brought timber or firewood to the burgh. This may have been intended to encourage rural dwellers in the burgh's hinterland, which included the timber-rich upper Tay catchment, to take supplies to the lowlands, but is more likely to have been aimed at the merchants who, by the early fourteenth century, were coming from southern Scandinavia and the eastern Baltic with cargoes of oak deals and spars (Crone and Watson, 2003: 74–75; but see also Stuart (ed.), 1878: 411). The almost treeless character of the estuarine districts is evident in the earliest existing maps of both areas, Timothy Pont's surveys of *c.* 1590. These maps show that substantial areas of woodland survived only as enclosed preserves associated with major royal or aristocratic residences (http://maps.nls.uk/pont/specialist/pont26.html and http://maps.nls.uk/pont/view/?id=pont32 [accessed 27 July 2011]), probably reflecting at the end of the sixteenth century the status quo that had been reached before 1400. Much of the current wooded character of the carses dates from the era of so-called Improvement agriculture after *c.* 1750. As General Roy's Military Survey of Scotland recorded for both districts in the middle of the eighteenth century, there had been little reafforestation yet undertaken beyond ornamental planting around aristocratic mansion houses or as shelter-belts and enclosures on recently reorganised farms (http://geo.nls.uk/roy-highlands/ [accessed 27 July 2011]). Later medieval record sources and Pont's maps confirm the existence of significant areas of oakwood around the headwaters of the Forth and Tay, but these were beyond the effective reach of consumers in the lowland zones around the river estuaries until the early sixteenth and seventeenth centuries (NAS GD22/3/706 and GD220/6/1846 (6); Hannay (ed.), 1915: 122, 129, 130, 139, 266). As a consequence, the urban communities in both estuarine zones became dependent on imports of foreign timber.

V. Wetland reclamation

Pressure on the Forth valley woodlands was matched on the Tay by substantial reclamation from the marshes and poorly drained clay soils of the Carse of Gowrie. This reclamation appears to have begun earlier and progressed faster on the Tay estuary than on the Forth, perhaps due in part to the political and settlement factors discussed above, but it is probable that the different salinity levels of the two river estuaries was a more important determinant in this differential development. The low salinity of the Tay estuary may have made reclamation easier and more rapid, as the significantly lower level of mineral salts in the estuarine silts would have flushed quicker from the drained ground and thus enabled it to be brought into productive cultivation sooner than in the carse of the saltier Forth estuary. Monastic records suggest misleadingly that the drainage activity was spearheaded by the Cistercians of Coupar Angus Abbey, who in the 1180s were developing 'Eddirpolles' (Gaelic = 'among the pools'), now known simply as Grange (Easson (ed.), 1947: documents III, IV, XXV). Agricultural exploitation here was not new, however, for in common with the Carse of Stirling the higher 'inches' and gravel ridges had long been cultivated or grazed (Maxwell, 1992; RCAHMS, 1994). In the 1230s the monks expanded the grange through direct purchases of blocks of already developed arable, bearing English names like Aithmuir and the Muirhouse, extending to around 120 acres that had been broken out of former pasture (Easson (ed.), 1947: documents XLII, XLVII, LVII). Agreements in the 1190s over tithes due from Eddirpolles suggest that these arable areas were established before the monks acquired Eddirpolles, as does reference to the pond of a mill there possessed by David de Hay, lord of Errol, *c.* 1200 (Easson (ed.), 1947: document III; Rogers (ed.), 1879: volume i, 337; Duncan, 1975: 320–321). Such lay ownership suggests that it was not just the monks who were actively expanding the arable acreage of settlements in the carse, but the scale of the monastic enterprise alone was nevertheless considerable. An indication of their labour investment can be seen by the mid-1200s in the causeway they constructed from the higher ground at Inchture across the wet land to their grange, 2.5 km south west. (Easson (ed.), 1947: document XXXVII).

Since the majority of the surviving record for land reclamation lies mainly in monastic cartularies, there has perhaps been a tendency to overstate the importance of the monasteries' role and understate the significance of lay reclaimers (Oram, 2011: 250–252). This situation is worsened by the poor survival of areas of pre-modern agricultural landscape that can be associated with any confidence to work undertaken by monastic or secular landlords in either of the study districts. Further fieldwork is necessary to determine whether it is at all possible to distinguish between areas of monastic and lay reclamation for agriculture in the modern landscape, as first the trend for monastic landlords to alienate areas of former demesne or grange agriculture to lay tenants and then the radical re-ordering of Scotland's rural exploitation systems which occurred in the century and a half after 1750 has obliterated much of the physical evidence. The earliest detailed 'modern' maps of both areas, produced *c.* 1750 just as that re-ordering was beginning to effect significant change in lowland areas, simply shows unenclosed blocks of rig-and-furrow cultivation interspersed with zones of rough grazing (http://geo.nls.uk/roy-highlands/ for the Carse of Gowrie [accessed 27 July 2011]). If there was any significant difference in the physical patterns of the regional social agro-systems arising from a contrast between medieval lay or ecclesiastical administrative or functional practice, those differences had already been excised before the eighteenth century.

As with woodland clearance, it is apparent that several monasteries benefited from the efforts of peasant cultivators whose landlords granted their properties to the church in a partly-developed state. Elsewhere, the monastic records reveal that other reclamation work was taking place alongside their own. For example, while Coupar Angus was developing Eddirpolles, neighbouring communities were equally active, and at Inchture the peasant tenants extended their arable fields from their *inch* of high ground onto the shore (*plaga*) (Easson (ed.), 1947: document XXV). It was, however, a piecemeal and protracted process; the scattered distribution of medieval settlement sites in the carse proper does not suggest any large-scale planned operations directed by a major landlord but rather a series of individual efforts by small tenants and minor landowners. While by 1400 the zones of settled agriculture in the Carse of Gowrie had extended south from the foot of the escarpment and radiated out from the inches, the lower ground along the sluggish, meandering streams or 'pows' which flowed across the carse remained as expanses of wetland exploited as water-meadow or for other resources, such as reeds and wildfowl. The drainage schemes which created the modern arable landscape of the region still lay 400 years in the future.

A similar picture of development emerges south of the Forth on the old march between Alba and Lothian. While the higher ground above the old raised beaches along this southern side of the inner tidal zone of Forth had seen settlement since the Neolithic period, the wetlands of the carse proper remained relatively unsettled into the twelfth century. The natural resources of this extensive wetland had attracted exploitation by hunter-gatherers since the Mesolithic period and, presumably, remained important to the inhabitants of the agricultural settlements along its southern edge into the Middle Ages, but there is no evidence of attempts to extend the cultivated ground by drainage and embankment operations before the mid-twelfth century. In the 1140s, when parts of the carse were granted to Newbattle Abbey, the district was referred to by the Norman-French speakers in the royal circle as 'Blankelande' (white waste ground) (Barrow (ed.), 1999: document 98), which is indicative of their perception of it as an undeveloped resource. Newbattle exploited their property in this 'waste' as a saltpan site, but they and the canons of Holyrood who

Figure 16.6 Mugdrum Island (Redinch) and sand-banks on the Tay estuary looking west from near Ballinbreich (Photograph: R. Oram)

received land for a pan at nearby Airth, were also draining the estuarine flats for pasture and agriculture. Newbattle's efforts were dwarfed by Holyrood, which was the principal agent in a process of reclamation which saw development of a major grange in the carse; by the sixteenth century it comprised of twenty-five townships and two salmon fisheries (Anonymous (ed.), 1840: Appendix, document 37). In contrast to the seemingly piece-meal developments in the Carse of Gowrie, the advance into the marshes along the Forth does appear to have been conceived as part of a planned programme. Some indication of a planned process of reclamation by Holyrood, Newbattle and their tenants may be obtained from Roy's mid-eighteenth-century survey of the district, which recorded a line of settlements running parallel with the shoreline almost continuously from Airth east to the mouth of the River Avon (http://geo.nls.uk/roy-lowlands/ [accessed 27 July 2011]). This distinctive feature appears to preserve the furthest advanced line of reclamation achieved in the later Middle Ages. Sadly, all opportunity for archaeological investigation of the main section of this reclaimed land on Holyrood's properties between the Carron and Avon has been lost in the twentieth-century development of Scotland's largest petro-chemical complex over the entire site. Similar reclamation operations to those organised by the abbeys were also undertaken by lay estate-owners into the fifteenth century and beyond, such as at Kinneil where peasant tenants extended their holdings from the higher ground into the salt-flats (Thomson (ed.), 1882: document 1178).

Other segments of the estuarine environment were also exploited for alternative agricultural operations, principally livestock management. Place-names at the east end of the Stirling carse point to grazing on water-meadows along the Forth's tributaries before the ninth century (Taylor, 2004: 15). On the Tay, the grass of upper estuary islands between Perth and Newburgh was in demand for pasturing oxen and cattle, and when Earl David granted Lindores the Redinch (now known as Mugdrum Island), he retained pasture rights for his animals there (Dowden (ed.), 1903: document II). More extensive pasture was provided on the links and salt-marshes of the dune systems at the mouth of the Tay, where Barry Links formed one of Balmerino Abbey's most valuable properties (Oram, 2008: 70). There, the 'salt griss' (salt-grass) of the coastal marshes was prized as sheep-pasture into the sixteenth century (NAS GD1/1039 (1)). The links along the East Lothian coast were also exploited for this purpose by the later 1100s; at Eldbotle, Dryburgh Abbey used its coastal pastures as a sheep-ranching operation (Morrison, Oram and Oliver, 2008: 42; Anonymous (ed.), 1847b: documents 104–108). Pressure for arable development, however, saw most grazing concentrated in upland districts by the thirteenth century, with the urban centres in the estuarine zones acting as conduits through which livestock-derived products were delivered to the domestic and international markets.

VI. Aquatic resource exploitation

Exploitation of the terrestrial resources of the estuarine zones was matched by exploitation of the aquatic resources. Shell-middens and flint-scatter sites demonstrate that the estuarine habitats have attracted human interest since the Mesolithic period (Wickham-Jones, 1996: 69). The diversity of the wildlife exploited in the Forth by the late eleventh century is underscored in various grants: to the monks of Dunfermline of seals caught at Kinghorn on the Fife coast; to the canons of Holyrood of every tenth whale caught – or beached – along the south side of the estuary (Barrow (ed.), 1999: documents 33, 147, 172). Malcolm IV (1153–65) awarded Dunfermline the heads of all whales and other large fish beached or caught in his demesne estates on the Fife coast, to which he later added half of the blubber and fish-oil from all whales and large fish caught anywhere in the estuary, while Kelso Abbey received the other half of the whale-blubber from the Forth (Barrow (ed.), 1960: documents 118, 131, 177, 214). By far the greatest resources of the two estuaries, however, were the fisheries.

There were significant differences between the fisheries on the two estuaries which stemmed from the greater volume of fresh water passing through the Tay and from the marine character of the outer estuary of the Forth. On the Tay, while other species were exploited in small numbers, the emphasis was on the salmon fishery. Scone Abbey was an early beneficiary, with two nets granted to it in the Tay at Perth by Alexander I, followed by a further two above the tidal limit in Scone parish and a tenth of all salmon taken between Scone and Kinfauns and Craigie on opposite sides of the river below Perth (Anon (ed.), 1843: documents 5, 12, 20, 85, 107). David I confirmed the draught of a net for salmon at Perth to the canons of St Andrews in 1144 and one to the monks of Dunfermline by 1150 (Barrow (ed.), 1999: documents 126, 172). Lindores received fishing rights from Earl David around Mugdrum Island on the south side (excluding a built fish-trap or *yair* which he retained for his own needs), and on the Errol side of the Tay at Clashbennie and

'Red Sands' from the Hay lords of Errol (Dowden (ed.), 1903: documents VI, LXXVI-LXXIX). Balmerino, however, was one of the biggest monastic beneficiaries of grants of fishing rights on the tidal reach from Perth to the rivermouth, salmon providing the monastery with its largest source of income (Oram, 2008: 71–73). The abundant evidence for monastic fishing stations flanking the river from Perth to just east of Dundee, however, must be balanced by recognition that these represented only a small percentage of the fishery, the majority of which remained in lay hands. The Hay lords of Errol, for example, only granted Lindores portions of their draw-net operations at Clashbennie and 'Red Sands' and specifically excluded their fixed stake-nets on the sandbanks along the north side of the river-channel, as did Earl David at Mugdrum, and the Hays actively acquired further fisheries into the late thirteenth century (NAS RH1/6/20; B66/25/7). The crown, too, retained extensive fishing interests on the Tay into the fourteenth century.

On the Forth, grants of salmon-fishing rights were awarded from the early twelfth century to various abbeys, including to Scone, Holyrood and Cambuskenneth (Anonymous (ed.), 1843: document 5; Barrow (ed.), 1999: documents 33, 147, 159). Most seem to have been draw-net operations, some using boats to make the sweep across the channel, but in the section of the Forth between Stirling and Avon-mouth stake-nets were fixed on the tidal flats on either side of the main stream. While these salmon fisheries were valuable, they were small-scale in comparison with those on the Tay and did not acquire the same level of economic importance (http://www.eh-resources.org/Fishing_results/). General fishing-rights, however, had been granted to Dunfermline at Musselburgh by the 1120s (Barrow (ed.), 1999: document 33), providing the first surviving record of the fishery in the outer firth. These fisheries were predominantly for herring, which Scotland was exporting by the thirteenth century (Gemmill and Mayhew, 1995: 319), and white fish. Large shoals of migratory herring entered the Forth estuary on a seasonal basis. Their presence gave the Forth fishery a distinctive profile, for the low salinity of the Tay deterred herring from entering its estuary. Again, there is a preponderance of record material relating to monastic-controlled fisheries in the outer Forth zone, such as the extensive rights of the monks of May Island in the waters at the outermost limits of the estuary (Barrow (ed.), 1960: documents 162, 169). Monasteries remote from the Forth actively sought to secure or expand their stake in this lucrative trade in the thirteenth and fourteenth centuries, Balmerino acquiring a fishing-station at Anstruther and Dryburgh at Anstruther and Kilrenny on the Fife coast (Oram, 2008: 73; Anonymous (ed.), 1847b: documents 98, 192). In common with the salmon fisheries, however, the main herring fishery remained in lay hands throughout the medieval period and it was principally on the profits of the export trade in dried herring that the twelfth- and thirteenth-century wealth of Crail in Fife was based (Gemmill and Mayhew, 1995: 319; Dowden (ed.), 1903: document LXXV).

VII. Salt manufacture

A second major divergence in the exploitation regimes between the two estuaries was salt-production, with the low salinity and strong flushing mechanism of the Tay deterring development while the Forth emerged as the principal centre of salt-manufacture in medieval Scotland (Oram, 2012). Saltpans are recorded from the 1140s in the Forth between

Avon-mouth and Stirling. It is likely that the earliest saltpans used brine obtained by sleeching, a method which extracts salt from inter-tidal silts by repeated rinsing; the low level of salts in the Carse of Gowrie estuarine mud would have made this a less economically viable method than on the salt-rich Forth estuary. The nature of the 'raw material' used meant that most sleeching sites were located on salt-marshes or tidal flats, close to a fuel-source. Place-names such as 'E Saltcotts', 'Mid Saltcotes' and 'W Salcots' on Timothy Pont's *c.* 1590s map and two 'Saltcotts' and 'Panstead' on William Roy's *c.* 1750 survey of the carse area between Airth and Bo'Ness on the Forth emphasise this positioning, all being located on the outer edge of the drained land at the margins of the coastal mudflats. The brine obtained by sleeching was boiled in metal pans in pan-houses until the water had evaporated (Cranstone, 2006: 4; Anonymous (ed.), 1840: Appendix, document 37); rapid boiling produced small, fine crystals of usually white salt (Gemmill and Mayhew, 1995: 323–324), while slow evaporation produced larger, darker crystals. Direct boiling, a method often claimed to have been developed in Northumberland in the earlier fifteenth century (Cranstone, 2006: 4), used sea-water obtained from coastal pools or built reservoirs. The favoured locations were on rockier coasts, possibly because sea-water there carried less sediment contaminants. The water was taken to metal pans in pan-houses above high-tide level where it was fast boiled over coal fires (other fuels were sometimes used) to produce fine, white salt (Cranstone, 2006: 4–5). Most probable later medieval direct boiling pan sites lie in the outer zone of the Firth of Forth, with Prestonpans in East Lothian dominating as an industrial centre on account of its access to the Lothian coal-field.

The earliest charter evidence for saltpans suggests a concentration of production in the upper tidal reach of the Forth. This distribution is an artificial result of the uneven survival of the documentation, but probably does reflect a focus of activity stimulated by suitable conditions for sleeching sites in combination with extensive fuel reserves. Salt-production was already pursued by inhabitants of royal estates around the Forth estuary by David I's reign. His grant in *c.* 1128 to Dunfermline of a tenth of all salt carried there for the king's use (Barrow (ed.), 1999: document 33) implies that bulk consumption of salt was already an established feature of the economic activity of the royal household. That this salt was the output of salt-pans on his estates is implicit in references to the king's pans in charters authorising the establishment alongside them of new pans by monastic communities (Barrow (ed.), 1999: documents 67, 157; Barrow (ed.), 1960: document 241; Barrow (ed.), 1971: document 197; Fraser (ed.), 1872: document 51). It may seem obvious, but it is worth emphasising that such references indicate that large-scale production and movement of salt was a feature of the economy of the Forth estuary before the founding of the ecclesiastical communities to which such activity has traditionally been attributed. Bulk salt supplies were needed by the royal household to preserve the meat from slaughtered stock, fish and dairy products of the king's estates, the volume of which is indicated by the grants of tenths of such by-products of the processing as hides, grease and tallow made to the monks of Dunfermline (Barrow (ed.), 1999: documents 18, 19, 33, 172). When David began to grant saltpans on the Forth to his newly-founded abbeys, the awards were of existing pans rather than vacant sites, and they came with access to both the salt-water and fuel. In the earliest surviving grants, wood was the designated fuel. At Callendar by Falkirk, for example, Newbattle had access to local royal woodland for their pan's needs (Barrow (ed.), 1960: document 109). King William's confirmation to Holyrood of David I's grant of a salt-pan at Airth also gave rights to unlimited fuel from his wood of Airth (Barrow (ed.), 1971: document 39). When he granted Arbroath a pan in the Carse of Stirling shortly after 1178, however, he gave them a share of the common pasture and fuel resources there alongside his own salt-workers (Barrow (ed.), 1971: document 226). Wood is not mentioned and it seems that the fuel intended was peat, indicating perhaps that local timber resources were depleted.

Access to dependable fuel supplies was a primary consideration behind the shift in areas of salt manufacture away from early centres on the upper part of the estuary to new locations on the Fife and Lothian coasts east of the Queensferry narrows (Oram, 2012). It was not simply a case of depletion of the peat reserves – that only presented serious problems from the sixteenth century – but may have been tied up more with the economics of peat-use in large-scale industrial production. Dependence on a fuel whose availability was determined by seasonal cycles of cutting, drying, transportation and stacking was exposed as a critical weakness in the fourteenth century, when first climate change and then population decline affected the ability to meet peat-demand. A switch to coal-use was perhaps initially a crisis response, but the ending of the seasonal constraints on production which peat entailed probably encouraged general adoption of coal and movement of salt-manufacturing operations closer to mines.

It is unlikely that only the king possessed salt-pans in the pre-1124 period but there is no surviving documentary evidence for wider lay ownership of pans before David I's reign. By the second half of the twelfth century, however, charter evidence indicates that salt-pans were important economic activities on lay estates. The tenants of the Earls of Dunbar were producing salt on the East Lothian coast before 1200 (Anonymous (ed.), 1837: documents 7, 8), and in the thirteenth century, important lordly families had salt-pans on the Forth carseland (NAS GD40/1/16; GD40/1/27). The majority of the pre-1400 record, however, relates to monastic proprietors, but not all monasteries controlled pans, especially those located on the Tay; neither Scone nor Balmerino nor Lindores possessed any, despite the pastoral basis of their estates and their substantial salmon fisheries (Oram, 2008). All had to satisfy their needs for salt through market purchases, but it is striking that none attempted to gain control of a production centre. This apparent disconnection between intensive involvement in fisheries which constituted a large and growing component in Scotland's export trade in the thirteenth and fourteenth centuries and salt-production is, however, less significant than it might seem. While Scottish salt may have been used for fish-curing domestically, fish for the export market was preserved in imported Bay salt; the large, slow-dissolving crystals of the much cheaper solar-evaporation produced French salt was preferred to the purer but more rapidly-dissolving Scottish product (Oram, 2012). The bulk of Scotland's salmon and herring was destined for the export market and was dry-salted in Bay salt for better preservation, especially after the development of new barrelling techniques in the later fourteenth century. For the main producers of fish, therefore, control of saltpans in Scotland mattered far less than secure access to French imports. Scottish saltpans, however, remained profitable ventures, supplying household needs in much of eastern Scotland for dairying and meat-processing, and they commanded a significant export market for their product in northern Europe.

Access to dependable sources of coal gave the Forth estuary salt-masters an effective monopoly on large-scale production. It also stimulated the development of new trading activity, with coals being shipped from mines on the Fife and Lothian coasts to fuel-poor coastal regions of eastern Scotland. By the early 1500s, elite households like that of the Bishop of Dunkeld were purchasing coal for their domestic needs, imported through Dundee and Perth (Hannay (ed.), 1915: 292), and it appears that by the fifteenth century the major burghs of the Tay estuary met most of their thermal energy needs through coal imports rather than continuing to depend on increasingly scarce wood and peat resources in their hinterlands. Rural communities around the Tay estuary remained dependent on their dwindling peat sources into the eighteenth century but around the Forth, most urban and rural areas close to the mine heads were largely coal consumers before the end of the 1400s. The rise of the coal trade, linked to salt export, also stimulated the growth of new port outlets for the commodities which began to challenge the mercantile monopolies of the older burghs. Culross and Dysart in Fife and Bo'ness and Prestonpans in Lothian, for example, rose in importance as Stirling and Crail – at opposite ends of the estuary – declined. The result was an intensification of commercial and proto-industrial activity around the central section of the Forth estuary and the laying of the foundations of the post-medieval economic prominence of this region. By the fifteenth century, the Forth ports and, increasingly, Edinburgh and its port of Leith in particular, had established a dominant hold over eastern Scotland's North Sea trade. By way of contrast, although Perth and its better-located rival and eventual supplanter Dundee remained important economic centres (Dundee emerging in the fifteenth century as second to Edinburgh in terms of volume of trade), their economic base rested on the export of traditional staples – wool, skins and leather – and cloth derived from the products of their rural agricultural hinterlands. Without the coal and mineral resources available to the Forth estuary communities, settlements around the Firth of Tay were destined to take a different path of economic development from the later medieval period.

VIII. Climate change and weather impacts

Although the two regions took divergent developmental routes at the close of the period under study here, a range of similar direct and indirect environmental consequences can be identified which relate to the intensification of exploitation around both estuaries in the medieval period. Long-term, while the clearance of woodland and the drainage of the carses around the two estuaries allowed significant population growth sustained on an expanded agricultural zone, there were also negative consequences; loss of woodland cover and the canalisation of 'pows' to ease drainage increased the levels of discharge into the main river channels, especially on the Tay. The results of this increased discharge were manifold, but sediment transportation was one of the more significant issues. It is an obvious fact that the larger catchment of the river basin increased the volume of discharge through the lower part of the system during the episodes of increased rainfall which occurred in the fourteenth and fifteenth centuries. Combined with the deforestation of the middle and lower reaches of the river system and the intensification of agricultural activity in those areas, a greater volume of sediment was carried down-river and deposited as sand-banks in its estuary and river-mouth. For the effectively inland or 'dry' port of Perth, this silting of the channel constituted a major obstacle to shipping and led to an acceleration of reorientation of regional trade links with the wider North Sea region in favour of Dundee, whose harbour was more regularly scoured and which lay downstream of the estuarine shallows. On the Forth, whilst its basin was still exposed to the same

Figure 16.7 Saltings and mudflats on the south-bank carse of the Forth looking east towards the petrochemical complex at Grangemouth (Photograph: R. Oram)

weather impacts as the Tay, the limited clearance of woodland west of Stirling and around its headwaters before the late seventeenth century resulted in only a marginal increase in sediment carriage during episodes of extreme rainfall. As a consequence, Stirling and the down-river ports did not experience the same silting as affected Perth.

A second consequence of increased discharge was heightened flood-risk exposure for reclaimed land on the estuarine margins – a situation which continues to the present – and also of inundations which threatened older settled areas. Heavy rainfall throughout Scotland in September 1209, for example, caused several of the major rivers to burst their banks and overwhelm adjoining arable land. More specifically, floodwaters from the rivers Almond and Tay devastated Perth, sweeping away not only the southern end of the bridge but also immersing the burgage plots close to the river (Watt (ed.), 1994: 457). Some people benefited from the misery of others caused by this event; in 1219, the canons of Scone purchased a toft in the burgh which they had formerly leased 'before the flood of the river carried it away in its violence', and acquired further land in Perth described as 'wasted by flood' (Anonymous (ed.), 1843: documents 82, 96). Perth, however, was almost uniquely exposed to such major flood risks – although even here analysis has shown that most of the early settlement areas and the major public buildings, including the parish church, occupied islands and ridges of higher ground that were rarely submerged (Bowler, 2004: 12–20) – with only Haddington of the major economic centres close to the Forth estuary ever suffering similar inundations. In 1358, the quarter of Haddington closest to the River Tyne was devastated by floodwater (Watt (ed.), 1996: 311–313). Major episodes of flooding of arable land occurred during the long phase of exceptionally poor weather in the mid-fourteenth century. The same extreme rain event which affected Haddington resulted in severe flooding through Lothian in September 1358. The Tyne around Haddington inundated a great expanse of agricultural land between the burgh and the coast near Dunbar, while the rain devastated growing crops and recently cut grain (Watt (ed.), 1996: 311–313). While the proximate cause of these events was weather-related in all cases, it was the human decision to concentrate arable production on flood-susceptible ground which turned a natural occurrence into an anthropogenic crisis.

Rising sea-levels were not a primary factor in increasing flood susceptibility in the estuarine hinterlands. In contrast to the southern part of the North Sea region and especially south-eastern England and the Netherlands where post-glacial isostatic adjustment has led to a downward movement in land level, the east coast of Scotland has experienced considerable uplift. Indeed, areas of the lower basins of the Forth and Tay which were submerged at the start of the Holocene in Scotland are now up to fifteen metres above sea level (Dawson, 2009: 83, 85). It was this process which created the circumstances for the formation of the carses along both river estuaries, where former shallow-water gulfs and estuarine mudflats were lifted above sea-level through isostatic rebound and lowland peat-mosses began to form on the exposed areas of former sea-bed. Even after the removal of the peat in the course of medieval fuel extraction and agricultural reclamation processes, the exposed clays remained above the medieval sea level by several centimetres. Although the favourable post-glacial isostatic adjustment ameliorated risk, the areas of reclamation remained vulnerable to inundation in extreme circumstances; the two major eastern Scottish estuaries are exposed to marine surges where easterly gales drive water up into the narrowing funnels of the firths. A major episode of such storm-related marine transgression affecting the whole of the east coast between the rivers Tay and Tweed was recorded in October 1267 (Watt (ed.), 1990: 359–361), but the loss of the Exchequer accounts for the half century after 1264 has rendered the impact of this event invisible in the record. In the inner reaches of the Forth, flood waters moving downstream meeting tidal surges probably driven up the estuary by strong easterly winds forced the development of new defences for land that had been won into cultivation or grazing from the estuarine marshes. At Stirling, earthwork dykes were constructed to protect suburban pasture and cultivated ground on the Queenshaugh. Lower down the river, dykes were built along the southern shore between Airth and Grangemouth by the tenants of Holyrood Abbey to defend land that had been taken from the estuarine marshes since the late twelfth century. By the early sixteenth century, however, Holyrood was disposing of large portions of the land that it had reclaimed in previous centuries from the marsh; in 1533, for example, they alienated their former saltpan site and arable property at Little Saltcots, 'since a great part of the lands … are submerged daily by inundation of the sea and the rivers known as Carron and Forth' (Balfour Paul and Thomson (eds.), 1883: document 3016). It was only in the later seventeenth century that this low-lying area was recovered by a major programme of draining and dyke-building, the course of which extending from the mouth of the Carron to the mouth of the Avon is shown on Roy's *c.* 1750 survey (http://geo.nls.uk/roy-lowlands/ [accessed 27 July 2011]).

Figure 16.8 Queenshaugh, Stirling: the remains of the post-medieval protective dyke can be seen curving approximately parallel with the river-bank (Photograph: R. Oram)

Flood was not the only threat, for climate change in the later Middle Ages may have brought additional weather-related risks to the Forth and Tay estuaries as atmospheric systems affecting northern mainland Britain became more volatile. There is still considerable debate surrounding the nature and indeed actual occurrence of significant climate change and associated extreme weather events in the late medieval period, but recent research has highlighted a series of episodes in which harvest-related crises in England between the early fourteenth and later fifteenth centuries can be tied directly to extraordinary runs of adverse weather (Campbell, 2009: 29–36). This deterioration appears to have been general across the British Isles, Scottish records revealing a striking correlation between the English crises and similar supply problems in Scotland (Dawson, 2009: 101–106). The potential socio-economic and cultural consequences of the impact of long-term climate change in that period has been better examined from the perspective of Highland Scotland and its Atlantic sea-board (Oram and Adderley, 2008; Oram and Adderley, 2010a; Oram and Adderley, 2010b), but while the easterly location of the catchments of the Tay and Forth may have buffered them from the worst weather-related effects it is likely that similar issues of shortened growing-seasons, reduced biomass production and increased water-logging of land affected the estuarine hinterlands. As the statistical evidence from England appears to indicate (Campbell, 2009: 23–29, 36–42), there was a significant impact on agricultural and industrial productivity with consequences for wages, living standards, human fertility and mortality rates, but there were also high-cost impacts on the built environment. The density of settlement and high concentration of monumental buildings around the Forth and Tay estuaries areas saw increased weather-damage to property, which had economic consequences. Increased storminess is recorded down the east coast in the later thirteenth century; in 1260, for example, summer thunder storms caused great damage to growing crops and trees, followed by heavy autumn rains and, in the early winter devastating gales; the 1267 storms have already been noted; the winter of 1270–1 was recorded as exceptionally severe, and caused extensive damage to the cathedral-church in St Andrews; in January 1272 a northerly gale and hail storms flattened houses, crushing those sleeping within, brought down tall buildings, and at Arbroath a lightning strike caused a fire which damaged the abbey church (Watt (ed.), 1990: 325, 385; Anderson (ed.), 1922: volume ii, 667; Dawson, 2009: 100).

A further problem being identified in both the archaeological and historical sources was episodes of severe wind-erosion of the dune systems around the outer reaches of the river estuaries. Most evidence for this has been identified on the Forth, where a combination of the stripping of the turf for soil-deepening activities, removal of the marram grass and whin-scrub for roofing material, fodder and fuel, and arable cultivation of the thin soils over the sand deposits, led to major incidents of sand-blow apparently from the later fourteenth century onwards. At Eldbotle in East Lothian, excavation has shown that the village contracted steadily from the later 1300s, losing significant portions of its fields to deep layers of wind-blown sand and with its abandoned buildings being filled rapidly as the dune systems were blown inland from the coast. Here, expansion into the sandy margins of the coastal strip had brought short-term gains but the settlement's success proved unsustainable and led to protracted decline through the later Middle Ages: Eldbotle was abandoned before the end of the seventeenth century.

IX. Conclusions

In conclusion, Scotland's two principal estuaries discharging into the North Sea experienced quite different processes of development driven by a combination of environmental and human factors. Post-glacial isotopic adjustment created large expanses of peat-covered carseland where former coastal clay mudflats and shallow-water inlets were raised above sea level. This geological process established superficial topographical similarities, but those physical similarities mask significant underlying differences in terms of resource potential, from the high fresh-water discharge and low salinity levels of the Tay affecting the nature of its fisheries and its viability as a salt-producing centre, to the availability of coal in the Forth basin providing an abundant source of thermal energy to power the proto-industrial development of communities around its estuary. Both estuaries presented significant opportunities for drainage and agricultural exploitation of expanses of estuarine marsh and, in the period under review here, increasing population levels in Lowland Scotland drove forward that process in the carses of both rivers. On the Tay, that development started earlier as a consequence of the importance of the lower Tay valley as a centre of Scottish royal power between the ninth and twelfth centuries, while similar development around the upper Forth may have been retarded by its status as a march between two rival powers. The absorption of Lothian into the Scottish kingdom from the eleventh century onwards saw a southward shift in the focus of royal activity and led to acceleration in the economic development of the territories around the Forth. In both areas, the processes of clearance and reclamation may have been aided by climatic amelioration during the so-called Medieval Warm Period, which encouraged encroachment into areas that were marginal in terms of sustainable agricultural potential; that marginality, however, may also have increased their vulnerability to weather-related impacts as climatic conditions deteriorated in the later Middle Ages. Mortality-related depopulation post-Black Death does not appear to have been a factor affecting settlement in either estuarine zone, with there being no evidence for settlement abandonment or significant contraction of the cultivated areas, but deteriorating weather conditions may have had significant negative consequences. Widespread deforestation and development of cultivation on former flood plains had led to the emergence of a suite of social, economic and environmental problems in the fourteenth century as the impact of climate change and deteriorating weather began to be felt.

Bibliography

Primary Sources

National Archives of Scotland, B66/25/7, GD1/1039 (1), GD22/3/706, GD40/1/16, GD40/1/27, GD220/6/1846 (6), RH1/6/20.

Edited Sources

Anderson, A.O. (ed.) (1922) *Early Sources of Scottish History*, 2 vols, Edinburgh.

Anonymous, (ed.) (1837) *Liber S Marie de* Melros, 2 vols, Edinburgh.

Anonymous, (ed.) (1840) *Liber Cartarum Sancte Crucis*, Edinburgh.

Anonymous, (ed.) (1843) *Liber Ecclesie de Scon*, Edinburgh.

Anonymous, (ed.) (1847a) *Libellus de Vita et Miraculis Sancti Godrici*, York.

Anonymous, (ed.) (1847b) *Liber S Marie de Dryburgh*, Edinburgh.

Anonymous, (ed.) (1848) *Liber S Thome de Aberbrothoc*, 2 vols, Edinburgh.

Balfour P.J. and Thomson, J.M. (eds) (1883) *Registrum Magni Sigilli Regum Scotorum*, vol. 3, Edinburgh.

Barrow, G.W.S. (ed.) (1960) *Regesta Regum Scotorum*, I, *The Acts of Malcolm IV*, Edinburgh.

Barrow, G.W.S. (ed.) (1971) *Regesta Regum Scotorum*, II, *The Acts of William I*, Edinburgh.

Barrow, G.W.S. (ed.) (1999) *The Charters of David I*, Woodbridge.

Dowden, J. (ed.) (1903) *The Chartulary of the Abbey of Lindores 1195–1479*, Edinburgh.

Easson, D.E. (ed.) (1947) *Charters of the Abbey of Coupar Angus*, 2 vols, Edinburgh.

Fraser, W. (ed.) (1872) *Registrum Monasterii de Cambuskenneth*, Edinburgh.

Hannay, R.K. (ed.) (1915) *Rentale Dunkeldense*, Edinburgh.

Rogers, C. (ed.) (1879–1880) *Rental Book of the Cistercian Abbey of Cupar Angus*, 2 vols, Edinburgh.

Skene, W.F. (ed.) (1872) *John of Fordun's Chronicle of the Scottish Nation*, 2 vols, Edinburgh.

Stuart, J. *et al.* (eds) (1878) *Exchequer Rolls of Scotland*, vol. 1, Edinburgh.

Thomson, J.M. (ed.) (1882) *Registrum Magni Sigilli Regum Scotorum*, vol. 2, Edinburgh.

Watt, D.E.R. (ed.) (1987–1997), Walter Bower, *Scotichronicon*, 9 vols, Aberdeen.

Secondary sources

Bowler, D. (2004) *Perth: The Archaeology and Development of a Scottish Burgh*, Perth (Tayside and Fife Archaeological Committee Monograph; 3).

Campbell, B.M.S. (2009) 'Four famines and a pestilence: harvest, price and wage variations in England, 13th to 19th centuries', in: B. Liljewall *et al.* (eds), *Agrarhistoria på många sätt, 28 studies on manniskan och jorden. Festskrift till Janken Myrdal på hans 60-årsdag*, Stockholm, pp. 23–56.

Crone, A. and Watson, F. (2003) 'Sufficiency to Scarcity: Medieval Scotland, 500–1600' in: T.C. Smout, (ed.), *People and Woods in Scotland*, Edinburgh, pp. 60–81.

Dawson, A. (2009) *So Foul and Fair a Day: A History of Scotland's Climate and Weather*, Edinburgh.

Duncan, A.A.M. (1975) *Scotland: The Making of the Kingdom*, Edinburgh.

Gemmill, E. and Mayhew, N. (1995) *Changing Values in Medieval Scotland: A Study of Prices, Money, and Weights and Measures*, Cambridge.

Graham, A. (1968–9) 'Archaeological notes on some harbours in eastern Scotland', *Proceedings of the Society of Antiquaries of Scotland*, 101, pp. 200–285.

Hall M., Hall, D. and Cook, G. (2005) 'What's cooking? New radiocarbon dates from the earliest phases of the Perth High Street excavations and the question of Perth's early medieval origins', *Proceedings of the Society of Antiquaries of Scotland*, 135, pp. 273–285.

Maxwell, G. (1992) 'Aerial Archaeology in South-East Perthshire', *Current Archaeology*, 131, pp. 451–454.

Morrison, J. Oram, R. and Oliver, F. (2008) 'Ancient Eldbotle Unearthed: Archaeological and Historical Evidence for a Long-Lost Early Medieval East Lothian Village', *Transactions of the East Lothian Antiquarian and Field Naturalists' Society*, 27, pp. 21–45.

Oram, R.D. (2008) 'A Fit and Ample Endowment? The Balmerino Estate, 1228–1603', in: *Life on the Edge: The Cistercian Abbey of Balmerino, Fife (Scotland), Cîteaux, Commentarii cistercienses*, vol. 50, fasc. 1–2, pp. 61–80.

Oram, R.D. (2011) *Domination and Lordship: Scotland 1070–1230*, Edinburgh.

Oram, R.D. (2012) 'The Sea-Salt Industry in Medieval Scotland', *Studies in Medieval and Renaissance History*, 9, pp. 209–232.

Oram, R.D. and Adderley, W.P. (2008) 'Lordship and environmental change in central Highland Scotland *c.*1300 to *c.*1450', *Journal of the North Atlantic*, 1, pp. 74–84.

Oram, R.D. and Adderley, W.P. (2010a) 'Lordship, land and environmental change in West Highland and Hebridean Scotland *c.*1300-*c.*1450', in: S. Cavaciocchi (ed.), *Economic and Biological Interactions in Pre-Industrial Europe from the 13th to the 18th Centuries*, Florence, pp. 257–268.

Oram, R.D. and Adderley, W.P. (2010b) 'Re Innse Gall: A Norse Colony in the Irish Sea and Hebrides?', in: S. Imsen (ed.), *The Norwegian Domination and the Norse World c.1100-c.1400*, Trondheim, pp. 125–148.

Perry, D. (2005) *Dundee Rediscovered: The Archaeology of Dundee Reconsidered,* Perth (Tayside and Fife Archaeological Committee Monograph; 4).

Royal Commission on the Ancient and Historical Monuments of Scotland [RCAHMS] (1994) *South-East Perth: an Archaeological Landscape*, Edinburgh.

Taylor, S. (2004) 'Celtic Place-names of Clackmannanshire', *History of Scotland*, 4, pp. 15–17.

Wickham-Jones, C. (1996) *Scotland's First Settlers*, London.

Unpublished Reports

Cranstone, D. (2006) *Solway Salt Project*, Unpublished report by Cranstone Consultants.

Hindmarch, E. (2006) *Renaissance Golf Club at Archerfield*, Evaluation and Data Structure Report. Unpublished report by AOC Archaeology Group.

Websites

Fawcett, R., Oram, R. and Luxford, J., *The Corpus of Scottish Medieval Parish Churches, Diocese of Dunblane, Abernethy Parish and Collegiate Church.* http://arts.st-andrews.ac.uk/corpusofscottishchurches/site.php?id=125608 Accessed 10 March 2010.

Oosthoek, J., *The Fishings of the Burgh of Stirling, 1700–1930.* http://www.eh-resources.org/Fishing_results/Accessed 10 March 2010.

Pont, T., *Maps of Scotland 1580s-1590s, Pont 26: Lower Angus and of Perthshire east of the River Tay*
http://maps.nls.uk/pont/specialist/pont26.html.

Pont 32: The East Central Lowlands (Stirling, Falkirk & Kilsyth)
http://maps.nls.uk/pont/view/?id=pont32 (for Abbot's Grange, Grange, East, Mid and West Saltcotts) Accessed 27 July 2011.

Roy, W., *Military Survey of Scotland, 1747–55.*
http://geo.nls.uk/roy-highlands/ and http://geo.nls.uk/roy-lowlands/Accessed 27 July 2011.

Tay District Salmon Fisheries Board, *The River Tay System.*
http://www.tdsfb.org/river-tay.htm Accessed 10 March 2010.

17 Storms, economics and environmental change in an English coastal wetland: The Thames Estuary *c.* 1250–1550*

James A. GALLOWAY, Institute of Historical Research, University of London

Abstract

This paper explores the interaction of natural and human forces in effecting change in the marshlands bordering the tidal river Thames and its estuary in south-east England between the thirteenth and sixteenth centuries[1]*. These lands lay at the heart of the hinterland of medieval London, England's capital and largest city, and were thus of considerable economic and strategic importance. They were also, however, subject to the repeated threat of inundation by storm flooding from the North Sea, which could overwhelm defences and ruin agricultural land. The paper will describe the context within which embankment occurred in the Thames area, and review the uses to which embanked and unembanked marshlands were put, before examining the impact of storm surges and other types of flooding upon these valuable but vulnerable lands. The changing economic parameters and social conditions of the later fourteenth and fifteenth centuries will be shown to have had a major impact upon human use of the marshes, and upon responses to storm flooding. Institutional responses to coping with these natural risks will also be reviewed, and the impact of changing ownership patterns in the first half of the sixteenth century will be discussed through a case-study of Erith in Kent. The paper will thus address some of the key concerns of this volume through its investigation of the dynamic forces, natural and human, which produced changes in the landscape, economy and ecology of the estuary over a period of three centuries.*

I. London, the Thames and the estuary

The Thames Estuary lies in the south-east of England, home to the country's greatest concentrations of wealth and highest population densities for most of recorded history. The river Thames is today tidal well upstream of London. In the Roman and medieval

* This paper arises from a research project carried out at Centre for Metropolitan History, part of the Institute of Historical Research (IHR) in the University of London, and an earlier pilot study based at the UK's National Maritime Museum in Greenwich. Funding was provided by a National Maritime Museum/Crown Estate Fellowship from January to June 2006 and by the UK Economic and Social Research Council (award no. RES-000-22-2693) from March 2008 to February 2010. For a fuller description of these projects see Galloway and Potts (2007) and Galloway (2009).

1 The Thames Estuary is here defined broadly to include both the tidal stretch of the river Thames from London Bridge to Gravesend ('the tidal river') and 'the outer estuary' from Gravesend to the open sea.

Landscapes or seascapes? The history of the coastal environment in the North Sea area reconsidered, ed. by Erik THOEN, Guus J. BORGER, Adriaan M.J. DE KRAKER, Tim SOENS, Dries TYS, Lies VERVAET & Henk J.T. WEERTS, Turnhout, 2013 (*CORN Publication Series*, 13), pp. 379–396.

BREPOLS PUBLISHERS DOI: 10.1484/M.CORN.1.101561

periods the tidal head was closer to the city, but still permitted sea-going vessels to anchor at London's harbours and wharfs (Milne, 2003: 144–146). The funnel-shaped estuary opens out towards the European mainland, and linked London to its most important trading partners in the Low Countries. The corridor of the Thames and the estuary formed the core of the London hinterland, characterised by the highest levels of price integration and the densest economic interaction (Galloway, 2000: 28–30). In the decades around 1300 London, with perhaps 80,000 inhabitants, constituted one of the largest cities in Europe, and exerted a strong influence upon patterns of production and marketing in its hinterland (Campbell *et al.*, 1993; Galloway, Keene and Murphy, 1996; Galloway, 2001). In the later fourteenth century that influence changed, with the decline of some agricultural and marketing specialisms and the rise of others, within an overall process of de-intensification.

The early commercialisation of the lands around the estuary is evident in the pattern of market foundation. A disproportionate number of places close to the tidal river and the estuary shore had acquired market rights by the second quarter of the thirteenth century, including West Thurrock (1207) Barking (1219) and Hadleigh (1228) in Essex and at a variety of locations in Kent (Letters, 2003). Britnell has calculated that one-third of all the markets established in Essex between 1200 and 1270 were located at small ports and other locations close to the coast (Britnell, 1981: 15–21). London's specialised grain dealers had strong contacts along the line of the river Thames and around the estuary, notably with the port of Faversham in north Kent. Faversham's grain hinterland was compact and highly productive . Yields of wheat and other grains, grown upon the embanked and drained alluvial soils adjoining the estuary and the Stour Levels between Faversham and the city of Canterbury, were among the highest in medieval England.. Faversham's grain market was held four days a week, drawing in both large and small-scale producers from the surrounding area and sometimes from further afield. Other important markets for wheat and oats around the Thames Estuary, frequented by the London cornmongers, included Canterbury, Rochester, Hoo, Dartford and Maidstone in Kent and Shoebury in Essex (Campbell *et al.*, 1993: 46–53, 91–95). This London-focused economic activity provides the vital context for understanding the history of the marshlands around the river Thames and its estuary.

The banks of the tidal Thames and the shores of its broad estuary are fringed by strips of alluvial soils, of varying width, but rarely extending more than a few km inland (Figure 17.1). Before the construction of river and sea-walls, these were regularly overflowed by the highest tides and thus exposed to salt-water – of high salinity in the outer estuary, of low but fluctuating salinity along the tidal river. The salt marsh and reed-bed plants which colonised these soils, and the creeks which bisected the marshes, were exploited by human beings for millennia before the marshes were embanked and cut off from tidal influence – for grazing, fishing and fowling and the cutting of reeds and rushes for thatch and flooring – while mudflats around the outer estuary were home to a seasonal salt-making industry (Rippon, 2000: 39–46; Keen, 1989).

The process of embanking the Thames marshes seems to have been largely accomplished in the twelfth and thirteenth centuries. Domesday Book, compiled in the late eleventh century, gives no indication that any of the marshes had at that time been embanked; the marshes feature as pasture for sheep, and the assumption is that this represents grazing on salt marsh, the rights to which were sometimes shared by inland parishes (Darby, 1977: 157–159). Deeds from the twelfth century begin to mention walls

Figure 17.1 Tidal River Thames and Thames Estuary, showing alluvial marshlands

and ditches in the marshes, however, and by the later thirteenth century detailed manorial extents and accounts show many of the Thames marshes to have been embanked and used for intensive agriculture (Galloway, 2009: 175–176). The most thoroughly embanked marshes appear to have been those along the tidal river, between London and Gravesend. To the east of this, around the outer estuary, the picture is more mixed. Extensive areas of unembanked salt marsh appear to have remained on Canvey Island in Essex, and in some of the neighbouring districts (Rippon, 2000: 203–207)[2]. Across the estuary, tracts of salt marsh also remained on Sheppey and the Hoo peninsula, but there they co-existed with extensive 'innings', while the marshes around the mouth of the River Medway had been in large part embanked and drained. The south-eastern Essex marshes of Foulness, and the other marshes lying between the Crouch and the Thames, were also mostly reclaimed by the thirteenth century, and put into intensive agricultural production (Smith, 1970: 9–12). Londoners were actively involved in the processes of reclamation, especially along the tidal river, both as landlords and as tenants of marshland holdings, the ease of water transport enhancing the utility and value of riverside and estuarine lands. An inquisition of lands seized in north-west Kent in the aftermath of the civil war of 1263–5 show that Nicholas Horn and John Capper of London held five and a half and eight acres of marsh land there respectively, while four other Londoners held various quantities of 'land' and rents in Erith and other locations (*CIM* vol. 1: 227). London religious houses including St Mary Graces, Holy Trinity, Aldgate and the great Abbey of St Peter's Westminster also held much land in the riparian and estuarine zones (Barron and Davies, 2007). The Bishops of London possessed much land around the tidal Thames, including extensive areas of marshland. The manor of Stepney, just to the east of London, incorporated the Walmarsh, which was embanked by the early thirteenth century. Arable and pastoral

[2] See also the paper by Stephen Rippon in this volume.

farming was practised there and in the sub-manor of Pomfret located on the Isle of Dogs, a marshland peninsula formed by a great loop of the Thames. Pomfret contained one of the few true marshland settlements within the Thames area (Croot, 1997).

The Thames marshes were of only limited inland extent compared to some of the larger English coastal wetlands like Romney Marsh and the Fenlands, and were dwarfed by the extensive wetlands of the Low Countries (Rippon, 2000; Van de Ven, 2004) This meant that they were comparatively expensive to reclaim per unit of area, as sea and river-frontage might only be protecting a kilometre-or-two wide strip of marshland. There needed, therefore, to be a significant economic inducement to undertake the costly business of building and maintaining the network of walls and ditches necessary to protect the marshes from the tides and to drain them of fresh water. The valuations of land in extents of the late thirteenth and early fourteenth century indicate how significant those inducements could sometimes be. On average arable land on reclaimed marsh was valued twice as highly as non-marsh or 'upland' arable, and in some areas the differential could be greater, with marsh arable valued four or five times higher than arable outside the marsh. The biggest differentials were found along the tidal river Thames, close to the city of London and its vast market for grain. Here some marsh arable achieved net annual values of 24d or 36d per acre, bearing comparison with the most fertile and productive areas of medieval England (Galloway, 2009: 175). For comparison, the average arable valuation for the ten counties around London was 4.8d per acre (Campbell, Galloway and Murphy, 1992: 16). In addition to the influence of the London market, it must be acknowledged that some of the 'upland' soils adjacent to the marshes were relatively poor and unproductive, and that the high marshland valuations therefore also reflect a local scarcity of good arable land.

Further east, around the outer estuary, a larger proportion of the reclaimed lands were used as pasture and meadow, providing improved grazing for both sheep and cattle, and often also for horses. Christ Church Priory Canterbury kept some 1300 sheep on their north Kent manors of Barksore, Elverton and Ham in the early fourteenth century, mostly grazed on embanked marshland, and a further 3–4000 on similar lands further east in the Stour Levels (Smith, 1943: 152). The existence of extensive tracts of good-quality arable land away from the coast meant that there was less pressure to convert these marshes to arable cultivation, one of the key differences between the tidal river and the outer estuary, and the contrast between marshland and upland values here were less extreme. Some arable farming was carried out on the outer estuary marshes, however, especially those on Foulness Island and in the marshes around the mouth of the river Medway. John de Rochford's manor of Rochford in south-east Essex was entirely marshland in 1309, including 300 acres of arable and an unspecified acreage of pasture (TNA: PRO C134/14(26)). At least some of this manor's lands were in Foulness.

The sea- and river-walls which protected the reclaimed or 'inned' lands were built using locally available materials, principally marsh earth or clay, timber and brushwood or rushes. The earth was heaped up, either free-standing or around a timber framework. The resulting bank was then faced or thatched with faggots made of brushwood or rushes, to give some added resistance to erosion by waves, spray and rainfall (Galloway, 2009: 178–179). Stone was not widely used as a building or facing material for sea-walls in the medieval period, due to the high cost of transport. However, in the Thames area chalk was readily available from quarries beside the Thames in north-west Kent, while ragstone could be sourced via the river Medway, and both these stones found some use in river defences. At Greenwich the remains of a probable medieval river-wall had ragstone blocks on the river side, and the wall was infilled with chalk rubble[3]. In addition to the principal sea- or river-walls, marshes were commonly subdivided by counter-walls, which helped to restrict the extent of any flood damage. Medieval Foulness Island, on the Essex shore of the outer Estuary, was divided up by counter-walls into as many as thirteen marshes, each independently protected against the sea (Smith, 1970: 26).

II. Storm flooding and the human response

Examination of Ordnance Survey maps of the nineteenth century, or other early detailed maps and plans pre-dating the eastwards sprawl of metropolitan London, reveals traces of historic flooding all around the tidal River Thames and the Estuary (Figure 17.2). Characteristic horse-shoe shaped 'set-backs' reveal where sea or river walls had been breached, and the line of the wall set back around the deep scour hole which had resulted from the movement of water in and out through the narrow gap. In Figure 17.2, the set-back wall to the south of the river dates from the Erith breach of 1530 or before (see also Figure 17.4 below). On the north bank, the large horseshoe repair (known into the nineteenth century as Horseshoe Corner, and used as a productive reed-cutting ground) may date from floods recorded in the early fifteenth century, while just to the east is the post-medieval Dagenham Breach of 1707. These breaches, and the resulting flooding of reclaimed marshland, could have a variety of contributory causes, including poor wall maintenance, or subsidence resulting from the building of a wall over an old salt-marsh creek. However, the immediate cause of most of the documented examples of breaching was exceptional storms, or extreme high tides associated with North Sea storm surges (Galloway and Potts, 2007). Some of these breaches could result in the formation of a permanent or semi-permanent lake, others to a new broad channel – often referred to as a *gut* or *gulf* – connecting to the river or the coast. There is reason to believe that some docks and basins in the London area may owe their origin to such flood-engendered features. It has recently been suggested that a large pond connected to the Thames in the Deptford Strand area, which is referred to in 1293 and which was probably the result of an earlier breach in the Thames bank, may later have been adapted to form the basin of the Royal Dockyard (Divers *et al.*, 2004: 21).

Local landholders and communities held the primary responsibility for defence against flooding in the medieval period, local custom or 'marsh law' obliging all those enjoying the protection of walls and drainage ditches to contribute to the cost of their maintenance. Early leases imply that tenants owed direct labour to maintain walls and ditches, as in the 1201-2 undertaking by a tenant at Rainham in Essex to 'defend the wall against the Thames' (Kirk, 1910: 26). Money payments in lieu of direct labour are commonly encountered, and typical thirteenth- and fourteenth-century levies in the Thames-side marshes amounted to between 2d and 6d per acre. It is clear, however, that on some of the major estates invest-

[3] London archaeological archive, site record TUM03, online at <http://www.museumoflondon.org.uk/laarc/catalogue/siteinfo.asp?id=17843&code=TUM03>, accessed 10 November 2008.

Figure 17.2 'Horseshoe' river wall repairs around historic Thames breaches near Barking and Erith in Ordnance Survey map of 1870–5 (© Crown Copyright and Landmark Information Group Limited 2010. All rights reserved)

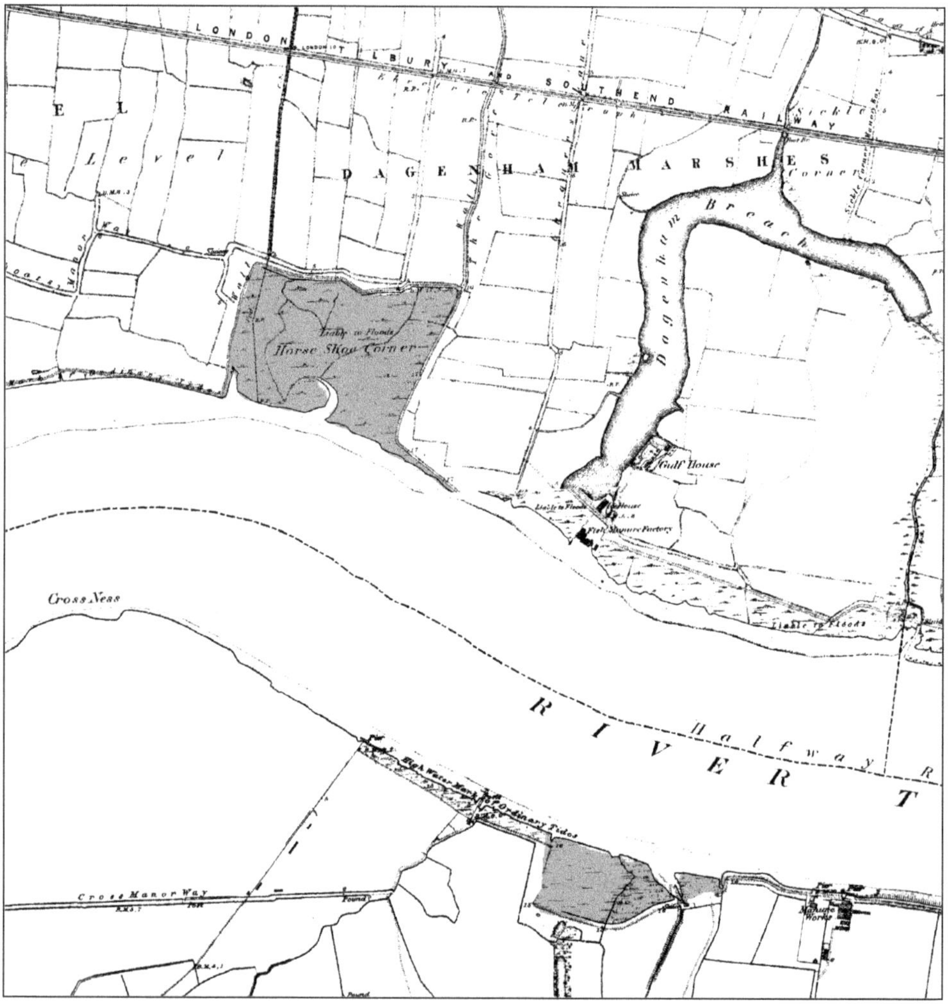

ment in reclamation and the maintenance of marsh defences required sustained central investment in addition to these levies on tenants (Smith, 1943: 170–1; Galloway, 2009: 178). These types of arrangement had presumably been in operation from an early date and must initially have worked adequately. From the mid- to later-thirteenth century onwards, however, increasing numbers of problems arose, many of them directly attributable to storm flooding. The accumulation of difficulties encouraged greater central supervision and legislation, with the Crown increasingly active in investigating problems and enforcing cooperation among landholders. The principal mechanism by which central intervention took place was the commission *de walliis et fossatis*, (of walls and ditches or dikes), later to be known as commission of sewers (Richardson, 1919; Owen, 1981: 9–23). These royal commissions were ad hoc, temporary bodies empowered to investigate problems in specific localities, or granted general oversight of flood protection and drainage throughout a county or along a defined stretch of coast. The principal local landholders were prominent among the membership of these commissions, and were often supplemented by trusted court officials and professional bureaucrats. Few records of the activities of early commissions survive, but it is clear that they were essentially reactive rather than proactive in their approach, responding to floods, failures of local organisation or petitions for aid from local landholders. The commissioners had power to distrain landholders who were found not to have performed their share of communal works (Owen, 1981: 14).

The impact of storm surges and other storm-related marine floods in the Thames area can be reconstructed from a variety of documentary sources, including the issuing of royal writs and commissions, manorial accounts, and the records of the City of London. The marshlands around London were flooded in the 1230s, but the Thames area seems to have escaped the worst effects of the mid-thirteenth-century storms. The concentrated episode of storm surge flooding in the years 1286–88 did, however, have a serious impact along the river and estuary and in adjacent areas of east Kent, and the decades around 1300 witnessed a number of serious inundations affecting London's southern suburbs (Galloway and Potts, 2007). Some river frontage was eroded away, and tidal mills, which processed much of the grain consumed in London, were damaged by extreme tides. The ex-Templar mills in Southwark were described as 'broken and ruined by the inundation of the water of the Thames' in 1309[4]. A tidal mill at Greenwich, excavated recently by Museum of London Archaeology, was built in the 1190s and later abandoned and partially dismantled, perhaps as a result of storm damage, the remains being overlain by a layer of estuarine silt (Goodburn and Davis, 2010: 6–12). Repeated issuing of commissions *de walliis et fossatis* for the Greenwich and Woolwich areas between 1315 and 1324 points to recurrent storm damage. At Stepney a river-wall breach on New Year's Eve 1323 marked the start of a sequence of floods which devastated extensive reclaimed marshlands, including much of the reclaimed Walmarsh which had been exploited as open-field arable since *c.*1200 (Galloway and Potts 2007; VCH Middx 11: 1–7).

The winter of 1334 witnessed one of the most severe storm surges to hit the Thames region in the medieval period, with the worst damage occurring around the outer estuary where many manors reported sea-wall breaches and livestock losses. Like many of the major storm floods affecting the Thames Estuary, this November 1334 event can be correlated with a storm surge affecting the Low Countries, from Flanders to North Holland (Gottschalk, 1971–8 vol. 1: 327–328). A significant difference, however, is that Thames area floods did not normally cause major loss of *human* life, as most settlements here were established on adjacent higher ground rather than in the marshlands themselves. The economic damage was severe, however. At Sharpness, a manor of Boxley Abbey with extensive embanked marshlands at the mouth of the river Medway in north Kent, expenditure on marsh walls and ditches increased from 19s.1d. in 1332–3 and £3.13s.5d. in 1333–4 to a substantial £29.16.7d

[4] Transcript of National Archives: PRO E 368/79, fol. 111 dorse by H.J. Nicholson, online at http://www.cf.ac.uk/hisar/people/hn/MilitaryOrders/MILORDOCS9.htm, accessed 15 June 2010.

Figure 17.3 Commissions *de walliis et fossatis* issued per decade for tidal River Thames and Thames Estuary, 1280–1499 (source: CPR)

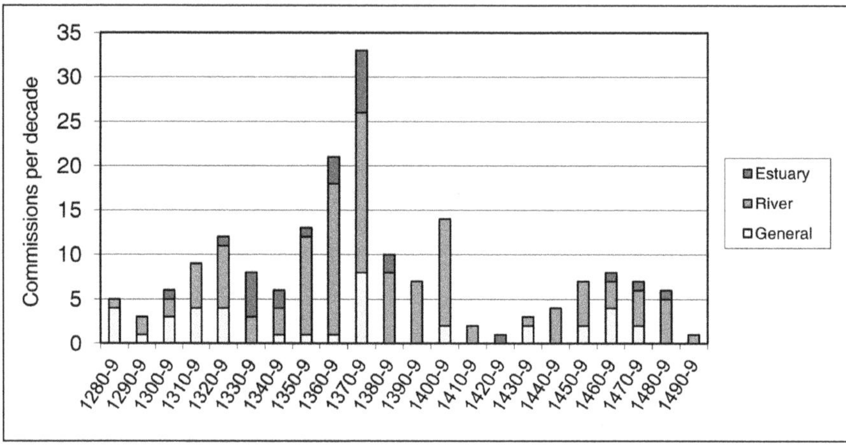

in 1334-5 (TNA: PRO SC6/896/17). Nearby, Barksore, a manor of Christ Church Priory, Canterbury, sustained a similar peak of expenditure on sea defences, while 150 demesne sheep were drowned along with an unknown number of tenants' livestock (Galloway, 2009: 184–187). At Great Wakering, west of Foulness in Essex, it was attested that lands submerged there would take more than seven years to return to their former state (CIPM 7: 474). Storms surges had multiple impacts upon marshland manors consequent upon the breaching of sea-walls. As well as direct damage to crops and livestock, storm flooding caused indirect damage through salinisation of soil and water-supplies. Consequently, and ironically, a shortage of water was one of the immediate short-term effects of flooding. At Sharpness new ditches and ponds had to be dug for fresh water, and 34 boatloads of fresh water had to be shipped to the marshes in the summer and autumn of 1335. The storm surge had also devastated the Sharpness eel fishery and 10 000 young eels had to be purchased to restock the waters there (TNA: PRO SC6/896/17).

It is clear that these major storm events wrought much short-term havoc in the Thames and estuarine marshlands. In the pre-Black Death period, however, they seem only rarely to have caused irreversible damage or long-term loss of land. It was the radical changes in economy and society wrought by the repeated visitations of plague from 1348–9 onwards that ushered in a new era for the marshes - one which in essence lasted for 200 years - in which long-term flooding and reversion to inter-tidal conditions became common. The 1370s appear to have been particularly calamitous, and may be seen to represent a turning point. Efforts to repair sea and river defences surged after 1350 and reached a peak in the 1370s as reflected in the issuing of royal commissions *de walliis et fossatis* (Figure 17.3). After the late 1370s the number of commissions fell away sharply, with some minor reversals in the first and middle decades of the fifteenth century. These trends appear to reflect thirty years of intensive efforts to deal with deterioration in the marshlands in the face of social and economic dislocation, followed by a sharp decline in activity reflecting a more pragmatic adaptation to economic and environmental change.

III. The crisis of the 1370s and its aftermath

It was in the 1370s that agricultural prices, which had remained generally buoyant since 1349, began a long-term downward trend. Concentrated demand for bread grains was collapsing along with London's population, which by 1377 had fallen to perhaps 40,000. The London cornmongers, a numerous and influential group in 1300, gradually declined in political and economic importance as corn-dealing passed increasingly into the hands of general merchants (Campbell *et al.*, 1993: 82). At the same time wages were pushing ever upwards, a trend most pronounced in the London region. Rising costs of labour and falling prices for agricultural produce had serious implications for the economies of the expensive to maintain riverine and estuarine marshlands, especially when they were hit by storm flooding and faced with major works to repair walls, ditches and sluices. The general labour shortage of the post-Black Death decades necessitated the granting of additional powers to commissioners, enabling them to compel labourers to work on marshland repairs, although paying for these works remained the responsibility of local landowners. In May 1353 a supplementary commission was issued for repair of the Thames banks between Greenwich and Plumstead in north-west Kent, empowering the commissioners to 'take in and impress such and so many labourers, as well within liberties as without, for competent wages, as they should have need of for the repair of those banks, ditches and sewers.' (Dugdale, 1772: 60). Similar powers were granted two years later to commissioners dealing with several sea-wall breaches in the hundred of Dengie in south-east Essex. The commissions attempting to cope with the widespread flooding of the 1370s, and their aftermath in the 1380s, were on several occasions granted such powers, often with the explicit power to arrest and imprison those who refused to work (Dugdale, 1772: 78–9; CPR, 1374–7: 126–7, 485; CPR, 1377–81: 447; CPR, 1381–5: 438). Even with such additional powers, which were not routinely granted but rather reserved for the most intractable cases, it does not seem that the medieval Thames commissions *de walliis et fossatis* ever attained the same degree of power and control over water management as the local and regional water boards of the Low Countries possessed (Soens, 2006). Their most fundamental weakness was undoubtedly their temporary character, commissions lapsing once short-term problems had been tackled and requiring new royal instructions to resume their duties subsequently.

The crisis of the 1370s played out all along the tidal river and far out into the estuary. A serious episode of flooding affected the Thames-side marshes in late winter 1373–74. Eight commissions were issued for the Thames area between February and May 1374, including one for Havering marsh, where the walls and dikes had been 'broken by the inundation of the water of Thames' (CPR, 1370–4: 473). These floods were closely followed by storm surges in the autumns of 1374 and 1375. Major expenditure was incurred in defending the estuarine marshlands around the mouth of the River Medway, including the exposed manor of Sharpness belonging to Boxley Abbey. A manorial account for Sharpness dating from 1374 describes 'the destruction of the marsh after the feast of St. Dionysius (9 October) by the great flood-tide' (TNA: PRO SC6/897/4). This same surge struck Zeeland and South Holland at this date. The following year expenditure on repairs to the marsh defences at Sharpness peaked at a remarkable £85, following another October storm surge (TNA: PRO SC6/897/5). In the following accounting year (1376–7) the Treasurers of Canterbury Cathedral Priory were spending large sums repairing sea-walls breached at their manor of Barksore, close to Sharpness (CCA DCc/MA 2).

A late fourteenth-century map of the Christ Church manor of Cliffe, on the Hoo peninsula in Kent, shows what can be interpreted as flood damage (Figure 17.4). The Thames Estuary is depicted at the top of the map, bounded by a sea-wall which appears to be breached at two points. The more easterly of these openings bears the inscription 'new *portus* (entry, port or breach) towards the church of Cliffe, made by (or on account of) the violence of the water', suggesting that this breach had recently occurred as a result of storm flooding. A large area of the marshlands is labelled '*allivio*', which may mean that it was subject to the tides. What may be sluices are indicated on the tidal inlets, adjacent to the lands of the Prior of Rochester. The style of the map suggests a working document drawn up rapidly for a specific purpose, perhaps in the aftermath of the surges of mid-1370s. An expert examination by Felix Hull concluded that it may have been prepared to accompany 'a hasty petition' explaining the disaster which had ensued following a breach in the sea-wall at Cliffe (Hull, 1986: 102–105).

At Barking, 15 km east of London, the surges of the mid-1370s breached the river walls and flooded extensive areas of marshland belonging to Barking Abbey. Large sums of money were expended by the Abbess over the following years and decades, and numerous grants of aid were made by the Crown. Special commissions with powers to impress labour were established in order to repair the breach and drain the marsh where, it was reported, 'a lake had formed on land once arable'. Gradually, however, the efforts faded away, the marsh remained subject to the tides, and serious efforts to recover it do not appear to have resumed for nearly a century and a half (Galloway and Potts, 2007: 374–375). This flooding had profound consequences, in the short, medium and long terms. The navigation of the Thames channel was harmed by 'logs, stone and sand' washed out of the marsh, probably associated with the formation or shifting of a shelf or sandbank in the river. Londoners expressed concern that the tides were not rising as high as formerly due to the overflowing of the Barking marshes. Fish stocks were said to be suffering damage through the installation of new fishing weirs on the marshes and in the marsh creeks, which at the ebb trapped fish and fry that had moved there at high tide. An enquiry in 1386 found that these weirs had been licensed by the Abbess of Barking and neighbouring landlords, suggesting that they had already realised that the marsh was not going to be recovered soon, and had sought to make what income they could from the marsh in its tidal, periodically inundated state (CPMR, 1381–1412: 116–117). The proximity of the London market created a major demand for fresh fish and no doubt added a further stimulus to these developments in the marshes. Indeed, London's demand for fresh rather than preserved fish may have been rising in absolute terms after 1349 despite population decline, due to rising living standards and related changes in diet. When in the 1390s the 'poor lieges of Lesnes, Plumstead, Erith, Woolwich and Greenwich' had petitioned the King to be allowed to continue their tradition of fishing with boats, nets and 'engines', they claimed that their activities supplied the City of London as well as 'all the surrounding countryside' (TNA: PRO SC/8/22/1061).

We may assume that some of the Barking fishermen were marshland tenants who had had to change their livelihoods, but otherwise we have no knowledge of what happened to the latter in the absence of court rolls or rentals from this period at Barking. Comparison with other Thames marshlands suggests that many of the tenants would have held acres in the marshland as part of a wider property holding, rather than being solely marshland farmers, and so they may not have been totally ruined by the inundation. Moreover, most would probably have lived in the town of Barking or other upland settlements, so few

Figure 17.4 Late fourteenth-century map of marshlands at Cliffe, Thames Estuary (source: Canterbury Cathedral Archives DCc/ChAnt/C/295, reproduced with the permission of the Dean and Chapter of Canterbury Cathedral)

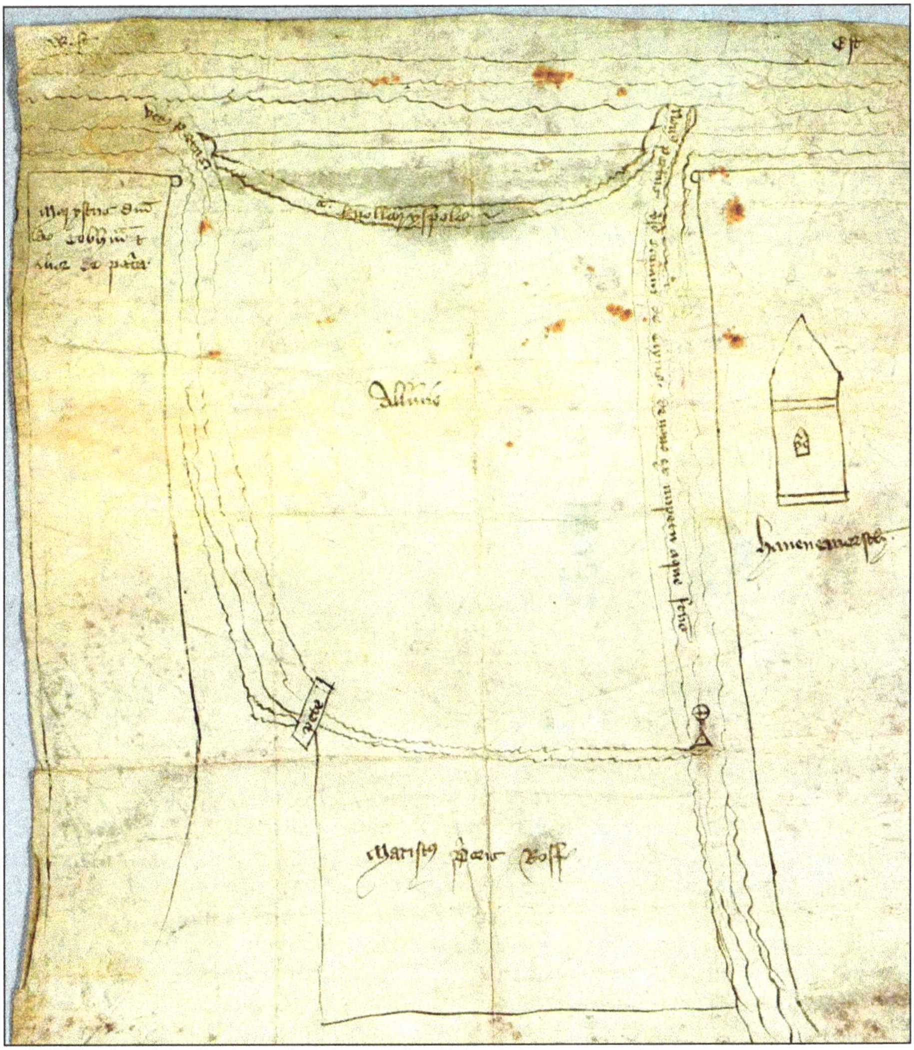

would have been directly made homeless as a result of the flooding of the marsh. The partial loss of livelihood, together with the repeated attempts to coerce labour to work on the riverside defences, must have caused widespread discontent, however, and perhaps it is not surprising to find that people from Barking and neighbouring Thames-side communities were prominent in the early phases of the great uprising of the English peasantry that began in May1381 (Oman, 1969: 32–34).

Examples of long-term abandonment of reclaimed marshlands in the Thames area could be multiplied. Along the north Kent shores of the Estuary, and around the mouth of the Medway, storm surge damage is recorded in the late fourteenth and early fifteenth centuries. A major surge struck in the winter of 1393–4, evident in large-scale stock losses and heavy expenditure at the manor of Ham (TNA: PRO SC6/892/1)[5]. Sharpness manor, which adjoined and was closely associated with Ham, must also have suffered badly from this storm and seems to have finally succumbed to recurrent flooding in the first quarter of the fifteenth century. The final straw may have been the 'St Elizabeth Day floods' of 1404 and 1421, both of which are known to have struck the Sheppey and Medway areas (Galloway and Potts 2007; Gottschalk, 1971–8, vol. 2: 18–19, 96–99). Close to London, Wapping marsh was recovered in the 1540s by Cornelius Wanderdelf, 'lately of Brabant', after being, in the words of an Act of Parliament, 'longe tyme surrounded and over flowen with water' (Statutes: 966).

As in the pre-Black Death period, loss of human life seems to have been rare in these floods, chroniclers and inquisition juries highlighting instead the drowning of livestock and the spoiling or long-term abandonment of farmland (Galloway, 2009: 183–4). The Thames marshes were thus spared the severe human tragedies which marked the later middle ages in the Low Countries (Van de Ven, 2004: 105–7). This must have been largely due to the fact that, as at Barking, the tenants who farmed marshland holdings usually lived in villages or farmsteads on the adjacent upland, the marshes themselves being largely devoid of settlement. A notable exception is provided by the settlement in the Isle of Dogs pertaining to the manor of Pomfret. Here mixed farming was practised in the fourteenth century, the arable being worked by the manor's villeins, and a manor house and chapel are documented. The Isle of Dogs was flooded by a major river-wall breach in the mid-fifteenth century, and the settlement, known simply as 'Marsh' was abandoned together with its chapel and manor-house, remains of which were still identifiable in the eighteenth century (Survey: 375–87). Fishing, fowling and reed-cutting replaced agriculture as sources of landlord income and peasant livelihood for several decades (VCH Middlesex, 11: 1–7, 52–63).

That the economic and environmental changes occurring in these marshes were characteristic of a more general process is suggested by legislation from the late fifteenth century. In 1489, parliament presented a petition to the crown, at the request of the citizens of London, concerning the state and custody of the River Thames[6]. It relates how 'tempests of weather and great abundance of water' had caused many breaches and had drowned 'pastures, meadows and grounds'. In these 'drowned lands' and their creeks most of the tidal river's fish fry (immature fish) were tending to congregate, where they were trapped by 'unlawful engines and nets', and used as bait or as pig-food. It was feared that this would cause the 'utter destruction' of fish stocks. By this date Londoners' main concern regarding flooding was thus its impact on fisheries, and the increasing use of fishing weirs both within and alongside the Thames – a threat to navigation as well as

[5] I am grateful to Elizabeth Eastlake for drawing my attention to this account and allowing me access to her digital photographs of the manuscript.

[6] 'Henry VII: January 1489', *Parliament Rolls of Medieval England*. URL: <http://www.british-history.ac.uk/report.aspx?compid=116566&amp> accessed 29 May 2010.

to fish stocks. Salt marsh creeks are recognised today as important fish nurseries, so the impact on fish stocks of these activities may have been real enough. The paradox is that an increase in the extent of brackish-water habitats in the later middle ages seems to have been associated with a *diminution* of fish-stocks.

IV. The Thames marshes at the close of the Middle Ages

Piecemeal recovery of parts of the 'drowned marshes' was underway by *c.*1500, at Barking, Stepney, East Ham, Wennington and other locations on the north bank of the Thames (VCH Essex 5 and 7; VCH Middx 11). The Isle of Dogs was partly recovered in 1488, but further floods meant that some parts were still under water in 1529. Subsequent secure recovery was accompanied by a transformation in land use and tenure, with pastoral farming based on consolidated holdings replacing the pre-flooding mixed farming system with its small tenant holdings (VCH Middx 11: 52–63). Similar developments were occurring in adjacent Wapping Marsh, together with peri-urban development of tenements and warehouses along the river wall (McDonnell, 1978: 98). These developments were spurred by the rapid re-growth of London's population during the sixteenth century, and the increasing demand of London's butchers for grazing lands around the city. However, on the south bank of the Thames institutional change, allied to storm surge activity, combined to bring about a new phase of flooding on parts of the tidal river, which left some marshes subject to the tides for much of the sixteenth century. The dissolution of the monastic orders in England, and the disposal of their estates, seems in some locations to have seriously interfered with the management of the marshlands. The transfer of church lands into royal and then into lay ownership, often passing through several hands within a decade or two, must have made it hard to sustain the traditional mechanisms for inspecting and carrying out repairs to walls, ditches and sluices, and for raising local taxes to pay for necessary works.

This problem is exemplified in the marshland peninsula which lay between the towns of Woolwich and Erith. Here some 1300 hectares of reclaimed marsh had been used as intensive and highly valued arable and pasture in the fourteenth century, and were retained largely intact throughout the fifteenth century, although the balance of use swung to the pastoral at that time (Galloway, 2010). Royal, lay and ecclesiastical lands lay intermixed in the marshes, the major landlords including Lesnes Abbey and St Augustine's Abbey (Canterbury). By the early sixteenth century the marshes at Lesnes and adjacent areas had a considerable history of embankment and flooding, and institutions existed to cope with the repeated environmental challenges. The tenants of Lesnes Abbey were obliged to make regular payments for maintenance of the marsh defences, at a rate of 4d per acre according to a surviving roll from 1462-3, while the Abbot of Lesnes played an active role in commissions *de walliis et fossatis* in the later fifteenth century (L&P Hen VIII vol. 4 pt. 2: 1593; CPR 1467–77: 463, 466).

On 26th December 1516 a damaging surge is recorded as affecting Zeeland and Flanders (Gottschalk, 1971–8 vol. 3: app. XI). There is evidence to show that this storm surge affected the tidal Thames, and specifically impacted upon the Erith area. A book of payments made by the controller of the English King's ships refers to a 'great and high tide that was upon St Stephen's Day (26 Dec) 1516', which wet ropes and cables in the storehouse at Erith. In response, expenditure was authorised on raising the doors of the storehouse 'for keeping out of the high tides' (L&P Hen VIII vol. 2 part 2: 1406). At the

Figure 17.5 Breaches in the Thames marshes near Erith, Kent, 1588. Note: this map was drawn with north at the bottom (source: Detail from British Library Maps Crace XVIII no. 17, reproduced with permission)

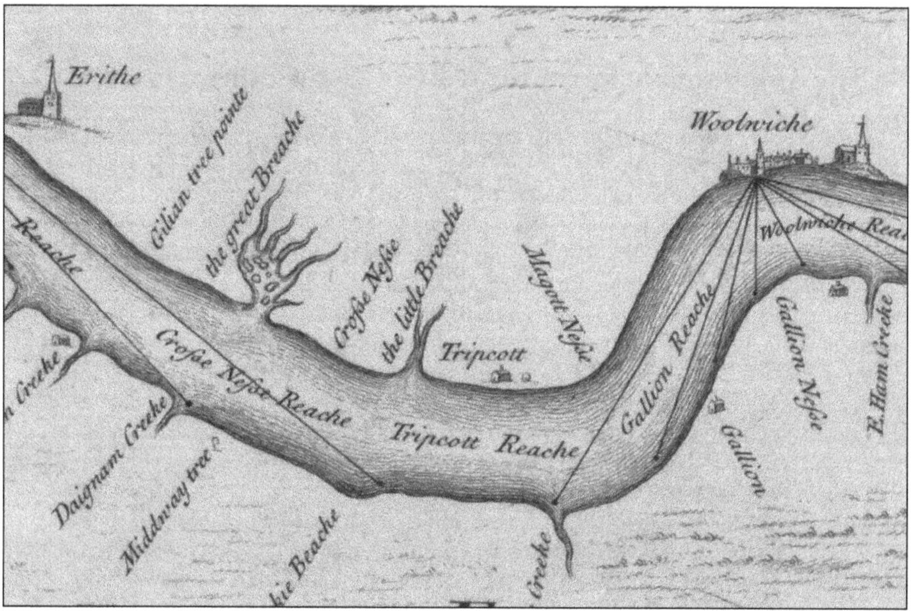

same time, records of maintenance of river defences on the Queen's manor of Erith cease abruptly, which together with other evidence suggests that the eastern part of the marshes may have been overwhelmed by this surge (Galloway, 2010: 18–19).

The central parts of the marsh, dominated by Lesnes Abbey, were seized on behalf of Henry VIII's minister Cardinal Wolsey in 1525, to fund the foundation of a new college in Oxford. Almost immediately, in 1526, a river-wall breach occurred, which was patched up, only for a worse one to follow in 1529. 'I have been at Lesnes,' wrote Wolsey's secretary Thomas Cromwell, 'where I saw one of the most piteous and grievous sights that ever I saw... concerning the breach out of the Thames into the marshes...which be all overflowen and drowned' (Merriman, 1902: 324–325). This seems to have been the 'little breach' shown on a map of 1588 (Figure 17.5). The cause of these particular incidents of flooding seems likely to have been institutional. The abbey's functions as landowner had been taken over by Wolsey's agents and with them the responsibility to maintain river defences, in cooperation with neighbouring landlords and tenants, and to contribute to general levies ordered by the Commissioners of Sewers for marsh defence work. As the prime reason for acquiring the abbey's lands was to sell or lease them as quickly as possible, it seems likely that these maintenance works were not regarded as a high priority, and local expertise and advise may have been disregarded. Letters written by Cromwell and others of Wolsey's officials certainly suggest that local people were resentful and reluctant to cooperate with their new masters (Galloway, 2010: 17).

Further damage resulted from the major storm surge of November 1530 – the 'St Felix's flood' in the Low Countries - which created or perhaps greatly enlarged an existing breach to the east, shown as the 'Great Breach' in Figure 17.5. The chronicler Holinshed wrote that on the 4th and 5th of November 'was a great wind that blew down many houses and trees, after which wind followed so high a tide that it drowned the marshes on Essex side and Kent, with the Isle of Thanet and other places, destroying much cattle' (Holinshed, 1808: 750–751). An 'Acte Concerning Plumsted Marsh' was passed by Parliament two months later, in January 1531, which in its preamble refers to 'the outrageousness' of the river Thames having drowned the marshes of Plumsted and Lesnes through breaches in the marsh walls of Plumstead, Lesnes and Erith. Repair work was being undertaken on the Plumstead breach, and a new cross-wall constructed, running from the Thames to the uplands, to prevent the penetration of the flood-waters which were entering the level at the more serious Erith breach (Statutes: 320). Institutional disruption, further changes of ownership after Cardinal Wolsey's fall - the Lesnes Abbey property passed through the hands of a series of lay owners during the sixteenth century - funding difficulties and further storm floods meant that despite Acts of Parliament and other measures these breaches remained open throughout the sixteenth century. The Great Breach developed into the lagoon-like feature with tributary creek network shown in the map. In the 1560s it was reported that attempted repairs were being harmed by fishing boats passing to and fro through the breach (Privy Council: 23). These fishermen were said, perhaps significantly, to be from Barking, as well as from Picardy hamlet in Erith parish. Barking was emerging as a major fishing port, and the early phases of its rise seem to owe much to the expertise of its fishermen in exploiting flooded marshlands – a process which had begun with the flooding of Barking marsh itself in the 1370s.

V. Conclusion

During the later Middle Ages shifting economic parameters – collapsing urban and regional populations, with falling prices and rapidly rising wages - allied to recurring damage from storm flooding, effected significant change in the wetlands bordering the tidal river and outer estuary of the Thames. Although many reclaimed marshes continued to be used as agricultural land, defended against the tides, others reverted to inter-tidal conditions and their fisheries and other natural resources came to assume greater importance. Changes in the structure of demand from the London market – including a decline in the bulk grain trade and rising demand for fresh fish as living standards rose – impacted upon the productive but vulnerable Thames marshes more intensely than elsewhere. The 1370s appears to represent a turning point in the history of the Thames marshes, after which the rapid repair of storm damaged sea- and river-defences was less likely to be the automatic response of marshland lords and communities. These conditions substantially persisted through the fifteenth century. A comparable partial retreat from coastal wetlands, with a similar but not identical chronology, may be seen in other parts of England such as coastal Sussex (Brandon, 1971). In the Low Countries, for instance around the Rhine-Meuse Delta, much more dramatic losses of land occurred in the later fourteenth and fifteenth centuries (Van de Ven, 2004: 105–7; Nienhuis, 2008: 232–48). The sixteenth century dissolution of the monastic orders in England added institutional disruption to the mix, and acted to prolong instances of flooding at some Thames-side locations. In all cases we see that the human element - the strength or weakness of institutions, and the economic incentives or disincentives to reclaim and defend coastal and estuarine wetlands - was decisive in determining the outcome of extreme natural events, and in shaping the landscape and environment of the productive but vulnerable coastal zone.

Archive Abbreviations

CCA = Canterbury Cathedral Archives

TNA: PRO = UK National Archives, Public Records Office, Kew.

Bibliography

Edited sources

CIM (1916–2003) *Calendar of Inquisitions Miscellaneous*, 8 vols., London.

CPR (1891–1911) *Calendar of Patent Rolls*, 44 vols., London.

Dasent, J.R. (ed.) (1895), Privy Council *Acts of the Privy Council of England,* New Series vol. X, *1577–8*, London.

Kirk, R.E.G (1910) *Feet of Fines for Essex*, Volume 1, A.D. 1182–A.D. 1272, Colchester.

L & P Hen VIII (1864–1920) *Letters and Papers, Foreign and Domestic, of the Reign of Henry VIII*, 21 vols., London.

Statutes (1817) *The Statutes of the Realm* vol. 3, London.

Survey (1994) *The Survey of London: volumes 43 and 44: Poplar, Blackwall and Isle of Dogs,* London.

Thomas, A.H. (ed.) (1932) CPMR *Calendar of Select Pleas and Memoranda of the City of London, 1381–1412*, Cambridge.

VCH Essex 5: Powell W.R. (ed.) (1966), *A History of the County of Essex* (Victoria County History), vol. 5, London.

VCH Essex 7: Powell W.R. (ed.) (1978) *A History of the County of Essex* (Victoria County History), vol. 7, London.

VCH Middx 11: Baker, T.F.T (ed.) (1998) *A History of the County of Middlesex* (Victoria County History), vol. 11, London.

Secondary sources

Brandon, P. (1971) 'Agriculture and the effects of floods and weather at Barnhorne, Sussex, during the Later Middle Ages', *Sussex Archaeological Collections*, 109, pp. 69–93.

Britnell, R.H. (1981) 'Essex markets before 1350', *Essex Archaeology and History*, 13, pp. 15–21.

Campbell, B.M.S., Galloway, J.A. and Murphy, M. (1992) 'Rural land-use in the metropolitan hinterland 1270-1339: the evidence of *inquisitions post mortem*', *Agricultural History Review*, 40, pp. 1–22.

Campbell, B.M.S., Galloway, J.A., Keene, D. and Murphy, M. (1993) *A Medieval capital and its grain supply: agrarian production and distribution in the London region* c. *1300*, London, (Historical Geography Research Group, Research Series; 30).

Croot, P. (1997) 'Settlement, tenure and land use in medieval Stepney: evidence of a field survey *c.* 1400', *London Journal*, 22, pp. 1–15.

Darby, H.C. (1977) *Domesday England*, Cambridge.

Divers, D. Jarrett, C., Philpotts C. *et al.* (2004) 'Excavations at Deptford on the site of the East India Company dockyards and the Trinity House almshouses, London', *Post-Medieval Arachaeology*, 38, pp. 17–132.

Dugdale, W. (1772) *The History of Imbanking and Draining*, London.

Galloway, J.A. (2000) 'One market or many? London and the grain trade of England', in: J.A. Galloway (ed.), *Trade, Urban Hinterlands and Market Integration* c. *1300–1600*, London, pp. 23–42 (Centre for Metropolitan History Working Papers Series; 3).

Galloway, J.A. (2001) 'Town and country in England, 1300–1570' in: S.R. Epstein (ed.), *Town and Country in Europe 1300–1800*, Cambridge, pp. 106–31.

Galloway, J.A. (2009) 'Storm flooding, coastal defence and land use around the Thames estuary and tidal river *c.* 1250–1450', *Journal of Medieval History*, 35, pp. 171–188.

Galloway, J.A. (2010) 'Piteous and grievous sights': the Thames marshes at the close of the Middle Ages', in: J.A. Galloway (ed.), *Tides and Floods: New research on London and the tidal Thames from the middle ages to the twentieth century*, London (Centre for Metropolitan History, Working Papers Series; 4).

Galloway, J.A., Keene, D. and Murphy, M. (1996) 'Fuelling the city: production and distribution of firewood and fuel in London's region 1290–1400', *Economic History Review*, 49, pp. 447–72.

Galloway, J.A. and Potts, J. (2007) 'Marine flooding in the Thames estuary and tidal river c.1250–1450: impact and response', *Area*, 39, pp. 370–9.

Goodburn, D. and Davis, S. (2010) 'Two new Thames tide mill finds of the 690s and 1190s and a brief up-date on archaeological evidence for changing medieval tidal levels', in: J.A. Galloway (ed.), *Tides and Floods: New research on London and the tidal Thames from the Middle Ages to the twentieth century*, London (Centre for Metropolitan History, Working Papers Series; 4).

Gottschalk, M.K.E. (1971–8) *Stormvloeden en Rivieroverstromingen in Nederland,* Assen.

Holinshed, R. (1808) *Holinshed's Chronicles of England, Scotland, and Ireland*, vol. 3, London.

Hull, F. (1986) 'Cliffe, Kent', in: R.A. Skelton and P.D.A. Harvey (eds), *Local Maps and Plans from medieval England*, Oxford, pp. 99–105.

Keen, L. (1989) 'Coastal salt production in Norman England', *Anglo-Norman Studies*, 11, pp. 133–79.

Letters, S. (2003) *Gazetteer of Markets and Fairs in England and Wales to 1516*, London, (List and Index Society Special Series; 32), available online at: http://www.history.ac.uk/cmh/gaz/gazweb2.html

McDonnell, K. (1978) *Medieval London Suburbs*, London-Chichester.

Merriman, R.B. (1902) *Life and Letters of Thomas Cromwell*, vol I: *Life, Letters to 1535,* Oxford.

Milne, G. (2003) *The Port of Medieval London*, Stroud.

Nienhuis, P.H. (2008) *Environmental History of the Rhine-Meuse Delta,* New York.

Oman, C. (1969) *The Great Revolt of 1381*, Oxford.

Owen, A.E.B. (1981) *The records of a commission of sewers for Wiggenhall 1319–24,* Norwich, (Norfolk Record Society; 48).

Richardson, H.G. (1919) 'The early history of the commission of sewers', *English Historical Review,* 34, pp. 385–93.

Rippon, S. (2000) *The Transformation of Coastal Wetlands*, Oxford.

Smith, R.A.L. (1943) *Canterbury Cathedral Priory: A Study in Monastic Administration,* Cambridge.

Soens, T. (2006) 'Explaining deficiencies of water management in the late Medieval Flemish coastal plain (13th-16th centuries)', *Jaarboek voor ecologische geschiedenis (2005/2006)*, pp. 35–62.

Ven, G.P. van de (2004) *Man-made Lowlands: History of water management and land reclamation in the Netherlands*, Utrecht.

18 Clio defeating Neptune: a pyrrhic victory? Men and their influence on the evolution of coastal landscapes in the North Sea area*

Erik Thoen, Ghent University

Abstract

In this article I tried to illuminate some of the most important aspects about the way human communities have influenced the natural coastal eco-system, often with disastrous long term results. It focuses especially on the human influence during the Middle Ages and the Old Regime. In its conclusion, it highlights arguments for a huge regional and temporal diversity of evolutions, since human influence was not everywhere identical. Moreover, because human activity often triggered natural phenomena and vice versa, and because we know from recent research (see e.g. the many other articles in this volume) that natural phenomena causing changes in marine environments were also much more regionally different compared to what was believed some decades ago (Verhulst and Gottschalk, 1980), it is logical that the regional component is crucial in the explanation of the evolution of coastal landscapes.

I. Introduction

Specialists increasingly agree that during the last few millennia the influence of mankind on the formation and change of coastal plains has been of huge importance. However, in most studies written by earth scientists, the human influence is rather superficially dealt with. Equally, in studies written by social historians and historical geographers is often only in a superficial way referred to the way humans influenced natural processes of land formation and/or degradation in coastal areas, often because on the one hand they do not know nor understand the natural processes, and on the other hand because many of them are not aware of the importance of changing social organisation on landscapes as a whole (Thoen, 2011).

II. Natural phenomena and human action

Many natural phenomena influenced the formation and changes of natural landscapes in coastal zones. Apart from the sediment supply and structure, the main natural actors for coastal changes were the relative sea and tidal amplitudes, as well as the (changing)

* I am very grateful to the colleagues Guus Borger, Tim Soens, Adriaan de Kraker and Henk Weerts for their interesting comments on the initial version of this text.

Landscapes or seascapes? The history of the coastal environment in the North Sea area reconsidered, ed. by Erik Thoen, Guus J. Borger, Adriaan M.J. de Kraker, Tim Soens, Dries Tys, Lies Vervaet & Henk J.T. Weerts, Turnhout, 2013 (*CORN Publication Series*, 13), pp. 397–428.

BREPOLS PUBLISHERS DOI: 10.1484/M.CORN.1.101562

height (and composition) of the adjacent land mass. This determined the formation of shore lines, the composition and relative height of the upper soils with the geometry, shape and depth of tidal channels and natural rivers for drainage of the hinterland, sedimentation processes, peat formation, its disappearance as well as its stratigraphy, and the structure and the resilience of the under lying stratigraphy (e.g. compaction of soft sediments and oxidation).

All these phenomena are to a large extent either determined by or themselves determine the changes/evolution of relative mean high water level (MHW).The extreme high water level having often has cumulative effects on soil formation or destruction. In the very long term and on a very large scale, these MHW changes had natural causes like the growing or melting of ice caps, regional glacio- and isostatic rebound, long-term climatic changes, and composition and volume of marine sediments etc. (see Weerts, this volume). Here human action was hardly present. However, if one looks at previous centuries and millennia only, regional differences in tide storage and MHW were huge and changed a lot over time and area. Consequently, the same also happened with extreme high water conditions during storms. In addition, the drainage conditions of the hinterland changed drastically over time. The formation of tidal channels, tidal flats, sedimentation, and soil structure was to a large extent influenced by these more regional factors which made a lot of difference to humans for using the land as resource. These (relatively) more short term but especially more regional and 'local' natural factors were much more influenced by human action. The mentioned phenomena were regional but could also influence neighboring regions! Human action influenced the natural ecosystems and changed/adapted natural processes but were still partly influenced themselves by these changing natural actions. This increases the complexity of a study of the evolution of coastal areas during the historical period.

For drainage reasons, rivers and channels discharged their water from the hinterland into the North Sea or in large deltas such as the Scheldt-Meuse or the Rhine. These tidal inlets had many moving side channels as well and created tidal flats mostly within a peat landscape that had formed since *c.* 7000 BP. The landscape of these flats was changed in a natural way: new deeper tidal channels could be created and sedimentation could take place. However, since human communities were active in these coastal areas, to a large extent they steered these natural phenomena. This was true at different geographic levels: in the mean tidal inlets but also more inland in the side channels and in the peat areas.

One could make a global distinction between human influence on natural phenomena: on the one hand human action causing *consciously wanted* changes and on the other hand *unconsciously created* changes in natural processes. However, such a division is largely artificial since most *consciously wanted* changes also caused natural but often unexpected long-term side effects.

Direct human influence in the coastal landscapes especially happened in two ways: via overexploitation and via direct changes in the hydraulic system.
First, changes could occur due to human overexploitation of the landscape. This was especially true in coastal landscapes where the upper soil formed under marine influence and became intensively cultivated. Indeed, in areas in the (wide) vicinity of the sea or tidal channels, drainage often resulted in compaction of soft sediments (particularly peat). Also the fact that soils were literally removed at the upper layer (peat digging), but also

by erosion and oxidation, caused subsidence of surface levels. Overexploitation and/or reclamation of much more inland situated areas, where the origin of the upper soils was often much older and not directly influenced by tidal erosive or sedimentation processes, could also cause changes in the coastal areas (see section 3).
Secondly, humans also influenced the coastal landscape directly through hydraulic systems: the reclamation of land, building of dikes, and canals or changing natural channels, the building of sluices, and the creation of an artificial possible 'man-made' accommodation space. This direct influence had also accumulative (often negative) effects on natural hydraulic phenomena, as will be discussed in section 4.

The social context determined why and at which pace people directly influenced the natural eco-system. It also determined the way they adapted and coped with the negative aspects of the changed ecosystem and how they could turn the balance in their favor. This will be dealt with in section 5.

III. Human action causing erosion and land subsidence

III.1. Drainage and farming causing land subsidence

One of the major natural phenomena influenced by mankind is **subsidence of surface levels.** This was particularly due to the compaction of soft sediments, especially peat but also underlying minerogenic sediments (Long, Waller, Stupples, 2006: 4). Compaction can have natural causes due to a changes in natural drainage via natural channels and the deposition of heavy sediments on top. Nowadays, human action during the Roman and even more during medieval periods is also thought to be a causal factor. Although written evidence is scarce, in the last few decades, archeology has demonstrated that at an early point in time peat and intertidal saltmarshes (not having any dikes) could have been intensively used for cattle and sheep breeding, and also for grain cultivation, from an early period. For example, traces and remnants of Roman drainage activities were excavated in the area near Vlaardingen in S. Holland as well as in Zealand Flanders (see e. g. Lascaris and de Kraker, this volume; see also Meier, this volume). In addition, large scale drainage took place during the Carolingian period (e.g. Vos *et al.* 2011 for the Netherlands, for coastal Flanders: Tys, 2003 and this volume; for Schleswig-Holstein: Meier, this volume). After embankments were built, drainage became even more intensive (see below). The systematic drainage which was necessary for cultivation caused land subsidence. It is well known that intensive drainage already occurred since the first occupation of these peaty areas (see many studies, inter alia Meier, this volume). The consequent subsidence must have been very important. In turn, it is obvious that land subsidence had far reaching societal consequences. It increased the amount of accommodation space and, after embankments, the risk of flooding! So, land subsidence was to a high degree caused by man and had accumulative negative societal aspects. During the last years and decades, both archeologists and geologists have increasingly underlined the importance of the process (Bungenstock and Weerts, 2009). According to Long, Waller and Stupples, 'this process is likely to have been a key driving mechanism behind rapid late Holocene coastal change, far exceeding the longer-term effects of eustatic change and uplift/subsidence'. In the Romney Marshes, compaction lowered the peat surface by at least 3m compared to the period before *c.* 700 AD when, in relation with this land subsidence, a saltmarsh was established in the area (Long, Waller and Stupples, 2006: 2). Earth scientists such as Bungenstock and Weerts (2009) as well as Van Asselen *et al.* (2011) recently stressed that peat compaction may have been underestimated by geologists, stating that subsidence due to peat compaction is highly variable in time and space, with local rates of not less than up to 15 mm per annum, depending on sedimentary sequence. Elsewhere, it was shown how important the phenomenon was for channel belt elevation, geometry, and configuration (Van Asselen, Stouthamer, Van Asch, 2009).

Moreover, due to drainage, gaseous CO^2 flux associated with **microbial oxidation** must have been very important too for the lowering of the surface level. It is only since the mid-1970s that this phenomenon became apparent to investigators researching the North Sea area (see e.g. Borger, 2007: 30 referring to his own work of 1975 about Western Friesland). Its impact must have been huge if we contextualise it with well-documented recent data. In the Sacramento-San Joaquin Delta of Southern California it seems to have been responsible for 75% of the current subsidence of surface level. There, subsidence made that the area, after the beginning of the farming practices in 1880, sink on average about 8m below a.s.l (Mount and Twiss, 2005: 4–5)!

Consequently, earth scientists' well described phenomenon of **land inversion** (the apparent uplift of former tidal channels where an underlying peat layer was absent), which ensured that less or no compaction took place, was also, in an indirect way, to a much larger extent influenced by man than previously thought since it was directly influenced by land subsidence. It is well recognised how important these inversed zones were for historical geography as they were very suitable for settlement.

The variation in land subsidence due to compaction and oxidation also caused and partly steered the natural phenomenon of **avulsion** (the constant reposition of tidal channels) (Kiden *et al.,* 2008: 59; Van Asselen *et al.,* 2009). This way, aversion was also partly caused by human action. It was demonstrated that aversion could lead to local changes in average mean high water levels (id: 60) and in that way also influence the sea and channels via sedimentation and storm surges.

Land subsidence was also influenced by drainage. Therefore, it is believed that subsidence was enhanced when drainage via windmills was introduced from the late fourteenth century onwards and in an accelerated pace from the sixteenth century when the practice became common (see e.g. van Tielhof and van Dam, 2006: 70). During the Early Modern Period, the intensive drainage of land via networks of mills in the polders sometimes caused problems in adjacent areas. This happened in parts of late eighteenth-century Groningen. Too intensive water discharge made it necessary to make polders in neighbouring lands too and created structural problems in the Wolden area which could only be resolved in the 1870s (Ligtendag, 1995: 249–50). Similar problems happened also in Rhineland and Noord-Holland in the Early Modern Period (an overview for the Netherlands in de Vleesschauwer, 2012:11–16).

Land subsidence increased the risk of flooding considerably. Not only were the consequences of flooding much higher due to the increased accommodation space in case of dike bursts, but the cumulative hydrostatic force acting on dikes is a function of the square of the depth of subsidence as has been showed by Mount and Twiss for the contemporary Sacramento-San Joaquin Delta (2005) (see also figure 18.3). In other words, due to land subsidence, the construction of dikes had to be more solid to resist increasing water pressure. This required new investment, which in many areas was only realised in the second half of the Early Modern Period after a period of problems had arisen (see below).

III.2. Peat cutting and salt making

Since the mid Holocene period in most coastal plains, a considerable peat layer had formed on the surface. Since *c.* 2500 BP, when an 'unstable' period of marine influence began, even larger parts were covered with marine sediments or were taken by the sea. In other areas peat could continue to grow, such as in the quiet areas of the inland Scheldt area; this even continued until the high Middle Ages (Kiden, 2001). However, much of the peat was not covered with marine sediments until the end of the medieval period and into the early modern period, not at least the areas which were situated further away from the sea and tidal channels. Although there are some indications that peat cutting for fuel was already practiced in Roman times, it was during the eleventh and twelfth centuries that it became very important in the North Sea Area. Across several centuries, vast amounts and large surface peat areas disappeared because of large scale peat cutting and the extent of areas where peat mining took place was enormous. Many studies have underlined the importance of peat digging for the economy of the Low Countries (Cornelisse, 2008; van Tielhof and van Dam, 2006; Soens and Thoen, 2009). Scholars were, and sometimes still are, divided in two groups: the 'minimalists' and the 'maximalists'; the first minimalise the former presence of peat, the others do the opposite. The former group also minimalise the presence of raised peat bogs which were very useful for peat cutting. However, thanks to the development of new research, a lot of written data becoming available, and the intensification of studies on peat cutting, in most areas the maximalists won the dispute in the end. The presence and discovery of *daliegaten* could prove the former presence of peat in some areas of the Northern Netherlands now covered with clay (e.g. Noord-Holland, Zuid-Holland, Utrecht and West-Friesland). As mentioned, at the beginning it was possible to cultivate crops on the well-drained peat soils. To improve the soil quality for arable farming, pits were dug through the peat to extract clay and sabulous clay from underneath. The clay was probably mixed with stable manure and used as fertiliser to improve the poor soil quality. However, although much of the peat has now disappeared due to erosion, oxidation or extraction, traces of these pits called *daliegaten* still can be found. Nowadays it is widely accepted that large amounts of peat were also once present in West-Friesland (Borger, 1975; Dekker 1981), the Campine area (Leenders, 1989) and deep in the Land van Waas as well as in the Meetjesland (for a state of the art of the discussions between geologists and historians in Flanders: Soens and Thoen, 2009). In all these areas the peat that was located at the surface is now completely gone.

The environmental consequences of peat digging are well known. Although it was mostly prohibited in the immediate vicinity of dikes, the resulting surface subsidence increased accommodation space in periods of flooding or when the dikes were broken. Vast areas, even as far as deep inland regions, increasingly became threatened by the sea and the possibility of being covered by marine sediments. Moreover, it was easy for sea water to penetrate an area where peat mining had previously taken place since such mining also created a network of ditches and canals for drainage and transporting peat. Exhausted peatlands also suffered from under-investment in the maintenance of drainage systems and flood protection, as the land value often was low (see below). Peat districts also became even more prone to flooding, when the practice of peat cutting below water lever via dredging became common from the later Middle Ages onwards. Many artificial lakes were created that way such as in Holland, Utrecht and the NW part of Overijssel and in Friesland and the Southern part of the Belgian coastal area (Van Tielhof, Van Dam, 2006: 120–151; Dekker, 1981: 63). From the late sixteenth century many of these lakes were again reclaimed and embanked using wind mills (these reclaimed lakes were called in Dutch *droogmakerijen*).

Since the eleventh century peat cutting for fuel was also carried out on a large scale and peat was sold in the many flourishing cities of the North Sea area. Salt making was also important. In the Middle Ages and the early modern times, salt was made in the North Sea area from salt peat. This was done in two stages: the burning of salt peat that was standing under the influence of marine salt outside the dikes (and sometimes covered with sediments that had to be first removed), and the process of refining for which also peat was used (Leenders, 2003). This practice was especially common in Zealand, South Holland, and West North-Brabant from the eleventh and twelfth centuries until the fifteenth century, when the import of foreign salt increased (id. and Mertens, 1963; de Kraker, 2007). There was some awareness of the environmental danger of peat digging outside the dikes, but prohibitive orders mostly were not effective (Leenders, 2003: 34; de Kraker *et al,* 2008).[1]

Peat mining happened at a different pace according the area. It can be demonstrated that in the northwestern part of Flanders (near the French border), peat cutting was undertaken in the twelfth and the first half of the thirteenth centuries, while along the Southern bank of the Western Scheldt it mainly took place between the thirteenth and fifteenth centuries (Soens and Thoen, 2009a). From the late sixteenth century large scale peat cutting stopped in the Southern Netherlands and the Western Scheldt area. In North Brabant and Holland and West Friesland (in both last mentioned regions, peat was especially used for farming) it started in the fourteenth century and reached its zenith in the seventeenth century. In Groningen and Drenthe, it was mainly a post-1600 activity that continued until recently (Borger, 2007; Cornelisse, 2008; Van Tielhof and Van Dam, 2006, Leenders, 1989).

These regional and local differences in peat cutting of course influenced the locally different environmental consequences of this activity. In combination with the described land subsidence it must have created locally different environmental fragility and possibilities for water to reshape the landscape into a marshy area. Together with the phenomenon of inversion due to channel silting and peat compaction, the intensity of peat digging created local differences in altitude in apparently flat areas which had far reaching environmental consequences for the stability and limited sustainability of coastal areas. Nevertheless, it is likely that the flooding of peat districts was never inescapable: even though land subsided, and water levels increased, peasant small-holders in the central Holland peat districts managed to survive well into the late seventeenth century, thanks to a well-balanced combination of dairy farming and peat dredging, and a careful land

[1] According to some, the salinity of peat – especially in not yet diked areas – probably also determined the date of reclamation (for arable). Low salinity could have made reclamation easier and more rapid (see Oram, this volume).

and water management. Drainage only started to fail significantly in the latter half of the seventeenth century, at a moment when the economy of the Dutch Republic witnessed a severe recession (Van Dam and Van Tielhof, 2006; Soens 2008).

III.3. Erosion and increased land surface fragility?

Erosion within the coastal areas

Drainage of peat areas was mostly done via a very dense network of canals and often rather deep parallel running ditches. This 'network' made the drained area very vulnerable to flooding. It is well known that due to storm surges large quantities of **peat were destroyed** and even large parts of the peat area were sometimes swept away to start drifting or were engulfed in the tidal canals or in the sea. As van Tielhof and van Dam could demonstrate, fragility of the peat surface caused by human intensive drainage was partly responsible for caving in of the river banks in the Rhineland area in the Netherlands during the later Middle Ages (van Tielhof and van Dam, 2006: 73). Where peat was located in areas with clay in the subsoil, this fragility also often increased during storm surges because deep canals had been made to mix this clay with the less fertile upper layer to increase land fertility (Dekker, 1981; de Kraker, 1997).

Human-caused **wind erosion** was also co-responsible for changing landscape conditions in the coastal areas. This was particularly true in the **dune barriers.** Initially these barriers were in most parts of the coastal North Sea area not as high as they are sometimes today but wider, more craggy, and at the same time more robust and covered with plants too. They were mostly displaced (land inward) between the post-Roman period and the Early Modern Period and sometimes even until the twentieth century (Van Haperen, 2009: 10–16). Despite natural elements, it is likely that this dune destruction was to a large extend caused by overexploitation. This was the consequence of too intensive grazing of cattle and sheep and arable farming (see e.g. van Dam, 1999, and the literature in the notes). Destruction by rabbits – introduced by humans in NW Europe for the fur trade – was also problematic (Augustijn, 1995; Rentenaar, 1978). Moreover, wood cutting and deforestation, the cutting of sods, and even the founding of settlements in the dunes was destructive for this eco-system (Labarque, 2007; de Ceuninck, 1992; Klijn, 1987: 57 ff. gives many data out of spread literature; see also Oram, this volume). The intensive use of the dune area was encouraged by the fact that dunes mostly had potable water in the subsoil (contrary to elsewhere in the coastal areas where salt water in the subsoil sometimes made arable farming and cattle breeding impossible) (Labarque, 2007: 83–84; Priester, 1998: 37–41). Destruction often made the dunes, more barren and blown up, covering clay areas. The measures taken to limit sand erosion made them higher and often narrower (Labarque, 2007: 25).

Human induced erosion of the hinterland influencing land and seascapes

Because (often far away situated) hinterlands were linked with the sea via tidal channels and/or tidal estuaries, erosion under the influence of reclamation activities and improper farming within the hinterlands often had a serious influence on sediment transportation. This could have caused the silting up of tidal channels, the creation of sand banks, and changed the sediment supply in a general way. Moreover, it changed the tidal prism in the channels with different physical consequences on the hydraulic system (see also below: the effects of direct human changes in the hydraulic system).

Figure 18.1 Manuscript showing rabbits destroying the dunes near the abbey of Ten Duinen in Koksijde at the Belgian coast, fifteenth century. Rabbits were introduced by men in NW Europe in the 12th century for fur production. Decades later it became wild animals that were damaging the natural eco-system of the dunes (source: Adriaan De Budt, Rapiarium, Royal Library Brussels, manuscripts 7978–79)

J. P. M. Syvitski *et al.* (2005), stressed that today this is still a huge environmental problem. Indeed, it has been estimated that in the twentieth century, world-wide, humans have simultaneously increased the sediment transport by global rivers through soil erosion by 2.3 ± 0.6 billion metric tons per year.[2]

In the first half of the Holocene soil erosion already seems to have been important (partly due to deforestation by hunter gatherers?) (Joosten, 2004). A recent study showed that erosion in the Rhine-Meuse delta caused by humans increased after 3000 BP. This caused a huge sediment supply and discharge from the hinterland (Gouw and Erkens,

[2] According to the same article referred to the flux of sediment reaching the world's coasts would have been reduced by 1.4 ± 0.3 billion metric tons per year because of retention within reservoirs. An opposite effect is caused by modern, large scale dredging techniques to deepen rivers. This way, sediments do not reach the sea anymore (personal communication H. Weerts).

2007) and meant that in the mouth areas of the Meuse and the Rhine, inland sedimentation could slightly counter the natural rise of the sea level (Vos *et al.,* 2011: 44).

Human influence was probably even larger in more hilly or mountainous areas. In Eastern Scotland, for example, R. Oram (this volume) suggests that deforestation and the loss of woodland cover, combined with an increased canalisation and intensification of agricultural activity between the twelfth and fourteenth centuries increased the levels of discharge into the main river channels, especially the River Tay, causing the silting up of navigable channels with associated problems for shipping.

A secondary consequence of the silting up of river mouths caused by humans was that erosion also created problems in relation to the discharge of inland water across large areas. This certainly was the case in Holland when the Rhine mouth silted up before *c.* 1200. The water in the peat areas behind the dunes could hardly be discharged. Therefore, already in the twelfth century, new canals needed to be constructed (see e.g. van Dam, 1999). Water discharge problems also enhanced flood risk due to heavy rainfall. This was the case in parts of Scotland during the fourteenth and fifteenth centuries (Oram, this volume).

IV. Changes of the hydraulic system directly triggered by men

In section 5.3, we deal with the chronology of embankments and their role in the – not necessary rectilinear – intensification process of land use. However, dike building and embankments themselves, as well as other infrastructural works, could have increased the vulnerability of a landscape and influenced natural phenomena. This is a complex matter and differs locally. However, here is a general outline along with some options.

IV.1 Outside the dikes

Outside the dikes, particularly in delta areas, embankments have an important influence on the tidal rivers and channels. The regionally changed tidal influence could have weakened the embanked hinterland on a local scale as we will describe – not only at ordinary tide, but also at spring tides, in particular coinciding with gales. It was the different geometry of the tidal river and side canals that was at the core of these regionally different influences (de Kramer, 2002: 16 ff.). That is why, for example, in the North Sea itself, the amplitude is much bigger at Vlissingen (Eng.: Flushing) (3.8 m) in the south than at Den Helder (*c.* 1.4 m) more to the north (id.:18). If this natural geometry was either entirely or only partly artificially changed because a natural process had been triggered, the tidal amplitude could also have changed. That amplitude was influenced by the geometry of the bottom. When depth was decreasing, a process of *shoaling* increased the tide amplitude. The (due to embankments) increased tidal range as well as the increased flow rates could have deepened the tidal channel due to erosion. This might have caused higher high water levels with an increasing risk of flooding as a consequence and a need for still higher dikes. This was the case in the lower Scheldt basin near Rupelmonde from the Middle Ages onwards (Kiden and Verbruggen, 2001: 27). Even more than depth, it is the frictional resistance of the borders that influences tidal amplitude. A narrowing shape of tidal channel can result in an increasing process of *reflection* (rebound) and higher tidal amplitudes (de Kramer, 2002: 20). This ensures that the tidal amplitude in the more

Figure 18.2 Water tax burden (in litre of wheat per hectare), important storm surges (number per period) and land losses (maps) in the Western Scheldt-Sealand (*c.* 1150 – *c.* 1850)

Important flood damages	Burden of water taxes (litre / ha)
1150-1250	
Little amount of important storm damage	small burden ('Verhoefslaging')
1250-1600	
c. 12 important storm damages and 2 periods of military inundations	Blankenbergse Watering c. 10-20l. Yvene Watering c. 50/80 l.
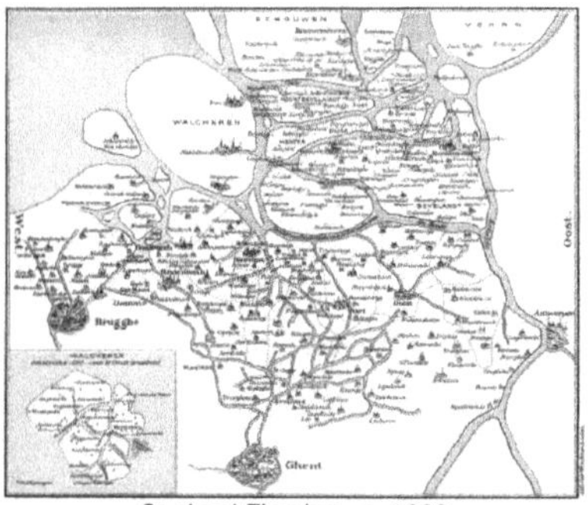 *Sealand Flanders c. 1300*	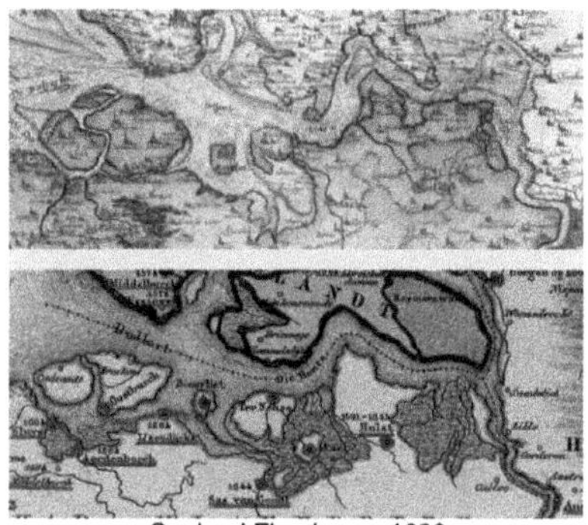 *Sealand Flanders c. 1600*
1600-1850	
c. 3 important storm damages	see table below
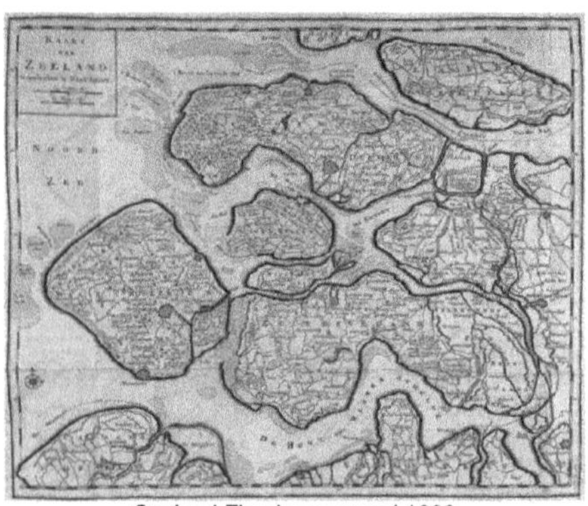 *Sealand Flanders a round 1800*	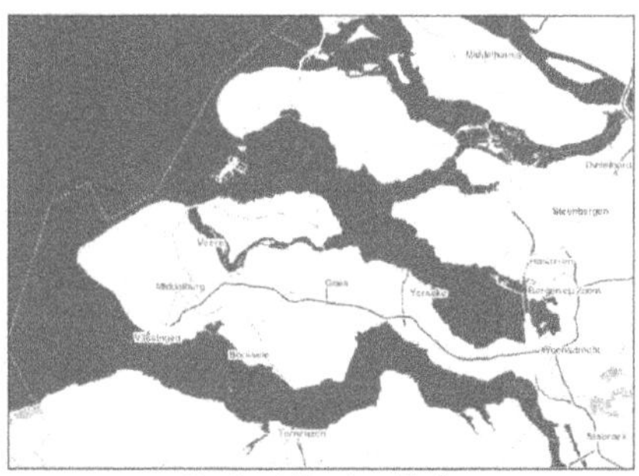 *Sealand Flanders Today*

	Blankenbergse W.	Polder Cadzand
1600-1650	c.32 l.	?
1650-1700	c.35 l.	c. 14 l.
1700-1750	c.40 l.	c.116 l.
1750-1800	?	c. 100 l.
1800-1850	?	c. 75 l.

inland part of the Western Scheldt is still much bigger than at its mouth (Vlissingen *c.* 4.6 m, Antwerp *c.* 5.9 m; de Kramer, 2002: 21). When the river banks are more parallel (e.g. due to embankments), reflection as well as the tidal amplitude can be lowered (de Kramer, ibidem). If this parallelism was due to embankments, flooding of one polder can change the whole geometry with new increases of tidal amplitudes as a result. More inward land frictional resistance can become so important that tidal amplitude progressively lowers again (the *tidal bowl effect*) (id.: 21), despite local differences still can occur due to local narrowing.

Changes in the geometry of the tidal channels can also cause changes in water speed during tidal activity, particularly in the volume of water displaced. Moreover, these elements determine tidal influence on the river banks and the adjacent polders.

The amount of water between the imaginary lines of high and low tides is called *tide storage area* (Dutch: 'komberging'). If one is only focusing on the acreage of a tidal area between the tides, it is called the *accommodation space.* When large scale embankments took place and side channels either disappeared or were closed with dams, this area was reduced. This could then trigger a deepening of the river bottom, an increased water speed, a larger tidal amplitude, and thus could result in more pressure on the river banks (de Kramer, 2002: 37). However, according to some this pressure would decrease again after a certain period (id: 42) due to the effect of deposition of marine sediments near the dikes in the second stage. This phenomenon occurring in intertidal environments is well known to earth scientists. The reduction of the tidal prism by embankments enhances sediment infilling and can even cause the closure of tidal inlets (Waller *et al.*, this volume). This phenomenon is reported a lot in medieval charters. However, after a certain period, these alluvial deposits were often embanked as well, this way again neutralising the risk reducing effect of the deposition of marine sediments.

Despite embankment or damming activities (of former tidal canals), human influence at the seaside of the dikes was also noticeable due to dredging activity in the tidal channels. Although this is a recent large scale phenomenon that became important from the late nineteenth century onwards, it was not uncommon in the Middle Ages either. So, we know that the former direct connection of the port of Bruges with the sea, the Zwin, was already dredged systematically during the fifteenth century (Augustyn, 1992). The influence could have been considerable because the activity might have changed the course of deeper tidal channels within the estuary since there was a natural under water difference in ebb tide and flood tide channels and flood tide channels were normally more dangerous for the neighboring river banks.

For the River Elbe, it has been shown that modifications of the estuary in the second half of the twentieth century increased storm surge levels in Hamburg by 70 cm. Experts estimate that about three-quarters of this increase was related to coastal defense measures and one quarter to the deepening of the shipping channel from less than 11 m to 14.5 m (von Storch and Woth, 2006).

Understanding the relation between the building of human infrastructure and the effects on natural phenomena is very complex because human influence may have triggered the natural 'morphodynamic' system in different ways: consolidating it, disturbing it, or triggering a new unexpected phenomenon in an adjacent area (for the last statement see

Figure 18.3 Drainage and land subsidence: conceptual diagram.
This diagram is illustrating the evolution of Delta islands in the the Sacramento-San Joaquin Delta due to levee (dike) construction and island subsidence in the nineteenth and twentieth centuries. The illustrated process is very similar to what has been happening in many coastal areas of the North Sea Area from the Roman period due to dry making and embankments (source: Mount and Twiss, 2005)

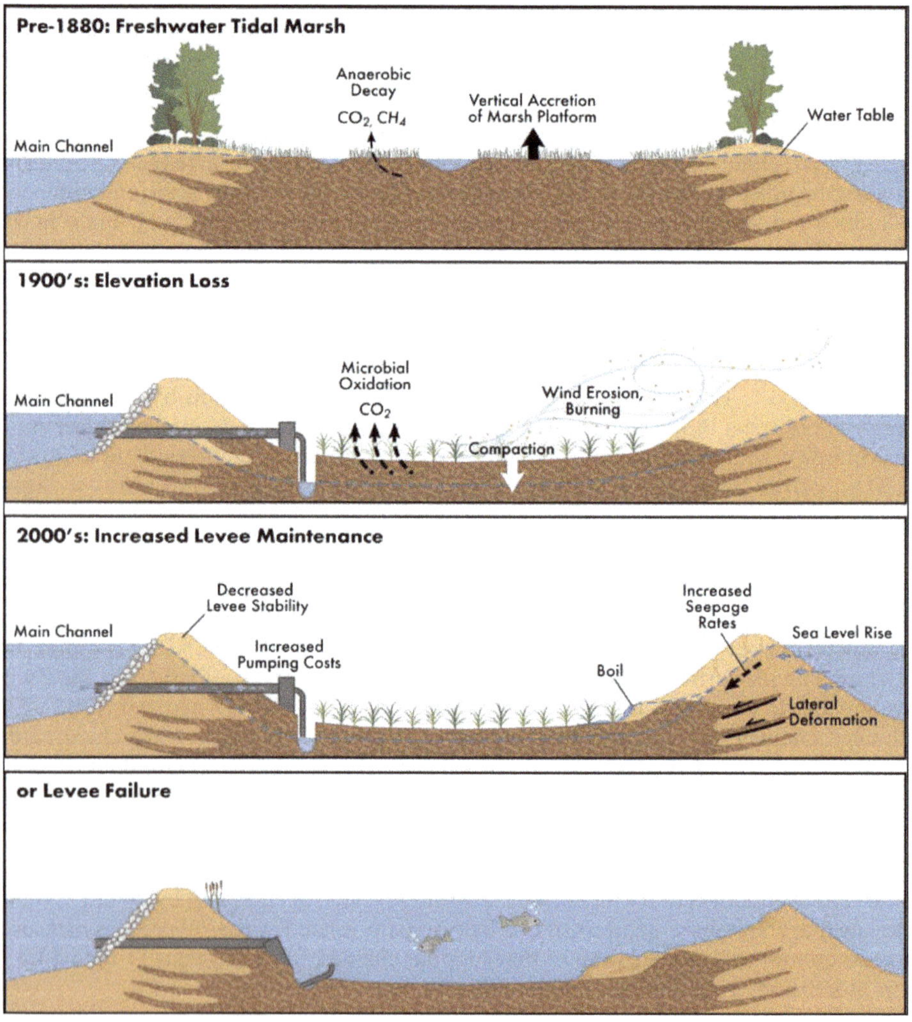

de Kramer, 2002: 47). The overview of some of the most important aspects of human influence on physical phenomena made by De Kramer and (in a simplified way) shown in figure 18.2 must therefore be interpreted with care since no 'consequence' is an absolute one. Everywhere and at any point in time the complete local context must be considered. This can be lead to an opposite result. (Idem: 48). However, history teaches that in most cases and especially since the medieval period, human influence has complicated occupation in the longer term.

IV.2. Inside the dikes

The previous section focused on the influence of embankments on the seaward part of the coastal landscape, dominated by the sea and the tidal channels. However, even more importantly, changing the inland landward part of the coastal landscape – the area inside the dikes – had a major influence on the future sustainability of the area. In some areas, the dike constructions themselves damaged the environment and made it more vulnerable: the constant digging of sods inside the dikes to repair them caused pits near the dikes to often fill up with salt mud. This weakened the area (de Vleesschauwer, 2012: 15).

However, land subsidence was a much greater risk! It has been already mentioned that (mostly) the process of peat compaction and land subsidence continued and even increased after embankments, as drainage and exploitation were intensified. This increased stress on dikes (see above) and asked for higher and more solid dikes and greater investment. Indeed, both the degree of investment and the degree to which the judicial and political organisations could ensure the maintenance of the infrastructure after embanking was very important for the sustainability of the area. The level of investment was determined by social organisation and this will be dealt with in the next section.

V. The social organisation as direct and indirect player determining the man-nature relation in coastal areas

Above, we described the most important direct effects of human occupation on natural phenomena and environmental changes. However this influence was very different, both locally and temporally. This is not only due to different natural conditions but also to the way people organised the landscape. In turn, this was determined by the evolving social organisation of society.

V.1. Why did social relations influence the geographical changes in coastal areas?

As everywhere (but to a larger degree in coastal areas because of the very high pressure of nature on living conditions), the man-nature relation is influenced by social organisation in three ways.

1. The social organisation determined the occupation process as well as the degree of intensification of land use. As we will see below, occupation processes and embankments –and also the possible natural consequences described above- were directly related to changing social relations!
2. Even in a society with apparently similar occupation phenomena, the influence of human settlement on the environment was different everywhere. This was due to different natural conditions. However, perhaps the different social organisation was even more important and the ability of this social organisation to cope (e.g. via effective social control systems) both with changing natural conditions and natural disasters. Essentially, some specific social organisations would better handle the

influence of natural changes than others. In the extreme, some types of social organisation even managed to almost eliminate natural hazards or developed strategies to turn natural conditions to their advantage.
3. A third point that needs to be made in this context is that the confrontation of certain social communities with negative natural phenomena and increased natural stress, could cause social changes that could have had in their turn (negative) cumulative effects on the influence of natural hazards.

For these reasons, it is necessary to study the social organisation of the communities living in or influencing the coastal areas to understand the evolution of these landscapes. For so many decades this was only superficially taken into account, but in the 1990s this changed when historical geographers made serious attempts in this direction, not at least for coastal areas (Thoen, 2004; van Tielhof and van Dam, 2006; and in this volume Tys and Soens), although even today historical geography as a discipline still too rarely deals with former social relations to explain landscape changes (Thoen, 2011).

However, in coastal areas different elements are complicating the study of the links between nature and social organisation:

1. Natural conditions, even on a regional scale, can have changed without direct and even indirect human influence. Climatic changes like the Medieval Warm Period and the Little Ice Age and their effect on the intensity of storm surges are good examples of this (de Kraker, 2008). For further causes we refer to the contribution of Weerts (this volume).
2. It is known that regional social causes for environmental change in the coastal area might have been caused by social or natural changes in a neighboring area, since the environmental cost of social or economic changes can be 'externalized' to neighboring areas. For example, bad maintenance within one polder can have consequences in neighboring polders as well, and land improvement through deep drainage in one area often results in increased drainage problems in the neighboring regions.
3. It is likely that humans were not always aware of the damaging nature of their interference in the natural ecosystem, at least not at the moment of interference itself (e.g. concerning land subsidence due to compaction of peat; the consequences of deforestation on erosion and silting up of rivers…; see above), since it often took some time for negative consequences to appear. Even if people did become aware of some of the negative consequences of their actions it is unlikely they understood the mechanisms underpinning them.

Despite these difficulties, a study of social organisation is necessary to understand the changes of many North Sea low coastal landscapes. In pre-industrial societies, the social organisation that determined survival mechanisms was regionally organised. Why and how did this happen? It brings us to some methodological considerations about the concept of social agro-systems, first raised in 2004 (Thoen, 2004).

V.2. Social agrosystems in coastal areas

As it has been stressed elsewhere, that during the Middle Ages and the old regime, regionally different 'social agro-systems' for decades determined the way people organised their production systems (Thoen, 2004). This was due to the fact that in a pre-industrial society distance was still more important and fragmentation in many senses was the rule (politically, geographically, in social and market relations etc...). Combined with a large pre-capitalist tendency towards path dependency, people organised their survival strategies to ensure their income through regionally organised social relations, both formal and informal.

Regional social relations are determined by many elements such as natural conditions, property and power structures, family and market structures, cultural traditions, and supra regional elements (Thoen, 2004). Since one of the elements is natural conditions there is a 'bounce'- effect between social organisation and natural environment. In particular, property and power structures are fundamental in vulnerable areas such as the coastal zones (van Bavel and Thoen, 2012). The social organisation of production processes, the use and formation of the land, the cultural landscape, and the environment where to a large extent shaped and directed by property arrangements. Property rights are understood as rights to land, such as the right of access, the right of sale and the right of inheritance, but also the right of use and the rights of profit, the rights of exclusion, the rights of management, and even the rights of prestige. These rights can be distributed between various persons and organisations (van Bavel and Thoen, 2012).

In pre-industrial societies, three types of social relations and property relations can be discerned. First, relations between local peasants who organised the production process locally. Second, relations between the local peasants 'in the fields' and their lords who did not cultivate land and often did not even live in the area under consideration. Third, there were the market relations inside and outside the region.

These described relations could be discerned on a regional basis. However, the social agro-systems in the coastal plain of the North Sea, also had a lot of similarities! Often rather identical changes in social relations happened, although not at the same pace. Indeed: in almost all areas there is a three stage evolution from a society and commons (and often with 'terps'), towards a (commercial) peasant economy with a majority of small holdings and no commons, and from there towards a more commercial society with larger and middle sized farms.

Within this context it is impossible – not at least because a lot of research still needs to be done – to describe in detail how the coastal social agro-systems determined landscape evolution. However, some important aspects and trends that are mostly the result of recent research or new hypotheses can be discussed.

V.3. Towards dike building and embankments: resulting from and resulting in new social organisation[3]

V.3.1. The social organisation of before the dikes and embankments. A peasant society with commons

During the early and high Middle Ages, in almost all documented coastal areas on both sides of the North Sea area, rather small and free peasants seem to have been the majority of the

[3] The original text of this section has been improved a lot thanks to important remarks of Tim Soens to which I am very grateful.

people who colonised the areas (Soens, 2011; Rippon, this volume). Although this was not an 'equal' society (differences in landed property and wealth existed), these peasants were relatively free and mostly possessed their own means of production. In most coastal areas, local lordship was scarce (see the recent overview of the property and power relations in the Low Countries in: van Bavel, van Cruyningen and Thoen, 2011). It has been suggested that the absence of lordship in the early Middle Ages was the result of the late integration of many coastal areas in the Merovingian-Carolingian empire but, more importantly, it can be caused by the retreat of lords due to plundering by the Vikings between the ninth and eleventh centuries (idem 2011:171) and maybe even earlier by other Germanic tribes from the late Roman period onwards (for these invasions see e.g. H. Thoen, 1978). The Vikings were especially interested in the wealthiest layers of the society (d'Haenens, 1967). It is worthwhile investigating whether the Viking invasions could also have triggered the creation of trading mechanisms within these relatively free peasant societies.

Until the 1980s /1990s, it was generally accepted that before *c.* 1000 the coastal plain was poorly occupied. Again in the eleventh century its occupation would have been considerably hindered, not at least due to the supposed increasing influence of the sea. More recent geological and historical research has shown, however, that in the coastal areas of the North Sea area, the Carolingian and post-Carolingian periods cannot be seen any more as a period of general abandonment and decline as one still thought in the last decades of the past millennium when this period still was seen as a period of 'transgression', land losses and settlement contraction (see e.g. Baeteman, this volume, Vos, 2011, Tys, 2003, van de Ven, 1993 *et al.*). Despite the fact that recent research is more focused on regional differences, the coastal area as a whole can be placed within the more pan-European optimistic vision on economic development of this period (McCormick, 2001). Lordship was in general weak and peasants could profit from it with large landed property rights, paying little or no rents. Settlements mostly coped well with the natural influence of the sea either via well-chosen *Flachsiedlungen* (un-raised settlements) (e. g. on river banks, see Meier, this vol.) or by smaller (single house) or larger settlements on artificial hills ('terps': see below). These coastal peasant societies could continue to grow relatively fast during the Middle Ages (van de Ven 1993; Tys, this volume; Soens, id.; Rippon, id.).

It is likely that in this social agro-system peasant society could still dispose of a huge amount of common lands: the still mostly un-diked freshwater peat lands and salt marshes. It has been demonstrated that peasant societies with commons were relatively stable. We know that also outside the coastal wetlands societies with a huge amount of common land to their disposal and dealing with only a limited amount of environmental and social changes, could survive very long (see e.g. Thoen, 2004)! Here, changes mostly occurred only when the communal system was changed which, as history has shown, occurred when new lordly (or governmentally) influence induced changes to property rights. This is actually what happened in many coastal areas in the course of the Middle Ages.

V.3.2. New social relations and embankments

Offensive or defensive dikes?

In older literature embankments were thought to have occurred in two stages: a defensive stage, and an offensive stage. During the offensive stage, land would have been 'conquered' from the sea. This chronological 'two stage model' (which is still partly apparent in the standard works of Gottschalk, 1983 and Verhulst, 1995) is impossible to maintain. We now know that dikes were made in marshland and peat land that was already used by men (so that in this sense the 'offensive 'character can be questioned), and that there were always aspects of 'defense' (against tidal influence) in dike making – even if in different and changing degrees to that seen previously. On the other hand, many embankments were caused by need of developing societies to move towards more intensive land use (e.g. more grain cultivation, more intensive cattle breeding (compare remarks of Rippon, this volume; see also Vervloet and van den Bergh, 2007: 21). So, in certain aspects even low dikes built for temporary protection, could be seen as 'offensive'. Moreover, as already mentioned, it was often the changed property rights of formerly un-diked areas that particularly triggered the earliest embankments. In that sense too, embankments can often be considered as 'offensive'. On the other hand, over exploitation of land between the twelfth to fourteenth centuries often meant that many (re-)embankments were progressively done for defense with higher and stronger dikes which again changed the property structures in the coastal lowlands. In sum, there is actually no clear cut periodisation apparent between 'offensive' and 'defensive' embanking.

Dike building and embanking as tools for population policy of lords and for agricultural intensification

Interventions in the hydraulic system of a tidal area initially occurred to protect settlements and newly reclaimed cultivated plots. Though some of these interventions date to the Roman period (Lascaris and de Kraker, this volume), most have occurred since the ninth and tenth century. The earliest dikes were built along and more or less parallel with the tidal channels in the neighborhood of settlements (and not as was thought along coast lines) as has been demonstrated by Henderikx for the Netherlands (1977) and by Tys for the Belgian coast (2003 and this volume).

Archeology showed that in coastal areas which were not yet protected with winter-resistant dikes (still used in a lesser intensive agrosystem), the oldest *Flachsiedlungen* (un-raised settlements) were easily replaced (see e.g. for Oost-Friesland, see van de Ven, 1993: 51; for Noord-Holland: Besteman and Guiran, 1986; for Flanders: Van Doorselaer and Verhaeghe, 1974), adapting to changing (often human influenced) natural conditions. Elsewhere, structures variously known as *terps, werven, vliedbergen,* and *Wurten* (artificial mounds), in some areas going back to the Roman period or even earlier (as in Friesland and Groningen – see van de Ven, 1993: 39 or Schleswig-Holstein, see e.g. Meier, this volume), were easily adaptable to counter change. Low dikes (also sometimes going back to the Roman period, see Lascaris *et al.,* this volume) had a number of uses. They could often be used as additional protection or to make the surrounding salt marshes more useful for cattle and sheep breeding. They could even have been used for new forms of arable farming necessary to feed a growing population. For example, the building of so-called 'summer dykes' would produce more hay for animal feed which, in turn, would increase manure production (see Bazelmans *et al.,* 1999: 61; compare also Kiden and Verbruggen, 2001: 26). Natural growing grasses in marine sediments are excellent fodder for cattle breeding (including winter fodder), which was a precondition for mixed farming. So, in certain areas, intensification did not yet make it necessary to generally build high and robust dikes and complex sluice and locks in this first stage. However, the growing land pressure in the course of the Middle Ages, stimulated by a new social organisation, must have created a higher need for the cultivation of bread cereals as well as a need for constant and protection from the sea. Tidal channels too would have required protection as well as the need for expensive sluice systems to drain the fresh water from embanked areas.[4] In sum, the earliest settlements probably did not really need much dikes at all.

So, we can see embankments as a stage towards a more intensive use of land. This intensification of land use began before dike building. Settlement creation on artificial mounds was stimulated in many areas. A further stage in the intensification process was dike building and embankments which was actually a logical stage in this evolution. This intensification had a cumulative effect on dike building since it often caused additional ecological problems, which in turn demanded more and stronger embankments. Logically, these stages happened in different areas according to a different chronology.

Property and power structures and their role in occupation of tidal marshlands and dike building

But what triggered this intensification process leading to a more intensive land use and especially to the earliest embankments? Was it the large landowners and lords, or was it the peasants? This requires a nuanced answer.

A major role has been attributed to the changed power relations in the coastal areas and the role of the lords. It has been stressed that in (parts of) coastal Flanders the count played an important role in the reclamation process of the coastal area from the tenth century onwards (Tys, 2003, 2004), while for Holland the role of the count and of the bishop of Utrecht have also been underlined from the eleventh century (H. Vanderlinden and others see for a recent state of the art in this respect: Borger, 2007: 59 and the literature there). In north-west Germany, the bishops of Bremen and Hamburg as well as the counts of Oldenburg played an important part in stimulating the creation of embankments (see Meier, this volume and the literature there)! In the English Fenlands, Glastonbury Abbey also had a strong role in shaping how landscapes were managed. However, for Rippon (this volume), who studied embankments in this area, it was locally different since in that neighborhood land shape and embankments were more influenced by local and peasant initiatives.

Certainly, the role of lords in early dike building can be proven. More than building dikes, their role was to encourage an intensification process. Therefore, collaboration with the local inhabitants and peasant land users was necessary.

Probably a major role must be attributed to changing property structures to explain the process of the earliest dike building. Lordship in preindustrial societies was built on two pillars: the *seigneurie foncière* (the possession of estates) and the (*seigneurie banale*) (the right to warrant and rule). These kinds of lordship were often in the hands of different groups. However, to change property structures in larger areas it was almost a pre-condition that the two pillars were in the same hands, or that the social groups controlling these

[4] Probably, in later stages cultivation of the cultivation of cereals did not play such a large part in the embankments and re-embankments. Some peasant areas could develop a market system to import cereals, as it may have been in Holland from the 14th century on. In other areas the peasant society developed towards a more commercial society. Here cattle breeding often played a larger part (see below).

pillars collaborated. To 'reclaim' land via dike building, one had to change the property structures of the peat areas and the marshlands that together comprised common lands. It was these groups who owned the *seigneurie banale* that could change the property rights. This was facilitated when they could collaborate with peasant property holders who were, since the early Middle Ages, found most frequently in coastal areas to encourage the reclamation process (or even better to stimulate intensification of that area).

How did this happen? Living outside or at the border of the (post-) Carolingian empire in a society with peasants possessing large property rights, a new class of (over-) lords could enhance their power so much that they probably managed to acquire (parts of) the formerly common lands which were huge in number. Some large estates were erected, but most of the land was given to *coloni* (peasants) who got huge property rights over the land they were now allowed to privately cultivate. The gains were reciprocal. The lords got more inhabitants (who could also be used for military service and indirect gains via increased commercial activities), while the peasants often only had to pay a symbolic rent. So the major role of the lords was to attract people: in other words, they pursued an active population policy which could enhance both their income and their power and prestige in the region. These lords could profit from decreasing central power structures (particularly in the former Carolingian areas) and were stimulated by changing (growing) needs for inhabitants as well as for cash which they needed more in a growing commercial society from the eight century and even more from the eleventh century on. This evolution happened everywhere (Thoen, 1993) but in the coastal wetlands the role of lords was reserved to a very small powerful elite who had to take the traditions of private peasant property rights into account. These peasant societies were also to a relatively large degree protected by their overlords from the ambitions of lesser lords through the creation of exclusive power structures within those societies.[5] To counter other seigniorial lords the overlords also collaborated with religious institutions who would not challenge their seigniorial power. These institutions were often gifted with land and rights over common lands where they could erect large holdings, priories, and they even got the right to stimulate colonisation of the former commons. So, an increasing destruction of the common lands was the result. All of which meant that peasant society could continue to grow and increase intensification (and embankments) in the twelfth and thirteenth centuries when land pressure became apparent.

These lords, who in the early Middle Ages were still interested in the direct management of land not only played an indirect role in the intensification process via a population policy. It is likely that they also played a direct role as well via the organisation of land use within these newly reclaimed lands, as the systematic and often geometric land divisions show. The reclamations named 'cope'– reclamations described by van der Linden for Holland and Utrecht (van der Linden, 2000 refers to all his earlier publications) as well as the more recent studies of Tys for Camerlinc Ambacht seem to indicate this as well (Tys, 2003; 2004), though there were also huge differences: while he count of Flanders seems to have been especially interested in the colonisation of marshlands, the count of Holland wanted to stimulate the intensive use and exploitation of peat lands. However,

[5] In Flanders for example small territorial administrative units called *ambachten* were erected under the lead of an officer from the count (Koch, 1951).

their general goals were similar. They both wanted to enhance population numbers and intensify land use through the privatisation of former common lands, probably and especially to increase their power position in the county since tenants could also be recruited for the army. In some areas (like in Holland, see van der Linden, 2000),only symbolic rents to recognise their seigniorial power needed to be paid.

Whether these lords not only stimulated but really financed the dike making itself cannot yet be proven. What we do know is that the peasants living in diked areas were (at least from the twelfth century) themselves responsible for the maintenance of this infrastructure and this could be an indication that they were actually co-responsible for the initial dike building as well. This suggestion does not preclude investment in these projects by their respective lords and without any doubt dike building in the reclaimed but former common lands was only possible with the consensus of the overlord who had appropriated the 'seigneurie banale' in these areas.

Towards a coastal area with a majority of small 'commercial peasants'

We mentioned already that the first dikes were often limited in size and height. They were often only protected parts of the areas adjacent to tidal channels. However, as a consequence of the tendency towards intensification, the first real 'polders' with a 'closed' ring dike mostly occurred between the eleventh and the thirteenth centuries (an early example can perhaps be found in tenth century Friesland, van de Ven, 1993). We now know that in almost all the embanked and un-embanked coastal areas of the North Sea area during 1100–1300 a society had developed with a growing population and a predominance of small peasant holders. Recent research demonstrates that the current polder landscape with a majority of large farms is to a large extend both a modern and early modern phenomenon. In the classic Middle Ages small scale farming was the rule, a social structure which almost everywhere goes back to the classic Middle Ages. These peasants were particularly interested in survival strategies. However, the growing needs of the overlords for money, caused by changing consumption patterns and growing urbanisation (Thoen, 1993), as well as the increasing network of waterways which was a precondition for drainage but was also very useful for transport of cash crops to the markets, stimulated these peasants to produce for the markets. Moreover, everywhere in the coastal areas towns of all sizes developed in the coastal areas (from Dorestad to Bruges and the many small towns (Stabel, 1996) erected in the coastal areas….) since the tidal channels and the man-made canals were of course also very useful for merchants. So, the growing commercialisation was very present in the coastal wetlands, which stimulated, together with the already mentioned lordly colonisation strategy, the fast colonisation between the eleventh and thirteenth century.

In sum, in many coastal areas the majority of peasants who owned small holdings were already producers of commercial products from an early stage in their existence. Salt making, peat digging, cloth production, as well as the cultivation of cash crops like dairy products, were huge. Accordingly, we could label this society at least from the Middle Ages onwards as a 'commercial peasant society', and this in a period when many inland sandy areas were still not as much commercially oriented (Thoen, 2004). Sometimes, like those from the fourteenth century in Holland, they already specialised in the production of dairy products and in peat extraction (van Tielhof, 2005).

New players in the occupation process...

In most areas, the role of important (over) lords to stimulate more intensive land occupation continued up to the twelfth century. The count of Holland, as well as the count of Flanders, are good examples of this. Later, in many areas, this kind of support disappeared since the overlords either re-aligned their interests towards supra-regional matters or suffered from impoverishment (like the Flemish counts in the thirteenth century).

In contrast, new social 'upper-class' groups could enhance their power in the area through typical medieval short-term profit seeking activities. This was the case with many religious institutions (see e.g. Mol, this volume), with new local lords (such as in parts of Zeeland, demonstrated by Dekker for Zuid-Beveland), but, often a bit later, also city burghers. In many areas they could also profit from the poverty of the peasants to acquire more property rights.

The role of the Cistercians and Praemonstratensians in the embankments is now clearer (see e.g. Verhulst, 1995: 48; Mol, 1992). For a long period it was thought that these institutions were the initial settlers in salt marshes, but now we know that the coastal area already had a high population density. Only through gifts of (already reclaimed) land, or when the peasants ran into financial difficulties, could they take over the initiative or create larger demesne centers, or even priories. Although active support from institutions and feudal lords most likely already dates from an earlier period, it was only apparent in Flanders since the twelfth century (Verhulst, 1995: 38), from the later Middle Ages in Zealand Flanders (van Cruyningen, 2012), and still later in the northern part of the Netherlands (Mol, 1992) that real entrepreneurs (private persons, institutions, lay lords) played a larger part in the embankments and that large scale embankment initiatives were taken. In particular, from the late thirteenth century onwards we can see more and more burghers of towns investing in the coastal rural estates, including peat digging centers (Thoen, 1988). The most important role of the increased influence of these large landowners was that the large estates that had often been created by them delivered, together with the increased peat digging activities, additional employment possibilities for the ever smaller peasant holdings (Thoen, 2003).

From the late thirteenth century on: the deficiency of a disrupted peasant society to cope with environmental problems and the change towards new social agro-systems

From the late Middle Ages until the eighteenth century relatively the same type of evolution of social agro-systems took place in most coastal areas around the North Sea, although there was a huge chronological differences between them.

The smaller peasant societies increasingly came under pressure. Gradually more and more entrepreneurs financed embankments in areas were peasants had previously lived and who had abandoned their properties due to financial problems. These investors profited from the insolvability of these people to make 'new land' through the re-embanking of certain areas.

In this context it is important to know that from the thirteenth-fourteenth centuries new institutional structures became necessary. The embankments themselves had provoked greater investments for protection against the sea and to counter other increased water management problems. Indeed, the growing population also needed more protection. Moreover, overexploitation had made the embanked areas more vulnerable for storm surges. To cope with these growing needs for maintenance, the first real water boards were installed. While, until the thirteenth century, peasants owning lands in the embanked areas and mostly living in the nearby holdings maintained the infrastructure themselves, from the thirteenth century onward it were organisations employing wage earners (paid through raising water taxes) who maintened the dikes, sluices, ditches and canals. The evolution of these institutions or waterboards (called: 'waterschappen', 'wateringues', 'wateringen','everingen', 'hoogheemraadschappen', 'polders', 'Wassergenossenschafte', developped between 1250 and 1600 all over the North Sea area (van de Ven, 1993: 97).[6] The judicial and social environment in which they functioned has been studied much more carefully during the last decades (Soens, 2009 and 2011; van Tielhof, van Dam, 2008 and others). Initially, these organisations mainly dealt with the supervision of maintenance and repair works which remained the task of individual landowners. Only gradually, the water boards started to organise maintenance works themselves, financing their activities by levying a *scot* or *geschot* (uniform land tax) to be paid by all landowners. As we will see, for many small landowners, these costs became too much of a burden to survive…

Within these boards the peasants gradually lost influence (Soens, 2008). This might even have created additional problems for the sustainability of many coastal areas. When, as in Flanders or in Romney Marsh, the government of the water boards fell into the hands of the large landowners in the course of the later Middle Ages, things got out of hand since peasants no longer controlled their own safety (Soens, 2006; 2009). A majority of 'feudally inspired' (read: interested in short term profit only) landowners took control of water management in wide coastal areas. It is no coincidence that in principalities where peasants remained influential on local water boards (like in Holland), water management seems to have been more effective and the peasant society could survive until the seventeenth century (van Tielhof, 2005).

In itself, the transition towards monetary taxes to finance the water control system had an important impact on coastal society. On average, water taxes were not so high but they often were highly irregular. Many smaller peasant landholders became bankrupt and their land was in turn taken over by larger peasants. Furthermore, the water taxes were levied in cash, which already presupposes a level of commercialisation of the rural economy. In difficult years, peasant landowners could be faced with high water taxes, instantly to be paid in cash. As a result, the monetarisation of water management was clearly linked to the commercialisation of coastal society, and with the transition from a society of peasant smallholders towards 'commercial' agriculture, where big landowners possessed the majority of the land. (Soens and Thoen, 2008). Natural conditions slowly became less favorable for small peasants because the cultivation of heavier clay lands required more capital in terms of horses and plows. Only larger and mid-sized farmers could survive since they were able to expand and could make enough economies of scale (for more details see e.g. Soens, 2012). In sum, there generally was a 'forced' evolution towards a more commercial society. Since the seventeenth century this evolution seems to have led towards a more sustainable landscape in most areas (see below). However, many regional differences occurred. The described evolution happened first in the Southern Low Countries and more later in the North. So while the deterioration of peasant society

[6] Today in the US these institutions are called, 'levee boards'.

and the evolution towards larger holdings had clearly begun by the thirteenth century in the Belgian coastal area and in the Dutch river area from the fourteenth century (van Bavel, 2010), in Holland the process seems to have been slower. Here, peasants adapted in the later Middle Ages. This was thanks to a new boost of the proto-industrialisation (peat digging, fishery, compare van Tielhof, 2005) and the increased demand for dairy products from booming towns. Peasants here specialised in such products and could profit from the grain supply of towns like Amsterdam. It was not until the seventeenth century that an evolution towards a more capitalist society with larger more specialised farms occurred here. Further north in the marine areas, this evolution did not take place until the eighteenth century (Groningen, parts of Friesland, Noord-Holland; Brusse *et al.*, 2010: 204). These evolutions were not the only reason for the man-made increased environmental stresses that changed social conditions. Other factors were important as well, not least the importance of the balance between the different social players (owners, land uses, power elites). However, in general the natural pressure in the long term was in favor of the more powerful social classes. So, contrary to the more sandy areas, many coastal areas experienced major social changes in the course of the later Middle Ages and the Old Regime. What is important in this context is that the deterioration of peasant society since the late Middle Ages and the change/development towards a more commercial society did NOT immediately generate higher investments! In many areas the opposite took place! It has been demonstrated that for some regions like Flanders and Sealand the new, often absentee, powerful landowners who succeeded in taking over the landed property (due to the impoverishment of the peasants) and part of the judicial property (since they got more and more control over the water boards, Soens, 2009), could not ensure the sustainability of the landscape for a long time between *c.* 1250 and 1600). Apparently, short term profit making was the main goal of these elites. Even the larger farmers who took over from the former peasants (who degraded towards wage earners), also struggled to survive and were not able to ensure the sustainability of the landscape. Farmer-elites who could support a more sustainable land use only emerged in the course of the early modern period in Sealand, Holland, Friesland, and Groningen.

Long term data series that can teach us more about the real investment level of the institutions are scarce. However, it is possible to get an idea of the investment rate for the Western Scheldt area (particularly the western part) from the later Middle Ages using water taxes as an index. As can be seen on figure 18.3, despite regional and temporal differences, in general the investment rate of the water boards was low (Thoen and Soens, 2001: 12 ff. and developed in Soens, 2009 and 2012, table 5.5)! In many coastal areas during the late Middle Ages and into the sixteenth century, damages from storm surges and land losses were huge and many villages completely disappeared (see for the Western Scheldt area figure 18.3; compare also Soens, this volume). However, the biggest land loss in this area occurred during the Eighty Years War where the deliberate flooding of lands became accepted military strategy for both the rebel side and the Spanish (de Kraker, 1997). However, the fact that land and villages were often lost forever proves that it this had been a vulnerable society before that war. For Sealand Flanders the loss of land cause by storm surges and warefare is estimated at roughly 75 per cent of the late medieval landscape (de Kraker, 2002).

In general, until the seventeenth century, investments by the water boards represented no more than 10–50l wheat per hectare, which can be estimated as only a few percent of the yearly income of arable land in the polders (estimated at about 20.000 litre per year of wheat per hectare).

V.4. Supra-regional support to regional social agro-systems in coastal areas and new attitudes of investment lead to increased sustainability in the Early Modern Period

Better techniques for sustainable landscapes?

A new stage in draining and the new embankments was linked to the application of new drainage technology, in particular the general use of *poldermolens/hoosmolens* (drainage mills). Wind driven drainage mills had spread slowly since the late fourteenth century in Holland. They only became generally used during the second part of the sixteenth century when they replaced horse driven mills (A. Bicker Caarten, 1990; Zeischka, 2008; Borger, 2007: 55). Initially, these (wind) mills were used for defensive reasons but their common use seems to be linked to a wider strategy of land reclamation. They also continued to be used for drainage via a pumping system of formerly flooded lands, as well as for making dry shallow lakes (in Dutch 'droogmakerijen'). which were often created by dredging peat under the water table, a practice which had become common from the late Middle Ages on in some areas. Although these mill-steered reclamations fundamentally increased the tax burden, in certain areas like Holland the peasants themselves seem to have been able to support these additional costs for centuries.

More sustainable investments

In the second part of the Old Régime, the investment rate of the water boards alone was clearly growing, more than doubling in general! In some areas, this increased investment had already begun by the end of the sixteenth century. Elsewhere, the investments only grew post-1650. The local water boards were not the only financers of investments anymore! Indeed, in many areas like Sealand Flanders, the polders that needed money for investments after flooding or dike collapse were now subsidised on a more regular basis by higher institutions, governments, and/or neighboring polders to overcome financial problems (de Vleeschauwer, 2012: 96). The same was true for Holland (van Tielhof and van Dam, 2008: 209). A new 'absentee' elite no longer just exploited the area (as seen in the period between 1200 and 1600); they also seem to have been more concerned with the general welfare of the landscape.

And more sustainable results…

From the second half of the eighteenth century onwards, provincial governments also took much more care in the construction and maintenance of 'sea walls' (de Vleeschauwer, 2012: 196)! It is very likely that this substantial growth of investments was responsible for the declining process of *dijkvallen* (dike erosion) – land losses due to storm surges and channel erosion caused by the ebb flow. While flooding caused by storms was very common in the late Middle Ages – a vast amount of land and many villages permanently disappeared – from the post-religious war period onwards these loses were relatively restrained. In the Western Scheldt and its adjacent tidal channels, a limited amount of polders still could be reclaimed. It is particularly important that a consolidation of the

Table 18.1 Natural versus human causes of historical change of the Scheldt river (Based on: de Kramer, 2002)

		Tide speed	Tide volume	Tide storage (Volume of water between high and low tide in a basin)	MHW	MLW	Height difference between low and high tide mode	Tidal flow rate (flood)	Tidal flow rate (ebb)	Maximum flow rates
Natural changes	silting	+	_	_	+	_	+	_	+	+
	Sea level up	+	+	+	+	±	+	_	+	+
	Storm surges (enlarging and diepening)	+	+	±	+	_	+	_	+	+
	Dike destruc-tion	_	+	+	_	0	_	+	_	+
	Sand import	_	_	_	_	+	_	+	_	_
	Sand export	+	+	+	+	_	+	_	+	±
Human changes	Embankments: diking	+	0	_	+	0	+	_	+	+
	Cutting of tidal channels with dams	+	0	_	+	_	+	_	+	+
	undiking	_	±	+	_	±	±	+	0	+
	Dredging on bars	+	+	±	+	_	+	_	+	+
	Sand winning	+/_	+	±	+	_	+	_	+	+
	River bank protection	0	0	0	0	0	0	0	0	+

border line and the river banks took place (see figure 18.3). Only a small number of storm surges still caused damage, despite a continuously growing tidal amplitude upstream of the river towards Antwerp between 1650 and 1850 (de Kramer, 2002: 21). Even the size and shape of the parcels cultivated since the seventeenth and eighteenth centuries is mostly still the same as today.

What is the long term balance?

Was anthropogenic influence in the coastal areas only negative for landscape evolution in the longer term? The above described influence of men on the nature of soils in the coastal area could give the impression that man's influence on the coastal area was negative. If we view it in the long term, taking into account only the available size of land, we can indeed say that from the Roman period until the end of the Old Regime (although the embankments made with the aid of windmills of the Early Modern period can be to a certain extend be seen as 'land conquests') the acreage of the totally available land had shrunk considerably and it is likely that to a very large extent this was due to human distortion of the natural ecosystem. Moreover, the valuable peat soils in many areas (although there value was declining from the eighteenth century with the growing importance of coal) disappeared too. One can say that probably gradually less people could live in the same area since population pressure was often reversed due to social changes. In the Middle Ages and the early modern times, the most commercially oriented rural areas, such as the 'forced' commercial areas in the coastal landscapes of the North Sea, had the lowest population densities!

However, particularly since the Early Modern Period, one could sometimes reverse an environmental problem into an advantage by integrating it into a new 'ecological balance'; this was, for example, the case in Holland, where many lakes which originated in the Middle Ages from peat dredging the underground water table and bank erosion were embanked again after the late sixteenth century. A large lake called the Haarlemmermeer also partly originated from peat digging got another function. It was transformed into a reservoir for drainage of the inland (van Tielhof and van Dam, 2006: 71).

More importantly, with the growing amount of clay soils as a result of 'the battle of the sea against men and *vice versa*', in general and in the longer term soils certainly became more valuable in terms of their nutrient content. The growing share of clay soils could produce the highest yields for arable farming, as well as for dairy farming In areas under marine influence, the reciprocal relations between social relation and environmental stress, as described, eventually resulted almost everywhere in coastal societies where only highly productive, more specialised, and commercially oriented farming could survive. In the longer term and after many difficult times, this commercial agriculture in which new elites emerged has contributed to the evolution of a more modern society, at least according to some scholars. But this is of course another subject.

VI. Epilogue

In another article of part 1 of this volume, H. Weerts brought together the major natural reasons for regionally different influence of the relative sea level movements. The text above focused on the influence of human social organisation particularly during the last millennium. This influence was huge and the text illustrates why human action caused landscape changes to be different in regions which experienced similar natural conditions. Moreover, we tried to illustrate how and according to which mechanisms the changes into a cultural landscape often enlarged or changed natural phenomena. We also underlined the importance of 'bounce effects' between the natural and human causes of environmental changes.

On the other hand, recent research demonstrates a larger link with specific general economic and social tendencies on a supra-regional level. Until the later Middle Ages, coastal areas seem to have more followed general economic and demographic tendencies contrary to what was previously before *c.* 1980, when researchers still believed in transgression and regression periods as the 'engines' for long term demographic movements. Today, with only minor differences compared to the rest of Europe, the Roman period until the third century is seen as a good economic and demographic period; the early Middle Ages as a period of low dynamics, while the economic revival seems to start in the Carolingian period until its zenith in the twelfth to thirteenth centuries. In older studies one considered the landscape in the period before 1000 (including the Roman period) as hardly being inhabited, and the period between 1000 and 1150 as one of low dynamics due to storm surges and sea level changes. Nowadays, the period between the eleventh and the thirteenth centuries is seen as a period of intensive growth in most coastal areas, with increasing amounts of small holders, just as it was in most areas outside marine influence.

From the late thirteenth century onwards, the evolution of many marine areas started to diverge from the rest of Europe. Since the coastal regions were more fragile environments, these landscapes were vulnerable to changes in social relations like state formation and to on-going deterioration processes. In their turn, the environmental processes that occurred triggered new social relations where only the upper part of rural society could survive and in which non-local elites could take over the power in the area. Although, these elites managed to transform many North Sea coastal wetlands into regions of intensive agricultural development and economic growth, they could not ensure longer term sustainability; on the contrary, the changed social relations mostly worsened the relationship between man and nature for a long time. Only when one moved from a 'feudal' society – where individual and short term interests were still more important than the general interest – towards a more modern society, where central institutions limited personal interests in favor of the inhabitants, and where economic actors again collaborated in a more productive way, could a greater sustainability be realised post-1600.

However, despite these described general tendencies, local and regional differences remained huge. This is actually due to the same mechanisms: social relations did not everywhere evolve in the same way and did not interact in an identical way with locally different natural conditions. Therefore, we hope that the text above helps to understand locally different evolutions of the coastal wetlands around the North Sea.

As well these general long term tendencies, regional differences still deserve to be underpinned with more 'in-depth' research. At the same time it must remain the task of the historical geographer to increasingly study in detail the larger scale mechanisms of man-nature relations. Therefore, much more real inter and trans disciplinary research is still very necessary to further analyze and understand '*the battle between Clio and Neptune*'.

Bibliography[7]

Asselen, S. van, Karssenberg, D., Stouthamer, E. (2011) 'Contribution of peat compaction to relative sea level rise within Holocene deltas', *Geophysical research letters,* 38, L24401.

Asselen, S. van, Stouthamer, E. and Asch, Th.W.J. van (2009) 'Effects of peat compaction on delta evolution: A review on responses, measuring and modeling', *Earth Sciences reviews,* 92, pp. 35–51.

Augustijn B. (1992), *Zeespiegelrijzing, transgressiefasen en stormvloeden in maritiem Vlaanderen tot het einde van de 16de eeuw,* Brussels.

Augustyn, B. (1995) 'De evolutie van het duinecosysteem in Vlaanderen in de Middeleeuwen: antropogene factoren versus zeespiegelrijzingstheorie', *Historisch Geografisch Tijdschrift,* 13, pp. 9–19.

Bavel van, B.J.P. (2010) *Manors and Markets. Economy and Society in the Low Countries, 500–1600,* Oxford.

Bavel, B.J.P. van and Thoen, E. (2013) 'Rural history and the environment. A survey of the relationship between property rights, social structures and sustainability of land use', in: B.J.P. van Bavel and E. Thoen (eds), *Rural societies and environments at risk. Ecology, property rights and social organisation in fragile areas (Middle Ages-Twentieth century),* Turnhout.

Bavel, B.J.P. van, Cruyningen, P. van and Thoen, E. (2010) 'The Low Countries, 1000–1750', in: B.J.P. van Bavel and R. Hoyle *et al.* (eds), *Social Relations: Property and Power. Rural Economy and Society in North-western Europe, 500–2000,* Turnhout, pp. 169–198.

Bazelmans, J., Gerrets, D., De Koning, J. and Vos, P. (1999) 'Zoden aan de dijk. Kleinschalige dijkbouw in de late prehistorie en protohistorie van noordelijk Westergo', *De Vrije fries,* 29, pp 8–71.

Bazelmans, J., Weerts, H.J.T., Meulen, M. van der (eds) (2011) *Atlas van Nederland in het Holoceen. Landschap en bewoning vanaf de laatste ijstijd tot nu,* Amsterdam.

Besteman, J.C. en Guiran, A.J. (1986) 'De middeleeuwse bewoningsgeschidenis van Noord-Holland boven het IJ en de ontginning van de veengebieden. Opgravingen in Assendelft in perspectief' in: M.C. van Trierum and H.E. Henkes (eds), *A contribution to prehistoric, roman and medieval archaeology,* pp. 183–212 (Rotterdam Papers; 5).

Borger, G.J. (2007) *Het verdwenen veen en de toekomst van het landschap. Rede,* UvA, Amsterdam.

Broeke, P.W. van den (1996) 'Turfwinning en zoutwinning langs de Noordzeekust. Een verbond sinds de Ijzertijd?', *Tijdschrift voor Waterstaatsgeschiedenis,* 5, pp. 78–59.

Brusse, P., Schuurman, A., Molle, L. Van, Vanhaute, E. (2010) 'The Low Countries, 1750–2000' in: B.J.P. van Bavel and R. Hoyle *et al.* (eds), *Social Relations: Property and Power. Rural Economy and Society in Noth-western Europe, 500–2000,* Turnhout, pp. 199–226.

[7] The chapters of this volume we refer to in the text are not mentioned in this bibliography. This bibliography is of course selective.

Bungenstock, F. and Weerts, H.J.T. (2010) 'The high resolution Holocene sea level curve for Northwest Germany: global signals, local effects or data-artefacts?', *International Journal of Earth Science,* 99, pp. 1687–1706.

Ceunynck, R. De (1992) 'Het duinlandschap, Ontstaan en evolutie' in: B. Billiau *et al.*(eds) *Tussen land en zee, het duingebied van Nieuwpoort tot De Panne,* Tielt, pp.18–45.

Cornelisse, C. (2008) *Energiemarkten en energiehandel in Holland in de late Middeleeuwen,* Hilversum.

Cruyningen, P. van (2003) 'Waterbeheer, landbouw en samenleving in West-Zeeuws-Vlaanderen in de 17[e] en 18[e] eeuw', *Jaarboek voor Ecologische Geschiedenis,* pp. 57–66.

Cruyningen, P. van (2012) 'Farmers' strategies and the West-Zeeland-Flanders grain trade, 1648–1794', in: Piet van Cruyningen and Erik Thoen (eds), *Food supply, demand and trade. Aspects of the economic relationship between town and countryside (Middle Ages-19th century*), Turnhout, pp. 161–172.

Dam, P. van (1999) 'Stuivend zand en stormende golven. De vorming van de Hollandse kust in de Middeleeuwen', *Madoc,* pp. 224–232.

Dam, P. van (2003) 'Schijven en beuken balken. Een sociaal-economische transformatie in de Riederwaard', in: B. Wouda (ed.), *Ingelanden als uitbaters.Sociaal-economische studies naar Oud- en Nieuw-Reijerwaard, een polder op een Zuid-Hollands eiland,* Hilversum, pp. 11–44.

Dekker L.W. (1981) 'Sporen van turfstekerijen in West-Friesland', *Boor en spade,* pp. 62–71.

D'Haenens, A. (1967) *Les invasions normandes en Belgique au IXe siècle,* Brussels.

Gottschalk, M.K.E. (1955–1958; 1983[2]) *Historische geografie van Westelijk Zeeuws Vlaanderen,* Assen.

Henderikx, P.A. (1977) 'De zorg voor de dijken in het baljuwschap Zuid-Holland en in de grensgebieden ten oosten daarvan tot het einde van de 13e eeuw', *Geografisch Tijdschrift,* 11, pp. 407–427.

Jongepier, J., Soens, T., Thoen, E., Eetvelde, V. Van, Crombé, Ph. and Bats, M. (2011) 'The brown gold: a reappraisal of medieval peat marshes in Northern Flanders (Belgium)', *Water history,* 3, pp. 76–93.

Kiden, P. and Verbrugggen, C. (2001) 'Het verhaal van een rivier: de evolutie van de Schelde na de laatste ijstijd', in: J. Bourgeois, Ph. Crombé, G. Mulder and M. Rogge (eds), *Een duik in het verleden. Schelde, Maas en rijn in de pre- en protohistorie,* pp. 11–35 (Publicaties van het Provinciaal Archeologisch Museum van Zuid-Oost-Vlaanderen-Site Velzeke, Buitengewone reeks; 4).

Kiden, P., Maraske, B. and Plassche, O. van de (2008) 'Waarom verschillen de zeespiegelreconstructies voor Nederland?', *Grondboor en hamer,* 3–4, pp. 54–61.

Klijn, J. A. (1981) *Nederlandse kustduinen, geologie en bodems,* Unpublished Dissertation, Wageningen University.

Kraker, A. de (1997) *Landschap uit balans. De invloed van de natuur, de economie en de politiek op de ontwikkeling van het landschap in de Vier Ambachten en het Land van Saeftinghe tussen 1488 en 1609,* Utrecht.

Kraker, A. de (2007) 'De centra van zoutproductie in Brabant, Vlaanderen en Zeeland tussen 1400 en 1600', in: Adriaan M.J. de Kraker en Guus J. Borger (eds), *Veen-Vis-Zout. Landschappelijke dynamiek in de zuidwestelijke delta van de Lage Landen,* Amsterdam, pp. 81–113 (Geoarchaeological and Bioarchaeological Studies; 8).

Kraker A. de and Borger, G.J. (2007) 'Veen-Vis-Zout: een multidisciplinaire uitdaging voor toekomstig onderzoek', in: Adriaan M.J. de Kraker en Guus J. Borger (eds), *Veen-Vis-Zout. Landschappelijke dynamiek in de zuidwestelijke delta van de Lage Landen,* Amsterdam, pp. 131–135 (Geoarchaeological and Bioarchaeological Studies; 8).

Kraker, A. de, Koning, D. de and Wielinga, R. (2008) 'Making salt the medieval way', in: *Experimentelle Archäologie in Europa. Bilanz 2008.* Heft 7, Oldenburg, pp. 75–90.

Kraker A. de (2008) 'Stormachtig weer in de Lage Landen tussen 1400 en 1625. Reconstructie van stormen langs de zuidoostelijke Noordzeekust, de wijze waarop hun invloed wordt bepaald en veranderingen in het stormpatroon', *Jaarboek voor Ecologische Geschiedenis,* pp. 1–33.

Koch, A.C.F. (1951) *De rechterlijke organisatie van het graafschap Vlaanderen,* Ghent.

Kramer, J. de (2002) *Waterbeweging in de Westerschelde. Een literatuurstudie,* Middelburg (ICG-rapport 02/6).

Labarque M. (2007) *Van duindorp aan zee tot verstedelijkt lint aan de kust. Onderzoek naar de verstedelijkingsprocessen van De Panne tot Nieuwpoort,* unpublished Masters Thesis, Ghent University.

Leenders, K.A.H.W. (2003) 'Middeleeuws zout uit de delta. Status questionis als basis voor nader onderzoek' (http://www.coenkd.dds.nl/Links_Info/tekst_darinkdelven.doc).

Leenders, K.A.H.W. (1989) *Verdwenen Venen. Een onderzoek naar de ligging en exploitatie van thans verdwenen venen in het gebied tussen Antwerpen, Turnhout, Geertruidenberg en Willemstad. 1250–1750,* Brussels/Wageningen.

Ligtendag, W. A. (1995) *De Wolden en het water. De landschaps- en waterstaatsontwikkeling in het lage land ten oosten van de stad Groningen vanaf de volle middeleeuwen tot ca. 1870,* Groningen.

Linden van der, H. (2000) 'De Koningsroede', *Jaarboek voor Middeleeuwse Geschiedenis,* pp. 7–43.

Long, A.J., Waller, M.P., Stupples, P. (2006) 'Driving mechanism of coastal peat compaction and the destruction of late Holocene coastal wetlands', *Marine geology,* 225, pp. 63–94.

McCormick, M. (2001) *Origins of the European economy. Communications and Commerce AD 300–900,* Cambridge.

Mertens, J.A. (1963) 'Biervliet, een laatmiddeleeuws centrum van zoutwinning, 1[e] helft XV[e] eeuw', *Handelingen der Maatschappij voor Geschiedenis en Oudheidkunde te Gent,* 17, pp. 105–107.

Mount, J. and Twiss, R. (2005) 'Subsidence, sea level rise and seismicity in the Sacramento-San Joaquin Delta', *San Francisco Estuary and Watershed Science,* 3, pp. 1–18.

Priester, P. (1998) *Geschiedenis van de Zeeuwse landbouw circa 1600–1910,* Wageningen (AAG Bijdragen; 37).

Rentenaar, R. (1978) 'De vroegste geschiedenis van het konijn in Holland en Zeeland', *Holland,* 10, pp. 2–16.

Soens, T. (2006) 'Explaining deficiencies of water management in late medieval Flemish coastal plain 13th-16th centuries', *Jaarboek voor Ecologische Geschiedenis,* 2005–6, pp. 35–62.

Soens, T. (2007) 'Dijkenbouwers of rustverstoorders? De Vlaamse graven en de strijd tegen het water in de laatmiddeleeuwse Vlaamse kustvlakte (veertiende-zestiende eeuw)', *Bijdragen en Mededelingen voor de Geschiedenis der Nederlanden,* 122, pp. 321–356.

Soens, T. (2009) *De spade in de dijk? Waterbeheer en rurale samenleving in de Vlaamse kustvlakte (1280–1580),* Ghent (Historische Economie en Ecologie).

Soens, T. (2011) 'Floods and money. Funding drainage and flood control in coastal Flanders (13th-16th centuries)', *Continuity and Change,* 26, pp. 333–365.

Soens, T. (2013) 'The social distribution of land along the north sea Coast: Flanders, Holland and Romney Marsh compared (c. 1250–1750)', in: B.J.P. van Bavel and E. Thoen (eds), *Rural societies and environments at risk. Ecology, property rights and social organisation in fragile areas (Middle Ages-Twentieth century),* Turnhout.

Soens T. and Thoen E. (2009) 'Mais où sont les tourbières d'antan? Géographie, chronologie et stratégies économiques du tourbage en Flandre maritime (XIIe-XVIe siècles)', *Aestuaria: Histoire et terres humides,* 14, pp. 45–60

Syvitski, J.P.M., Vörösmarty, Ch.J., Kettner, A.J., Green, P. (2005) 'Impact of Humans on the Flux of Terrestrial Sediment to the Global Coastal Ocean', *Science,* 308, pp. 376–380.

Thoen, E. (1993) 'The Count, the Countryside and the Economic development of the Towns in Flanders from the eleventh Century to the thirteenth Century. Some Provisional Remarks and Hypotheses, in: E. Aerts, B. Henau, P. Janssens and R. Van Uytven (eds), *Studia Historica Oeconomica – Volume 1: Liber amicorum Herman Van der Wee,* Louvain, pp. 259–278.

Thoen, E. (2004) '"Social agrosystems" as an economic concept to explain regional differences. An essay taking the former county of Flanders as an example (Middle Ages-19th century)', in: B.J.P. van Bavel and P. Hoppenbrouwers (eds), *Landholding and land transfer in the North Sea Area (Late Middle Ages-19th century),* Turnhout, pp. 47–66.

Thoen, E. (2011) 'La géographie historique, discipline en crise: plaidoyer pour une nouvelle approche. Un regard subjectif sur l'étude historique du paysage', *Revue Belge d'Histoire et de Philology,* 89, pp. 803–820.

Thoen, E. and Soens, T. (2001) 'Waterbeheer in de Vlaamse kustvlakte in de Late Middeleeuwen en het Ancien Régime', *Jaarboek voor Ecologische Geschiedenis,* pp. 1–23.

Thoen, H. (1978) *De Belgische kustvlakte in de Romeinse tijd: bijdrage tot de studie van de landelijke bewoningsgeschiedenis,* Brussels.

Tielhof, M. van (2005) 'Turfwinning en proletarisering in Rijnland 1530–1670', *Tijdschrift voor Sociale en Economische Geschiedenis,* 2, pp. 95–121.

Tielhof, M. van and Dam, P.J.E.M. van (2006) *Waterstaat in stedenland. Het hoogheemraadschap van Rijnland voor 1857,* Utrecht.

Tys, D. (2003) *Een middeleeuws landschap als materiële cultuur: de interactie tussen macht en ruimte in het kustgebied en de wording van een laatmiddeleeuws tot vroegmodern landschap. Kamerlingsambacht, 500–1200/1600,* Unpublished Dissertation, Free University Brussels.

Tys, D. (2004) 'Domeinvorming in de "wildernis" en de ontwikkeling van de vorstelijke macht – Het voorbeeld van het bezit van de graven van Vlaanderen in het IJzerestuarium tussen 900 en 1100', *Jaarboek voor Middeleeuwse Geschiedenis,* 7, pp. 31–83.

Ven, G.P. van de (ed.) (1993) *Leefbaar laagland. Geschiedenis van de waterbeheersing en landaanwinning in Nederland,* Utrecht.

Verhulst, A. (1995) *Landschap en landbouw in Middeleeuws Vlaanderen,* Brussels.

Verhulst A. and Gottschalk, M.K.E. (eds) (1980) *Transgressies en occupatiegeschiedenis in de kustgebieden van Nederland en België. Colloquium 5–7 september 1978.* Handelingen, Ghent.

Vervloet, J. and Bergh, S. van den (eds) (2007) *Eemland in verandering. Ontginning en ruilverfkaveling in het gebied van de Eem,* Utrecht.

Vleesschauwer, M.L.M. de (2012) *Van water landt, van landt weder water. Waterstaat en waterhuishouding in midden Zeeuws-Vlaanderen, 1600–1999,* Amsterdam (Geoarchaeological and Bioarchaeological Studies; 14).

Von Storch, H. and Woth, K. (2006) 'Storm surges. The case of Hamburg, Germany'. *OSC panel session on 'GEC, natural disasters and their implications for human security in coastal urban areas'* (http://www.safecoast.nl/editor/databank/File/hamburg-storms.pdf.).

Vos, P.C., Bazelmans, J., Weerts, H.J.T., Meulen, M.J. van der (eds) (2011) *Atlas van Nederland in het Holoceen. Landschap en bewoning van de laatste ijstijd tot nu,* Amsterdam.

Weerts, H.J.T., Westerhoff, W.E., Cleveringa, P., Bierkens, M.F.P., Veldkamp, J.G. and Rijsdijk, K.F. (2005) 'Quaternary geological mapping of the lowlands of the Netherlands, a 21st century perspective', *Quaternary International,* 133–134, pp. 159–178.

Weerts, H., Cleveringa, P., Westerhoff,W. and Vos, P. (2006) 'Nooit meer: Afzettingen van Duinkerke en Calais', *Archeobrief (methoden en technieken),* pp. 28–34.

Zeischka, S. (2008) *Minerva in de polder. Waterstaat en techniek in het hoogheemraadschap van Rijnland, 1500–1856,* Utrecht.

www.ingramcontent.com/pod-product-compliance
Ingram Content Group UK Ltd.
Pitfield, Milton Keynes, MK11 3LW, UK
UKHW021825070126
466728UK00002B/18

9 782503 540580